SECOND EDITION

Make: Electronics

Charles Platt

Make: Electronics

by Charles Platt

Printed in Canada.

Published by Maker Media, Inc., 1160 Battery Street East, Suite 125, San Francisco, CA 94111.

Maker Media books may be purchased for educational, business, or sales promotional use. Online editions are also available for most titles (*http://safaribooksonline.com*). For more information, contact our corporate/institutional sales department: 800-998-9938 or *corporate@oreilly.com*.

Editor: Brian Jepson
Production Editor: Nicole Shelby
Proofreader: Charles Roumeliotis
Indexer: Charles Platt
Interior Designer: David Futato
Cover Designers: Marc de Vinck and Charles Platt
Interior Photographs and Illustrations: Charles Platt

September 2015: Second Edition
December 2009: First Edition

Revision History for the Second Edition

2015-08-07: First Release
2015-10-23: Second Release

See *http://oreilly.com/catalog/errata.csp?isbn=9781680450262* for release details.

978-1-680-45026-2

[TI]

Dedication

To readers of the first edition of *Make: Electronics* who contributed many ideas and suggestions for this second edition. In particular: Jeremy Frank, Russ Sprouse, Darral Teeples, Andrew Shaw, Brian Good, Behram Patel, Brian Smith, Gary White, Tom Malone, Joe Everhart, Don Girvin, Marshall Magee, Albert Qin, Vida John, Mark Jones, Chris Silva, and Warren Smith. Several of them also volunteered to review the text for errors. Feedback from my readers continues to be an amazing resource.

Acknowledgments

I discovered electronics with my school friends. We were nerds before the word existed. Patrick Fagg, Hugh Levinson, Graham Rogers, and John Witty showed me some of the possibilities.

It was Mark Frauenfelder who nudged me back into the habit of making things. Gareth Branwyn facilitated *Make: Electronics*, and Brian Jepson enabled the sequel and this new edition. They are three of the best editors I have known, and they are also three of my favorite people. Most writers are not so fortunate.

I am also grateful to Dale Dougherty for starting something that I never imagined could become so significant, and for welcoming me as a participant.

Russ Sprouse and Anthony Golin built and tested the circuits. Technical fact checking was provided by Philipp Marek, Fredrik Jansson, and Steve Conklin. Don't blame them if there are still any errors in this book. It's much easier for me to make an error than it is for someone else to find it.

Table of Contents

What's New in the Second Edition

All of the text from the first edition of this book has been rewritten, and most of the photographs and schematics have been replaced.

Single-bus breadboards are now used throughout (as in *Make: More Electronics*) to reduce the risk of wiring errors. This change entailed rebuilding the circuits, but I believe it was worthwhile.

Diagrams showing component placement are now used instead of photographs of breadboarded circuits. I think the diagrams are clearer.

Internal views of breadboard connections have been redrawn to match the revisions noted above.

New photographs of tools and supplies have been included. For small items, I have used a ruled background to indicate the scale.

Where possible, I have substituted components that cost less. I have also reduced the range that you need to buy.

Three experiments have been completely revised:

- The Nice Dice project that used LS-series 74xx chips in the first edition now uses 74HCxx chips, to be consistent with the rest of the book and with modern usage.
- The project using a unijunction transistor has been replaced with an astable multivibrator circuit using two bipolar transistors.
- The section on microcontrollers now recognizes that the Arduino has become the most popular choice in the Maker community.

In addition, two projects involving workshop fabrication using ABS plastic have been omitted, as many readers did not seem to find them useful.

All the page layouts have been changed to make them easily adaptable for handheld devices. The formatting is controlled by a plaintext markup language, so that future revisions will be simpler and quicker. We want the book to remain relevant and useful for many more years to come.

--Charles Platt, 2015

[illegible]

[illegible]

[illegible] and [illegible] that you need [illegible] electronics should not cost [illegible] significantly more than [illegible], and you don't need a [illegible]. Everything can be done on a table top.

[illegible]

[illegible]

I don't believe that the concepts will be hard to understand. Of course, if you want to study electronics more formally [illegible] circuit design, that can be [illegible]

[illegible]

Most [illegible] begin by [illegible] and [illegible] concepts [illegible]

[illegible] this as [illegible]

This book works the other way around. I want you to dive right in and start putting components together without necessarily knowing what to expect. As you see [illegible]

Preface: How to Have Fun with This Book

Everyone uses electronic devices, but most of us don't really know what goes on inside them.

You may feel that you don't need to know. You can drive a car without understanding the workings of an internal combustion engine, so why should you learn about electricity and electronics?

I think there are three reasons:

- By learning how technology works, you become better able to control your world instead of being controlled by it. When you run into problems, you can solve them instead of feeling frustrated by them.
- Learning about electronics can be fun, so long as you approach the process in the right way. It is also very affordable.
- Knowledge of electronics can enhance your value as an employee, or perhaps even lead to a whole new career.

Learning by Discovery

Most introductory guides begin by using definitions and theory to explain some fundamental concepts. Circuits are included to demonstrate what you have been told.

Science education in schools often follows a similar plan. I think of this as *learning by explanation*.

This book works the other way around. I want you to dive right in and start putting components together without necessarily knowing what to expect. As you see what happens, you will figure out what's going on. This is *Learning by Discovery*, which I believe is more fun, more interesting, and more memorable.

Working on an exploratory basis, you run the risk of making mistakes. But I don't see this as a bad thing, because mistakes are a valuable way to learn. I want you to burn things out and mess things up, to see for yourself the behavior and limitations of the parts that you are dealing with. The very low voltages used throughout this book may damage sensitive components, but will not damage you.

The key requirement of Learning by Discovery is that it has to be hands-on. You can derive some value from this book merely by reading it, but you will enjoy a much more valuable experience if you perform the experiments yourself.

Fortunately, the tools and components that you need are inexpensive. Hobby electronics should not cost significantly more than a recreation such as needlepoint, and you don't need a workshop. Everything can be done on a tabletop.

Will It Be Difficult?

I assume that you're beginning with no prior knowledge. Consequently, the first few experiments will be extremely simple, and you won't even use prototyping boards or a soldering iron.

I don't believe that the concepts will be hard to understand. Of course, if you want to study electronics more formally and do your own circuit design, that can be challenging. But in this book I have kept theory to a minimum, and the only math you'll need will be addition, subtraction, multiplication, and division. You may also find it helpful (but not absolutely necessary) if you can move decimal points from one position to another.

How This Book Is Organized

An introductory book can present information in two ways: in tutorials or in reference sections. I decided to use both of these methods.

You'll find the *tutorials* in sections headed as follows:

- Experiments
- What You Will Need
- Cautions

Experiments are the heart of the book, and they have been sequenced so that the knowledge you gain at the beginning can be applied to subsequent projects. I suggest that you perform the experiments in numerical order, skipping as few as possible.

You'll find *reference sections* under the following headings:

- Fundamentals
- Theory
- Background

I think the reference sections are important (otherwise, I would not have included them), but if you're impatient, you can dip into them at random or skip them and come back to them later.

If Something Doesn't Work

Usually there is only one way to build a circuit that works, while there are hundreds of ways to make mistakes that will prevent it from working. Therefore the odds are against you, unless you proceed in a really careful and methodical manner.

I know how frustrating it is when components just sit there doing nothing, but if you build a circuit that doesn't work, please begin by following the fault-tracing procedure that I have recommended (see "Fundamentals: Fault Tracing" on page 73). I will do my best to answer emails from readers who run into problems, but it's only fair for you to try to solve your problems first.

Writer–Reader Communication

There are three situations where you and I may want to communicate with each other.

- I may want to tell you if it turns out that the book contains a mistake which will prevent you from building a project successfully. I may also want to tell you if a parts kit, sold in association with the book, has something wrong with it. This is *me-informing-you* feedback.
- You may want to tell me if you think you found an error in the book, or in a parts kit. This is *you-informing-me* feedback.
- You may be having trouble making something work, and you don't know whether I made a mistake or you made a mistake. You would like some help. This is *you-asking-me* feedback.

I will explain how to deal with each of these situations.

Me Informing You

If you already registered with me in connection with *Make: More Electronics*, you don't need to register again for updates relating to *Make: Electronics*. But if you have not already registered, here's how it works.

I can't notify you if there's an error in the book or in a parts kit unless I have your contact information. Therefore I am asking you to send me your email address for the following purposes. Your email will not be used or abused for any other purpose.

- I will notify you if any significant errors are found in this book or in its successor, *Make: More Electronics*, and I will provide a workaround.
- I will notify you of any errors or problems relating to kits of components sold in association with this book or in *Make: More Electronics*.
- I will notify you if there is a completely new edition of this book, or of *Make: More Electronics*, or of my other books. These notifications will be very rare.

We've all seen registration cards that promise to enter you for a prize drawing. I'm going to offer you a much better deal. If you submit your email address, which may only be used for the three purposes listed above, I will send you an unpublished electronics project with complete construction plans as a two-page PDF. It will be fun, it will be unique, and it will be relatively easy. You won't be able to get this in any other way.

The reason I am encouraging you to participate is that if an error is found, and I have no way to tell you, and you discover it later on your own, you're likely to get annoyed. This will be bad for my reputation and the reputation of my work. It is very much in my interest to avoid a situation where you have a complaint.

- Simply send a blank email (or include some comments in it, if you like) to *make.electronics@gmail.com*. Please put REGISTER in the subject line.

You Informing Me

If you only want to notify me of an error that you have found, it's really better to use the "errata" system maintained by my publisher. The publisher uses the "errata" information to fix the error in updates of the book.

If you are sure that you found an error, please visit:

http://shop.oreilly.com/category/customer-service/faq-errata.do

The web page will tell you how to submit errata.

You Asking Me

My time is obviously limited, but if you attach a photograph of a project that doesn't work, I may have a suggestion. The photograph is essential.

You can use *make.electronics@gmail.com* for this purpose. Please put the word HELP in the subject line.

Going Public

There are dozens of forums online where you can discuss this book and mention any problems you are having, but please be aware of the power that you have as a reader, and use it fairly. A single negative review can create a bigger effect than you may realize. It can certainly outweigh half-a-dozen positive reviews.

The responses that I receive are generally very positive, but in a couple of cases people have been annoyed over small issues such as being unable to find a part online. I would have been happy to help these people if they had asked me.

I do read my reviews on Amazon about once each month, and will always provide a response if necessary.

Of course, if you simply don't like the way in which I have written this book, you should feel free to say so.

Going Further

After you work your way through this book, you will have grasped many of the basic principles in electronics. I like to think that if you want to know more, my own *Make: More Electronics* is the ideal next step. It is slightly more difficult, but uses the same "Learning by Discovery" method that I have used here. My intention is that you will end up with what I consider an "intermediate" understanding of electronics.

I am not qualified to write an "advanced" guide, and consequently I don't expect to create a third book with a title such as "Make: Even More Electronics."

If you want to know more electrical theory, *Practical Electronics for Inventors* by Paul Scherz is still the book that I recommend most often. You don't have to be an inventor to find it useful.

Safari® Books Online

Safari Books Online is an on-demand digital library that delivers expert content in both book and video form from the world's leading authors in technology and business.

Technology professionals, software developers, web designers, and business and creative professionals use Safari Books Online as their primary resource for research, problem solving, learning, and certification training.

Safari Books Online offers a range of plans and pricing for enterprise, government, education, and individuals.

Members have access to thousands of books, training videos, and prepublication manuscripts in one fully searchable database from publishers like O'Reilly Media, Prentice Hall Professional, Addison-Wesley Professional, Microsoft Press, Sams, Que, Peachpit Press, Focal Press, Cisco Press, John Wiley & Sons, Syngress, Morgan Kaufmann, IBM Redbooks, Packt, Adobe Press, FT Press, Apress, Manning, New Riders, McGraw-Hill, Jones & Bartlett, Course Technology, and hundreds more. For more information about Safari Books Online, please visit us online.

How to Contact Us

Please address comments and questions concerning this book to the publisher:

Make:
1160 Battery Street East, Suite 125
San Francisco, CA 94111
877-306-6253 (in the United States or Canada)
707-829-0515 (international or local)

Make: unites, inspires, informs, and entertains a growing community of resourceful people who undertake amazing projects in their backyards, basements, and garages. Make: celebrates your right to tweak, hack, and bend any technology to your will. The Make: audience continues to be a growing culture and community that believes in bettering ourselves, our environment, our educational system—our entire world. This is much more than an audience, it's a worldwide movement that Make: is leading—we call it the Maker Movement.

For more information about Make:, visit us online:

Make: magazine: *http://makezine.com/magazine/*
Maker Faire: *http://makerfaire.com*
Makezine.com: *http://makezine.com*
Maker Shed: *http://makershed.com/*

We have a web page for this book, where we list errata, examples, and any additional information. You can access this page at *http://bit.ly/make_elect_2e*.

To comment or ask technical questions about this book, send email to *bookquestions@oreilly.com*.

The Basics

1

Chapter One of this book contains Experiments 1 through 5.

In Experiment 1, I want you to get a taste for electricity—literally! You'll experience electric current and discover the nature of electrical resistance, not just in wires and components but in the world around you.

Experiments 2 through 5 will show you how to measure and understand the pressure and flow of electricity—and finally, how to generate electricity with everyday items on a tabletop.

Even if you have prior knowledge of electronics, I encourage you to try these experiments before venturing into subsequent parts of the book. They're fun, and they clarify some basic concepts.

Necessary Items for Chapter One

Each chapter of this book begins with pictures and descriptions of the tools, equipment, components, and supplies that will be required. After you have learned about them, you can flip to the back of the book where your buying options are summarized for quick reference.

- To buy tools and equipment, see "Buying Tools and Equipment" on page 324.
- For components, see "Components" on page 317.
- For supplies, see "Supplies" on page 316.
- If you prefer to get a prepackaged set of the components that you need, you have a choice of kits. See "Kits" on page 311 for more information.

I classify *tools and equipment* as items that should be useful indefinitely. They range from pliers to a multimeter. *Supplies*, such as wire and solder, will gradually be consumed in a variety of projects, but the quantities that I am recommending should be sufficient for all the experiments in the book. *Components* will be listed for individual projects, and will become part of those projects.

The Multimeter

Figure 1-1 *This kind of analog meter is inadequate for your purposes. You need a digital meter.*

I'm beginning my instructional overview of tools and equipment with the multimeter, because I consider it the most essential piece of equipment. It will tell you how much voltage exists between any two points in a circuit, or how much current is passing through the circuit. It will help you to find a wiring error, and can also evaluate a component to determine its electrical resistance—or its capacitance, which is the ability to store an electrical charge.

If you're starting with little or no knowledge, these terms may seem confusing, and you may feel that a multimeter looks complicated and difficult to use. This is not the case. It makes the learning process easier, because it reveals what you cannot see.

Before I discuss which meter to buy, I can tell you what not to buy. You don't want an old-school meter with a needle that moves across a scale, as shown in Figure 1-1. That is an *analog* meter.

You want a *digital* meter that displays values numerically —and to give you an idea of the equipment available, I have selected four examples.

Figure 1-2 shows the cheapest digital meter that I could find, costing less than a paperback novel or a six-pack of soda. It cannot measure very high resistances or very low voltages, its accuracy is poor, and it does not measure capacitance at all. However, if your budget is very tight, it will probably see you through the experiments in this book.

Figure 1-2 *The cheapest meter that I could find.*

The meter in Figure 1-3 offers more accuracy and more features. This meter, or one similar to it, is a good basic choice while you are learning electronics.

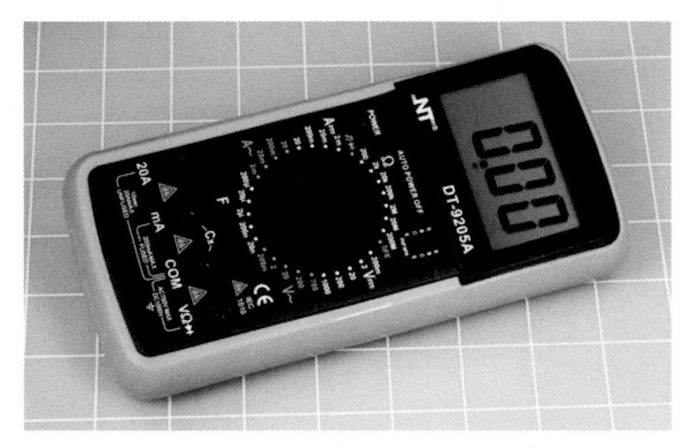

Figure 1-3 *Any meter similar to this one is a good basic choice.*

The example in Figure 1-4 is slightly more expensive but much better made. This particular model has been discontinued, but you can find many like it, probably costing two to three times as much as the NT brand in Figure 1-3. Extech is a well-established company trying to maintain its standards in the face of cut-price competitors.

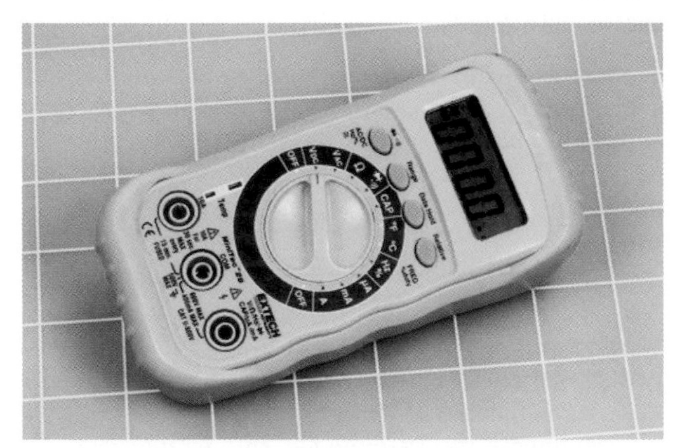

Figure 1-4 *A better-made meter at a somewhat higher price.*

Figure 1-5 shows my personal preferred meter at the time of writing. It is physically rugged, has all the features I could want, and measures a wide range of values with extremely good accuracy. However, it costs more than twenty times as much as the lowest-priced, bargain-basement product. I regard it as a long-term investment.

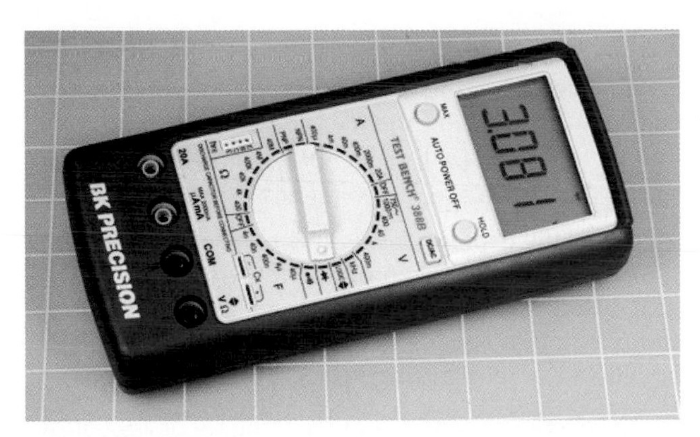

Figure 1-5 *A high-quality product.*

How do you decide which meter to buy? Well, if you were learning to drive, you wouldn't necessarily need a high-priced car. Similarly, you don't need a high-priced meter while you are learning electronics. On the other hand, the absolute cheapest meter may have some drawbacks, such as an internal fuse that is not easily replaceable, or a rotary switch with contacts that wear out quickly. So here's a rule of thumb if you want something that I would regard as inexpensive but acceptable:

- Search eBay for the absolute cheapest model you can find, then double the price, and use that as your guideline.

Regardless of how much you spend, the following attributes and capabilities are important.

Ranging

A meter can measure so many values, it has to have a way to narrow the range. Some meters have *manual ranging*, meaning that you turn a dial to choose a ballpark for the quantity that interests you. A range could be from 2 to 20 volts, for instance.

Other meters have *autoranging*, which is more convenient, because you just connect the meter and wait for it to figure everything out. The key word, however, is "wait." Every time you make a measurement with an autoranging meter, you will wait a couple of seconds while it performs an internal evaluation. Personally I tend to be impatient, so I prefer manual meters.

Another problem with autoranging is that because you have not selected a range yourself, you must pay attention to little letters in the display where the meter is telling you which units it has decided to use. For example, the difference between a "K" or an "M" when measuring electrical resistance is a factor of 1,000. This leads me to my personal recommendation:

- I suggest you use a manual-ranging meter for your initial adventures. You'll have fewer chances to make errors, and it should cost slightly less.

A vendor's description of a meter should say whether it uses manual ranging or autoranging, but if not, you can tell by looking at a photograph of its selector dial. If you don't see any numbers around the dial, it's an autoranging meter. The meter in Figure 1-4 does autoranging. The others that I pictured do not.

Values

The dial will also reveal what types of measurements are possible. At the very least, you should expect:

Volts, amps, and ohms, often abbreviated with the letter V, the letter A, and the ohm symbol, which is the Greek letter omega, shown in Figure 1-6. You may not know what these attributes mean right now, but they are fundamental.

Your meter should also be capable of measuring milliamps (abbreviated *mA*) and millivolts (abbreviated *mV*). This may not be immediately clear from the dial on the meter, but will be listed in its specification.

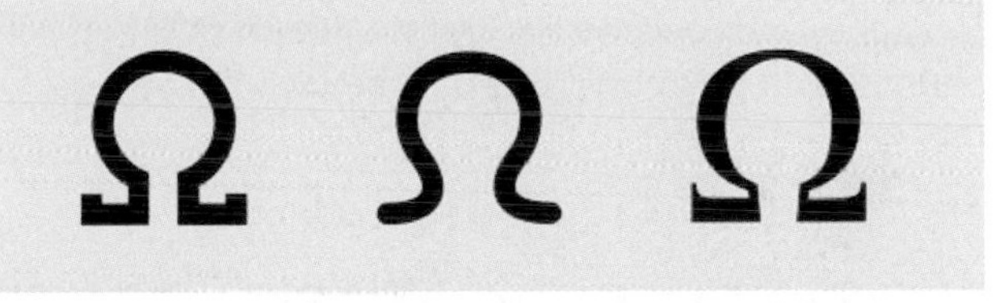

Figure 1-6 *Three samples of the Greek symbol omega, used to represent electrical resistance.*

DC/AC, meaning direct current and alternating current. These options may be selected with a DC/AC pushbutton, or they may be chosen on the main selector dial. A pushbutton is probably more convenient.

Continuity testing. This useful feature enables you to check for bad connections or breaks in an electrical circuit. Ideally it should create an audible alert, in which case it will be represented symbolically with a little dot that has semicircular lines radiating from it, as shown in Figure 1-7.

Figure 1-7 *This symbol indicates the option to test a circuit for continuity, with audible feedback. It's a very useful feature.*

For a small additional sum, you should be able to buy a meter that makes the following measurements. In order of importance:

Capacitance. Capacitors are small components that are needed in the majority of electronic circuits. Because small ones usually don't have their values printed on them, the ability to measure their values can be important, especially if some of them get mixed up or (worse) fall on the floor. Very cheap meters usually cannot measure capacitance. When the feature exists, it is usually indicated with a letter F, meaning farads, which are the units of measurement. The abbreviation CAP may also be used.

Transistor testing, indicated by little holes labeled E, B, C, and E. You plug the transistor into the holes. This enables you to verify which way up the transistor should be placed in a circuit, or if you have burned it out.

Frequency, abbreviated Hz. This is unimportant in the experiments in this book, but may be useful if you go further.

Any features beyond these are not significant.

If you still feel unsure about which meter to buy, read ahead a little to get an idea of how you will be using a meter in Experiments 1, 2, 3, and 4.

Safety Glasses

For Experiment 2, you may want to use safety glasses. The cheapest plastic type is satisfactory for this little adventure, as the risk of a battery bursting is almost nonexistent, and probably would not occur with much force.

Regular eyeglasses would be an acceptable substitute, or you could view the experiment through a little piece of transparent plastic (for instance, you can cut out a piece of a water bottle).

Batteries and Connectors

Because batteries and connectors become part of a circuit, I am categorizing them as components. See "Other Components" on page 319 for details about ordering these parts.

Almost all the experiments in this book will use a power source of 9 volts. You can obtain this from a basic nine-volt battery sold in supermarkets and convenience stores. Later I'll suggest an upgrade to an AC adapter, but you don't need that right now.

For Experiment 2, you will need a couple of 1.5-volt AA batteries. These have to be the alkaline type. You must not perform this experiment with any kind of rechargeable battery.

To transfer the power from a battery into a circuit, you need a connector for the 9-volt battery, as shown in Figure 1-8, and a carrier for a single AA battery, as shown in Figure 1-9. One carrier will be enough, but I suggest you get at least three 9-volt connectors for future use.

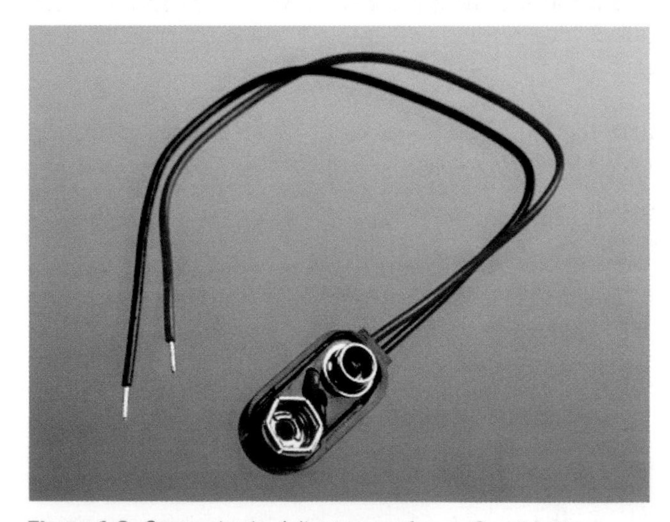

Figure 1-8 *Connector to deliver power from a 9-volt battery.*

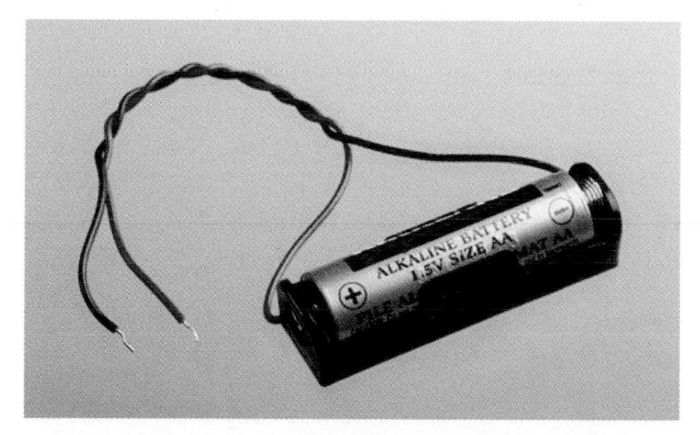

Figure 1-9 *You need a carrier like this for a single AA battery. Don't buy the type of carrier that holds two batteries (or three, or four).*

Test Leads

You will use test leads (pronounced "leeds") to connect components with each other in the first few experiments. The type of leads I mean are *double-ended*. Surely, any piece of wire has two ends, so why should it be called "double-ended"? The term usually means that each end is fitted with an *alligator clip* as shown in Figure 1-10. Each clip can make a connection by grabbing something and gripping it securely, freeing you to use your hands elsewhere.

You don't want the kind of test leads that have a plug at each end. Those are sometimes known as *jumper wires*.

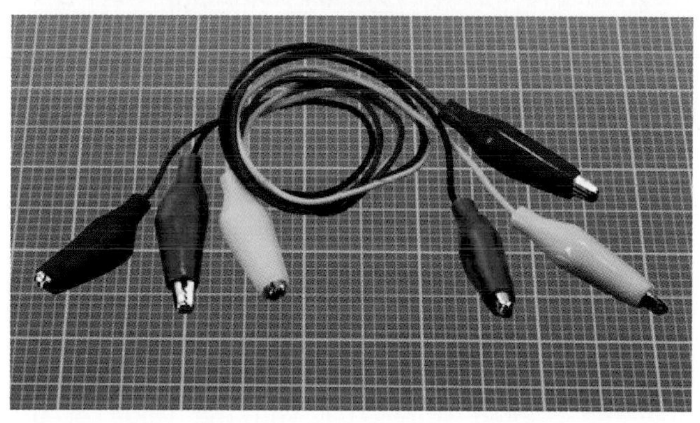

Figure 1-10 *Double-ended test leads with an alligator clip at each end.*

Test leads are classified as equipment for the purposes of this book. See "Buying Tools and Equipment" on page 324 for more information.

Potentiometer

A potentiometer functions like the volume control on an old-fashioned stereo. The kinds shown in Figure 1-11 are considered large by modern standards, but large is what you need, because you'll be gripping the terminals with the alligator clips on your test leads. A 1" diameter potentiometer is preferred. Its resistance should be listed as 1K. If you are buying your own, see "Other Components" on page 319 for details.

Figure 1-11 *Potentiometers of the general type required for your first experiments.*

Fuse

A fuse interrupts a circuit if too much electricity passes through it. Ideally you'll buy the type of 3-amp automotive fuse shown in Figure 1-12, which is easy to grip with test leads, and clearly reveals the element inside it. Automotive fuses are sold in a variety of physical sizes, but so long as you use one rated for 3 amps, the dimensions don't matter. Buy three to allow for destroying them intentionally or accidentally. If you don't want to use an auto parts supplier, a 2AG-size 3-amp glass cartridge fuse of the kind shown in Figure 1-13 will be available from electronics component suppliers, although it is not quite so easy to use.

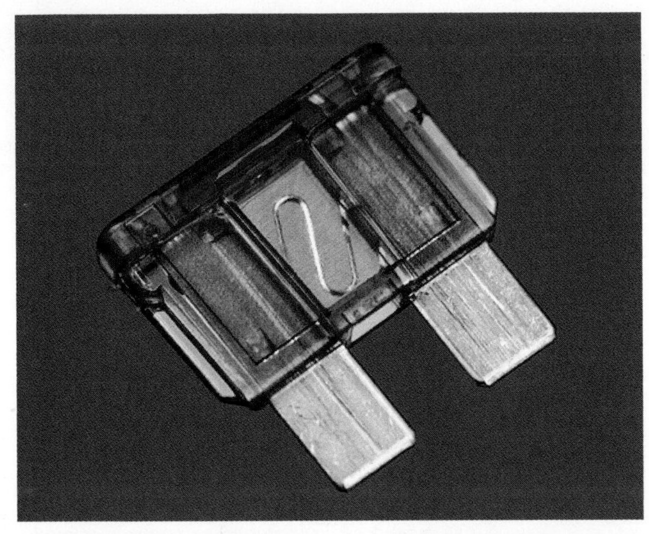

Figure 1-12 *This type of automotive fuse is easier to handle than the cartridge fuses used in electronics hardware.*

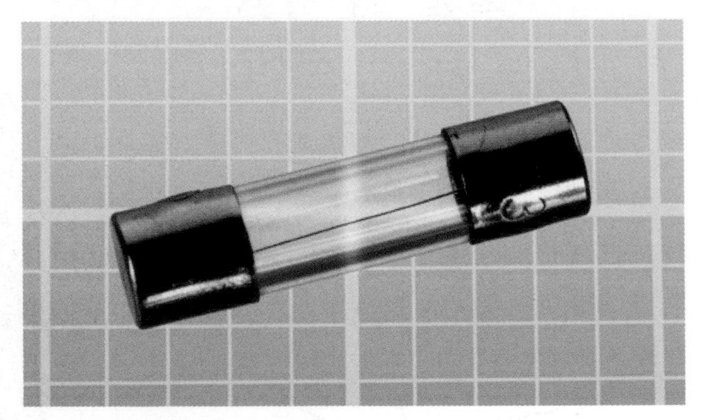

Figure 1-13 *You can use a cartridge fuse like this, although your alligator clips won't grip it so easily.*

Light-Emitting Diodes

More commonly known as *LEDs*, they come in various shapes and forms. The ones we will be using are properly known as *LED indicators*, and are often described as *standard through-hole LEDs* in catalogs. A sample in Figure 1-14 is 5mm in diameter, but 3mm is sometimes easier to fit into a circuit when space is limited. Either will do.

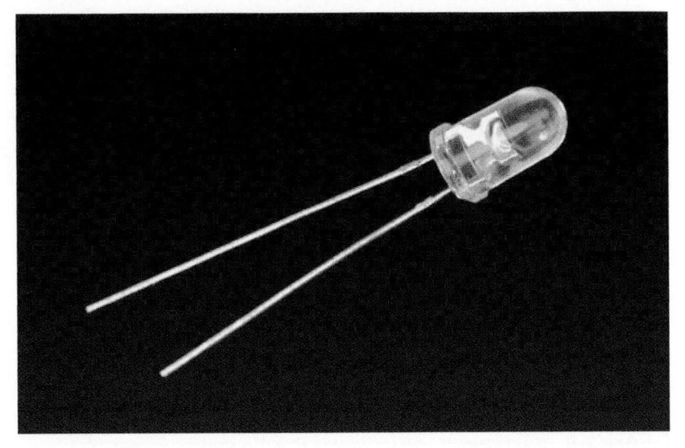

Figure 1-14 *A light-emitting diode (LED) approximately 5mm in diameter.*

Throughout this book I will refer to *generic LEDs*, by which I mean the cheapest ones that don't emit a high-intensity light and are commonly available in red, yellow, or green. They are often sold in bulk quantities, and are used in so many applications that I suggest you buy at least a dozen of each color.

Some generic LEDs are encapsulated in "water clear" plastic or resin, but emit a color when power is applied. Other LEDs are encapsulated in plastic or resin tinted with the same color that they will display. Either type is acceptable.

In a few experiments, *low-current LEDs* are preferred. They cost slightly more, but are more sensitive. For example, in Experiment 5, where you will generate a small amount of current with an improvised battery, you'll get better results with a low-current LED. See "Other Components" on page 319 for additional guidance, if you are not using components that were supplied in a kit.

Resistors

You'll need a variety of resistors to restrict the voltage and current in various parts of a circuit. They should look something like the ones in Figure 1-15. The color of the body of the resistor doesn't matter. Later I will explain how the colored stripes tell you the value of the component.

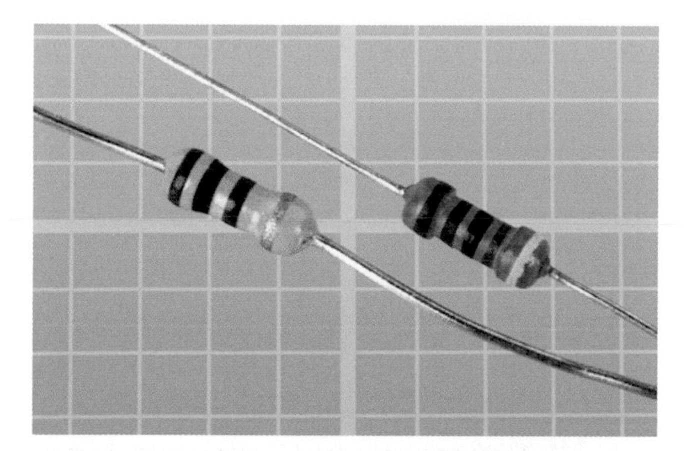

Figure 1-15 *Two resistors of the type you need, all rated for 1/4 watt.*

If you are buying your own resistors, they are so small and cheap, you would be foolish to select just the values listed in each experiment. Get a prepackaged selection in bulk from surplus or discount sources, or eBay. For more information about resistors, including a complete list of all the values used throughout this book, see "Components" on page 317.

You don't need any other components to take you through Experiments 1 through 5. So let's get started!

Experiment 1: Taste the Power!

Can you taste electricity? It feels as if you can.

What You Will Need

- 9-volt battery (1)
- Multimeter (1)

That's all!

Caution: No More than Nine Volts

This experiment should only use a 9-volt battery. *Do not* try it with a higher voltage, and *do not* use a bigger battery that can deliver more current. Also, if you have metal braces on your teeth, be careful not to touch them with the battery. Most important, never apply electric current from any size of battery through a break in your skin.

Procedure

Moisten your tongue and touch the tip of it to the metal terminals of a 9-volt battery, as shown in Figure 1-16. (Maybe your tongue isn't quite as big as the one in the picture. Mine certainly isn't. But this experiment will work regardless of how big or small your tongue may be.)

Figure 1-16 *An intrepid Maker tests the characteristics of an alkaline battery.*

Do you feel that tingle? Now set aside the battery, stick out your tongue, and dry the tip of it very thoroughly with a tissue. Touch the battery to your tongue again, and you should feel less of a tingle.

What's happening here? You can use a meter to find out.

Setting Up Your Meter

Does your meter have a battery preinstalled? Select any function with the dial, and wait to see if the display shows a number. If nothing is visible, you may have to open the meter and put in a battery before you can use it. To find out how to do this, check the instructions that came with the meter.

Meters are supplied with a red lead and a black lead. Each lead has a plug on one end, and a steel probe on the other end. You insert the plugs into the meter, then touch the probes on locations where you want to know what's going on. See Figure 1-17. The probes detect electricity; they don't emit it in significant quantities. When you are dealing with the small currents and voltages in the experiments in this book, the probes cannot hurt you (unless you poke yourself with their sharp ends).

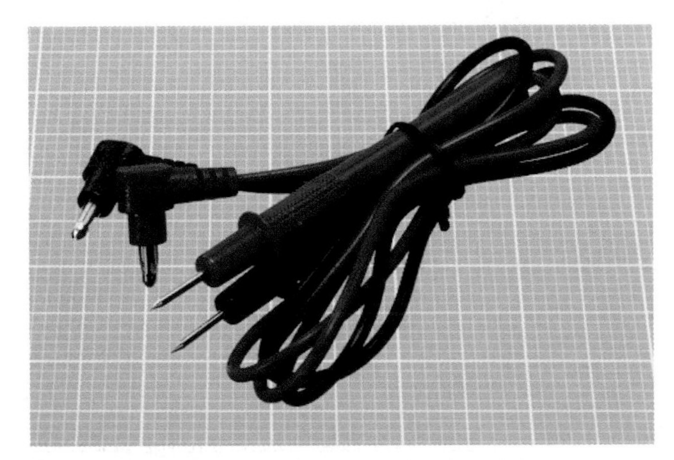

Figure 1-17 *Leads for a meter, terminating in metal probes.*

Most meters have three sockets, but some have four. See Figure 1-18, Figure 1-19, and Figure 1-20 for examples. Here are the general rules:

- One socket should be labelled COM. This is *common* to all your measurements. Plug the black lead into this socket, and leave it there.
- Another socket should be identified with the ohm (omega) symbol, and the letter V for volts. It can measure either resistance or voltage. Plug the red lead into this socket.
- The voltage/ohms socket may also be used for measuring small currents in mA (milliamps) . . . or you may see a separate socket for this, which will require you to move the red lead sometimes. We'll get to that later.
- An additional socket may be labelled 2A, 5A, 10A, 20A, or something similar, to indicate a maximum number of amps. This is used for measuring high currents. We won't be needing it for projects in this book.

Figure 1-18 *Note the labeling of sockets on this meter.*

Figure 1-19 *Socket functions are split up differently on this meter.*

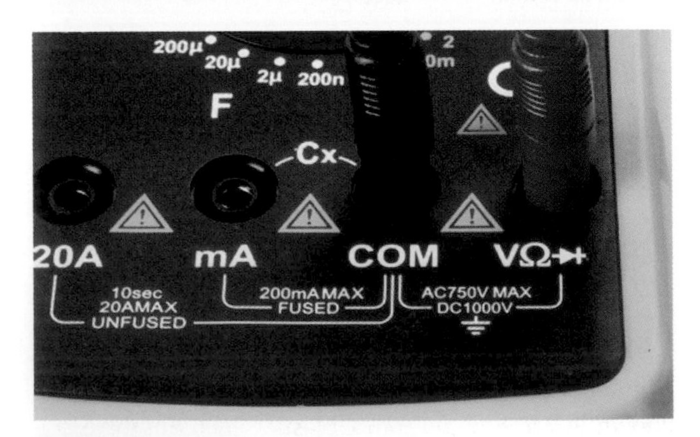

Figure 1-20 *Sockets on one more meter.*

Fundamentals: Ohms

You're going to evaluate the resistance of your tongue, in ohms. But what is an ohm?

We measure distance in miles or kilometers, mass in pounds or kilograms, and temperature in Fahrenheit or Centigrade. We measure electrical resistance in ohms, which is an international unit named after Georg Simon Ohm, who was an electrical pioneer.

The Greek omega symbol indicates ohms, but for resistances above 999 ohms the uppercase letter K is used, which means *kilohm*, equivalent to a thousand ohms. For example, a resistance of 1,500 ohms will be referred to as 1.5K.

Above 999,999 ohms, the uppercase letter M is used, meaning *megohm*, which is a million ohms. In everyday speech, a megohm is often referred to as a "meg." If someone is using a "two-point-two meg resistor," its value will be 2.2M.

A conversion table for ohms, kilohms, and megohms is shown in Figure 1-21.

Ohms	Kilohms	Megohms
1 Ω	0.001K	0.000001M
10 Ω	0.01K	0.00001M
100 Ω	0.1K	0.0001M
1,000 Ω	1K	0.001M
10,000 Ω	10K	0.01M
100,000 Ω	100K	0.1M
1,000,000 Ω	1,000K	1M

Figure 1-21 *Conversion table for the most common multiples of ohms.*

- In Europe, the letter R, K, or M is substituted for a decimal point, to reduce the risk of errors. Thus, 5K6 in a European circuit diagram means 5.6K, 6M8 means 6.8M, and 6R8 means 6.8 ohms. I won't be using the European style here, but you may find it in some circuit diagrams elsewhere.

A material that has very high resistance to electricity is known as an *insulator*. Most plastics, including the colored sheaths around wires, are insulators.

A material with very low resistance is a *conductor*. Metals such as copper, aluminum, silver, and gold are excellent conductors.

Measuring Your Tongue

Inspect the dial on the front of your meter, and you'll find at least one position identified with the ohm symbol. On an autoranging meter, turn the dial to point to the ohm symbol as shown in Figure 1-22, touch the probes *gently* to your tongue, and wait for the meter to choose a range automatically. Watch for the letter K in the numeric display. Never stick the probes *into* your tongue!

Figure 1-22 *On an autoranging meter, just turn the dial to the ohm (omega) symbol.*

On a manual meter, you must choose a range of values. For a tongue measurement, probably 200K (200,000 ohms) would be about right. Note that the numbers beside the dial are maximums, so 200K means "no more than 200,000 ohms" while 20K means "no more than 20,000 ohms." See the close-ups of the manual meters in Figure 1-23 and Figure 1-24.

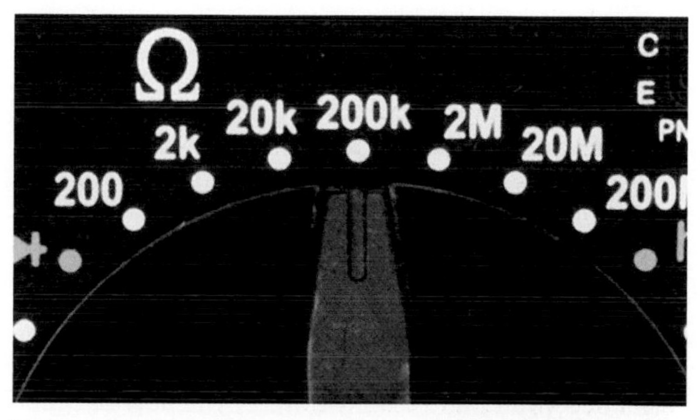

Figure 1-23 *A manual meter requires you to select the range.*

Figure 1-24 *A different manual meter dial, but the principle is the same.*

Touch the probes to your tongue about one inch apart. Note the meter reading, which should be around 50K. Put aside the probes, stick out your tongue, and use a tissue to dry it carefully and thoroughly, as you did before. Without allowing your tongue to become moist again, repeat the test, and the reading should be higher. Using a manual ranging meter, you may have to select a higher range to see a resistance value.

- When your skin is moist (for instance, if you perspire), its electrical resistance decreases. This principle is used in lie detectors, because someone who knowingly tells a lie, under conditions of stress, may tend to perspire.

Here's the conclusion that your test may suggest. A lower resistance allows more electric current to flow, and in your initial tongue test, more current created a bigger tingle.

Fundamentals: Inside a Battery

When you used a battery for the original tongue test, I didn't bother to mention how a battery works. Now is the time to rectify that omission.

A 9-volt battery contains chemicals that liberate *electrons* (particles of electricity), which want to flow from one terminal to the other as a result of a chemical reaction. Think of the cells inside a battery as being like two water tanks—one of them full, the other empty. If the tanks are connected with each other by a pipe and a valve, and you open the valve, water will flow between them until their levels are equal. Figure 1-25 may help you to visualize this. Similarly, when you open up an electrical pathway between the two sides of a battery, electrons flow between them, even if the pathway consists only of the moisture on your tongue.

Electrons flow more easily through some substances (such as a moist tongue) than others (such as a dry tongue).

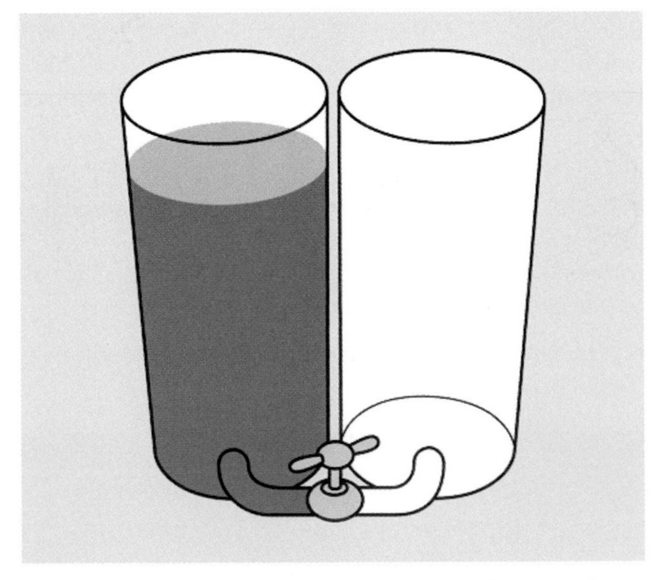

Figure 1-25 *You can think of a battery as being like a pair of interconnected water reservoirs.*

Further Investigation

The tongue test was not a very well-controlled experiment, because the distance between the probes might have varied a little between one trial and the next. Do you think that may be significant? Let's find out.

Hold the meter probes so that their tips are only 1/4" apart. Touch them to your moist tongue. Now separate the probes by 1" and try again. What readings do you get?

When electricity travels through a shorter distance, it encounters less resistance. As a result, the current will increase.

Try a similar experiment on your arm, as shown in Figure 1-26. You can vary the distance between the probes in fixed steps, such as 1/4", and note the resistance shown by your meter. Do you think that doubling the distance between the probes doubles the resistance shown by the meter? How can you prove or disprove this?

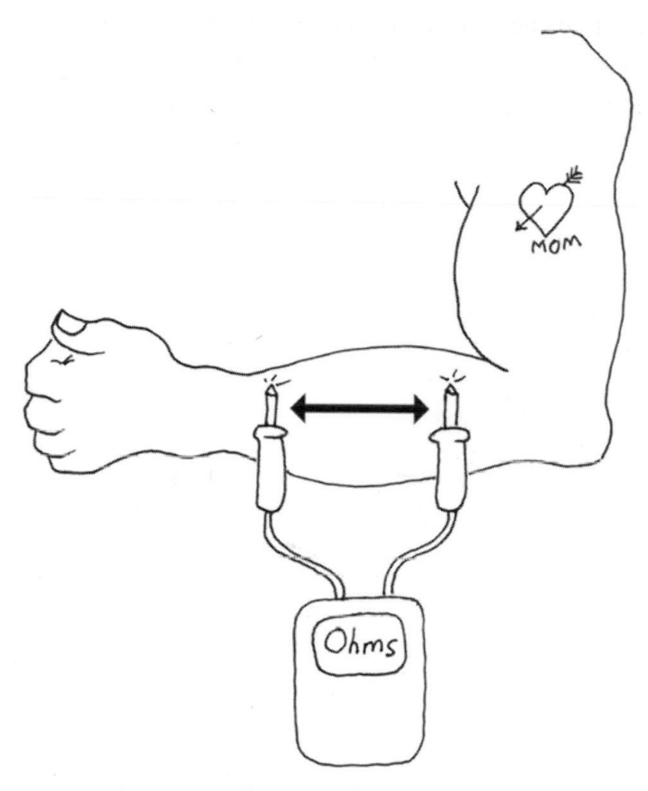

Figure 1-26 *Vary the distance between the probes, and note the reading on your meter.*

If the resistance is too high for your meter to measure, you will see an error message, such as L, instead of some numbers. Try moistening your skin, then repeat the test, and you should get a result. The only problem is, as the moisture on your skin evaporates, the resistance will change. You see how difficult it is to control all the factors in an experiment. The random factors are properly known as *uncontrolled variables*.

There is still one more variable that I haven't discussed, which is the amount of pressure between each probe and the skin. If you press harder, I suspect that the resistance will diminish. Can you prove this? How could you design an experiment to eliminate this variable?

If you're tired of measuring skin resistance, you can try dunking the probes into a glass of water. Then dissolve some salt in the water, and test it again. No doubt you've heard that water conducts electricity, but the full story is not so simple. Impurities in water play an important part.

What do you think will happen if you try to measure the resistance of water that contains no impurities at all? Your first step would be to obtain some pure water. So-called *purified water* usually has minerals added after it was purified, so, that's not what you want. Similarly, *spring water* is not totally pure. What you need is *distilled water*, also known as *deionized water*. This is often sold in supermarkets. I think you'll find that its resistance, per inch between the meter probes, is higher than the resistance of your tongue. Try it to find out.

Those are all the experiments relating to resistance that I can think of, right now. But I still have a little background information for you.

Background: The Man Who Discovered Resistance

Georg Simon Ohm, pictured in Figure 1-27, was born in Bavaria in 1787 and worked in obscurity for much of his life, studying the nature of electricity using metal wire that he had to make for himself (you couldn't truck on down to Home Depot for a spool of hookup wire back in the early 1800s).

Despite his limited resources and inadequate mathematical abilities, Ohm was able to demonstrate in 1827 that the electrical resistance of a conductor such as copper varied in inverse proportion with its area of cross-section, and the current flowing through it is proportional to the voltage applied to it, so long as temperature is held constant. Fourteen years later, the Royal Society in London finally recognized the significance of his contribution and awarded him the Copley Medal. Today, his discovery is known as Ohm's Law. I'll explain more about this in Experiment 4.

Figure 1-27 *Georg Simon Ohm, after being honored for his pioneering work, most of which he pursued in relative obscurity.*

Cleanup and Recycling

Your battery should not have been damaged or significantly discharged by this experiment. You can use it again.

Remember to switch off your meter before putting it away. Many meters will beep to remind you to switch them off if you don't use them for a while, but some don't. A meter consumes a very small amount of electricity while it is switched on, even when you are not using it to measure anything.

Experiment 2: Let's Abuse a Battery!

To get a better feeling for electrical power, you're going to do what most books tell you not to do. You're going to short out a battery. (A *short circuit* is a shortcut between two sides of a power source.)

Caution: Use a Small Battery

The experiment that I am going to describe is safe, but some short circuits can be dangerous. Never short out a power outlet in your home: there'll be a loud bang, a bright flash, and the wire or tool that you use will be partially melted, while flying particles of melted metal can burn you or blind you.

If you short out a car battery, the flow of current is so huge that the battery might even explode, drenching you in acid. Just ask the guy in Figure 1-28 (if he is able to answer).

Figure 1-28 *Dropping a wrench between the terminals of a car battery will be bad for your health. Short circuits can be dramatic, even at a "mere" 12 volts, if the battery is big enough.*

Lithium batteries are often found in power tools, laptop computers, and other portable devices. Never short-circuit a lithium battery: it can catch fire and burn you. Lithium batteries have been known to catch fire even if you don't short them out, as shown in Figure 1-29. After some early laptops self-destructed, lithium battery packs were modified to prevent this kind of thing. But short-circuiting them is still a very bad idea.

Figure 1-29 *Never fool around with lithium batteries.*

Use only an alkaline battery in this experiment, and only a single AA cell. You may also want to wear safety glasses in case you happen to have a defective battery.

What You Will Need

- 1.5-volt AA battery (2)
- Battery carrier (1)
- 3-amp fuse (2)
- Safety glasses (regular eyeglasses or sunglasses will do)
- Test leads with alligator clips at each end (2)

Generating Heat with Current

Use an alkaline battery. Do not use any kind of rechargeable battery.

Put the battery into a battery carrier that terminates in two thin wires, as shown in Figure 1-9. Twist the bare ends of the wires together, as shown in Figure 1-30. At first it seems that nothing happens. But wait one minute, and you'll find that the wires are getting hot. Wait another minute, and the battery, too, will be hot.

Figure 1-30 *Shorting out an alkaline battery can be safe if you follow the directions precisely.*

The heat is caused by electricity flowing through the wires and through the *electrolyte* (the conductive fluid) inside the battery. If you've ever used a hand pump to force air into a bicycle tire, you know that the pump gets warm. Electricity behaves in much the same way. You can imagine the electricity being composed of particles (electrons) that make the wire hot as they push through it. This isn't a perfect analogy, but it's close enough for our purposes.

Where do the electrons come from? Chemical reactions inside the battery liberate them, creating electrical pressure. The correct name for this pressure is *voltage*, which is measured in volts and is named after Alessandro Volta, another electrical pioneer.

Going back to the water analogy: the height of the water in a tank is proportional with the pressure of the water, and is similar to voltage. Figure 1-31 may help you to visualize this.

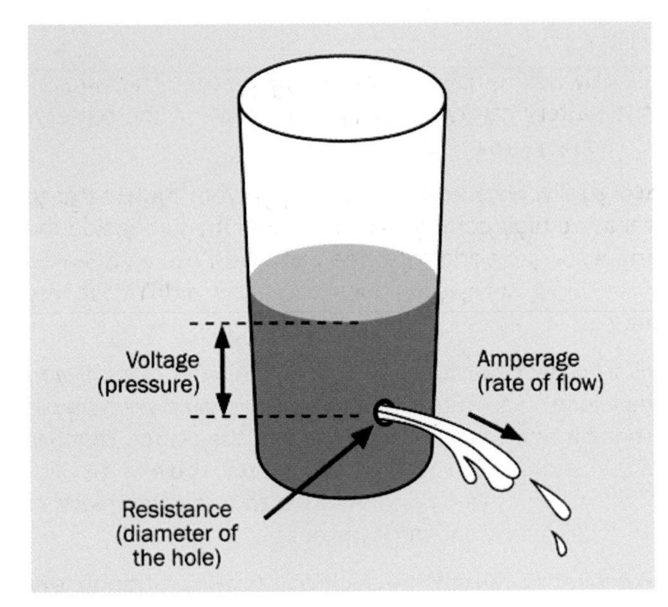

Figure 1-31 *The pressure in a source of water is analogous to the voltage in a source of electricity.*

But volts are only half of the story. When electrons flow through a wire, the amount of flow during a period of time is known as *amperage*, named after yet another electrical pioneer, André-Marie Ampère. That flow is also generally known as *current*. The current—the amperage—generates the heat.

- Think of voltage as pressure.
- Think of amperes as the rate of flow, properly known as current.

Background: Why Didn't Your Tongue Get Hot?

When you touched the 9-volt battery to your tongue, you felt a tingle, but no perceptible heat. When you shorted out a battery, you generated a noticeable amount of heat, even though you only used a 1.5-volt battery. How can you explain this?

Your meter showed you that the electrical resistance of your tongue is very high. This high resistance reduced the flow of electrons.

The resistance of a wire is very low, so if there's a wire connecting the two terminals of the battery, more current will pass through it than passed through your tongue, and it will create more heat. If all other factors remain constant:

- Lower resistance allows more current to flow.
- The heat generated by electricity is proportional with the amount of electricity (the current) passing through a conductor in a period of time. (This relationship is no longer exactly true if the resistance of the wire changes as the wire gets hot. But it remains approximately true.)

Here are some other basic concepts:

- The flow of electricity per second is measured in amperes, very commonly abbreviated as *amps*.
- The pressure of electricity that causes the flow is measured in *volts*.
- The resistance to the flow is measured in *ohms*.
- A higher resistance restricts the current.
- A higher voltage is better able to overcome resistance and increase the current.

The relationship between voltage, resistance, and amperage (pressure, resistance, and flow) is illustrated in Figure 1-32.

Figure 1-32 *Resistance blocks pressure and reduces flow—in water, and in electricity.*

Fundamentals: Volt Basics

The volt is an international unit, represented by an uppercase letter V. In the United States and much of Europe, AC power for domestic use is supplied at 110V, 115V, or 120V, with separate circuits for heavy-duty appliances delivering 220V, 230V, or 240V. Solid-state electronic components traditionally required DC power ranging from 5V to about 20V, although modern surface-mount devices may use less than 2V. Some components, such as a microphone, deliver voltage measured in millivolts, abbreviated mV, each millivolt being 1/1,000th of a volt. When electricity is distributed over long distances, it is measured in kilovolts, abbreviated kV. A few long-distance lines use megavolts. A conversion table for millivolts, volts, and kilovolts is shown in Figure 1-33.

Millivolts	Volts	Kilovolts
1mV	0.001V	0.000001kV
10mV	0.01V	0.00001kV
100mV	0.1V	0.0001kV
1,000mV	1V	0.001kV
10,000mV	10V	0.01kV
100,000mV	100V	0.1kV
1,000,000mV	1,000V	1kV

Figure 1-33 *Conversion table for the most common multiples of volts.*

Fundamentals: Ampere Basics

The ampere is an international unit, represented by an uppercase letter A. Household appliances may draw several amps, and a typical circuit breaker in the United States is rated for 20A. Electronic components often are rated in milliamps, abbreviated mA, each milliamp being 1/1,000th of an amp. Devices such as liquid-crystal displays may draw microamps, abbreviated μA, each microamp being 1/1,000th of a milliamp. A conversion table for amps, milliamps, and microamps is shown in Figure 1-34.

Microamps	Milliamps	Amps
1μA	0.001mA	0.000001A
10μA	0.01mA	0.00001A
100μA	0.1mA	0.0001A
1,000μA	1mA	0.001A
10,000μA	10mA	0.01A
100,000μA	100mA	0.1A
1,000,000μA	1,000mA	1A

Figure 1-34 *Conversion table for the most common multiples of amps.*

How to Blow a Fuse

Exactly how much current flowed through the wires of the battery carrier, when you shorted out the battery? Could we have measured it?

Not easily, because if you try to use your multimeter to measure high current, you may blow the fuse inside the meter. So set aside the meter, and we'll use your 3-amp fuse, which we can sacrifice because it didn't cost very much.

First inspect the fuse, using a magnifying glass, if you have one. In an automotive fuse, you should see a tiny S-shape in the transparent window at the center. That S is a thin section of metal that melts easily. You saw it in Figure 1-12. In a glass cartridge fuse, there's a thin piece of wire that serves the same purpose.

Remove the 1.5-volt battery from its carrier. The battery is no longer useful for anything, and should be recycled if possible. Separate the two wires that you twisted together, and use two test leads to connect the carrier with a fuse, as shown in Figure 1-35 or Figure 1-36. Watch the fuse as you insert a new battery in the carrier. A break should occur in the center of the fuse element, where the metal melted. Figure 1-37 and Figure 1-38 show what I mean.

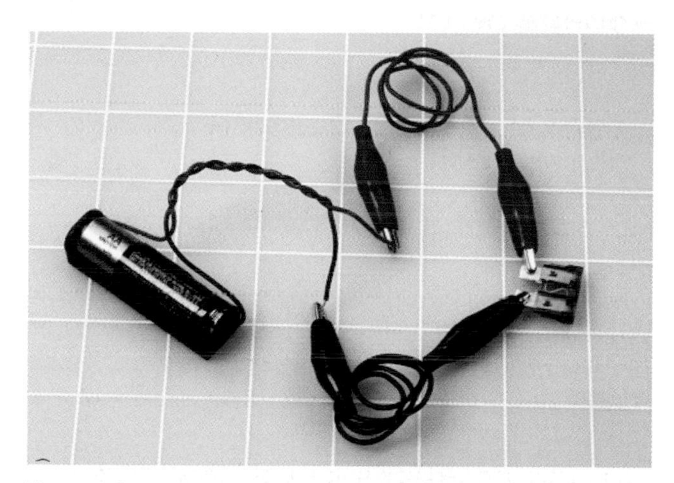

Figure 1-35 *How to short out an automotive fuse.*

Some 3-amp fuses blow more easily than others, even though they all have the same rating. If you think your fuse has not been affected, try applying the wires from the battery to it directly, instead of passing the current through the test leads. If you are not using a fresh AA battery, you may have to wait a few seconds for the fuse to respond. If all else fails, you can apply a C cell or a D cell, which have the same voltage as an AA battery but can deliver more current. But this shouldn't be necessary.

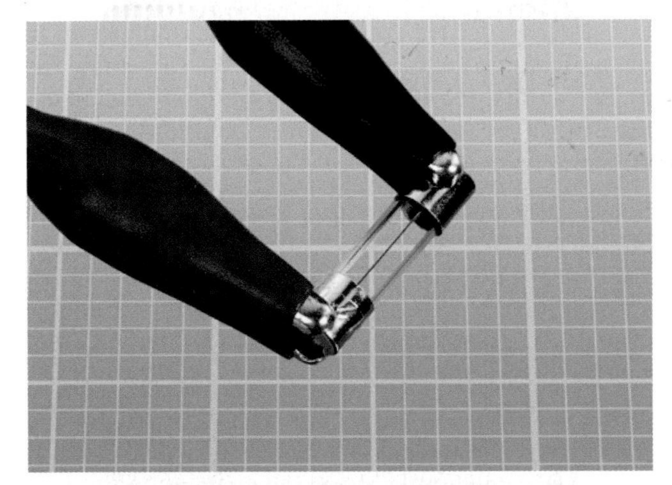

Figure 1-36 *How to apply your test leads to a small cartridge fuse.*

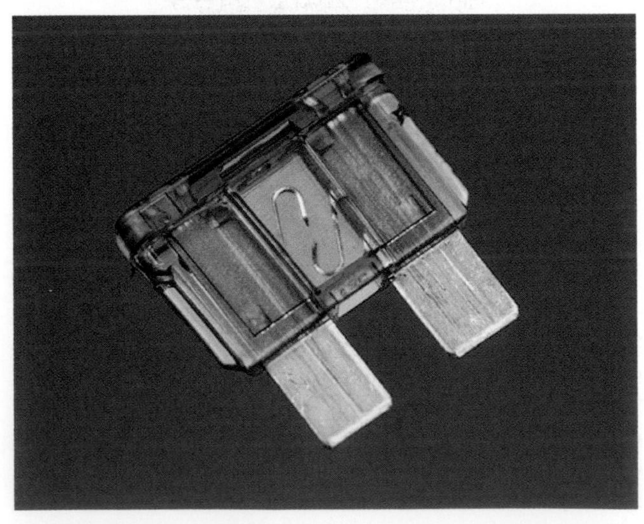

Figure 1-37 *Note the break in the element.*

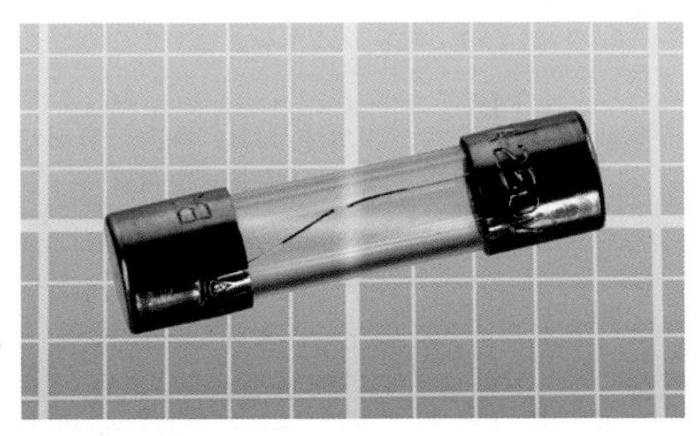

Figure 1-38 *In a shorted cartridge fuse, a similar break appears.*

This is how a fuse works: it melts to protect the rest of the circuit. That tiny break inside the fuse stops any more current from flowing.

Fundamentals: Direct and Alternating Current

The flow of current that you get from a battery is known as *direct current*, or DC. Like the flow of water from a faucet, it is a steady stream that flows in one direction.

The flow of current that you get from the power outlet in your home is very different. The "live" side of the outlet changes from positive to negative, relative to the "neutral" side, at a rate of 60 times each second (in many foreign countries, including Europe, 50 times per second). This is known as *alternating current*, or AC, which is more like the pulsing flow you get when you use a power washer to wash a car.

Alternating current is essential for some purposes, such as cranking up voltage so that electricity can be distributed over long distances. AC is also useful in motors and domestic appliances. The parts of a power outlet are shown in Figure 1-39. This style of outlet is found in North America, South America, Japan, and some other nations. European outlets look different, but the principle remains the same.

In the figure, socket A is the "hot" or "live" side of the outlet, supplying voltage that alternates between positive and negative relative to socket B, which is the "neutral" side. If an appliance develops a fault such as an internal loose wire, it should protect you by sinking the voltage through socket C, the ground.

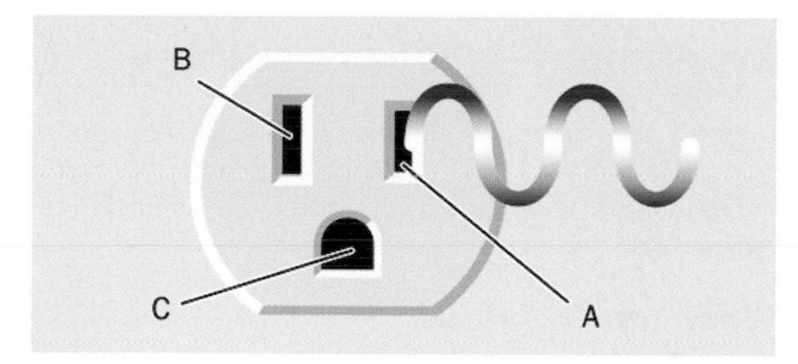

Figure 1-39 *The parts of a power outlet.*

In the United States, the outlet shown in the diagram is rated for 110 to 120 volts. Other configurations of outlet are used for higher voltages, but they still have live, neutral, and ground wires—with the exception of three-phase outlets, which are used primarily in industry.

For most of this book I'm going to be talking about DC, for two reasons: first, most simple electronic circuits are powered with DC, and second, the way it behaves is much easier to understand.

- I won't bother to mention repeatedly that I'm dealing with DC. Just assume that everything is DC unless otherwise noted.

Background: Inventor of the Battery

Alessandro Volta, shown in Figure 1-40, was born in Italy in 1745, long before science was broken up into specialties. After studying chemistry (he discovered methane in 1776) he became a professor of physics and developed an interest in the so-called galvanic response, whereby a frog's leg will twitch in response to a jolt of static electricity.

Using a wine glass full of salt water, Volta demonstrated that the chemical reaction between two electrodes (one made of copper, the other of zinc) will generate a steady electric current. In 1800, he refined his apparatus by stacking plates of copper and zinc, separated by cardboard soaked in salt and water. This "voltaic pile" was the first electric battery in Western civilization.

Figure 1-40 *Alessandro Volta discovered that chemical reactions can create electricity.*

Background: Father of Electromagnetism

Born in 1775 in France, André-Marie Ampère (shown in Figure 1-41) was a mathematical prodigy who became a science teacher, despite being largely self-educated in his father's library. His best-known work was to derive a theory of electromagnetism in 1820, describing the way that an electric current generates a magnetic field. He also built the first instrument to measure the flow of electricity (now known as a *galvanometer*), and discovered the element fluorine.

Figure 1-41 *André-Marie Ampère found that an electric current running through a wire creates a magnetic field around it. He used this principle to make the first reliable measurements of what came to be known as amperage.*

Cleanup and Recycling

You can dispose of the battery that you damaged when you shorted it. Putting batteries in the trash is not a great idea, because they contain heavy metals that should be kept out of the ecosystem. Your state or town may include batteries in a local recycling scheme. (California requires that almost all batteries be recycled.) Check your local regulations for details.

The blown fuse is of no further use, and can be thrown away.

The second battery, which was protected by the fuse, should still be OK.

The battery carrier can be reused later.

Experiment 3: Your First Circuit

Now it's time to make electricity do something more useful. To achieve this, you'll be experimenting with components known as resistors, and a light-emitting diode, or LED.

What You Will Need

- 9-volt battery (1)
- Resistors: 470 ohm (1), 1K (1), 2.2K (1)
- Generic LED (1)
- Test leads with alligator clips at each end (3)
- Multimeter (1)

Setup

It's time to get acquainted with the most fundamental component we'll be using in electronic circuits: the humble resistor. As its name implies, it resists the flow of electricity. As you might expect, its value is measured in ohms.

If you bought a bargain-basement assortment of resistors, they may be delivered to you in unmarked bags. No problem; we can determine their values easily enough. In fact, even if the packages are clearly labeled, I want you to check the resistors as we go along, because it's easy to get them mixed up. You have two alternatives:

- Use your multimeter, after setting it to measure ohms.
- Learn the color codes that are printed on most resistors. I'll explain them immediately below.

After you check the values of resistors, it's a good idea to sort them into labeled compartments in little plastic parts boxes. Personally, I like the boxes sold at the Michael's chain of crafts stores in the United States, but there are many alternatives. You can also use miniature plastic bags, which you can find by searching eBay for:

```
plastic bags parts
```

Fundamentals: Decoding Resistors

Some resistors have a value clearly stated on them in microscopic print that you can read with a magnifying glass, as shown in Figure 1-42.

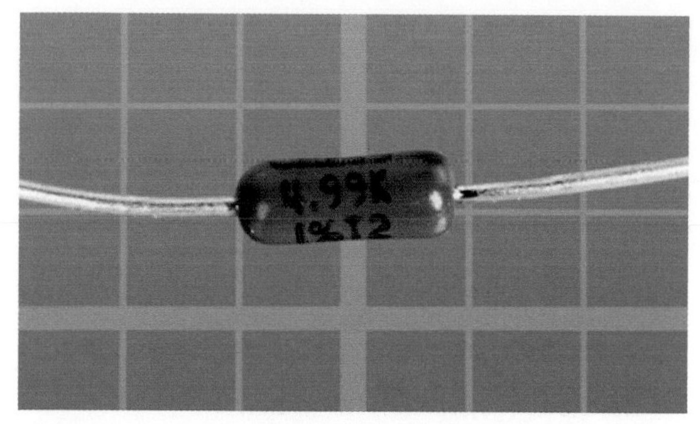

Figure 1-42 *Only a minority of resistors have their values printed on them.*

However, most resistors are color-coded with stripes. Figure 1-43 shows the coding scheme.

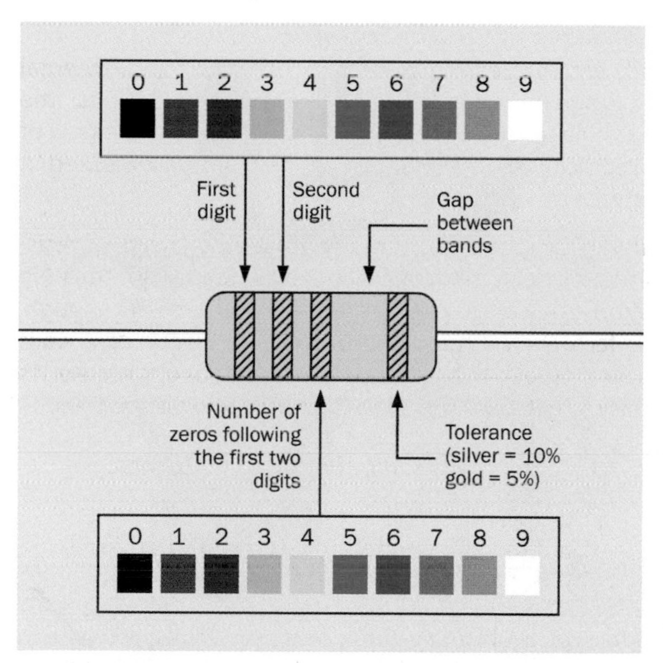

Figure 1-43 *The coding scheme that shows the value of a resistor. Some resistors have four stripes on the left instead of three, as explained in the text.*

Figure 1-44 shows you some examples. From top to bottom: 1,500,000 ohms (1.5M) at 10% tolerance, 560 ohms at 5%, 4,700 ohms (4.7K) at 10%, and 65,500 ohms (65.5K) at 5%.

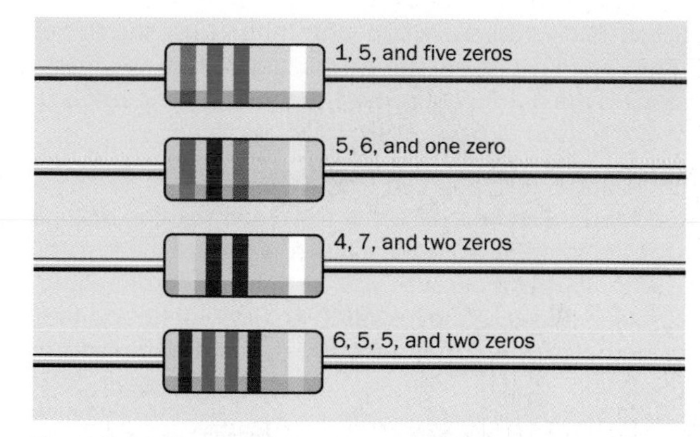

Figure 1-44 *Four examples of color-coded resistors.*

The code can be summarized in words like this:

- Ignore the color of the body of the resistor. (The exception to this rule would be a white resistor that may be fireproof or fused, and must be replaced with the same type. But you are unlikely to encounter one of these.)
- Look for a silver or gold stripe. If you find it, turn the resistor so that the stripe is on the right-hand side. Silver means that the value of the resistor is accurate within 10%, while gold means that the value is accurate within 5%. This is known as the *tolerance* of the resistor.
- If you don't find a silver or gold stripe, turn the resistor so that the colored stripes are clustered at the left end. Usually there are three of them. If there are four, I'll deal with those in a moment.
- The colors of the first two stripes, from left to right, tell you the first two digits in the value of the resistor. The color of the third stripe from the left tells you how many zeros follow the two numbers. The color values are shown in Figure 1-43.

If you run across a resistor with four stripes instead of three, the first *three* stripes are digits and the *fourth* stripe is the number of zeros. The third numeric stripe allows a resistor to have an intermediate value.

Confusing? Well, you can still use your meter to check the values. Just be aware that the meter reading may be slightly different from the claimed value of the resistor. This can happen because your meter isn't absolutely accurate, or because the resistor is not absolutely accurate, or both. Small variations don't matter in the projects in this book.

Lighting an LED

Now take a look at one of your generic LEDs. Old-fashioned lightbulbs used to waste power by converting a lot of the power into heat. LEDs are much smarter: they convert almost all their power into light, and they last almost indefinitely—provided you treat them right!

An LED is quite fussy about the amount of power it gets, and the way it gets it. Always follow these rules:

- The *longer* wire sticking out of the LED must receive a *more positive* voltage relative to the shorter wire.
- The positive voltage difference that you apply between the long wire and the short wire must not exceed the limit stated by the manufacturer. This is known as the *forward voltage*.
- The current passing into the LED through the long wire and out through the short wire must not exceed the limit stated by the manufacturer. This is known as the *forward current*.

What happens if you break these rules? You'll see for yourself in Experiment 4.

Make sure you have a fresh 9-volt battery. You could use a connector with the battery, as shown in Figure 1-8, but I think it's easier just to clip a couple of test leads directly to the battery terminals, as in Figure 1-45.

Select a 2.2K resistor. Remember, 2.2K means 2,200 ohms. Why is it 2,200 and not a nice round amount such as 2,000? I'll explain that shortly. See "Background: Puzzling Numbers" on page 21 if you want to know right now.

The colored stripes on your 2.2K resistor should be red-red-red, meaning 2 followed by another 2 and two more zeros. You will also need a 1K resistor (brown-black-red) and a 470-ohm resistor (yellow-violet-brown), so have them ready.

Wire the 2.2K resistor into the circuit shown in Figure 1-45. Make sure you get the battery the right way around, with its positive terminal on the right.

- The "plus" symbol always means "positive."
- The "minus" symbol always means "negative."

Be sure that the long lead of the LED is on the right, and be careful that none of the alligator clips touches each other. You should see the LED glow dimly.

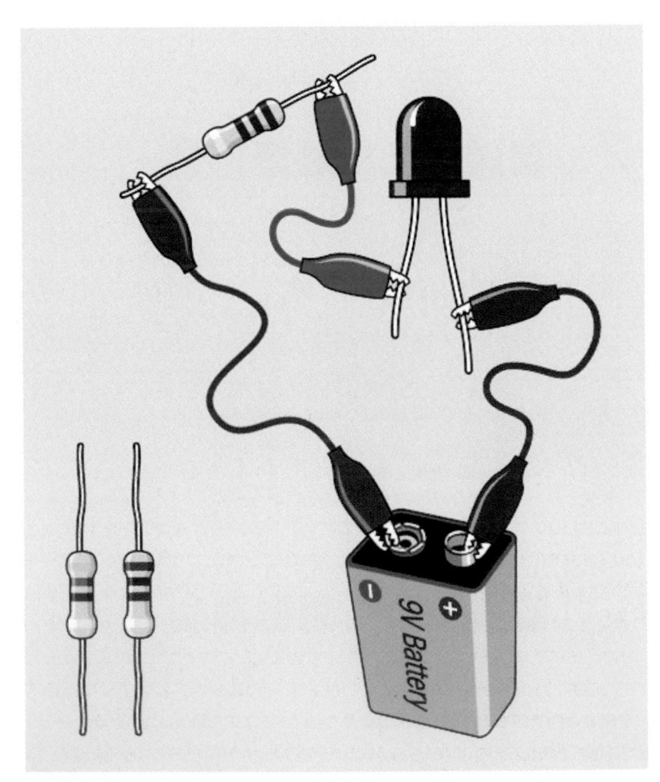

Figure 1-45 *Your first circuit, to light an LED.*

Now swap out your 2.2K resistor and substitute the 1K resistor. The LED should glow more brightly.

Swap out the 1K resistor and substitute a 470-ohm resistor, and the LED should be brighter still.

This may seem obvious, but it makes an important point. The resistor blocks a percentage of the current in the circuit. A higher-value resistor blocks more current, leaving less for the LED.

Checking a Resistor

I mentioned that you can use your meter to check the value of a resistor. This is really very easy. The procedure is shown in Figure 1-46. First, don't forget to set your meter to ohms. Disconnect the resistor from any other components, and apply the probes of your meter. If you have a manual-ranging meter, you must set the meter to a higher value than you expect to find. Otherwise, you'll get an error message.

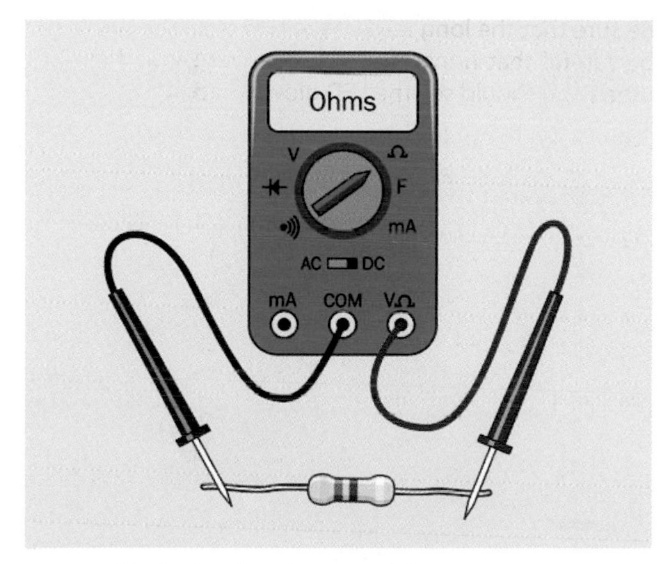

Figure 1-46 *Testing the value of a resistor.*

One thing to bear in mind is that you will get a more accurate reading if you press the probes firmly against the leads of the resistor. Don't hold the resistors and probes between your fingers—you don't want to measure the resistance of your body along with the resistance of the resistor. Place the resistor on an insulating surface, such as a nonmetallic desktop. Hold the probes by their plastic handles, and press down hard with the metal tips.

Alternatively, you can use a couple of test leads. Clip one end of each lead to each end of the resistor, and clip the other ends of the leads to the meter probes. Now you can do hands-free resistor testing, which is much easier.

Background: Puzzling Numbers

After you check a few resistors (or shop for them online) you'll notice that the same pairs of digits keep recurring. In thousands of ohms, we often find 1.0K, 1.5K, 2.2K, 3.3K, 4.7K, and 6.8K. In tens of thousands, we find 10K, 15K, 22K, 33K, 47K, and 68K.

The pairs of digits are known as *multipliers*, because you can multiply them by 1, or 1,000, or 10,000, or 100, or 10 to get basic resistor values in ohms.

There is a logical reason for this. Long ago, many resistors had an accuracy of plus-or-minus 20%, and therefore a 1.0K resistor could have an actual resistance as high as 1 + 20% = 1.2K while a 1.5K resistor could have a resistance as low as 1.5 – 20% = 1.2K. Therefore, it was pointless to have any values between 1K and 1.5K. Similarly, a 68 ohm resistor could have a value as high as 68 + 20% = just over 80 ohms, while a 100 ohm resistor could have a value as low as 100 – 20% = 80 ohms; so, it was unnecessary to have a value between 68 and 100.

In the top row of the table in Figure 1-47, the white numbers were the original multipliers for resistors. These numbers are still the most widely used today, even though modern resistor values are plus-or-minus 10% or better.

If you include the numbers in black type with the numbers in white type, you get all the possible multipliers for 10% resistors. If you then include the values in blue type, you have all the possible multipliers for 5% resistors.

1.0	1.5	2.2	3.3	4.7	6.8
1.1	1.6	2.4	3.6	5.1	7.5
1.2	1.8	2.7	3.9	5.6	8.2
1.3	2.0	3.0	4.3	6.2	9.1

Figure 1-47 *Traditional multipliers for resistor and capacitor values. See text for details.*

I have only used the original six multipliers for the projects in this book, to minimize the range of resistors that you will require. If accuracy is important (in Experiment 19, for example, where a circuit measures the speed of your reflexes) you can use a potentiometer to fine-tune the output—as I will show you in the very next experiment.

Cleanup and Recycling

You'll use the battery and the LED in the next experiment. The resistors can be reused in the future.

Experiment 4: Variable Resistance

You can vary the resistance in a circuit by inserting a *potentiometer*, which will control the current. The potentiometer in this experiment will enable you to learn more about voltage, amperage, and the relationship between them. You'll also learn how to read a manufacturer's datasheet.

What You Will Need

- 9-volt battery (1)
- Resistors: 470 ohms (1) and 1K (1)
- Generic LEDs (2)
- Test leads with alligator clips at each end (4)
- Potentiometer, 1K linear (2)
- Multimeter (1)

Look Inside Your Potentiometer

The first thing I want you to do is see for yourself how a potentiometer works, and the best way to accomplish this is to open it. This is why I asked you to acquire two potentiometers for this experiment—in case you can't put the first one back together again.

Some readers of the first edition of this book complained that it's wasteful to risk destroying a potentiometer by prying it open. But almost any learning experience consumes some resources, from pens and paper to whiteboard markers. If you really don't want to risk the future of your potentiometer, you can leave it untouched while you study the photographs that follow.

Most potentiometers are held together with little metal tabs. You need to bend the tabs upward. One way to do this is to slide in a knife and use it as a lever. Another way is to use a screwdriver—or maybe some pliers. I haven't specified any tools for this experiment, because I am hoping you already have a knife, or a screwdriver, or pliers in your home.

Figure 1-48 shows the tabs circled in red. (A fourth one is hidden behind the shaft of the component.) Figure 1-49 shows the tabs bent upward and outward.

Figure 1-48 *The tabs that hold a potentiometer together.*

Figure 1-49 *The tabs have been bent upward and outward.*

After you have pried up the tabs, very carefully pull up on the shaft while holding the body of the potentiometer in your other hand. It should come apart as shown in Figure 1-50.

Figure 1-50 *The wiper of the potentiometer is circled.*

Inside the shell you will find a circular *track*. Depending on whether you have a really cheap potentiometer or a slightly more high-class version, the track will be made of conductive plastic or will have thin wire wrapped around it, as shown in the photograph. Either way, the principle is the same. The wire or the plastic possesses some resistance (a total of 1,000 ohms in a 1K potentiometer), and as you turn the shaft, a *wiper* rubs against the resistance, giving you a shortcut to any point from the center terminal. The wiper is circled red in Figure 1-50.

You can probably put it back together, but if necessary, use your backup potentiometer.

Testing the Potentiometer

Set your meter to measure resistance (at least 1K, on a manual-ranging meter) and touch the probes to the two adjacent terminals shown in Figure 1-51. You should find that when you turn the shaft of the potentiometer clockwise (seen from above), the resistance diminishes to almost zero. When you turn the shaft counterclockwise, the resistance increases to about 1K. Now keep the black probe where it is, and touch the red probe on the opposite terminal. The behavior of the potentiometer will be reversed.

Do you think, maybe, the middle terminal connects with the wiper inside the potentiometer? Do you think the other two terminals connect with the ends of the track?

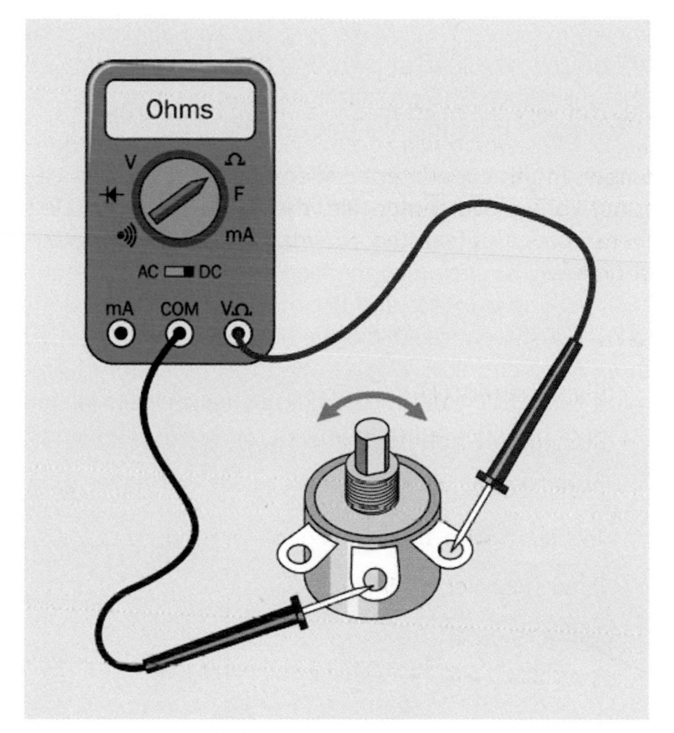

Figure 1-51 *Procedure for testing the behavior of a potentiometer.*

If you move the red probe to where the black probe is, and move the black probe to where the red probe was, the resistance between them will not change. It's the same in both directions. Unlike an LED, which has to be connected the right way around, a potentiometer has *no polarity*.

Caution: Don't Add Power

Don't apply power to a circuit while trying to measure resistance. Your meter uses a small amount of voltage from its internal battery when you are measuring resistance. You don't want that voltage to fight with voltage that you are applying from a battery.

Caution: Destructive Experiment Ahead

I have performed the next procedure many times uneventfully, but one reader reports that his LED fractured. You may wish to use safety glasses, if you want to be cautious. Regular eyeglasses will be acceptable.

Dimming Your LED

Now you can use the potentiometer to control the brightness of your LED. Connect everything exactly as shown in Figure 1-52. Make sure the two alligator clips are on the terminals shown. You are now using a variable resistance (the potentiometer) where the fixed resistor was in Experiment 3 (see Figure 1-45).

Begin with the shaft turned all the way *counterclockwise* (seen from above), otherwise you'll burn out the LED before we even get started. Now turn the shaft clockwise, very slowly, as shown by the blue arrow. You'll notice the LED glowing brighter, and brighter, and brighter—until, oops, it just went dark! You see how easy it is to destroy modern electronics? When I titled this procedure "Dimming your LED," you probably didn't realize I was talking about dimming it permanently.

Set aside that LED. I'm sorry to say, it will never glow again.

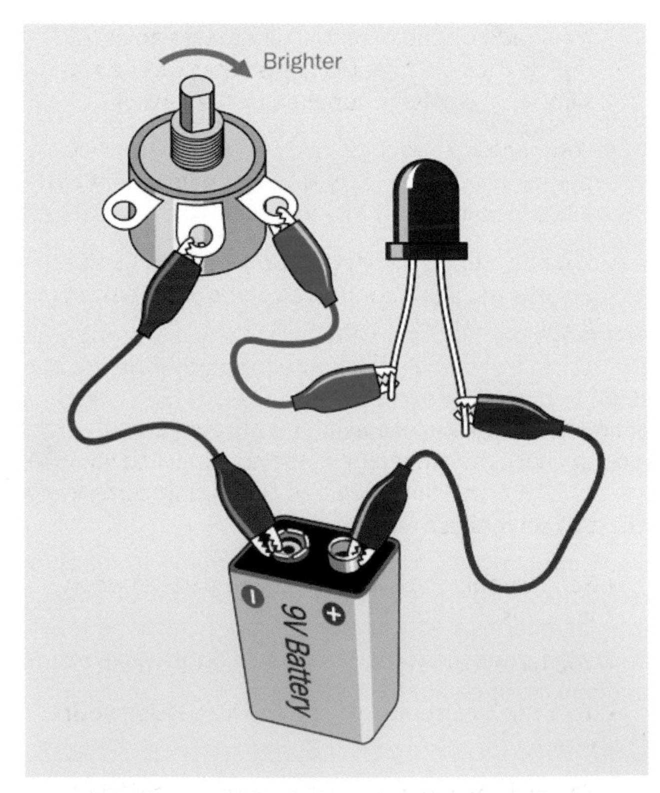

Figure 1-52 *Adjusting the brightness of an LED with a potentiometer.*

Substitute a new LED, and this time, let's protect it. Add a 470-ohm resistor, as shown in Figure 1-53. Electricity now passes through the 470-ohm resistor as well as the potentiometer, so that the LED will be protected even if the potentiometer's resistance diminishes to zero. You can turn the shaft of the potentiometer without worrying about destroying anything.

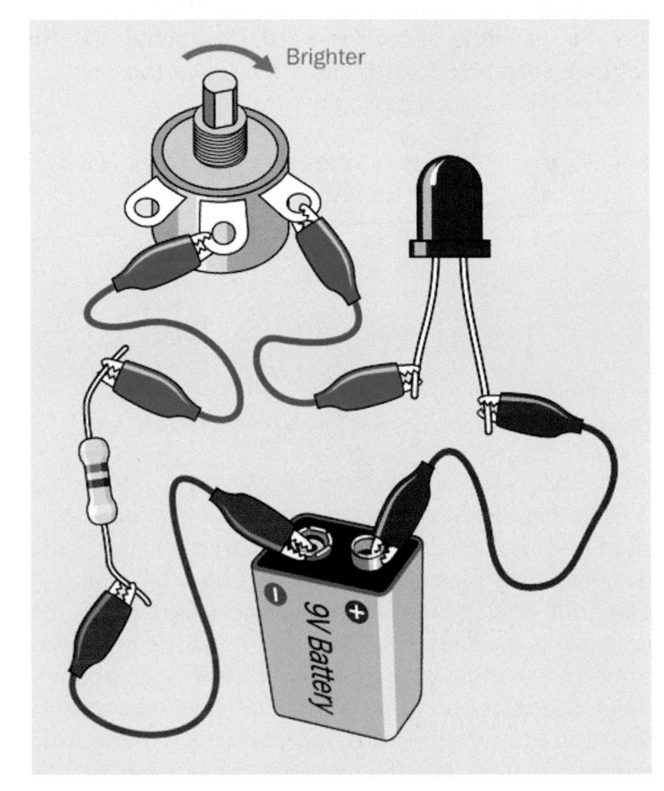

Figure 1-53 *Protecting the LED.*

The lesson that I hope you have learned is that an LED is too sensitive to be connected directly with a 9-volt battery. It must always be protected by some extra resistance in the circuit.

Could you power an LED directly from a single 1.5-volt battery? Try it. You may get a dim glow, but 1.5 volts is below the *threshold* for the LED. Let's find out how much voltage an LED needs.

Measuring Potential Difference

While the battery is connected in the circuit, set the dial of your meter to measure volts DC. You can leave the red lead plugged into the meter where it was before, because the socket for measuring volts is the same as the socket for measuring ohms.

If your meter uses manual ranging, set the voltage higher then 9 volts. Remember, the numbers beside the dial on the meter are the maximum in each range.

Now touch the probes to the terminals of the potentiometer that you are using, as shown in Figure 1-54. Try to hold the probes in place while you turn the potentiometer up a little, and down a little. You should see the voltage changing accordingly. We call this the *potential difference* between the two probes.

- "Potential difference" means the same as voltage between two points.

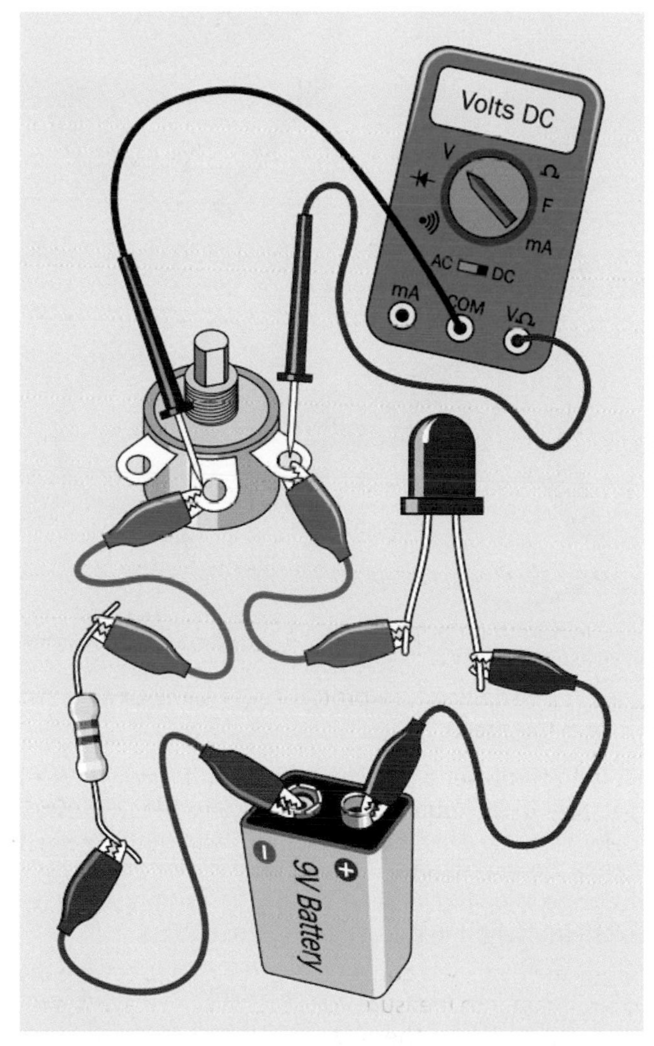

Figure 1-54 *Measuring the potential difference across an LED.*

If you measure the potential across the LED, this will also change when you adjust the potentiometer, although not as much as you might expect. An LED self-adjusts to some extent, modifying its resistance as the voltage and current fluctuate.

What if you swap the positions of the red and black probes? A minus sign should appear in the meter's display. You won't damage the meter this way, but it's less confusing if you always measure voltage with the red probe more positive than the black probe.

Lastly, touch the probes across the fixed-value resistor, and once again the potential difference will change when you adjust the potentiometer. Voltage from the battery is shared by all the components in this simple circuit. When the potentiometer reduces its share, a larger voltage difference is available for the fixed-value resistor and the LED. In addition, when the potentiometer has a lower resistance, the total resistance in the circuit is lower, allowing more current to flow.

A few things to keep in mind:

- If you add up the potential differences across all the devices in the circuit, the total will be the same as the voltage supplied by the battery.
- You measure voltage *relatively* between two points in a circuit. This is what their potential difference means.
- When measuring voltage, apply your meter like a stethoscope, without disturbing or breaking the connections in the circuit.

Checking the Flow

Now I want you to make a different measurement. I want you to check the amperage in the circuit, using your meter set to mA (milliamps). When measuring current, you must observe these rules:

- You can only measure current (amperage) when it passes *through* the meter.
- You have to insert your meter into the circuit.
- Too much current will blow the fuse inside your meter.
- You will have to use a socket on the meter labelled mA. This may be the same as the socket that you have used so far, or it may be different.

Make sure you turn the dial on your meter to measure mA, not volts, before you try this.

Caution: Meter Overload

Be careful when measuring current. For instance, if you put the probes of your meter directly against the terminals of a battery, and the meter is set to measure mA, you will create an instant overload, and the meter will blow its internal fuse. A cheap meter will not have any fuses supplied with it, so you'll have to open the case, check the value of the fuse, and search around online until you find an exact replacement. This is really annoying (I have been through it myself, more than once). A really cheap meter may not even have an easily replacable fuse.

- Always measure current when there are components in the circuit to restrict the flow.
- As a precaution, if your meter has a separate socket for measuring current, plug the red lead into that socket only while you are actually performing that task. Move the red lead back to the volts/ohms socket afterward.

Checking the Current

Insert the meter between the LED and the potentiometer, as shown in Figure 1-55. As you adjust the potentiometer up and down a little, you should find that the varying resistance in the circuit changes the flow of current—the *amperage*. The LED burned out in the previous experiment because too much current made it hot, and the heat melted it inside, just like a fuse. A higher resistance limits the amperage.

Now here's an interesting test. Turn the potentiometer all the way counterclockwise. Make a note of the current that you measure.

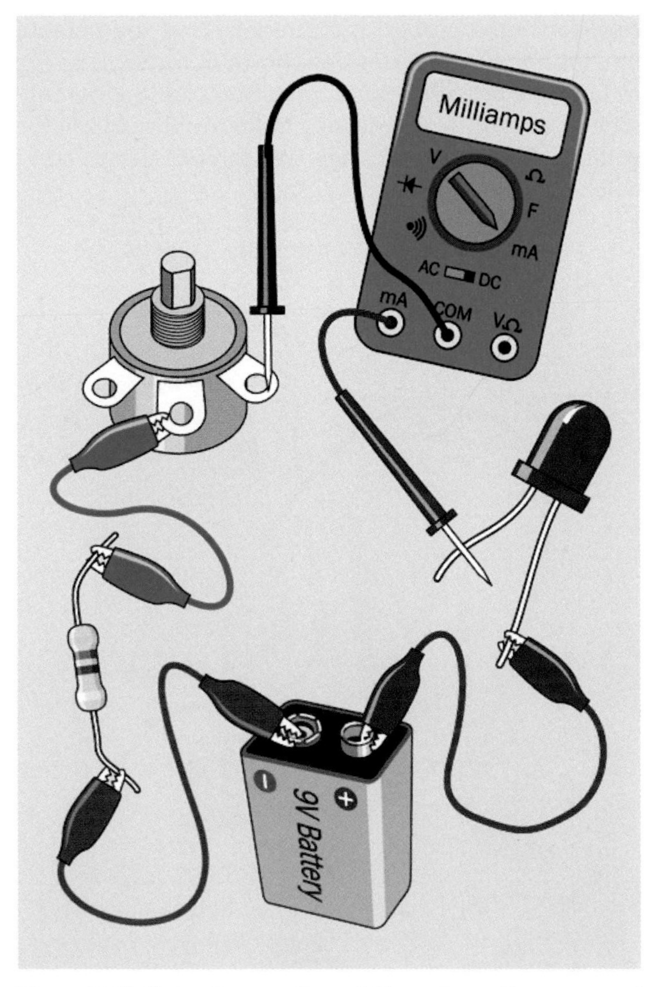

Figure 1-55 *Current passes through the meter on its way around the circuit.*

Without readjusting the potentiometer, move the meter and insert it between the battery and the LED, as shown in Figure 1-56. What's the value of the current now? It should be exactly the same as before—or very nearly the same, allowing for tiny changes in resistance that resulted from shifting the alligator clips.

- The current is the same at all points in a simple circuit. It has to be, because the flow of electrons has no place else to go.

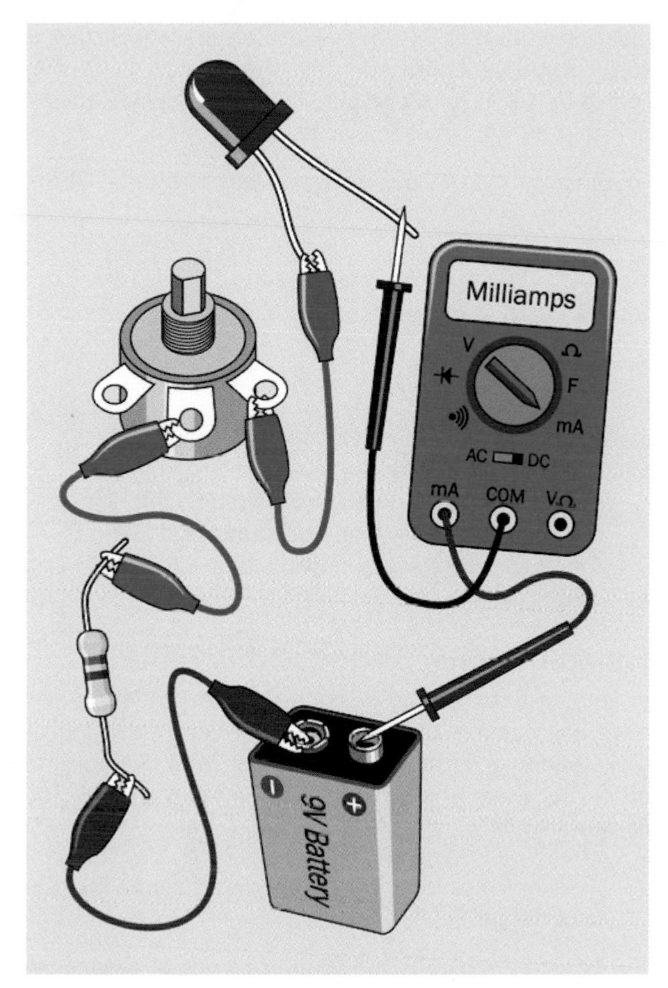

Figure 1-56 *The current flowing through a simple circuit is always the same throughout the circuit, regardless of where you measure it.*

Making Measurements

It's time now to nail this down with some numbers. This will enable you to establish the most fundamental rule in all of electronics.

Remove the LED from the circuit, and put the meter directly between the battery and the potentiometer. Remove the 470-ohm resistor, and substitute a 1K resistor (with colors brown-black-red), as shown in Figure 1-57. Now the only resistance in the circuit is provided by the 1K potentiometer plus the 1K resistor. (Your meter also has some resistance, but it's so low, we can ignore it. The wires and alligator clips also have some tiny amount of resistance, but this is even lower than the resistance of the meter.)

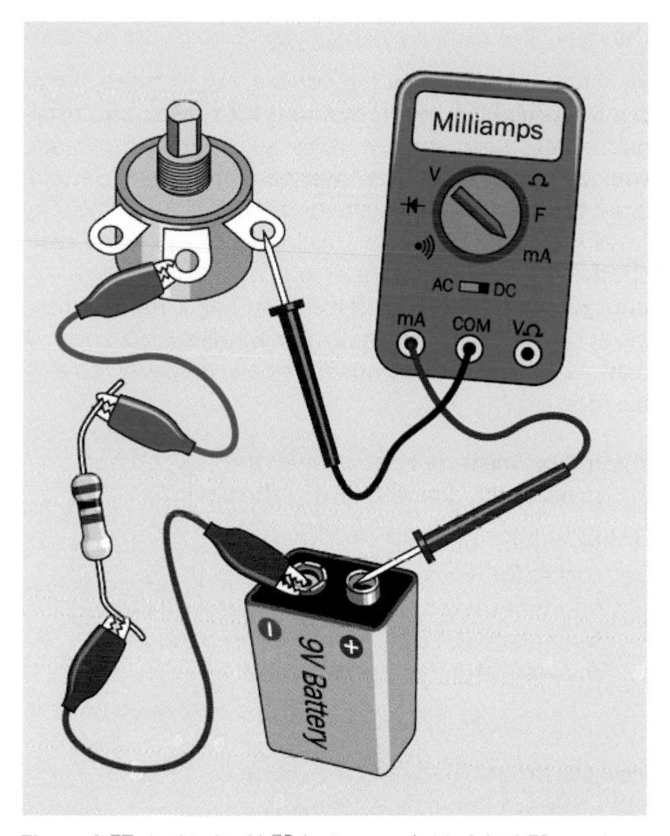

Figure 1-57 *In this final LED test, you get rid of the LED.*

Turn the potentiometer all the way clockwise, so that it provides almost-zero resistance. You now have only 1,000 ohms in the circuit, from the resistor. How much current does your meter show flowing?

Turn the potentiometer halfway, so it creates about 500 ohms resistance. The total resistance in the circuit is now 1,500 ohms, approximately. How much current does your meter show, now?

Turn the potentiometer all the way counterclockwise, so its full value is in the circuit, plus the resistor, making 2,000 ohms total. What's the amperage now?

When I tried this, I got the values shown below. Yours should be about the same.

```
9mA with 1K total resistance

6mA with 1.5K total resistance

4.5mA with 2K total resistance
```

Do you notice something interesting? On each line, if you multiply the number on the left by the number on the right, the result is always 9. And 9 volts just happens to be the voltage of the battery.

We only have three measurements, but if you ran a more detailed test using a series of resistors with fixed values, I'll bet that the result would be the same. I can summarize it like this:

```
battery voltage = milliamps × kilohms
```

But wait a minute: 1K is 1,000 ohms, and 1mA is 1/1,000 of an amp. Therefore, using the fundamental units of volts, amps, and ohms, our formula should really look like this:

```
voltage = (amps / 1,000) × (ohms × 1,000)
```

(I'm using the / symbol, which is often referred to as the "slash" symbol, to mean "divide by.")

The two factors of 1,000 cancel each other out, so we get this:

```
volts = amps × ohms
```

This is known as *Ohm's Law*. It is absolutely fundamental.

Fundamentals: Ohm's Law

The general way to express Ohm's Law is:

```
voltage = current × resistance
```

which is usually abbreviated like this:

```
V = I × R
```

Letter I represents the flow of current, because originally current was measured by its *inductance*, meaning the ability to induce magnetic effects. Perhaps it would be more helpful if some other letter, such as C, was used to represent current, but it's too late to persuade everyone to do that. You just have to remember that I means current.

By shifting the terms around, you get these versions of the formula:

```
I = V / R
```

```
R = V / I
```

To apply the formula, you have to make sure that the units are consistent. If V is measured in volts, and I is measured in amps, then R must be measured in ohms.

What if you have measured a current in milliamps? You must express it in amps. For instance, a current of 30mA must be written as 0.03 in the formula, because 0.03A = 30mA. If you get confused, use a calculator to divide milliamps by 1,000 to get a value in amps. Likewise, divide millivolts by 1,000 to get a value in volts.

To minimize the risk of errors, you can memorize Ohm's Law using the actual units, like this:

```
volts = amps × ohms
```

```
amps = volts / ohms
```

```
ohms = volts / amps
```

But remember:

- Volts are measured as a *difference in voltage* between two points in a simple circuit. Ohms are the resistance between the same two points. Amps are the current flowing through the circuit.

Fundamentals: Series and Parallel

In your test circuit, the resistor and the potentiometer have been connected in *series*, meaning that the electricity had to go through one before it went through the other. The alternative would have been to put them side by side, in *parallel*.

- Resistors in series are positioned so that one follows the other.
- Resistors in parallel are positioned side by side.

When you put two equal-valued resistors in series, you double the total resistance, because electricity has to pass through two barriers in succession. This is shown in Figure 1-58.

When you put two equal-valued resistors in parallel, you cut the total resistance in half, because you're giving the electricity two paths of equal resistance instead of one. This is shown in Figure 1-59.

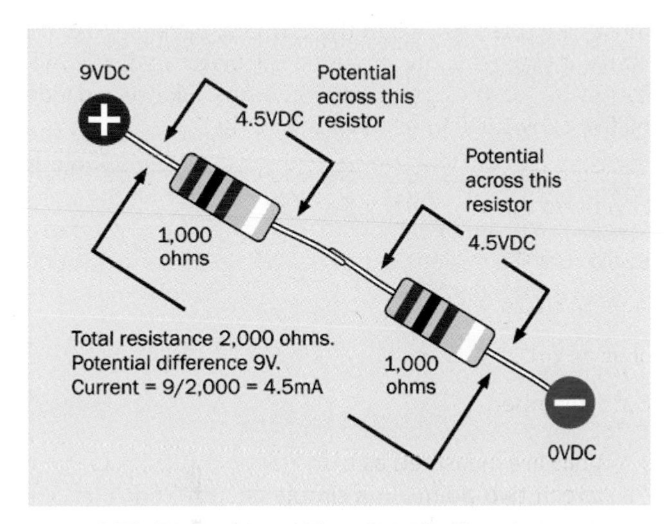

Figure 1-58 *Two resistors of the same value in series.*

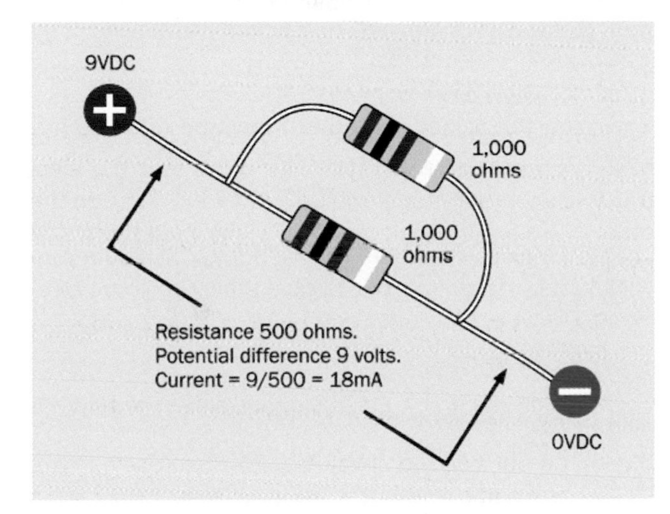

Figure 1-59 *Two resistors of the same value in parallel.*

In both figures, the current in milliamps has been calculated using Ohm's Law.

In reality we don't normally need to put resistors in parallel, but we often put other types of components in parallel. All the lightbulbs in your house, for instance, are wired in parallel across the main supply. So, it's useful to understand that resistance in a circuit goes down if you keep adding components in parallel. At the same time, as you add more pathways for electricity to follow, the total current through the circuit goes up.

Using Ohm's Law

Ohm's Law is extremely useful. For example, it can tell you precisely what resistance to put in series with an LED, to protect it adequately while generating as much light as possible.

The first step is to discover the specification for the LED established by its manufacturer. This information is easily available in a datasheet that you can locate online. Suppose you have an LED made by Vishay Semiconductors. You know that its part number is TLHR5400, because the number was printed on a label when you received a bag of LEDs in the mail, and you snipped out the label and stored it with the LEDs. (At least, that's what you should have done.)

All you have to do is Google the part number and the manufacturer's name:

```
vishay tlhr5400
```

The very first hit is the datasheet maintained by Vishay. Scroll down, and you see the information that you need. I've included the left side and right side of a screen capture in Figure 1-60. I outlined the part number of the component in red on the left, and I outlined two types of forward voltage on the right. "Typ" means typical and "Max," as I'm sure you guessed, is maximum. So, the LED typically should run with a 2V potential difference. But what does "at I_F (mA)" mean? Well, remember that the letter I is used to represent the current passing through a circuit. The letter F means "Forward." So the forward voltage in the table is measured at a forward current of 20mA, which is the recommended value for this LED.

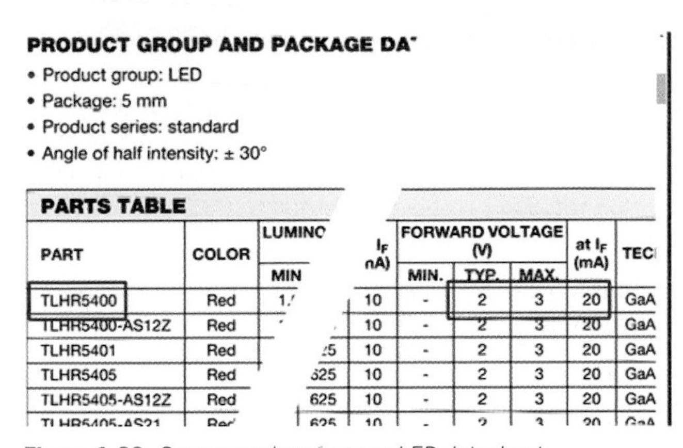

PRODUCT GROUP AND PACKAGE DA

- Product group: LED
- Package: 5 mm
- Product series: standard
- Angle of half intensity: ± 30°

PARTS TABLE

PART	COLOR	LUMINO MIN		I_F (mA)	FORWARD VOLTAGE (V) MIN.	TYP.	MAX.	at I_F (mA)	TEC
TLHR5400	Red	1.		10	-	2	3	20	GaA
TLHR5400-AS12Z	Red			10	-	2	3	20	GaA
TLHR5401	Red		5	10	-	2	3	20	GaA
TLHR5405	Red		25	10	-	2	3	20	GaA
TLHR5405-AS12Z	Red		625	10	-	2	3	20	GaA

Figure 1-60 *Screen capture from an LED datasheet.*

How about if you have a Kingbright WP7113SGC? This time, the second hit from a Google search takes you to the appropriate datasheet, where the second page specifies a typical forward voltage of 2.2V, maximum

2.5V, and a maximum forward current of 25mA. The layout of the Kingbright datasheet is different from that of the Vishay datasheet, but the information is still easy to find.

Let's go with the Vishay LED. Now that you know that it works well with 2V and 20mA, Ohm's Law can tell you the rest.

How Big a Resistor?

In the simple circuit shown in Figure 1-61, you want to know the correct value for the resistor. Begin by recalling the rule I mentioned before:

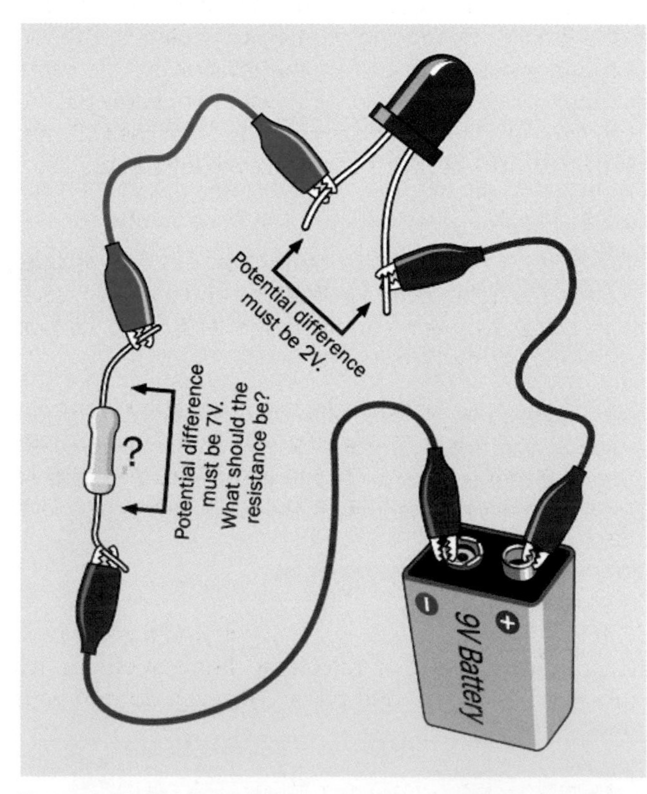

Figure 1-61 *This basic circuit enables you to calculate the value of the resistor.*

- If you add up the potential differences across all the devices in the circuit, the total will be the same as the voltage supplied by the battery.

The battery is 9V, of which we want the LED to take 2V. Therefore, the resistor must drop the voltage by 7V. What about the current? Remember another rule that I mentioned previously:

- The current in a simple circuit is the same at all locations in the circuit.

Therefore, the current through the resistor will be the same as the current through the LED. Your target is 20mA, but Ohm's Law requires you to make all the units match. If you are dealing with volts and ohms, you have to express the current in amps. Well, 20mA is 20 / 1,000 amps, which is 0.02 amps.

Now you can write down what you know, which is always the first step:

```
V = 7
```

```
I = 0.02
```

Which version of the Ohm's Law formula should you use? The one where the value that you don't know, and want to know, is on the left. That would be this one:

```
R = V / I
```

Now plug the values for V and I into the formula, like this:

```
R = 7 / 0.02
```

There's a trick that I will tell you for doing calculations involving decimal points, but to save time, use your calculator to show you the answer:

```
7 / 0.02 = 350 ohms
```

That is not a standard value for resistors, but 330 ohms is. Alternatively, just in case you use a more sensitive LED, you could move up to the next higher standard resistance value, which is 470 ohms. You may remember, I used a 470-ohm resistor in Experiment 3. Now you know why: I did the math.

Some people make the mistake of thinking that when they divide volts by amps to find the correct value for a series resistor, they should use the supply voltage (9V in this case). This is not correct, because the supply voltage is being applied to the resistor *and* the LED. To find the value for the resistor, you have to consider just the potential difference across it alone, which is 7V.

What happens if you use a different power supply? Later in the book, you'll be using a 5V supply in several experiments. How will this change the appropriate value of the resistor?

The LED still has to take 2V. The power supply is 5V, so the resistor has to drop 3V. The current should still be the same, and the calculation now looks like this:

```
R = 3 / 0.02
```

Therefore the resistance value is 150 ohms. But you don't need the LED to deliver its maximum light output, and you might happen to use an LED that has a lower limit than 20mA. Also, if a battery is powering the circuit, you might want to reduce the power consumption to make the battery last longer. Bearing this in mind, you could use the next higher standard value for the resistor, which would be 220 ohms.

Background: Hot Wires

I mentioned that wires have a very low resistance. Is it so low that you can always ignore it? Actually, no. If a large current is flowing through a wire, the wire will get hot, as you saw yourself when the 1.5-volt battery was shorted out in Experiment 2. And if the wire gets hot, you can be sure that some voltage is being blocked by the wire, leaving less voltage available for any device attached to the wire.

Once again, you can use Ohm's Law to establish some numbers.

Suppose that a very long piece of wire has a resistance of 0.2 ohms. You want to run 15 amps through it, to run a device that consumes a lot of power.

You begin by writing down what you know:

```
R = 0.2 (resistance of the wire)
```

```
I = 15 (amperage through the circuit)
```

You want to know V, the voltage drop between one end of the wire and the other. So you should use the version of Ohm's Law that places V on the left side:

```
V = I × R
```

Now plug in the values:

```
V = 15 × 0.2 = 3 volts
```

Three volts is not a big deal if you have a high-voltage power supply, but if you are using a 12-volt car battery, this length of wire will take one-quarter of the available voltage.

Now you know why the wiring in automobiles is relatively thick—to waste as little of the 12V power as possible.

Fundamentals: Decimals

Legendary British politician Sir Winston Churchill is famous for complaining about "those damned dots." He was referring to decimal points. Because Churchill was Chancellor of the Exchequer at the time, supervising all government expenditures, his difficulty with decimals was a bit of a problem. Still, he muddled through in time-honored British fashion, and so can you.

Suppose you have decimals in a division sum. You can make it easier by moving the decimal points in the top half and the bottom half of the division by the same number of places. So, when you wanted to know the answer to 7 / 0.02 to find the series resistor for the LED, you could move the decimals two spaces to the right:

```
7 / 0.02 = 700 / 2
```

This is much easier. Notice that if you move a decimal point to the right beyond a digit, you add a zero for each extra place. So, when you move the decimal in 7.0 two places to the right, you get 700.

What if you have decimals in a multiplication sum? For example, you need to multiply 0.03 by 0.002. Because you are now multiplying instead of dividing, you have to move the decimal points in opposite directions. Like this:

```
0.03 × 0.002 = 3 × 0.00002
```

So the answer is 0.00006. Again, if you find this too confusing, you can use a calculator. But sometimes it's quicker to use a pen and paper—or even do it in your head.

Theory: The Math on Your Tongue

I'm going back to the question I asked in the previous experiment: why didn't your tongue get hot?

Now that you know Ohm's Law, you can figure out the answer in numbers. Let's suppose the battery delivered its rated 9 volts, and your tongue had a resistance of 50K, which is 50,000 ohms. As always, begin by writing down what you know:

```
V = 9
```

```
R = 50,000
```

You want to know the current, I, so you use the version of Ohm's Law that puts this on the left:

```
I = V / R
```

Plug in the numbers:

```
I = 9 / 50,000 = 0.00018 amps
```

Move the decimal point three places to convert from amps to milliamps:

```
I = 0.18 mA
```

That's a very small current. It will not produce much heat.

What about when you shorted out the battery? How much current made the wires get hot? Well, suppose the wires had a resistance of 0.1 ohms (probably it's less, but I'll start with 0.1 as a guess). Write down what you know:

```
V = 1.5
```

```
R = 0.1
```

Once again I'm trying to find I, the current, so I use:

```
I = V / R
```

Plug in the numbers:

```
I = 1.5 / 0.1 = 15 amps
```

That's almost 100,000 times the current that may have passed through your tongue. It generated substantial heat in a thin wire.

A room heater or a large power tool, such as a table saw, might draw 15 amps. Maybe you're wondering if that little AA battery really could deliver as much current as that. The answer is . . . I'm not sure. I couldn't measure the current with my meter, because 15A would blow the meter's fuse, even if I plug the probe into the high-current socket labeled 10A. But I did try the experiment with a 10-amp fuse instead of a 3-amp fuse, and the 10-amp fuse survived.

Now, why was that? Ohm's Law says that the current should be 15A, but for some reason it was less. Maybe the resistance of the wire on the battery carrier was actually higher than 0.1 ohms? No, I think it was probably lower. So—what was restricting the current to a lower value than Ohm's Law predicted?

The answer is that everything in the everyday world has some electrical resistance, *even including a battery.* Always remember that a battery is an active part of a circuit.

Do you remember that when you shorted out the battery, it became hot, as well as the wires? Definitely, the battery has some *internal resistance*. You can ignore it when you are dealing with small currents in milliamps, but for large currents, the battery is actively involved.

This is why I cautioned against using a larger battery (especially a car battery). Larger batteries have a much lower internal resistance, allowing much higher currents, which can generate explosive amounts of heat. A car battery is designed to deliver literally hundreds of amps when it turns a starter motor. That's quite enough current to melt wires and cause nasty burns. In fact, you can weld metal using a car battery.

Lithium batteries also have low internal resistance, making them very dangerous when they're shorted out. Here's the take-home message:

- High current is not dangerous in the same way as high voltage. But it is still dangerous.

Background: The Watt

So far I haven't mentioned a unit that everyone is familiar with: watts.

A watt is a unit of power, and when power is applied throughout a period of time, it performs work. An engineer might say that work is done when a person, an animal, or a machine pushes something to overcome mechanical resistance. Examples would be a car cruising along a level stretch of road (overcoming friction and air resistance) or a person walking upstairs (overcoming the force of gravity).

When one watt of power is applied for one second, the work done is one *joule*, usually represented by letter J. If P is used to represent power:

```
J = P × s
```

Or if the formula is turned around:

```
P = J / s
```

When electrons push their way through a circuit, they are overcoming a kind of resistance, and so they are doing work.

The electrical definition of a watt is easy:

```
watts = volts × amps
```

Or, using the units customarily assigned, with W meaning watts, these three formulas all mean the same thing:

```
W = V × I (watts = volts × amps)
```

```
V = W / I
```

```
I = W / V
```

The terms milliwatts (mW), kilowatts (kW), and megawatts (MW) are commonly used in different situations—megawatts being usually reserved for heavy-duty equipment such as generators in power stations. Be careful not to confuse the lowercase m in the abbreviation for milliwatts with the uppercase M in the abbreviation for megawatts. A conversion table for milliwatts, watts, and kilowatts is shown in Figure 1-62.

Milliwatts	Watts	Kilowatts
1mW	0.001W	0.000001kW
10mW	0.01W	0.00001kW
100mW	0.1W	0.0001kW
1,000mW	1W	0.001kW
10,000mW	10W	0.01kW
100,000mW	100W	0.1kW
1,000,000mW	1,000W	1kW

Figure 1-62 *Conversion table for the most common multiples of watts.*

Old-style incandescent light bulbs are calibrated in watts. So are stereo systems. The watt is named after James Watt, inventor of the steam engine. Incidentally, watts can be converted to horsepower, and vice versa.

Resistors are commonly rated as being capable of dealing with 1/4 watt, 1/2 watt, 1 watt, and up. For all of the projects in this book, you can use 1/4-watt resistors. How do I know this?

Go back to the first LED circuit, using a 9V battery. Remember you wanted the resistor to drop the voltage by 7 volts, at a current of 20mA. How many watts of power would the resistor have to deal with?

Write down what you know:

```
V = 7 (potential difference at the resistor)
```

```
I = 20mA = 0.02 amps
```

You want to know W, so use this version of the formula:

```
W = V × I
```

Plug in the numbers:

```
W = 7 × 0.02 = 0.14 watts
```

This is the power being dissipated by the resistor.

Because 1/4 watt is 0.25 watts, a 1/4 watt resistor will have no trouble dealing with 0.14 watts. In fact you could almost use a 1/8 watt resistor, but in future experiments we may need resistors that can handle 1/4 watt, and there's no penalty for using a resistor that is rated for more watts than necessary. They just cost slightly more and are slightly larger.

Background: The Origins of Wattage

James Watt, shown in Figure 1-63, is known as the inventor of the steam engine. Born in 1736 in Scotland, he set up a small workshop in the University of Glasgow, where he struggled to perfect an efficient design for using steam to move a piston in a cylinder. Financial problems and the primitive state of the art of metal working delayed practical applications until 1776.

Figure 1-63 *James Watt's development of steam power enabled the industrial revolution. After his death, he was honored by having his name applied to the basic unit of power in electricity.*

Despite difficulties in obtaining patents (which could only be granted by an Act of Parliament in those times), Watt and his business partner eventually made a lot of money from his innovations. Although he predated the pioneers in electricity, in 1889 (70 years after his death) his name was assigned to the basic unit of electric power that can be defined by multiplying amperes by volts.

Cleanup and Recycling

The dead LED can be thrown away. Everything else is reusable.

Experiment 5: Let's Make a Battery

Long ago, before the Web existed, kids were so horribly deprived, they tried to amuse themselves with kitchen-table experiments such as making a battery by pushing a nail and a penny into a lemon. Hard to believe, but true!

Now that modern LEDs will emit light when just a few milliamps flow through them, the old lemon-battery experiment is more interesting. If you've never tried it, the time is right.

What You Will Need

- Lemons (2) or squeeze-bottle of pure lemon juice (1)
- Copper-plated coins, such as US pennies (4)
- One-inch (or larger) zinc-plated steel brackets from a hardware store (4)
- Test leads with alligator clips at each end (5)
- Multimeter (1)
- Low-current LED (1). (See "Light-Emitting Diodes" on page 6 for a reminder of the difference between generic and low-current LEDs)

Setup

A battery is an *electrochemical* device, meaning that chemical reactions create electricity. Naturally, this only works if you have the right chemicals working for you, and the ones I'm going to use are copper, zinc, and lemon juice.

The juice should be no problem. Lemons are cheap, or you can buy one of those little yellow plastic squeeze-bottles of concentrated juice. Either will work.

Pennies are not made of copper anymore, but they are still plated with a thin layer of copper, which is good enough. Just make sure that your pennies are new and bright. If the copper has oxidized, it will be a dull brown, and the experiment won't work as well.

Zinc is a bit more of a problem. What you need is a metal part that is *galvanized*, meaning it is coated in zinc to prevent rust. Small galvanized steel brackets should be available at your local hardware store, and they don't cost much. Brackets that measure about one inch along each side will be fine.

Lemon Test: Part One

Cut a lemon in half, and push a penny into it. As close as possible to the penny (but not touching it), push in your galvanized bracket. Now set your multimeter so that it can measure up to 2V DC, and hold one probe against the penny while you hold the other probe against the bracket. You should find that your meter detects between 0.8V and 1V.

To power a typical LED, you need more voltage. How can you get it? By putting batteries in series. In other words, more lemons! You'll use test leads to link the batteries, as shown in Figure 1-64. Notice that each lead connects a bracket to a penny. Do not connect pennies to pennies or brackets to brackets.

If you set things up carefully, keeping the pennies and brackets close to each other but making sure they don't actually touch each other, you should be able to illuminate your LED with three lemon-juice batteries in series.

Another option is to use a little parts box divided into small sections, as shown in Figure 1-65. When everything is nicely aligned, squeeze in some concentrated lemon juice. Vinegar or grapefruit juice may also work.

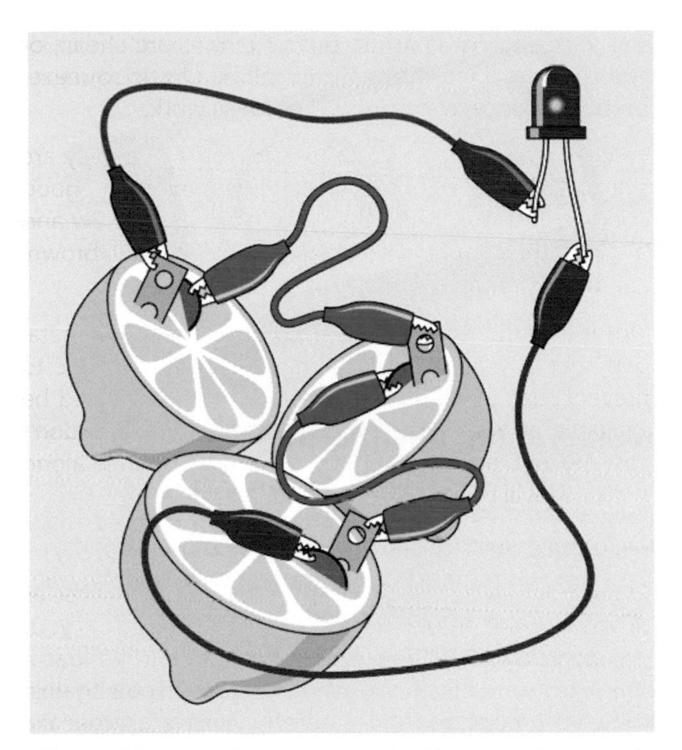

Figure 1-64 *A three-lemon battery should generate just enough voltage to drive a low-current LED.*

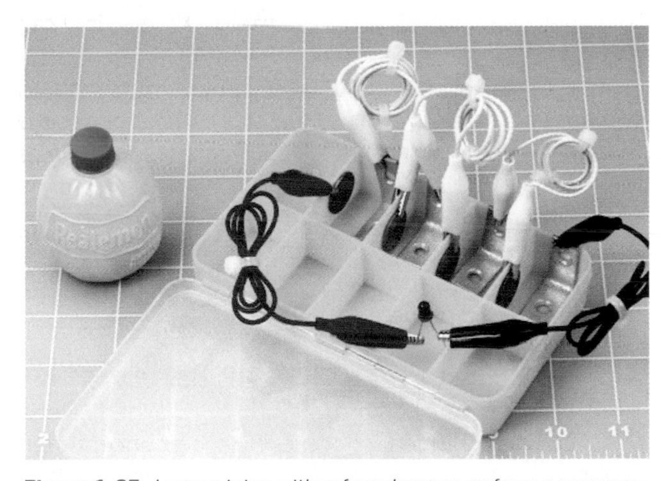

Figure 1-65 *Lemon juice, either from lemons or from a squeeze-bottle, will produce reliable results, even though the equipment doesn't look so nice. Here a box intended for parts has been repurposed as a four-cell juice battery.*

I decided to have four cells in my juice battery, because the LED pulls the voltage down somewhat, and the battery is not capable of delivering enough current to damage the LED. The setup in the photograph worked immediately.

Theory: The Nature of Electricity

To understand why the lemon battery works, you have to start with some basic information about atoms. Each atom consists of a nucleus at the center, containing particles called protons, which have a positive charge. The nucleus is surrounded by electrons, which carry a negative charge.

Breaking up the nucleus of an atom requires a lot of energy, and can also liberate a lot of energy—as happens in a nuclear explosion. But persuading a couple of electrons to leave an atom (or join an atom) can take very little energy. For instance, when zinc reacts chemically with an acid, this can liberate electrons.

The reaction soon stops if the zinc-plated part isn't connected with anything, as electrons accumulate with nowhere to go. They have a mutual force of repulsion, and you can imagine them like a crowd of hostile people, each one wanting the others to leave, and refusing to allow new ones to join them, as shown in Figure 1-66.

Figure 1-66 *Electrons on an electrode have a bad attitude known as mutual repulsion.*

Now consider what happens when a wire connects the zinc electrode, which has a surplus of electrons, to another electrode, made from a different material (such as copper), which contains "holes" for the electrons to occupy. The electrons can pass through the wire very easily by jumping from one atom to the next. As soon as we open up this pathway, mutual repulsion makes the electrons try to escape from each other to their new home as quickly as possible. This is how an electric current is created. See Figure 1-67.

Figure 1-67 *Electrons escaping from a zinc electrode to a copper electrode.*

Now that the population of electrons on the zinc electrode has been reduced, the zinc–acid reaction can continue, replacing the missing electrons with new ones—which promptly imitate their predecessors and try to get away from each other by running away down the wire. They move with such force, we can divert them through an LED, and they will liberate some of their energy by making it light up.

The process continues until the zinc–acid reaction grinds to a halt, usually because it creates a layer of a compound such as zinc oxide, which won't react with acid and prevents the acid from reacting with the zinc underneath. (This is why your zinc electrode may have looked sooty when you pulled it out of the acidic electrolyte.)

This description applies to a *primary battery*, meaning one that is ready to generate electricity as soon as a connection between its terminals allows electrons to transfer from one electrode to the other. The amount of current that a primary battery can generate is determined by the speed at which chemical reactions inside the battery can liberate electrons. When the raw metal in the electrodes has all been used up in chemical reactions, the battery can't generate any more electricity and is dead. It cannot easily be recharged, because the chemical reactions are not easily reversible, and the electrodes may have oxidized.

In a rechargeable battery, also known as a *secondary battery*, a smarter choice of electrodes and electrolyte does allow the chemical reactions to be reversed.

Background: Positive and Negative

I've told you that electricity is a flow of electrons, which have a negative charge. In that case, why have I been talking as if electricity flows from the positive terminal to the negative terminal of a battery, in the experiments that you have performed so far?

The story started with an embarrassment in the history of research into electricity. When Benjamin Franklin was trying to understand the nature of electric current by studying phenomena such as lightning during thunderstorms, he believed he observed a flow of "electrical fluid" from positive to negative. He proposed this concept in 1747.

In fact, Franklin had made an unfortunate error that remained uncorrected until physicist J. J. Thomson announced his discovery of the electron in 1897. Electricity really is a flow of negatively charged particles, from an area of greater negative charge to some other location that is "less negative" or "more positive." In a battery, electrons originate from the negative terminal and flow to the positive terminal.

You might think that when this fact was established, everyone should have discarded Ben Franklin's idea of a flow from positive to negative. But people had been thinking in those terms for 150 years. Also, when an electron moves through a wire, you can think of an equal positive charge flowing in the opposite direction. When the electron leaves home, it takes a small negative charge with it; therefore, its home becomes a bit more positive. When the electron arrives at its destination, its negative charge makes the destination a bit less positive. This is pretty much what would happen if an imaginary positive particle traveled in the opposite direction. Moreover, all of the mathematics describing electrical behavior are still valid if you apply them to the imaginary flow of positive charges.

As a matter of tradition and convenience, Ben Franklin's erroneous concept of flow from positive to negative survived, because in the end, it makes no difference.

Figure 1-68 *In some weather conditions, the flow of electrons during a lightning strike can be from the ground, through your feet, out of the top of your head, and up to the clouds. Benjamin Franklin would have been surprised.*

In the symbols that represent components such as diodes and transistors, you will find arrows reminding you which way these components should be placed—and the arrows all point from positive to negative, even though that's not the way things work at all.

When Ben Franklin studied lightning, he saw this as an electric charge moving from a positive domain (the clouds in the sky) to a negative reservoir (the planet Earth). Well, it's true that the clouds are more positive, but this simply means that in reality, lightning is a transfer of electrons from the ground up to the sky. That's right: someone who is "struck by lightning" may be hurt by *emitting* electrons rather than by receiving them, as shown in Figure 1-68.

Theory: Basic Measurements

I'm going to backtrack, now, to the kinds of definitions you would normally find at the beginning of an electronics text.

Electrical potential is measured by adding up the charges on individual electrons. The basic unit is the *coulomb*, equal to the total charge on 6,241,509,629,152,650,000 electrons.

If you know how many electrons pass through a piece of wire each second, you can calculate the flow of electricity, which can be expressed in amperes. In fact:

```
1 ampere = 1 coulomb/second
```

```
(about 6.24 quintillion electrons/second)
```

Even if you could see inside a wire carrying electric current, electrons are smaller than the wavelength of visible light, so you would have no way to observe them, and there are far too many, moving much too fast. However, we have indirect ways of detecting them. For instance, the motion of an electron creates a wave of electromagnetic force. More electrons create more force, and this force can be measured. We can calculate the amperage from that. The electric meter installed at your home by the utility company functions on this principle.

The force required to push electrons through a conductor is voltage, and it creates a flow that can create heat, as you saw when you shorted out a battery. (If the wire that you used had zero resistance, the electricity running through it would not have created any heat.) We can use the heat directly, as in an electric stove, or we can use the electrical energy in other ways—to run a motor, for instance. Either way, we are taking energy from the electrons to do some work.

One volt can be defined as the amount of pressure that you need to create a flow of 1 ampere, which does 1 watt of work. As previously defined, 1 watt = 1 volt × 1 ampere, but the definition actually originated the other way around:

```
1 volt = 1 watt / 1 ampere
```

It's more meaningful this way, because a watt can be defined in nonelectrical terms. Just in case you're interested, we can work backward through the units of the metric system like this:

```
1 watt = 1 joule / second

1 joule = 1 newton force acting through 1 meter

1 newton accelerates 1kg by 1m / sec each second
```

On this basis, the electrical units can all be anchored with observations of mass, time, and the charge on electrons.

Practically Speaking

For practical purposes, I think an intuitive understanding of electricity can be more useful than the theory. I like to go back to the water analogies that have been used for decades in guides to electricity.

In Figure 1-31 I showed how the rate at which water leaks from a hole in a tank can be compared to amperage, while the height of the water in the tank creates pressure, comparable to voltage, and the size of the hole is equivalent to resistance.

Where's the wattage in this picture? Suppose you place a little water wheel where it is hit by the flow from the hole, as shown in Figure 1-69. You could attach some machinery to the water wheel. Now the flow would be doing some work. (Remember, wattage is a measurement of the rate at which work is done.)

Maybe this looks as if you would be getting something for nothing, extracting work from the flow of water without putting any energy back into the system. But remember, the water level in the tank is falling. As soon as I include some helpers hauling the waste water back up to the top of the tank, it becomes obvious that you have to put work in to get work out. See Figure 1-70.

Figure 1-69 *If a wheel extracts energy from the flow of water, the flow is now doing some work, which could be measured in watts during a period of time.*

Similarly, a battery may seem to be giving power out without taking anything in, but the chemical reactions inside it are changing pure metals into metallic compounds, and the power we get out of a battery is enabled by this change of state. If it's a rechargeable battery, we have to push power back into it during the charging process, to reverse the chemical reactions.

Going back to the tank of water, suppose we can't get enough power out of it to turn the wheel. One answer could be to increase the height of the water, to create more force, as in Figure 1-71.

Figure 1-70 *To continue getting work out of the system, we have to put work into it.*

This would be the same as putting two batteries end to end, positive to negative, in series (as I suggested with lemons, in the lemon battery). Two batteries in series will double the voltage, as shown in Figure 1-72. As long as the resistance in the circuit remains the same, greater voltage will create more amperage, because amperage = voltage / resistance.

Thinking again about the tank analogy, what if we want to run a wheel for twice as long, and we've run out of tank capacity? Maybe we should build a second tank, and pipe their outputs through the same hole. Similarly, if you wire two batteries side by side, in parallel, you get the same voltage, but the batteries should last twice as long. Alternatively, the two batteries may be able to deliver more current than if you just used one. See Figure 1-73.

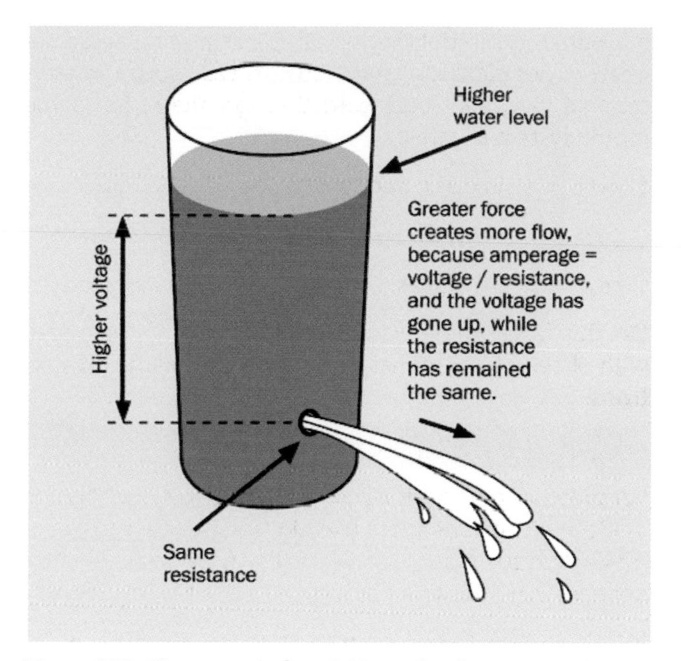

Figure 1-71 *The amount of available work will increase with greater water pressure.*

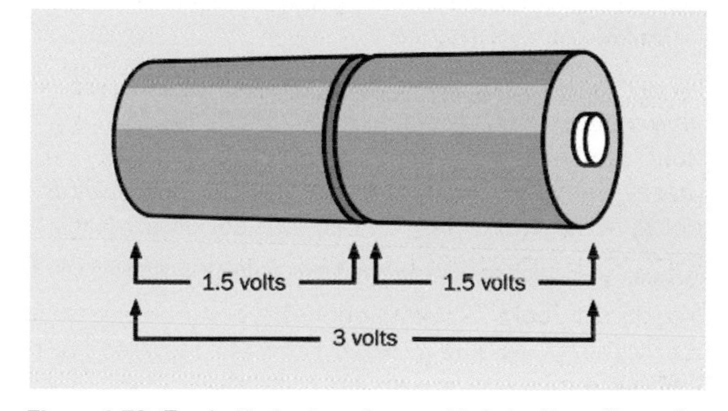

Figure 1-72 *Two batteries in series provide twice the voltage of a single battery, provided they are both fully charged.*

Summing up:

- Two batteries in series deliver twice the voltage.
- Two batteries in parallel can deliver the same current for twice as long, or twice the current for the same amount of time.

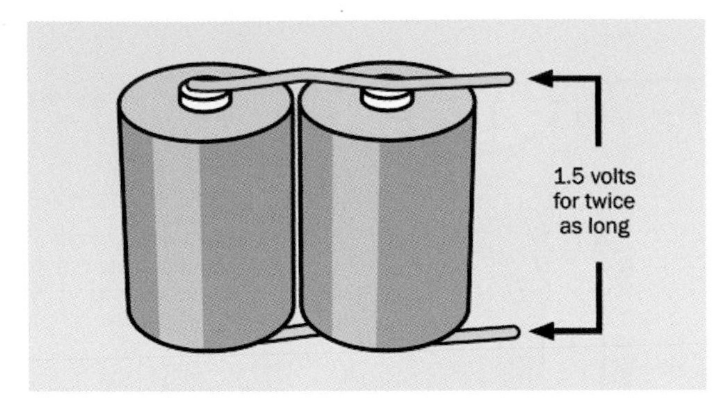

Figure 1-73 *Batteries in parallel, powering the same load as before, will run it for about twice as long. Alternatively, they can provide twice the current for the same time as a single battery.*

That's enough theory for now. In the next chapter, I'll continue with some experiments that will build on the foundations of knowledge about electricity, to take you gradually toward gadgets that can be fun and useful.

Cleanup and Recycling

The hardware that you immersed in lemons or lemon juice may be discolored, but it is reusable. Bearing in mind that some zinc ions may have been deposited in the lemons, eating them might not be such a good idea.

Switching

2

This chapter of the book contains Experiments 6 through 11, which explore the seemingly simple topic of switching—by which I mean not just manual control, but using one flow of electricity to switch, or control, another. It's such an important principle, no digital device can exist without it.

Today, switching is mostly done with transistors. I will deal with them in detail, but before that I'll back up and illustrate the concept by introducing you to relays, which are easier to understand, because you can see what's going on inside them. And before I get to relays, I'll show how manually activated switches can demonstrate some of the concepts that will follow. So, the sequence will be switches—relays—transistors.

Also in this chapter, I'll deal with capacitance, because capacitance is almost as fundamental as resistance in electronic circuits.

Necessary Items for Chapter Two

As before, when buying tools and equipment, see "Buying Tools and Equipment" on page 324 for a shopping list. If you want kits containing components and supplies, see "Kits" on page 311. If you prefer to buy your own components and supplies from online sources, see "Components" on page 317.

Essential: Small Screwdrivers

A set such as the one made by Stanley (part number 66-052) is shown in Figure 2-1. Screwdrivers that you may already have in your home will be too big for most of the little screws that you find on components.

You can buy cheaper screwdriver sets that look very similar to the one in Figure 2-1, but I think the quality of steel is better in name-brand screwdrivers.

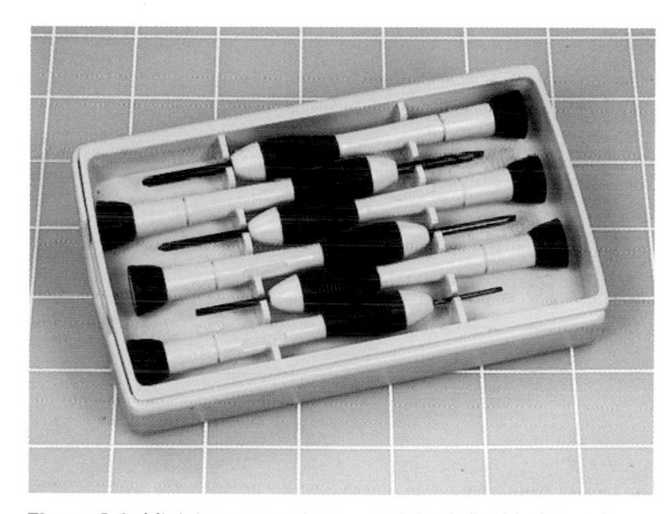

Figure 2-1 *Miniature screwdrivers, with both flat blades and Phillips blades. The white lines are spaced at intervals of 1".*

Essential: Small Long-Nosed Pliers

The type of *long-nosed pliers* that you need are no more than five inches from end to end. You'll be using them to bend wire precisely, or to pick up small parts where a finger and thumb are too big and clumsy. For these kinds of operations I don't think you benefit much from spending extra money on high-quality tools, so feel free to buy the cheapest. An example is shown in Figure 2-2. These have spring-loaded handles, which some people don't like, but you can pull out the springs—if you have a second pair of pliers to do so.

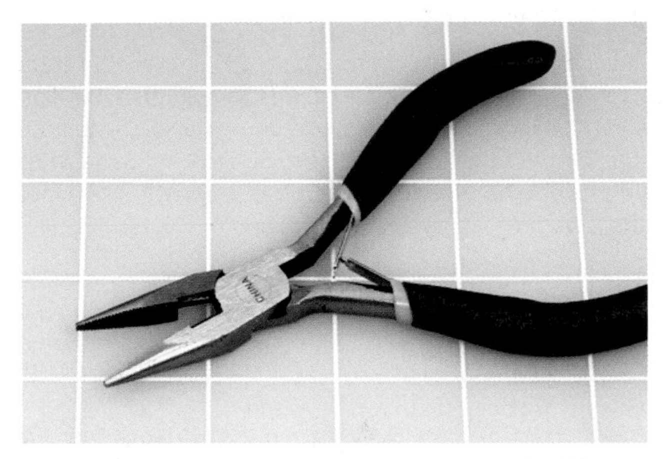

Figure 2-2 *Appropriate pliers for electronics work should be no more than 5" long.*

Optional: Sharp-Nosed Pliers

These are like small long-nosed pliers, but with very precise, pointed jaws. I like them for accessing tightly packed components on a breadboard. Your best source is a website or store that specializes in craft work such as beading. Be careful, however, that you don't buy beading pliers that have rounded noses for making loops in wire. For our purposes, the inner surfaces of the jaws should be flat, as shown in Figure 2-3.

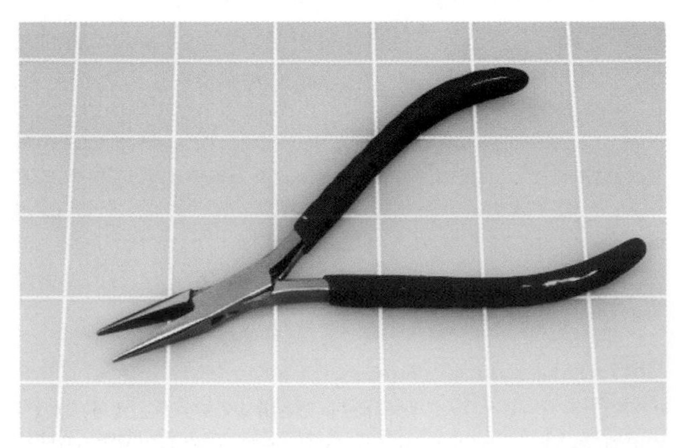

Figure 2-3 *Sharp-nosed pliers enable very precise work on a small scale.*

Essential: Wire Cutters

Pliers usually have cutting edges near the joint, which you can use for chopping wire. Often, however, a piece of wire will be attached to something else, and you won't be able to reach it with your pliers. You really need *wire cutters* such as those shown in Figure 2-4. They should be no longer than five inches. So long as you use them to cut thin, soft copper wire, they do not have to be of high quality.

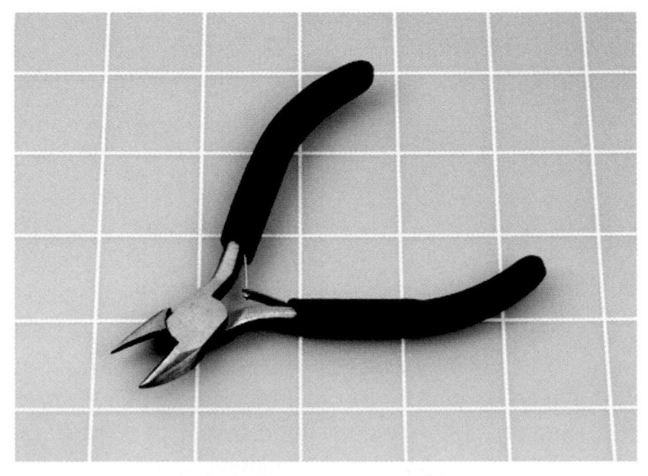

Figure 2-4 *Wire cutters should be no more than 5" long.*

Optional: Flush Cutters

Flush cutters, shown in Figure 2-5, are similar to wire cutters, and do the same job, but they are thinner, smaller, and better able to get into small spaces. However, they are less robust. Whether you use them or wire cutters is a matter of personal preference. Personally I like wire cutters.

Figure 2-5 *Flush cutters can get into smaller spaces than wire cutters.*

Essential: Wire Strippers

Wire of the type that you will be using has a coating of plastic insulation. *Wire strippers* are specifically designed to remove a short section of insulation to expose the conductor inside it. Macho hobbyists may claim that they don't need any tool to do this job, but I have two corners missing from the insides of my front teeth to testify that this is a really bad idea (see Figure 2-6).

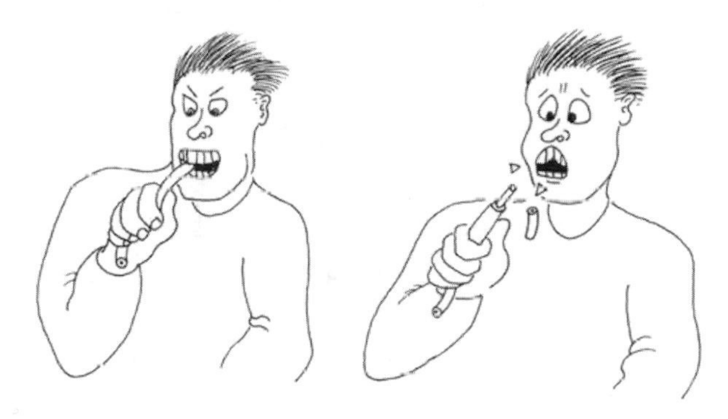

Figure 2-6 *In a hurry? Can't find your wire strippers? The temptation is obvious, but not a good idea.*

Another option is to use wire cutters, as shown in Figure 2-7. You hold the wire in one hand, close the jaws of the cutters gently with the other hand, and lever your two hands apart. However, this is a skill that requires practice. Sometimes the cutters slip off without doing anything, or they may cut the wire instead of stripping it.

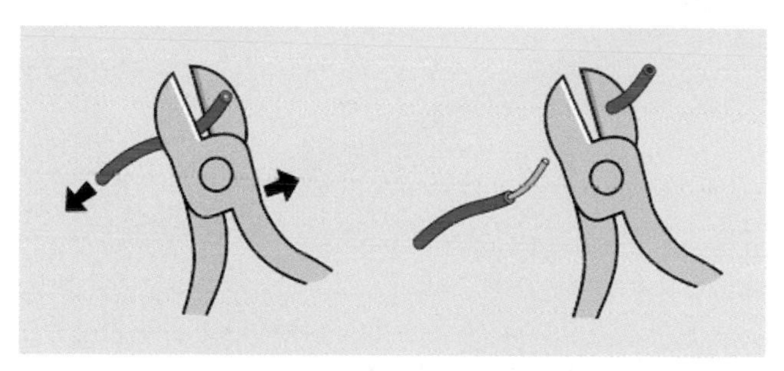

Figure 2-7 *How to use wire cutters to strip insulation. Wire strippers are easier to use.*

For a few extra dollars, a pair of wire strippers makes the job so much easier.

The first edition of this book included an option to buy so-called automatic wire strippers, which can be used with only one hand. Unfortunately they are significantly more expensive than other types of wire strippers, and many brands don't work well with the 22-gauge hookup wire required for all the circuits in this book. Therefore, I don't recommend them anymore.

The type of tool shown in Figure 2-8 is available from many manufacturers. Some brands have angled handles, some have straight handles, and some have curved handles. I don't think it matters. They all work the same way: you insert the wire in a hole of appropriate size, close the jaws, and pull away the insulation.

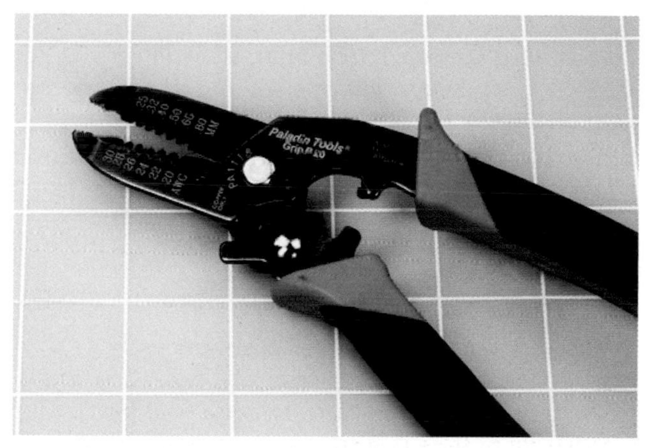

Figure 2-8 *The recommended wire strippers are designed for 20-gauge to 30-gauge wire.*

You do have to be careful, though, that they are suitable for the right size of wire.

Wire gauge is a measure of the thickness of a conductor. A higher number indicates a thinner wire. As it happens, 20-gauge wire is slightly too thick for our purposes, while 24-gauge wire is slightly too thin. The optimal thickness is 22 gauge, and you will have a much easier time if you buy a tool that is calibrated for that specific size. As you can see in Figure 2-8, a range from 20 gauge to 30 gauge includes a little hole for 22 gauge. This is the right tool for the job.

Essential: Breadboards

Breadboards are not required until Experiment 8, but I will provide a brief introduction here. The breadboard is a little plastic slab perforated with holes at intervals of 0.1". You push components and wires into the holes. Hidden beneath the plastic are conductors that make connections along the rows of holes.

A breadboard allows components to be connected together more neatly than with the test leads that you have used so far, and more easily (and reversibly) than if you join them with solder.

- A breadboard is properly known as a *solderless breadboard* and is sometimes referred to as a *prototyping board*.

The brand or source is not important, but you should be careful to buy the same configuration that I have used throughout this book. You have three options, only one of which is correct:

Breadboard Option 1: The Mini, shown in Figure 2-9. Often sold as being "suitable for Arduino," it doesn't have enough holes in it for our purposes, so, don't buy one of those.

Figure 2-9 *A mini-board is not big enough for many of the projects in this book.*

Breadboard Option 2: Single-Bus, shown in Figure 2-10. The term "bus" refers to a long column of holes alongside the short, numbered rows of holes. There is a single bus on each side, outlined in red in the photograph. This is the kind of board that you want. Check a photograph of the product that you are buying, to make sure. Also, note that it should have 60 rows of holes and 700 connection points (also known as tie points). If you are buying your own, search Amazon or eBay for:

```
solderless breadboard 700
```

If you prefer, you can use a dual-bus breadboard and just ignore the extra holes.

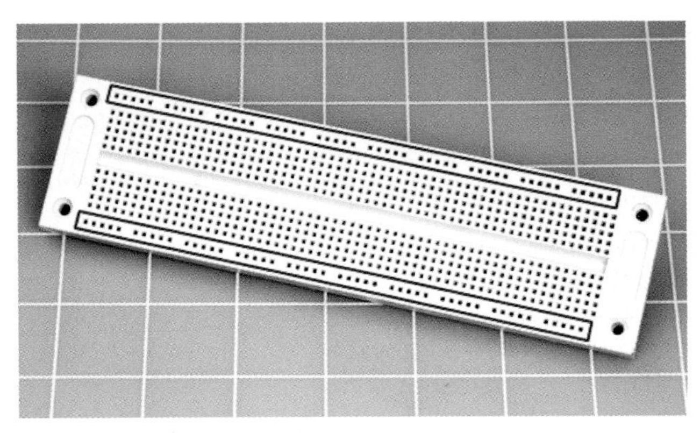

Figure 2-10 *A single-bus breadboard has a single long row of holes along each side.*

Breadboard Option 3: Dual Bus, shown in Figure 2-11. This has two long rows of holes on each side, which are the dual buses, outlined in red in the photograph. I used this type of board in the first edition of this book, because it can be more convenient. Then I saw how it encouraged wiring errors when people used it for the first time, so I don't recommend dual-bus breadboards anymore.

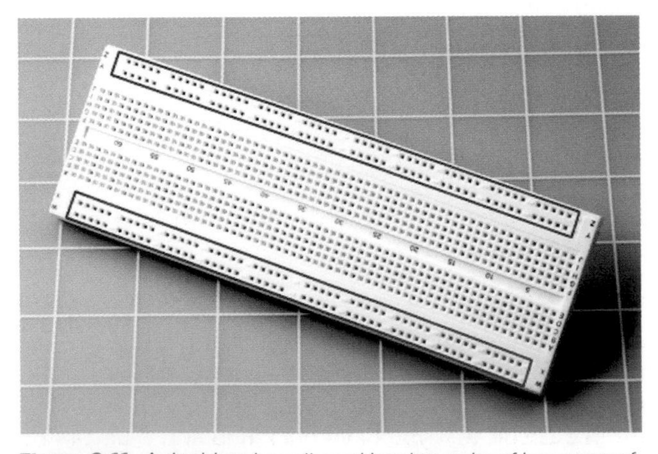

Figure 2-11 *A dual-bus breadboard has two pairs of long rows of holes, outlined in red in this photograph. This style of board is no longer recommended.*

Now that I've established the type of board that I recommend, how many of them do you need? I used to say, "Only one," bearing in mind that they are reusable, but their price has been driven down to the point where I think you should consider buying two or three. This way, you can prototype new circuits without having to disassemble old ones first.

Supplies

If you want to buy a kit containing components and supplies, see "Kits" on page 311. When buying your own supplies, see "Supplies" on page 316.

Essential: Hookup Wire

The type of wire you need for making connections on a breadboard is often called *hookup wire* although it can be found, sometimes, under the general category of *bulk wire*. Either way, it must have a solid core (not stranded), and must be 22 gauge in size.

Often it is sold in 25-foot and 100-foot lengths on plastic spools, as shown in Figure 2-12.

Figure 2-12 *Hookup wire is available on spools holding 25 feet and 100 feet, shown here.*

Wire is cheaper, per foot, if you buy a 100-foot spool, but I suggest you should buy smaller quantities in at least three different colors of insulation. The reason is that wire colors are helpful when you are trying to find an error in a circuit that you have built. Red and blue can be used for connections with positive and negative power, and one other color for other connections.

When the insulation is stripped away, it reveals a solid conductor, as shown in Figure 2-13. Compare this with stranded wire, shown in Figure 2-14. Stranded wire does have some uses, which I will talk about in a moment, but if you try to push it into holes in a breadboard, you are likely to become frustrated quite quickly. You really need solid wire.

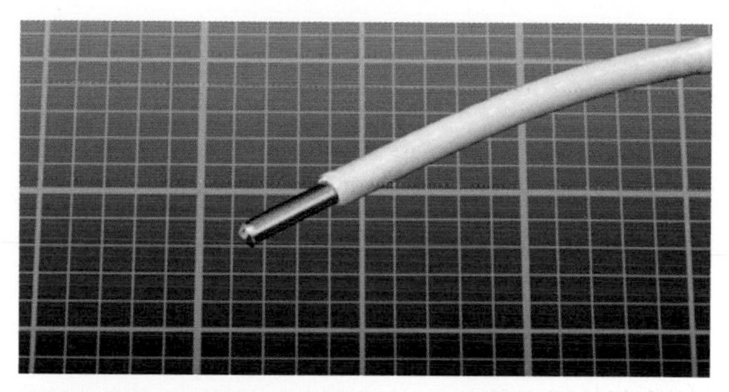

Figure 2-13 *Inside the plastic insulation, there should be a single conductor of solid wire.*

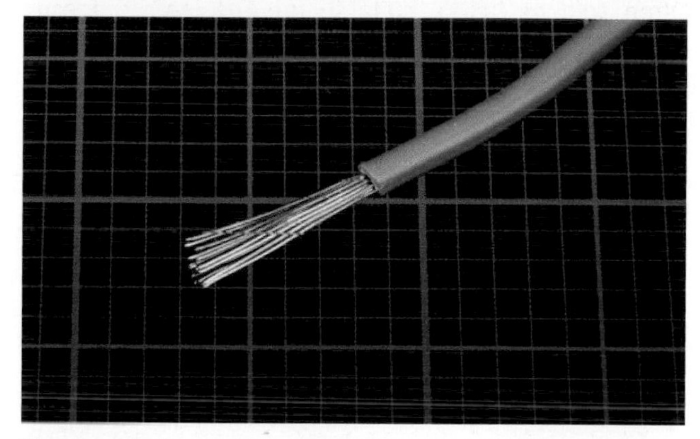

Figure 2-14 *For specific purposes (described in the text) stranded wire can be a useful option.*

Just as I emphasized the need for wire strippers that are specifically designed for 22-gauge wire, I will now emphasize that the wire should be 22-gauge, not 20-gauge, and not 24-gauge. You will find that the 24-gauge wire doesn't fit tightly in a breadboard, and may not make a reliable connection, while 20-gauge wire is just a little too thick, so that when you try to push it in, it can bend instead of sliding into place—and if you do manage to insert it, pulling it out can be difficult.

Some copper wire has a silver-colored coating, when you strip the insulation away. This is said to be "tinned." Other wire is plain copper, and I don't have any opinion about which type is better.

How much will you need? For assembling the circuits in this book, 25 feet of each color will be more than enough. However, Experiments 26, 28, and 29 require you to make coils of wire, to explore the relationship between electricity and magnetism, and to build your own crystal-set radio. If you want to pursue these projects (which I think are worthwhile), you'll need 200 feet of wire. This is your choice, as none of the kits for this book contains that much wire. See "Supplies" on page 316 for information about buying wire.

Jumpers

If you cut a section of wire, strip at least 1/4" but no more than 3/8" insulation from each end, bend the ends down, and push them into holes in a breadboard, you have just made a *jumper*, which creates a connection by jumping across some intervening holes in the board. Jumpers of this type result in a neat circuit where you can find errors relatively easily.

The trouble is, stripping insulation and making right-angle bends is tedious, even with the right tools for the job. Therefore you may be inclined to purchase *precut wire* which has been made into jumpers for you. An assortment looks like the one shown in Figure 2-15. For guidance about finding a selection like this, see "Supplies" on page 316.

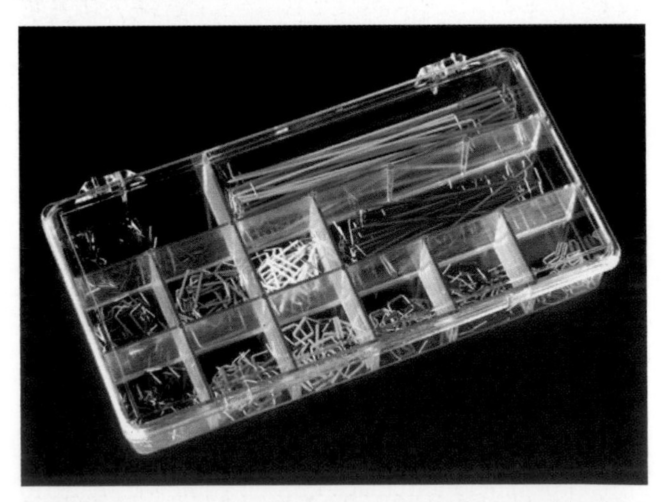

Figure 2-15 *An assortment of precut wire, stripped and bent for use in a breadboard.*

I used to use precut wire myself, but gave up on it because the wire segments are colored according to their length rather than their function. Red wires are all 0.2" long, yellow wires are all 0.3" long, and so on.

I want wires to be colored according to what they do in a circuit. Thus, red wires should always be connected to the positive side of the power supply, regardless of how long they are.

The only way to achieve this is to cut your own, which is what I do. You are welcome to use the precut wire if you so choose—but in addition to being confusingly colored, it costs more.

There is one more issue I have to clarify regarding jumpers. Many people like to use a different type of jumper wire that has a little plug at each end, just the right size to push into holes in a breadboard. These "plugged jumpers" are sold in bundles, and are probably the first thing you will find if you search for jumper wires online.

Because they are flexible, and they are usually about three inches long, you can use them to make almost any connection you are likely to need on a breadboarded circuit. They are reusable, and appear to be the simplest, quickest, cheapest option.

So far, so good; but if you make a mistake, you are going to have a problem finding it. Figure 2-16 shows a small circuit for an application outside of this book, using flexible jumpers that have plugs on each end. Figure 2-17 shows exactly the same circuit with hand-cut jumper connections using solid-core 22-gauge wire. Each of these circuits contains one wiring error. On the one using hand-cut wire, I can see it within a few seconds. On the one using flexible jumpers, I would have to dig around for quite a while, probably using a meter to search for the fault.

Figure 2-16 *A circuit on two mini-breadboards using flexible jumper wires with a plug at each end.*

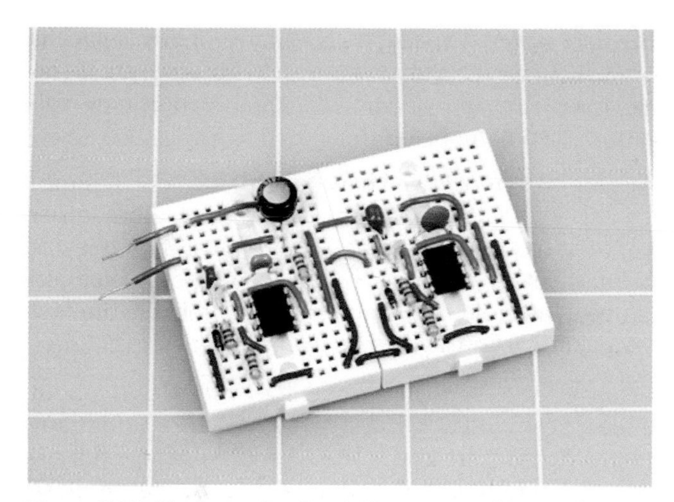

Figure 2-17 *The same circuit as in the previous figure, using hand-cut solid jumper wires.*

To make matters worse, the plugs on flexible jumpers are sometimes defective, and may contain loose connections. This can make fault-tracing almost impossible. Therefore:

- I don't recommend flexible jumpers that have plugs on each end.

Optional: Stranded Wire

Getting back to stranded hookup wire, it does have one advantage. It is much more flexible than solid wire, which is useful if you are running it out from a circuit board to a switch or a potentiometer. Flexibility can be essential if the wire is making a connection with an object that moves or vibrates.

While flexible wire is not essential for any of the projects in this book, 25 feet of 22-gauge stranded will come in handy occasionally. If you buy some, I suggest you choose a color that is different from your colors of solid wire, so that you won't confuse it with them.

Components

Again I must remind you that component kits for the projects in this book are available. See "Kits" on page 311. If you prefer to buy your own components from online sources, see "Components" on page 317.

Essential: Toggle Switch

A full-size *toggle switch* is an old-fashioned device, but useful for your switching experiments. You will need two. They should be described as SPDT, meaning single-pole double-throw. I'll explain this in detail a little later. Double-pole double-throw, described as DPDT, can also be used, but may cost slightly more.

A toggle switch with *screw terminals* will reduce the inconvenience of attaching it to hookup wire, but other types of terminals are acceptable.

A typical full-size toggle switch is shown in Figure 2-18. E-Switch ST16DD00 is an example, but cheaper generic alternatives can be found on eBay.

Figure 2-18 *A full-size toggle switch.*

Essential: Tactile Switch

Confusingly, a tactile switch is not what you would think of as a switch. It is a very tiny pushbutton. If you plug it into your breadboard, it provides a convenient way for a circuit to receive user input.

The most commonly used tactile switches have four little legs to push into a board, and can be annoying, because the legs often don't engage properly. The component is liable to spring out like a baby grasshopper at unexpected moments. I suggest you use a tactile switch with two pins spaced 0.2" apart. The Alps SKRGAFD-010 will be used in projects throughout this book, and is shown in Figure 2-19. Any other tactile switch with two pins spaced 0.2" apart can be substituted, such as the Panasonic EVQ-11 series.

Figure 2-19 *The recommended tactile switch for breadboarded projects in this book.*

Essential: Relay

Because pin functions are not standardized among manufacturers, you have to be cautious about making substitutions when you buy a *relay*. I am recommending the Omron G5V-2-H1-DC9, shown in Figure 2-20, which should minimize confusion, because it has its pin functions printed on it.

Figure 2-20 *The relay recommended for use with this book.*

Omron is a large manufacturer of relays, so I have some hope that the one I am recommending will be around for a while. You may also use the Axicom V23105-A5006-A201 or the Fujitsu RY-9W-K. All of them are are 9VDC DPDT relays, with pins spaced as shown on the left side of Figure 2-21. If the spacing is shown in millimeters, 5mm or 5.08mm are acceptable substitutes for 0.2", and 7.5mm or 7.62mm can substitute for 0.3".

If a diagram is printed on the relay, it should look like the one on the right side of Figure 2-21. Datasheets for relays almost always contain this information. You can use relays with different pin functions, but they will cause you some inconvenience, because they won't match the schematics that I will be providing.

The relays I have recommended are high-sensitivity type, meaning that they consume less current. You can substitute others, but they will draw more current. Whichever type of relay you use, it must have the same 9VDC coil voltage and pin spacing.

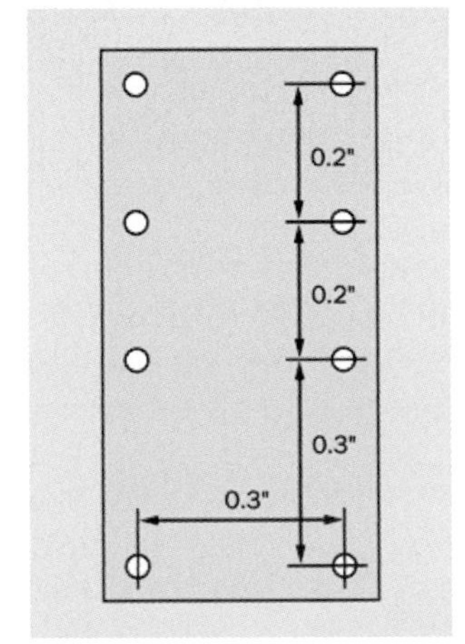

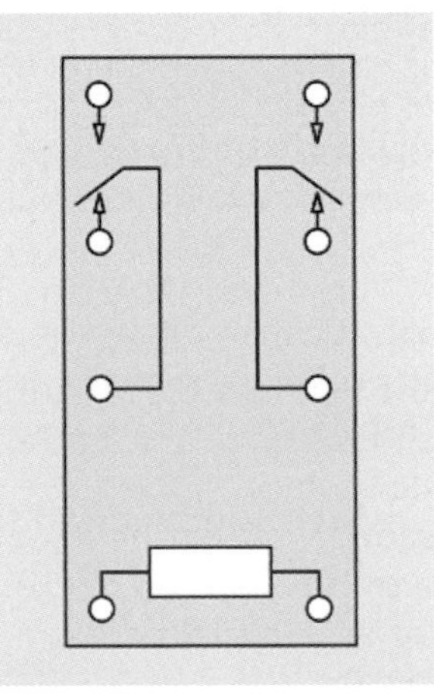

Figure 2-21 *Pin spacing and internal connections in a relay should be as shown.*

One thing to watch out for, when buying relays, is their polarity, meaning a requirement to apply current in one specified direction, because the relay won't work when current flows through its coil in the other direction. I have recommended relays that do not require polarity. Many Panasonic relays do, so read the datasheet carefully before buying one of them.

Lastly, any relay that you buy must be of the *nonlatching* type.

If this seems confusing and overly technical, you can put off buying the relay until you read Experiment 7, which describes how to use it. You will need two relays to carry out that experiment fully.

Essential: Trimmer Potentiometer

Instead of the big clunky potentiometer that you used in Experiment 4, you'll be using a *trimmer potentiometer*, which is smaller, cheaper, and plugs into a breadboard.

Examples (with various arbitrary values) are shown in Figure 2-22.

Figure 2-22 *Trimmer potentiometers.*

The trimmers at left and at right in this photograph are the type that I have chosen to use in this book. They will sit flush with a breadboard when their leads are pushed down into it. The only difference between these two samples is that one is slightly larger than the other. They are also available in variants that stand up at an angle of 90 degrees with the breadboard, but these are less accessible.

The one in the center of the photograph is a *multi-turn trimmer*, which allows much finer adjustment via a brass screw connected with a worm gear inside the component. This is less convenient, more expensive, and is not necessary for our purposes, as you won't require that degree of accuracy.

Essential: Transistors

Only one type of *transistor* is used in this book. The generic part number is 2N2222, but unfortunately, not all 2N2222 transistors are alike.

If you are using a kit, you should have no problems. If you do your own shopping, you absolutely must avoid any part that has P2N preceding the 2222 number. When the P2N2222 was introduced, the manufacturers reversed the pin functions of the 2N2222 that had been standardized for decades. (Why would they do this? I do not know.)

Here's the rule.

- Part numbers 2N2222 or PN2222 or PN2222A are okay. PN2222 has become a more common designation than 2N2222, but either will work.
- Part numbers P2N2222 or P2N2222A are *not* okay.

The trap is that if you are searching for 2N2222, you will be offered a P2N2222 because the search engine will help you out by showing parts that have extra letters preceding the number. So—shop with care! And if you have a meter that tests transistors, check each one. If the transistor has the traditional pin functions, the meter should tell you that it has an amplification ratio exceeding 200. If you have the wrong type of transistor, your meter will show an error or an amplification value lower than 50.

2N2222 transistors used to be packaged in tiny metal cans. These days, they are almost always packaged in black plastic. Examples of both types are shown in Figure 2-23. Plastic and metal packages both work equally well—so long as the transistor's part number does not begin with P2N.

Figure 2-23 *Two 2N2222 transistors. Either can be used.*

Essential: Capacitors

Capacitors are not quite as cheap as resistors, but still cheap enough for you to consider buying an assortment of small ones in bulk. Capacitor values in the range that we will be using are most often measured in microfarads, abbreviated μF. I'll explain this in detail when you start using capacitors in your circuits.

For small values, *ceramic* capacitors are recommended. For larger values, *electrolytics* are cheaper. For additional guidance about purchasing them, see "Components" on page 317. Various capacitors are shown in Figure 2-24. The cylindrical ones are electrolytic, while the others are ceramic.

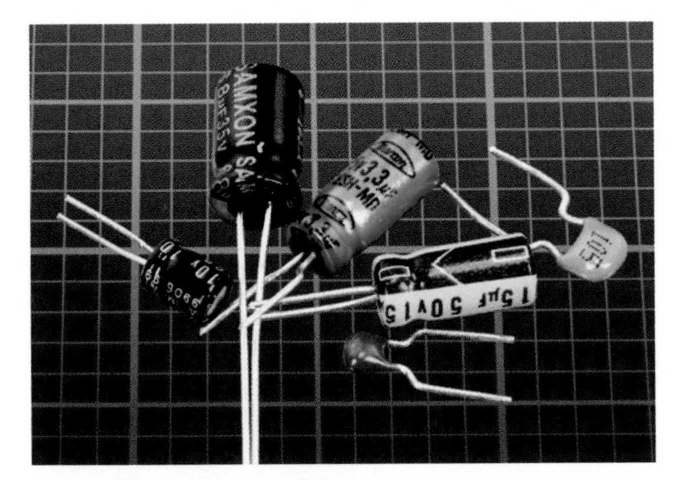

Figure 2-24 *A variety of capacitors.*

Essential: Resistors

If you are shopping for your own components, I am assuming that you bought a good selection of resistors as I suggested for Experiment 1.

Essential: Loudspeaker

Minimum size for a *loudspeaker* is 1" diameter, but 2" diameter is good. A 3" diameter is maximum. The impedance should be 8 ohms or higher. When searching for a loudspeaker, bear in mind that some suppliers call them "speakers." This may cause a search for "loudspeaker" to yield no results.

We will not be dealing with high-fidelity sound, so any cheap loudspeaker will do. A couple of samples are shown in Figure 2-25.

Figure 2-25 *Two speakers, one measuring 1" diameter, the other measuring 2".*

And More?

By now you may be thinking that I have specified quite a lot of components. Rest assured that almost everything that I have listed here will be reusable, and you will not need many additional parts for the remaining chapters of the book.

Experiment 6: Very Simple Switching

This experiment will acquaint you with the function of manually operated switches. You may feel that you already know how to use a switch, but when two double-throw switches are combined in a circuit, the subject becomes a little more interesting.

What You Will Need

- Screwdriver, wire cutters, wire strippers
- Hookup wire, 22-gauge, no more than 12"
- 9-volt battery (1)
- Generic LED (1)
- Toggle switches, SPDT or DPDT (2)
- 470-ohm resistor (1)
- Test leads with alligator clips at each end (2)

Assemble the parts as shown in Figure 2-26. You'll need to practice your abilities at wire stripping, here, to bare the ends of the two sections of black wire. To secure them in the screw terminals of the switches, try using your pliers to curl the end of each wire so that it looks like a letter J. Then feed it under the screw from the left, so that the screw will draw in the wire when tightened in a clockwise direction.

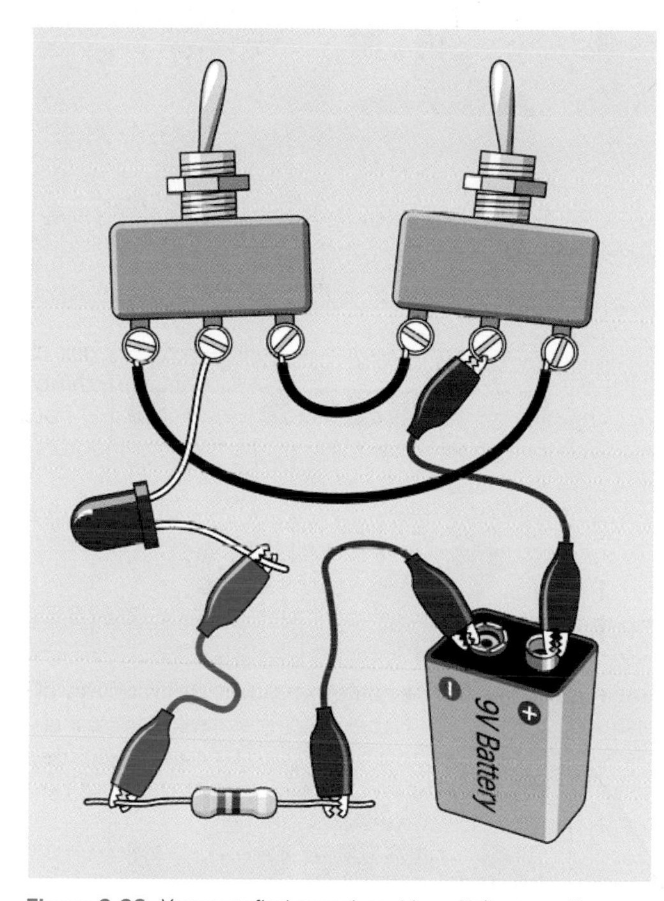

Figure 2-26 *Your very first experiment in switch connections.*

The long lead on the LED also fits one of the screw terminals. Don't use the short lead of the LED by mistake. Remember, the long lead must always be more positive than the short lead.

If your switches don't terminate in screw terminals, you will have to use a couple of alligator test leads instead of the black wires, and one more test lead to connect the LED with the center terminal of the lefthand switch.

After you connect the battery, experiment, flipping the switches. What do you find?

If the LED is on, flipping either of the switches will turn it off. If the LED is off, either of the switches will turn it on. I'll explain this interesting behavior shortly (see "Introducing Schematics" on page 55), but I have to tell you some fundamentals and background information first.

Fundamentals: All About Switches

The *toggle* in a toggle switch is the part that you flip with your finger. In the type of switch shown in Figure 2-26, flipping the toggle connects the center terminal with one of the terminals on either side of it, as shown in Figure 2-27.

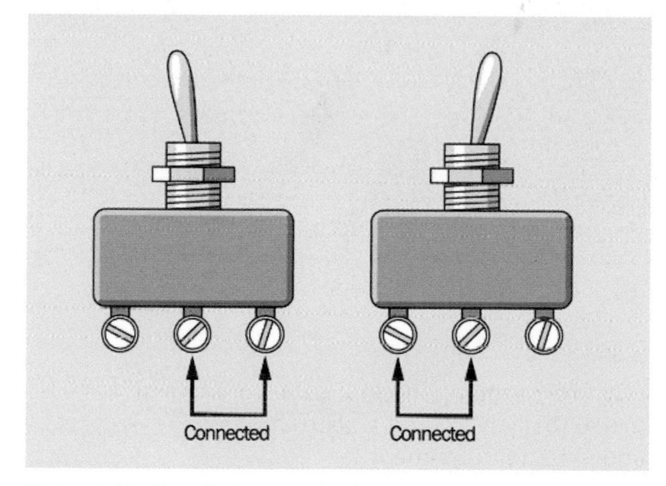

Figure 2-27 *Usually, but not absolutely always, toggle switches work like this.*

The center terminal is called the *pole* of the switch. Because you can flip, or throw, this switch to make two possible connections, it is called a *double-throw* switch, abbreviated DT (or, sometimes, 2T). A single-pole, double-throw switch is abbreviated *SPDT* (or, sometimes, 1P2T).

Some switches only have two terminals instead of three. They are on/off, meaning that if you throw them in one direction they make a contact, but in the other direction they make no contact at all. Most of the light switches in your house are like this. They are known as *single-throw* switches. A single-pole, single-throw switch is abbreviated *SPST* (or, sometimes, 1P1T).

Some switches have two entirely separate poles, so you can make two separate connections simultaneously when you flip the switch. These are called *double-pole* switches, abbreviated DP (or, sometimes, 2P). Check Figure 2-28, Figure 2-29, and Figure 2-30 for photographs of old-fashioned "knife" switches, which are still used sometimes to teach electronics to kids in school. You wouldn't use a switch like this for any practical purpose, but they illustrate very clearly the differences between SPST, SPDT, and DPST connections.

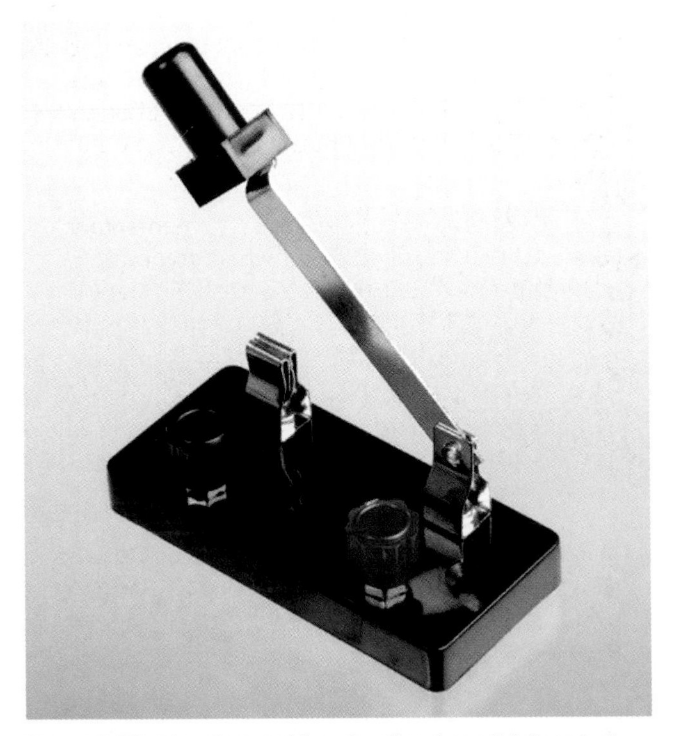

Figure 2-28 *Manufactured for educational use, this is a single-pole, single-throw (SPST) switch.*

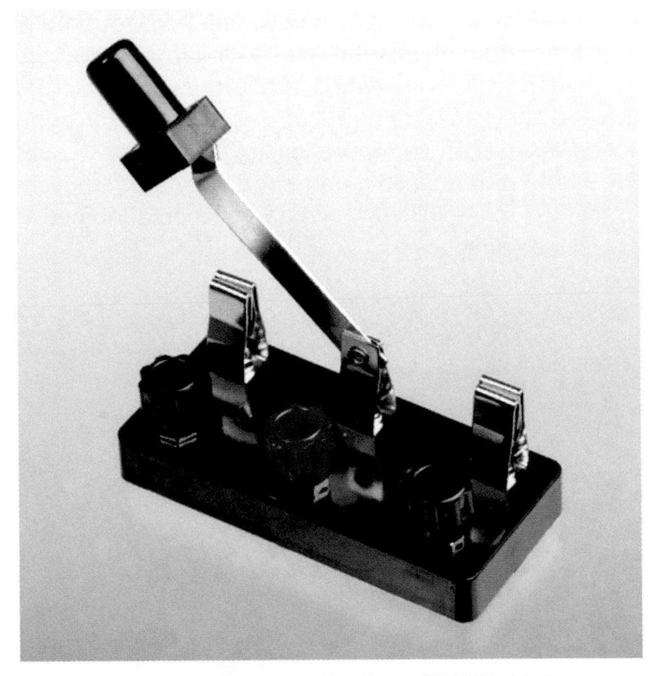

Figure 2-29 *A single-pole, double-throw (SPDT) switch connects one pole with a choice of contacts.*

The only place you're likely to see a knife switch being used for serious purposes is in a horror movie. In Figure 2-31, a mad scientist is powering up his experiment with a single-pole, double-throw knife switch, conveniently mounted on the wall of his basement laboratory.

Figure 2-30 *A double-pole, single-throw (DPST) switch has two poles that are completely isolated from each other. Each pole can connect with only one contact.*

Figure 2-31 *Left: Mad scientist. Right: SPDT knife switch.*

To make things more interesting, you can buy switches that have three or four poles. (Some rotary switches have even more, but we won't be using them.) Also, some double-throw toggle switches have an additional "center off" position.

Putting all this together, I made a table showing some possible types of switches and the abbreviations that describe them. Pushbuttons use the same abbreviations. See Figure 2-32. If you're reading a parts catalog, you can check this table to remind yourself what the abbreviations mean.

	Single Pole	Double Pole	Three Pole	Four Pole
Single Throw	SPST (or 1P1T) on-off	DPST (or 2P1T) on-off	3PST (or 3P1T) on-off	4PST (or 4P1T) on-off
Double Throw	SPDT (or 1P2T) on-on	DPDT (or 2P2T) on-on	3PDT (or 3P2T) on-on	4PDT (or 4P2T) on-on
Double Throw with Center Off	SPDT (or 1P2T) on-off-on	DPDT (or 2P2T) on-off-on	3PDT (or 3P2T) on-off-on	4PDT (or 4P2T) on-off-on

Figure 2-32 *This table summarizes various options for toggle switches (and pushbuttons).*

Some switches are spring-loaded, so that they snap back to a default position when you release pressure on them. When you see ON or OFF in parentheses, you know that you have to maintain pressure on the switch to keep it in that position.

Here are some examples:

- OFF-(ON): Because the ON state is in parentheses, it's the momentary state. Therefore, this is a single-pole switch that makes contact only when you push it, and flips back to make no contact when you let it go. It is also known as a "normally open" momentary switch, abbreviated "NO." Most pushbuttons also work this way.
- ON-(OFF): The opposite kind of momentary switch. It's normally ON, but when you push it, you break the connection. So, the OFF state is momentary. It is known as a "normally closed" momentary switch, abbreviated "NC."
- (ON)-OFF-(ON): This switch has a center-off position. When you push it either way, it makes a momentary contact, and returns to the center when you let it go.

Other variations are possible, such as ON-OFF-(ON) or ON-(ON). As long as you remember that parentheses indicate the momentary state, you should be able to figure out how these switches behave.

Sparking

When you make and break an electrical connection, it tends to create a spark. Sparking is bad for switch contacts. It erodes them until the switch doesn't make a reliable connection anymore. For this reason, you must use a switch that is appropriate for the voltage and amperage that you are dealing with.

The electronic circuits in this book are low-current and low-voltage, so you can use almost any switch; but if you are switching a motor, it will tend to suck an initial surge of current that is at least double the rating of the motor when it is running constantly. For instance, you should probably use a 4-amp switch to turn a 2-amp motor on and off.

Checking Continuity

You can use your meter to check a switch. Doing this enables you to find out which contacts are connected when you turn a switch one way or the other. It's also useful if you have a pushbutton and you can't remember whether it's the type that is normally open (you press it to make a connection) or normally closed (you press it to break the connection).

It's convenient to set your meter to measure "continuity" when checking a switch. The meter will beep (or show a visual indication) if it finds a connection, and will do nothing if it doesn't. See Figure 2-33, Figure 2-34, and Figure 2-35 for examples of meters that are set to measure continuity. Remember that in Experiment 1, I showed you the symbol that is used on meters to represent continuity. See Figure 1-7.

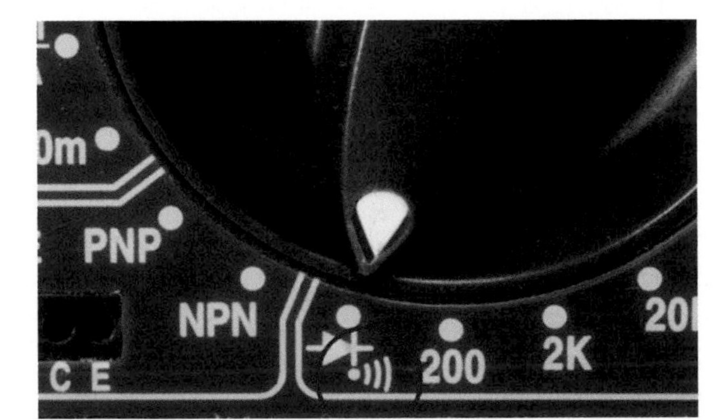

Figure 2-33 *A meter dial that has been turned to measure continuity.*

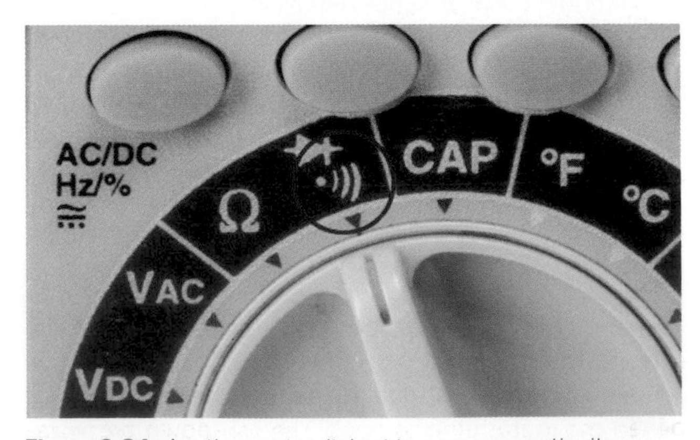

Figure 2-34 *Another meter dial set to measure continuity.*

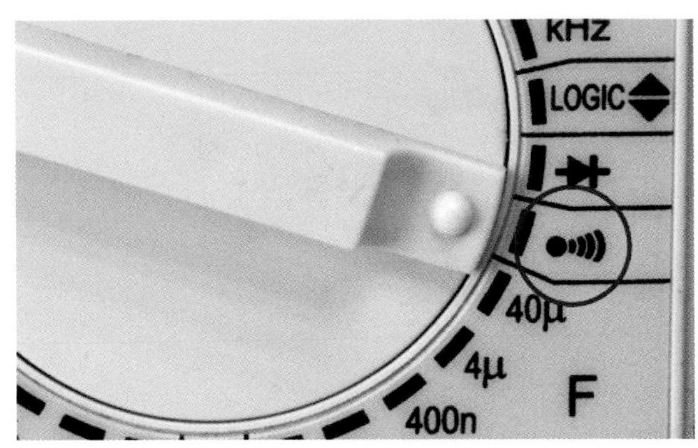

Figure 2-35 *A third meter measuring continuity.*

Background: Early Switching Systems

Switches seem to be such a fundamental feature of our world, and their concept is so simple that we can easily forget that they went through a gradual process of evolution. Primitive knife switches were quite adequate for pioneers of electricity who simply wanted to connect and disconnect some apparatus in a laboratory, but a more sophisticated approach was needed when telephone systems began to proliferate. Typically, an operator at a "switchboard" needed to connect any pair of 10,000 lines on the board. How could it be done?

In 1878, Charles E. Scribner (shown in Figure 2-36) developed the "jack-knife switch," so called because the part of it that the operator held looked like the handle of a jack knife. Protruding from it was a plug, and when the plug was pushed into a socket, it made contact inside the socket. The socket, in fact, contained the switch contacts.

Figure 2-36 *Charles E. Scribner invented the "jack-knife switch" to satisfy the switching needs of telephone systems in the late 1800s. Today's audio jacks still work on the same principle.*

Audio connectors on guitars and amplifiers still work on the same principle, and when we speak of them as being "jacks," the term dates back to Scribner's invention. Switch contacts are still mounted inside a jack socket.

Today, of course, telephone switchboards have become as rare as telephone operators. First they were replaced with relays—electrically operated switches, which I'll talk about later in this chapter. And then the relays were superceded by transistors, which made everything happen without any moving parts. In Experiment 10, you'll be switching current with transistors.

Introducing Schematics

In Figure 2-37, I've redrawn the circuit from Figure 2-26 in a simplified style known as a "schematic." From this point onward, I will be illustrating circuits with schematics, because they make the connections easier to understand. You only need to know a few symbols to interpret them.

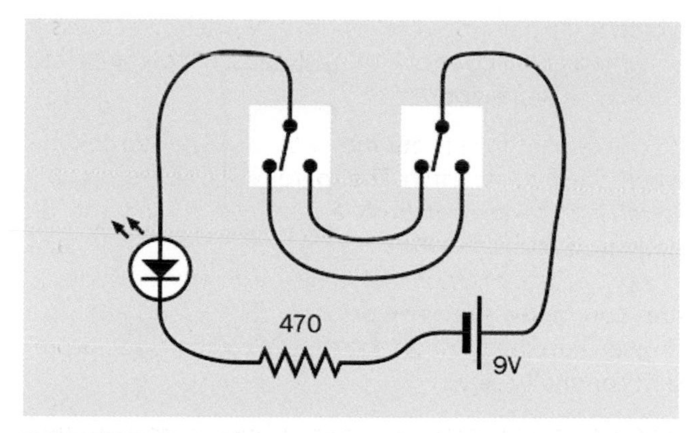

Figure 2-37 *The circuit with two switches is redrawn here as a schematic.*

Figure 2-26 and Figure 2-37 both show the same components and connections between them. In the schematic, the zigzag thing is the resistor, the symbol with two diagonal arrows is the LED, and the battery is shown as two parallel lines of unequal length.

The large triangle in the LED symbol indicates the direction of flow of *conventional current*, meaning the imaginary flow that runs from positive to negative. The pair of diagonal arrows tells you that this is the type of diode that is *light-emitting* (I'll get to other kinds of diodes later). In the battery symbol, the longer of the two lines identifies the positive side.

Trace the path that electricity can take through the circuit and imagine the switches turning one way or the other. You should see clearly, now, why either switch will reverse the state of the LED from on to off or off to on.

Figure 2-38 shows the same schematic cleaned up a bit. The lines are straight, and the power supply is now shown with the positive side at top-left, and the negative side at bottom-right. You tend to see conventional current moving from top to bottom, in schematics, while signals of some kind (such as the audio input to an amplifier) move from left to right. The top-down organization of a circuit makes it easier to understand.

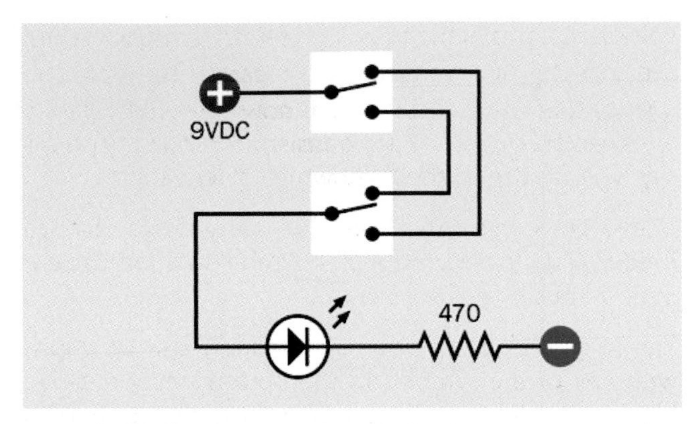

Figure 2-38 *The previous schematic has been reorganized along more conventional lines.*

The important thing to grasp is that the two schematics show exactly the same circuit, even though they look different. The type of components, and the way in which they are connected, are all that matters. The exact locations of the components are irrelevant.

- A schematic doesn't tell you where to put the components. It just tells you how to join them together.

Incidentally, you probably have an example of the circuit in Figure 2-38 in your home, especially where two light switches are placed at the top and at the bottom of a flight of stairs, and either switch can be used to turn a light on and off. This is shown in Figure 2-39, where the live and neutral wires of the AC supply enter the picture at bottom-left. The live wire is switched, while the neutral wire runs alongside it to the lightbulb (the white circle with a curly line representing the element of an old-fashioned incandescent bulb).

The only problem with schematics is that some symbols are not standardized. You may see several variants that all mean the same thing. I will be explaining them as we go along.

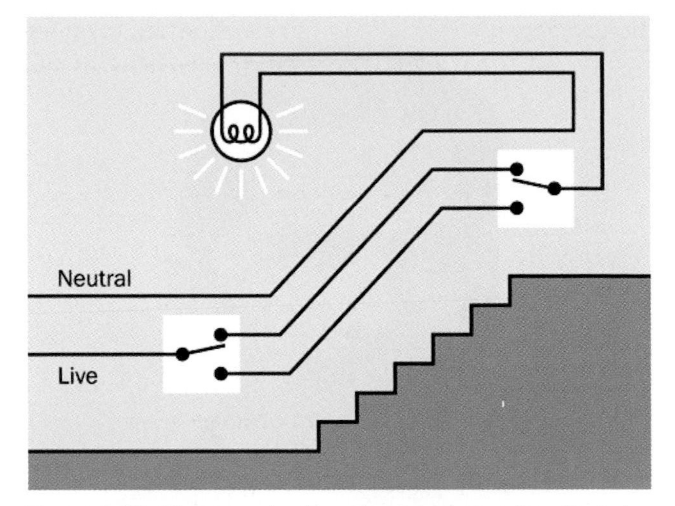

Figure 2-39 *The same circuit as shown in the previous figure is used in homes where two switches control a single light.*

Fundamentals: Basic Schematic Symbols

1. The Switch. Figure 2-40 shows five variations of that most basic component, a single-pole, single-throw switch. In each case, the pole happens to be on the right, while the contact is on the left—although with a SPST switch, this does not make a significant difference. In this book I have chosen to include a white rectangle around each switch, to emphasize that although it has two parts, they constitute one component.

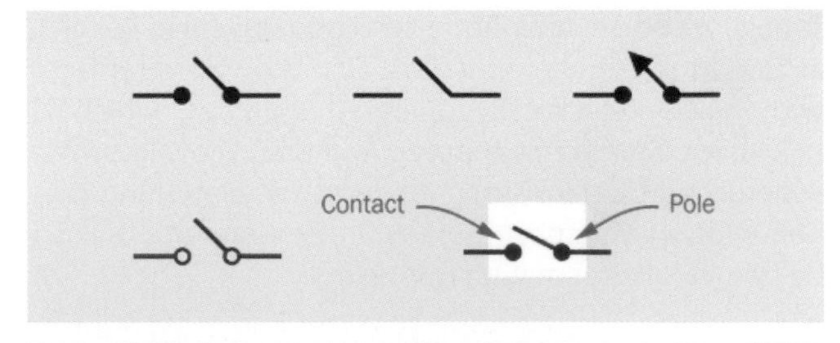

Figure 2-40 *Five variations in the schematic symbol for a SPST switch. They are all functionally identical.*

Figure 2-41 shows how things get a little more complicated when you have a double-throw, double-pole switch. A dashed line indicates that both segments of the switch move together when the switch is turned, even though each pole and set of contacts is electrically isolated from the other. The variation at center is sometimes found in large schematics where the layout makes it difficult to put the sections of a switch close to each other. Each set of contacts is identified by an abbreviation ending in A, B, C . . . and you are expected to understand that the contacts are actually contained in one switch.

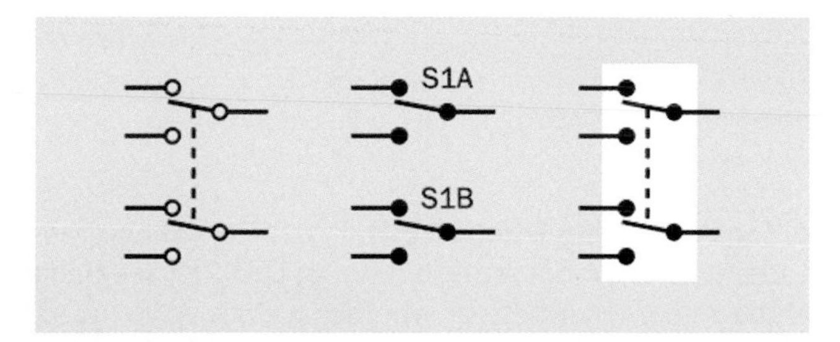

Figure 2-41 *Three variations on a DPDT switch symbol.*

2. Power Supply. The power supply for a DC (direct-current) circuit can be indicated in several ways. The top section of Figure 2-42 shows symbols for a battery. A short line indicates the negative end, while a longer line indicates the positive end. Traditionally, a single pair of lines represented a single 1.5-volt cell, two pairs indicated a 3-volt cell, and so on. But when circuits used high voltage with vacuum tubes, the person drawing the schematic would usually show a dashed line between cells instead of drawing dozens of them in a row.

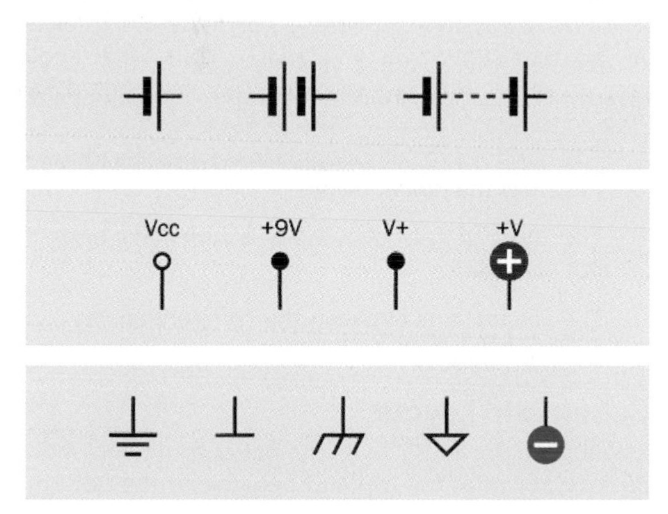

Figure 2-42 *Various ways to show positive and negative DC power in a circuit.*

A battery symbol may still be used in a simple schematic, but more often positive and negative DC power is indicated by separate symbols, shown in the center section and the bottom section of Figure 2-42. Positive is applied in one location of a circuit labelled Vcc, V_{CC}, or V+, or +V, or +V with a number added to indicate the voltage. Originally the term V_C referred to the voltage at the collector of a transistor. V_{CC} meant the supply voltage for the whole circuit, and is now used regardless of whether the circuit has any transistors in it. Many people say "vee cee cee" without knowing its derivation.

In this book, because I have the luxury of full-color reproduction, I have used a plus symbol in a red circle to indicate positive power input.

The negative side of the power supply can be shown with any of the symbols in the bottom section of Figure 2-42. It may be referred to as "negative ground" or simply "ground." Because many parts of a circuit may share a negative potential, multiple ground symbols may be found scattered around a schematic. This is more convenient than drawing lines linking them all together.

In this book, I have chosen to use the minus sign in a blue circle because it is so intuitively obvious. You will not see it used often in schematics generally.

My discussion so far has referred to battery-powered devices. In a gadget that uses AC power from a wall outlet, the situation is more complicated, because the outlet has three sockets in it for live, neutral, and ground connections. A schematic typically shows the AC source as an S shape turned on its side, as in Figure 2-43. Often the value of the power supply is shown, and in the US it is usually 110, 115, or 120 volts. Elsewhere in the circuit, the symbols shown on the righthand side of Figure 2-43 refer to the chassis of the device in which the electronics are mounted.

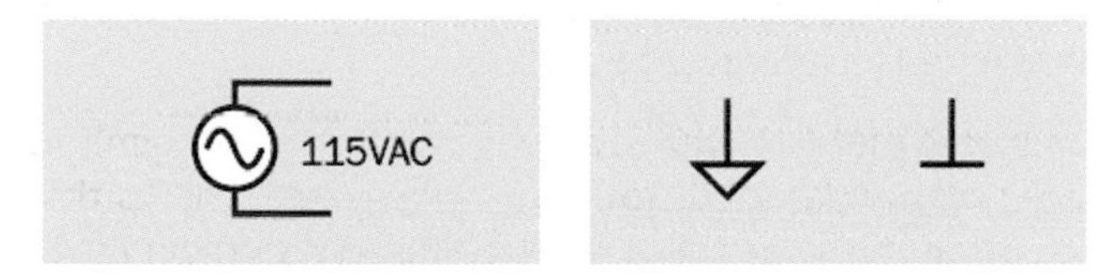

Figure 2-43 *Symbolic representation of AC power (left) and chassis in an AC device (right).*

Note that the ground pin in an AC power outlet in the home really does connect with the ground outside the building. An electronic device with a metal chassis that connects with that pin is "grounded." In a battery-powered circuit without any high voltages, grounding "in the ground" is unnecessary, but a ground symbol may still be used.

In the UK, a grounded device is sometimes referred to as being "earthed."

3. Resistor. Only two variants of the symbol for a resistor exist, shown in Figure 2-44. The symbol on the left is used in the United States, with a number beside it indicating its resistance in ohms. Alternatively the resistor may be identified as R1, R2, R3 . . . with a separate parts list showing the values. The righthand symbol in Figure 2-44 originated in Europe, and here again the number is the value of the resistor in ohms. The value of 220 ohms in the figure was selected arbitrarily.

Figure 2-44 *Symbol for a resistor in the United States (left) and Europe (right).*

Remember that where a resistance value includes a decimal point, Europeans omit the point and substitute a K or M, while resistances less than 1K are shown as one or more digits followed by letter R.

4. Potentiometer. In Figure 2-45 the lefthand symbol is used in the United States, while the righthand symbol originated in Europe. In both symbols, the arrow identifies the wiper in the potentiometer. The value of 470 ohms was selected arbitrarily.

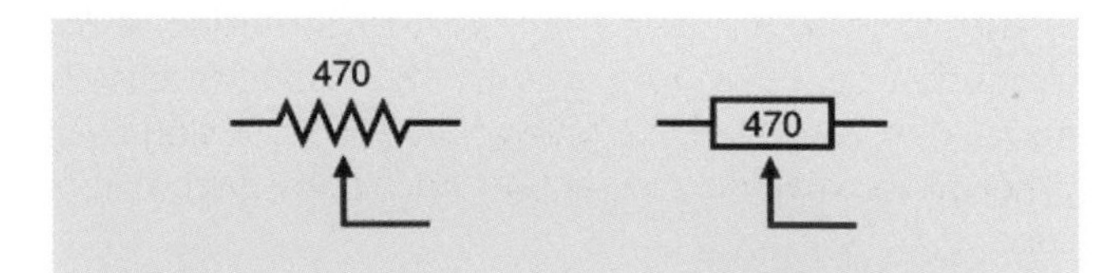

Figure 2-45 *The lefthand symbol for a potentiometer is used in the United States, while the righthand symbol originated in Europe.*

5. Pushbutton. Three possible pushbutton symbols are shown in Figure 2-46. These symbols represent the most common type of normally open pushbutton or momentary switch, where pressure closes two contacts and release of the pressure opens the circuit. In more complex pushbuttons, where a single button press closes or opens numerous contacts, the symbol for a multi-pole switch may be used.

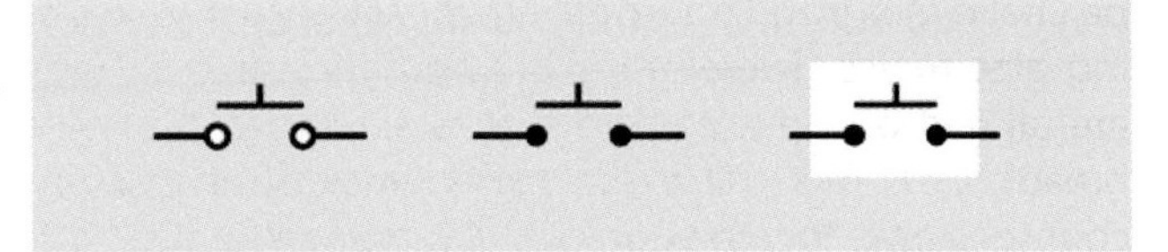

Figure 2-46 *Three variants of the pushbutton symbol. The white rectangle is added in this book for clarity, but is not used elsewhere.*

6. Light-Emitting Diode (LED). Figure 2-47 shows four variants of a symbol representing an LED. The meaning is the same, regardless of whether a circle is included, and regardless of whether the triangle is solid or open. The white highlight in the circle is added in this book for clarity, but is not used elsewhere. LED symbols may be oriented in any direction, for convenience in drawing a circuit. Arrows also may point in any direction.

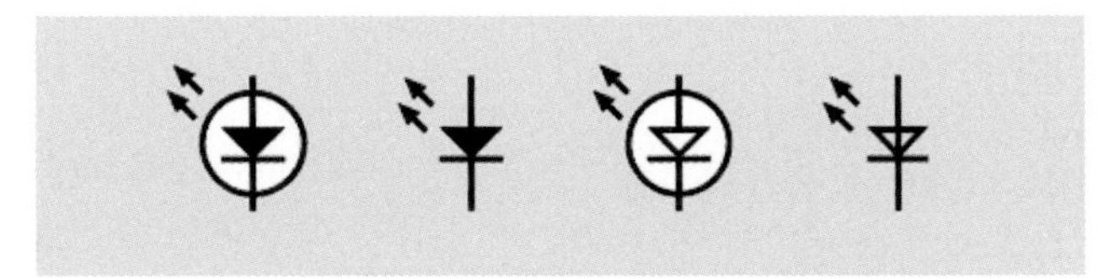

Figure 2-47 *Four ways of representing an LED. They are functionally identical.*

I'll explore other symbol variants later in the book. Meanwhile, the most important things to remember are:

- The positions of components in a schematic do not affect the functionality.
- The styles of symbols used in a schematic are not important.
- The connections between the components are extremely important.

Schematic Layout

I mentioned previously that schematics commonly show positive power at the top and negative at the bottom. This convention is a great help in understanding how the circuit works, but is not helpful at all when you actually want to build the circuit, because almost certainly you will begin by using a breadboard that imposes a completely different geometry.

Almost all the electronics books that I have seen expect you to change a circuit from the way it looks as a schematic to the way it has to be on a breadboard. This can be challenging, and may be a significant barrier to learning electronics. Therefore, the schematics in this book are all laid out in a pattern that is similar to that of a breadboard. This will make more sense after you encounter a breadboard yourself in Experiment 8.

Crossovers

The last topic I have to mention about schematics concerns the way in which they show two wires crossing each other. In the simple circuits that you have built so far, no crossovers occurred, but as circuits become more complicated, wires must pass over each other without making an electrical connection. How can a schematic illustrate this?

In the first edition of this book, I used a style in which one wire crossing another had a little semicircular "jump" in it. This is identified as the "Old Style" in Figure 2-48. I still prefer this style, because you can see so clearly that the wires are not making an electrical connection. However, a few decades ago, rendering the little jump became more troublesome as circuits were created with graphics software instead of pen and ink. At that point, jumps were used less frequently.

An alternative, identified as "Newer Style" in Figure 2-48, showed one of the wires with a break in it. This was confusing, and was not easily rendered by automated circuit-drawing software. Consequently it, too, has become rare.

The third style, labelled "Conventional," is now extremely common. In this edition of *Make: Electronics* I decided I should conform with the conventional style that is used in the rest of the world, even though I think it isn't as clear as the old style.

Perhaps you are wondering—if two lines crossing each other are not electrically connected, how do you draw them when they are connected? The answer is, you use a dot, and to avoid confusion, the dot should be large, not just a little pin-prick. The lower half of Figure 2-48 shows what I mean. This leads to the general rule:

- Two lines crossing each other do not indicate an electrical connection.
- Where lines intersect in a dot, there is an electrical connection.

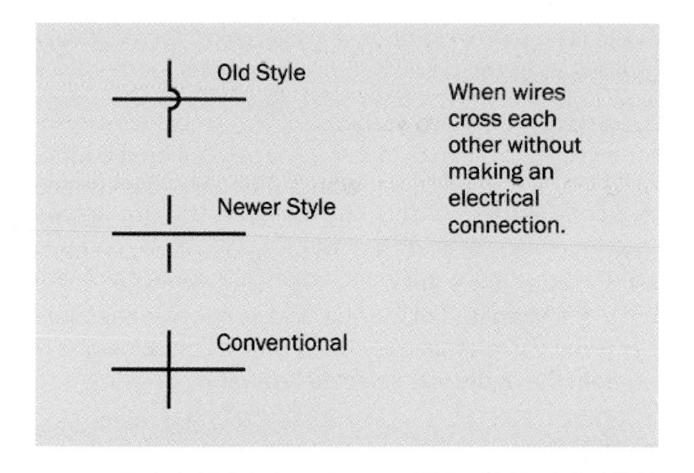

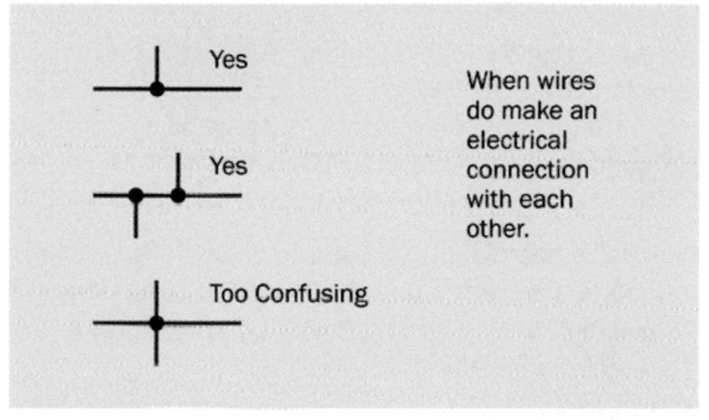

Figure 2-48 *Various styles for depicting wires that do or do not connect. See text for details.*

There's still one more note to add. In the interests of clarity, I think it's good practice to avoid the style shown at the bottom of Figure 2-48. It's too confusing. If we avoid that configuration, and use the one immediately above it instead, then we know that wires crossing each other never make a connection under any circumstances.

Colored Conductors

Did I say that crossovers would be my last topic about schematics? Actually there is one more small matter. Because I never want you to get confused between the positive and negative sides of a power supply, I'm going to be coloring all the positive conductors red in schematic drawings in the rest of the book, while the negative/ground side will be blue. Readers have told me that this was very helpful where I used it occasionally in the past, so, now I'm going to use it consistently throughout.

Black is a more commonly used color for negative/ground (as in the black wire on your meter, or the black wire from a battery connector). But blue is still sometimes used, and is more distinctive.

Just bear in mind that schematics that you encounter in the world outside of this book will not use this helpful coloring convention. All the wires will be black, and you will have to figure out which ones are connected with the power supply.

Experiment 7: Investigating a Relay

The next step in your exploration of switching is to use a remote-controlled switch. By "remote-controlled," I mean that it turns on or off in response to a signal that you send to it. This kind of switch is known as a *relay*, because it relays an instruction from one part of a circuit to another.

- Often a relay is controlled by a low voltage or small current, and switches a larger voltage or higher current.

This can be very useful. When you start your car, for instance, a relatively small, cheap ignition switch sends a small signal down a thin, inexpensive piece of wire to a relay that is near the starter motor. The relay activates the motor through a shorter, much thicker, more expensive piece of wire, capable of carrying as much as 100 amps.

Similarly, if you raise the lid on an old-fashioned top-loading washing machine during its spin cycle, you close a small switch that sends a small signal down a thin wire to a relay. The relay handles the bigger task of switching off the large motor spinning the drum full of wet clothes.

What You Will Need:

- 9-volt battery (1)
- DPDT 9VDC relays (2)
- Tactile switch, SPST (1)
- Test leads with alligator clips at each end (5)
- Utility knife (1)
- Multimeter (1)

The Relay

The type of relay that I want you to use has two pins at one end and six at the other. The six are clustered in two lines of three, as in Figure 2-49 (where the relay is upside-down with its pins in the air). If you buy two relays, you can use one for investigational purposes, meaning that you'll be cutting it open to take a look inside. If you do this very, very carefully, the relay should still be usable afterward. If not—well, you will have a spare.

Caution: Polarity Problems

Some relays are fussy about the way that you apply voltage to the coil that's hidden inside. Everything works fine with the electricity flowing one way through the coil, but if you reverse the positive and negative connections (in other words, if you reverse the polarity) the relay stops working.

This is especially annoying when the relay datasheet doesn't make it clear. The relays that I have recommended do not require a particular polarity. See "Essential: Relay" on page 48.

Procedure

Attach test leads and a tactile switch (pushbutton) as shown in Figure 2-49. (Note that the parts in these drawings are not drawn to scale.) When you press the button to apply 9 volts to the pair of pins that are separate from the others, you should hear a very faint click. Let go of the button, and you may hear another click. (If your hearing is not good, touch the relay gently with your fingertip, and you should feel a faint vibration when the click occurs.)

What's happening here? Your meter will help you to investigate. Set it to measure continuity, and verify that it's working by touching the two probes together. If it doesn't beep, you haven't set it to measure continuity, or the battery is dead, or one of the probes is plugged into the wrong socket.

Now hold the probes against the pins as shown in Figure 2-50, and press the button. The meter should beep while the button is pressed.

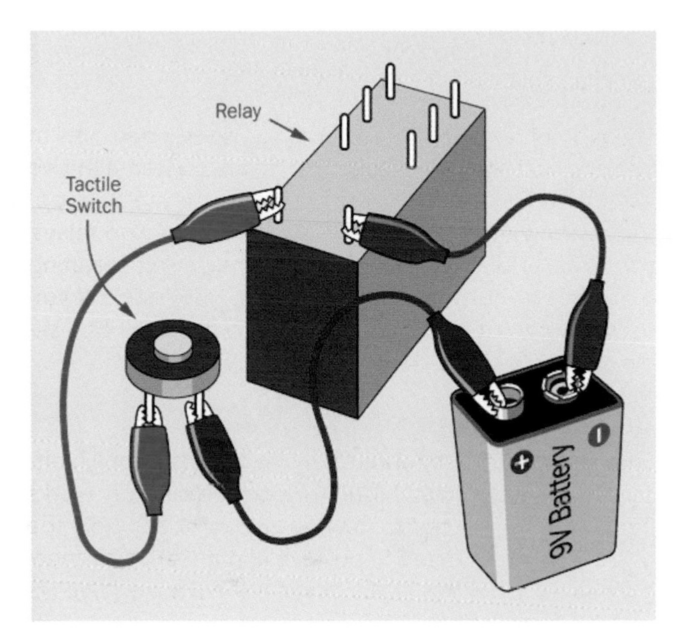

Figure 2-49 *The first step in figuring out what's happening inside a relay.*

This test tells you that some contacts must close inside the relay when you apply voltage to the pair of terminals at the end nearest you. But maybe you're having difficulty holding the meter probes against the pins at the same time as you press the button. In that case, try using a couple of test leads as shown in Figure 2-51. With one end of each lead clipped to a meter probe, and the other end of each lead clipped to a relay pin, you have your hands free.

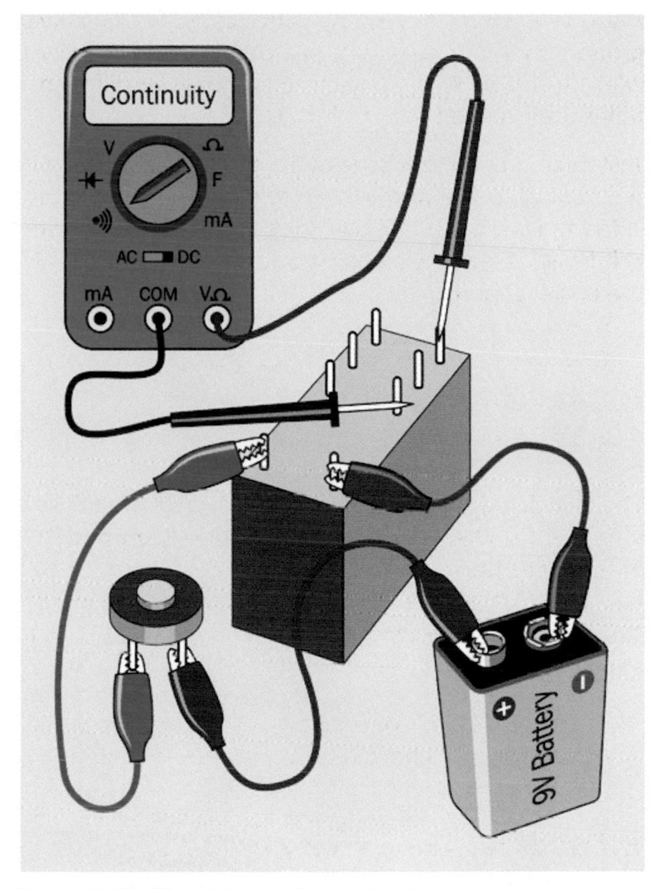

Figure 2-50 *Step 2: Measuring continuity.*

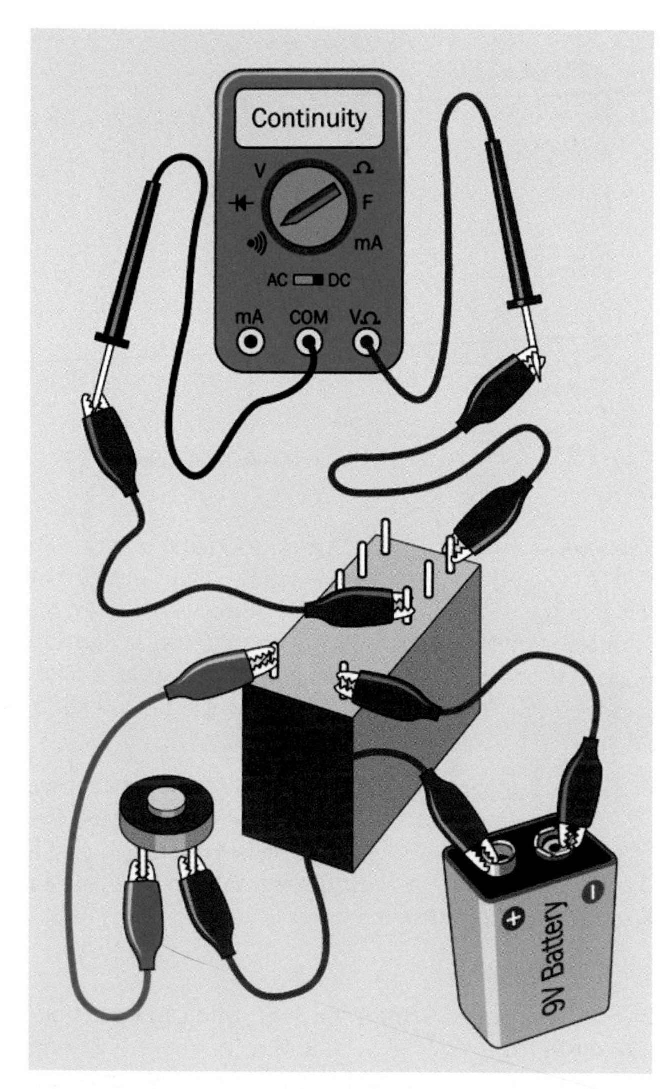

Figure 2-51 *You can extend your meter probes with test leads, to free yourself from holding the probes.*

Now try moving the red test lead from the relay pin farthest away from you to the vacant pin next to it. You should find that the behavior of the meter is reversed, so that it beeps when you don't press the button, and then stops beeping when you do press it.

What's Going On Inside

Figure 2-52 shows an x-ray view of the interior of the relay when you pressed the button. The relay contains a coil at the bottom, which generates a magnetic field, which moves a pair of internal switches. The coil moved the switch on the right to connect pins A and C internally, so the meter beeped.

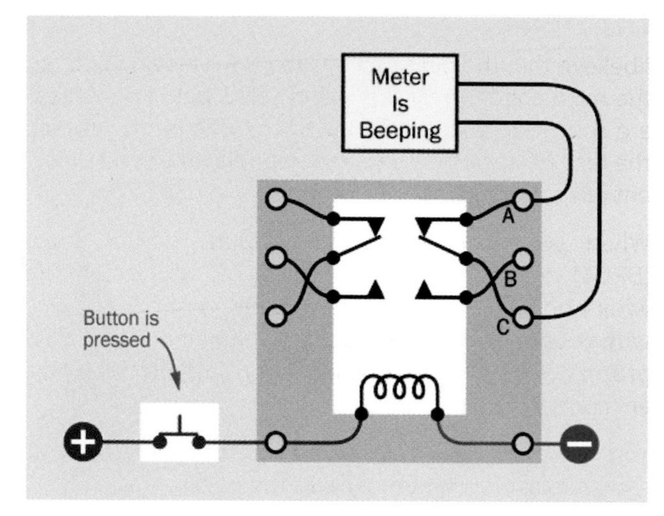

Figure 2-52 *Inside the relay, when the button is pressed, the meter starts beeping.*

You may be wondering why the coil in the relay seems to push the internal switch away from it. The reason is that there is a mechanical linkage inside the relay which converts a pulling force to a pushing force. You'll be able to inspect this when I get to the point of opening up the relay, later in this experiment.

Figure 2-53 shows what happened when you were not pressing the button. The switch contacts relaxed into their opposite state, breaking the connection between A and B and making the connection between B and C. The contacts remain in this position when no power is flowing through the relay coil.

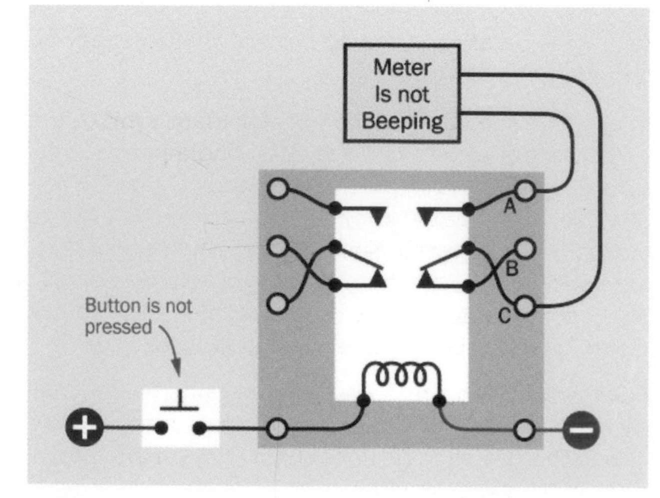

Figure 2-53 *Inside the relay, when the button is not pressed and the meter is not beeping.*

Other Relays

I believe that the pin functions that I have described are the most common for this size of relay, but I know there are some exceptions that do things differently. In fact, the first edition of this book used a relay that had different pin functions.

When you encounter a double-pole, double-throw (DPDT) relay for the first time, how can you determine what's going on inside it? You test different pairs of pins with your meter while you apply voltage to the coil. By a process of elimination, you can figure out how the pins are connected.

You can also read the manufacturer's datasheet. It should contain diagrams such as the one in Figure 2-21.

Is this all you need to know about relays? No, I've hardly scratched the surface.

- Some relays are *latching*, meaning that the internal switches remain in either position when the power is off. Latching relays usually have *two coils* to move the switches each way. I will not be using them in this book.
- Some relays have two poles, some have only one; some are double-throw, and some are single-throw.
- Some coils switch with AC, some switch with DC, and as I've mentioned previously, some DC coils require you to apply DC voltage with correct polarity.

As always, datasheets should provide the necessary information.

Figure 2-54 shows a selection of schematic symbols for various types of relays. Type A is single-pole, single-throw. Type B is single-pole, double-throw. Type C is single-pole, single-throw, drawn in the style that I like to use, with a white rectangle reminding you that the parts are enclosed in a single component. Type D is single-pole, double-throw. Type E is double-pole, double throw. Type F is single-pole, double-throw, latching.

Relay schematics are always drawn with the internal switch in its relaxed position, when power is not applied—with the exception of the latching relay, where the position of the switch is arbitrary.

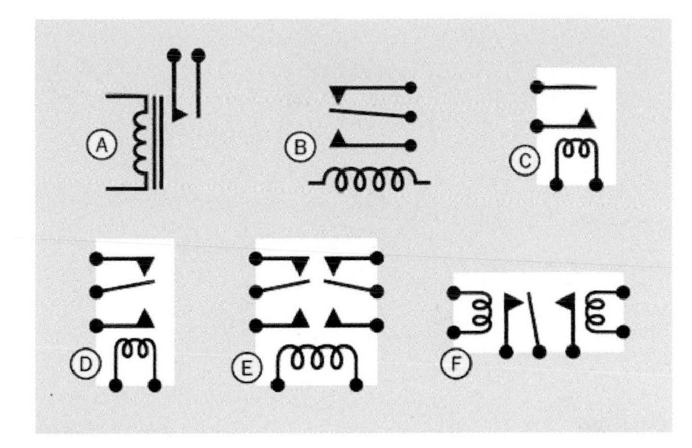

Figure 2-54 *Various schematic symbols for relays. See text for details.*

The type of relay you have been testing is a *small-signal relay*, meaning that it can't switch a lot of current. Larger relays may be capable of switching many amperes. It's important to choose a relay that is rated for the maximum current in your circuit (or higher), because overloading a relay will cause sparking that quickly erodes its contacts.

In future experiments you'll discover some practical uses for a relay—for example, in a home security system. Before you get to that, I'm going to show you how to turn a relay into an oscillator that buzzes. But first, I think it's time to take a look inside.

Opening It Up

If you have an impatient disposition, you can open your relay using methods such as those in Figure 2-55 and Figure 2-56. Generally, though, you may be better off using a most mundane piece of equipment: a box cutter or utility knife.

Figure 2-55 *Option 1 for opening a relay (probably not recommended).*

Figure 2-56 *Option 2 for opening a relay (definitely not recommended).*

Figure 2-57 and Figure 2-58 illustrate the technique that I like to use. You shave the edges of the plastic shell, beveling them until you see just a hair-thin opening. Don't go any farther; the parts inside are very, very close to your knife blade. Now pop the top off. Repeat this procedure with the remaining edges of the shell, and if you were really careful, the relay will be exposed but will still work when you energize its coil.

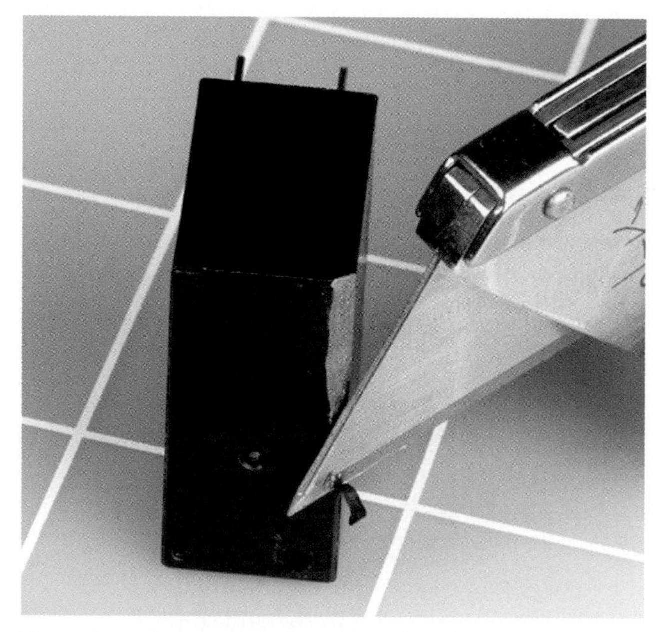

Figure 2-57 *Shaving the edges of the plastic box of a relay is a first step to opening it. Always cut away from you and downward toward your work bench.*

Figure 2-58 *After shaving the edges, you should be able to pry open one section of the case.*

What's Inside?

Figure 2-59 shows a simplified view of the parts in a typical relay. The coil, A, generates a magnetic attraction pulling lever B downward. A plastic extension, C, pushes outward against flexible metal strips and moves the poles of the relay, D, between the contacts. (This is a slightly different configuration from the relay recommended for experiments in this book, but the general principle is the same.)

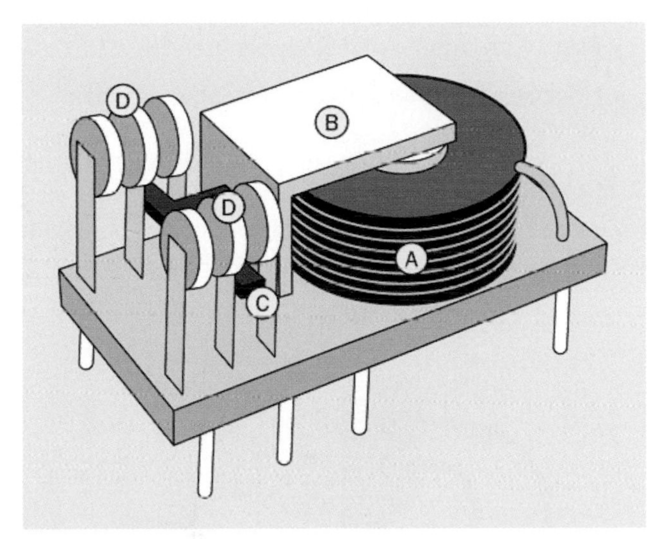

Figure 2-59 *Simplified view of the interior of a relay. See text for details.*

You can compare the diagram with an actual relay that I opened up, in Figure 2-60.

Various sizes of relays are shown with their cases removed in Figure 2-61. All of them happen to be designed for 12 volts DC. The automotive relay at far left is the simplest and easiest to understand, because it is designed without much concern for the size of the package. Smaller relays are more ingeniously designed, more complex, and more difficult to figure out. Usually, but not always, a smaller relay is designed to switch less current than a larger one.

Figure 2-60 *An actual relay, exposed. The squares on the cutting mat are 1" x 1".*

Figure 2-61 *A variety of 12-volt relays. See text for details.*

Fundamentals: Relay Terminology

Coil voltage is the voltage that the relay is supposed to receive when you energize it. This may be AC or DC.

Set voltage is the minimum that the relay will accept, to close its switch. This will be a bit less than the ideal coil voltage. In practice, a relay will probably work with even less power than the set voltage, but the set voltage tells you the minimum at which it is guaranteed to work.

Operating current is the power consumption of the coil, usually in milliamps, when the relay is energized. Sometimes the power is expressed in milliwatts.

Switching capacity is the maximum amount of current that the contacts inside the relay can switch without damage. Usually this is specified for a *resistive load*, meaning a passive device such as an incandescent lightbulb. When you use a relay to switch on a motor, it imposes an *inductive load*, which takes a big initial surge of current before it gets up to speed. Switching the motor off creates another surge. If the datasheet for a relay doesn't rate its ability to handle an inductive load, a rule of thumb is to assume that a motor may draw twice as much current when it starts, compared with when it is running.

Experiment 8: A Relay Oscillator

When you used test leads with alligator clips in previous experiments, they had two big advantages: you could assemble a circuit quickly, and you could see its connections easily.

Sooner or later, though, you have to get acquainted with a quicker, more convenient, more compact, and more versatile method for building circuits, and that time has come. I'm referring to the most widely used prototyping device: a *solderless breadboard*.

In the 1940s, circuits were built on a platform that really did look like a board that could be used for slicing bread. Wires and components were nailed, stapled, or screwed into place, because this was a whole lot easier than the alternative, which was mounting them on pieces of sheet metal. Remember, plastic barely existed back then. (A world without plastic—can you imagine it?)

Today the term "breadboard" is used for a little slab measuring about 2" by 7", and no more than 1/2" thick, as pictured in Figure 2-10. This is a wonderfully quick and easy system for assembling components. The only problem is that it creates internal connections between the components that are difficult to visualize—although I have ways to help you deal with that.

The best way to learn breadboarding is to assemble a circuit, which is exactly what you are about to do, taking the previous experiment with a relay one step further.

What You Will Need

- 9-volt battery (1)
- Battery connector (1)
- Breadboard (1)
- DPDT 9VDC relay (1)
- Generic LEDs (2)
- Tactile switch (1)
- Resistor, 470 ohms (1)
- Capacitor, 1,000µF (1)
- Pliers, wire cutters, wire strippers (1 each)
- Hookup wire, at least two colors, no more than 12" each

A Beginner's Board

Figure 2-62 shows the top part of your breadboard with the components that I would like you to plug into it.

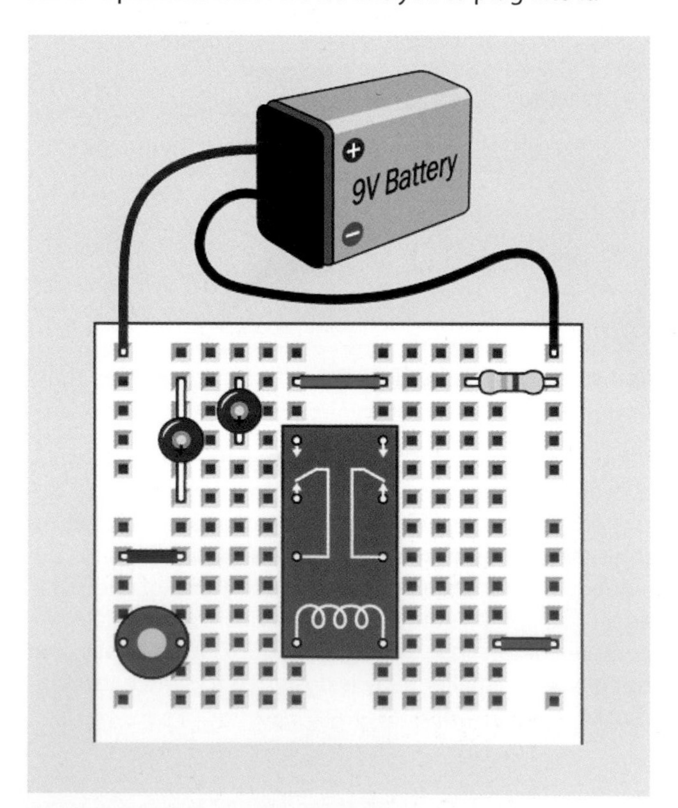

Figure 2-62 *A relay test circuit mounted in a breadboard.*

In case you're wondering exactly what some of these components are, Figure 2-63 shows all the pictorial symbols that will be used in the rest of the book in breadboard diagrams. You have not encountered most of these components yet, but you can refer back to this diagram for reference.

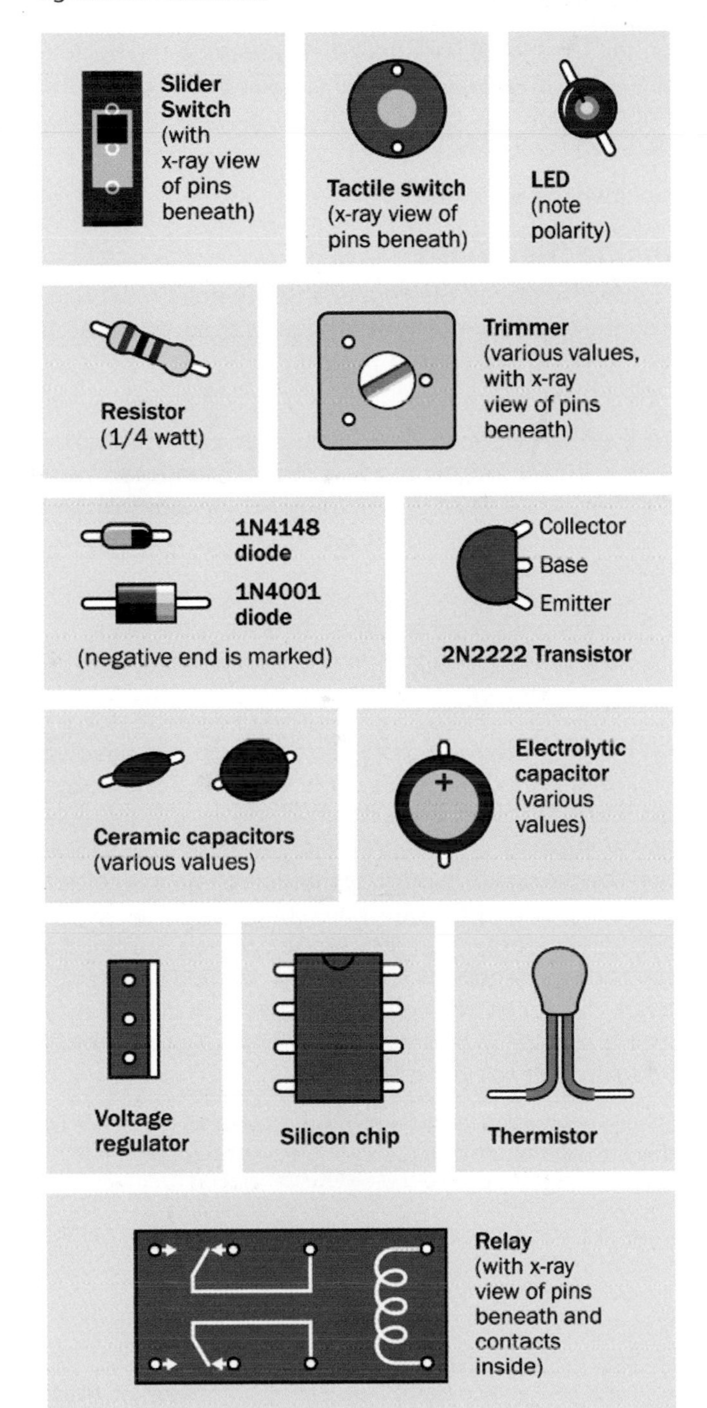

Figure 2-63 *Representation of breadboarded components.*

In Figure 2-62 the relay from the previous experiment is in the center. You won't be able to see the pins from above, because they are inserted into the board underneath. I have shown their positions so that you know which way around the relay should be (that is, with its coil pins at the bottom). I have also shown the connections that exist inside the relay, just to remind you how they are configured. The switch is in the position where it rests when there is no power. This is the "relaxed" position.

The gray circular object is your pushbutton, more properly known as a tactile switch. I have shown an x-ray view of its pin positions, too, so you know how it should be oriented.

The two red circular objects are LEDs. Make sure the long lead of each one is on the side where the plus sign is shown.

The resistor value is 470 ohms—but you could have figured that out by looking at the colors of its stripes.

The red, green, and blue segments that look like pieces of wire, plugged into the breadboard, really are pieces of wire, plugged into the breadboard. My next task is to tell you how to make them.

Making Jumpers

If you bought an assortment of precut hookup wires, also known as jumpers, you can just go ahead and push them into the breadboard in the positions shown, although they won't be the same colors as the ones in my illustration.

As I mentioned previously, I advocate making your own. The exact procedure that I use is shown in Figure 2-64. First, strip off a couple of inches of insulation from some hookup wire. To do this, hold the wire in your left hand (or your right hand, if you are lefthanded). Hold the wire strippers in your other hand. Close the wire strippers so that they grip the wire in the hole marked "22" on the blade of the strippers. Pull the strippers away from your other hand, and they should take the insulation with them. (If you are wondering why you use the hole marked "22," it's because you are using 22-gauge wire. At least, I hope that's the size you are using.)

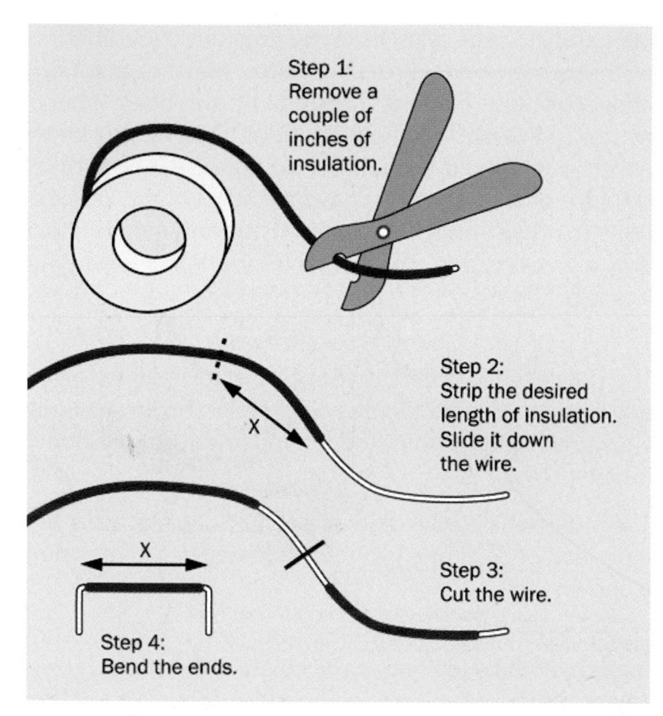
Step 1:
Remove a
couple of
inches of
insulation.
Step 2:
Strip the desired
length of insulation.
Slide it down
the wire.
X
X
Step 3:
Cut the wire.
Step 4:
Bend the ends.

Figure 2-64 *Procedure for making jumper wires. See text for details.*

Next assess how long the visible part of your wire is supposed to be when it is installed on the board. I will call that distance X inches. Measure X inches of the insulation that remains on your hookup wire. Drag a section of insulation that is X″ long until it is about 3/8″ from the end of the wire.

Using your wire cutters, or the blades built into your wire strippers, snip the wire 3/8″ behind the X″ section of insulation that you just dragged along the wire.

Finally, use your pliers to make a neat right-angle bend at each end, and push it into the board. Wait—does it not quite fit? If you practice a little, you'll soon get to the point where you can make jumpers of the right length just by visual estimation.

Power Up

Lastly you need to apply power from your 9-volt battery. You should find that the wires attached to the connector terminate in little bare soldered ends that will poke into the holes in the breadboard. If you have trouble with them, try feeding them in with the tips of your pliers. If you still have trouble, you may need to strip off a couple more millimeters of insulation, using your wire strippers.

After inserting the wires into the breadboard, snap the connector onto the battery, as suggested in Figure 2-62. As soon as you have power on the breadboard, the LED on the left should light up. When you press the button, the switch inside the relay will close and the LED on the right will light up. Congratulations! You just breadboarded your first circuit.

Now, why does it work?

Inside the Board

Figure 2-65 reveals the copper strips that are hidden inside the breadboard. The little squares show where each lead of a component can poke through and make contact with the strips inside.

Each of the two long, vertical strips is known as a *bus* (plural: buses). This type of bus does not transport people, but does transport electrons, as the positive and negative sides of the power supply are typically connected with the buses.

- In this book, I am consistently placing the positive side of the supply on the left bus and negative ground on the right bus.

An important thing to notice is that each bus has two breaks in it. Not all breadboards have this feature, but many do. The purpose is to allow you to use multiple power supplies of different voltages at different locations on the board. In practice this doesn't happen often, and the breaks in the buses are annoying, because you may tend to forget that they're there. When you build a circuit that extends down the board, and you find a mysterious lack of power around the halfway mark, you may realize eventually that you forgot to add jumper wires bridging the gaps in the buses.

Where necessary, I will be reminding you to take care of this little detail.

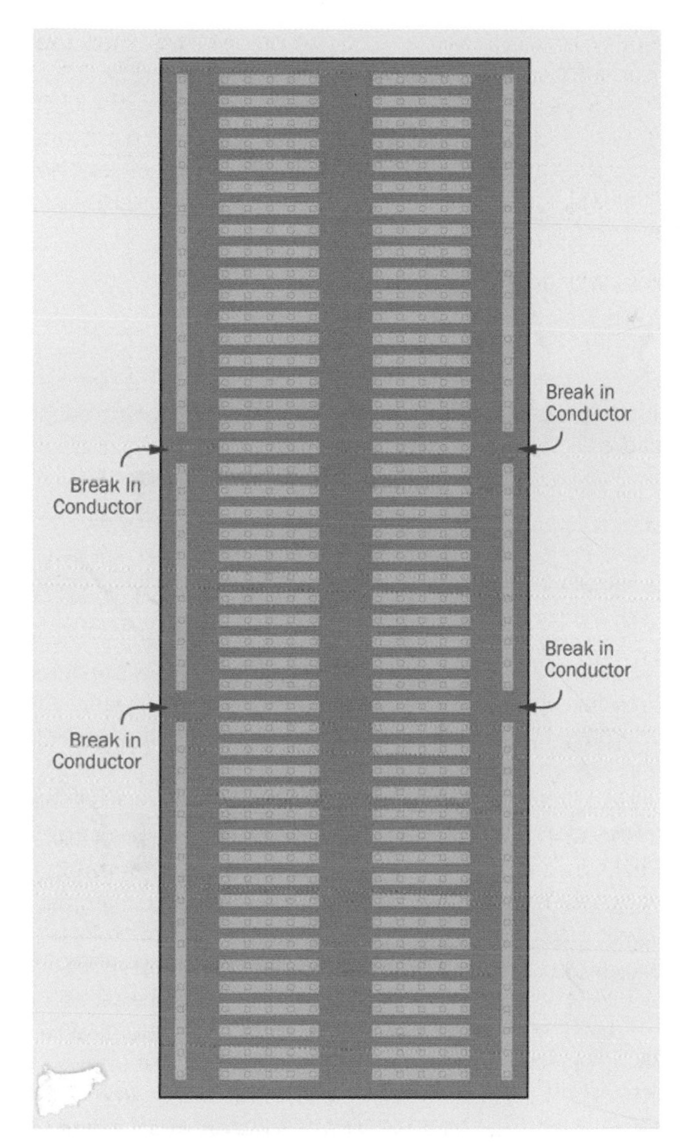

Figure 2-65 *A single-bus breadboard contains copper connecting strips in this configuration.*

Relay Circuit, Revealed

Figure 2-66 shows you the copper strips hidden inside your breadboard. They make connections between components plugged into the breadboard. The electricity takes a zigzag path, but the resistance of the copper strips is so low, the length of the path doesn't matter.

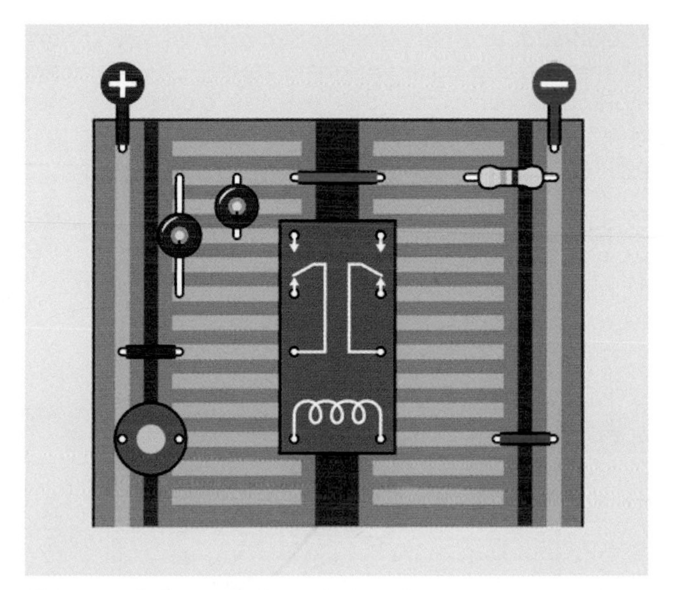

Figure 2-66 *Components on the breadboard are connected through the copper strips inside it.*

Maybe the diagram is easier to understand if I hide the copper strips that aren't doing anything, and just show the ones that are part of the circuit, as shown in Figure 2-67.

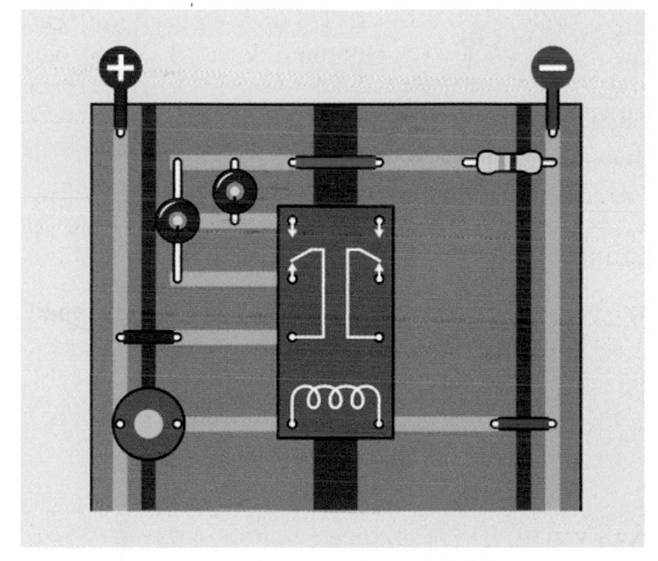

Figure 2-67 *The previous diagram is modified so that strips in the breadboard that are not an active part of the circuit have been omitted.*

Now take a look at the schematic for the same circuit, in Figure 2-68. I laid out the schematic to resemble the breadboard, to emphasize the similarity. As you continue through the book I'm going to rely on schematics more, and will expect that you can create your own breadboard layouts. But I'll take a while to get to that point.

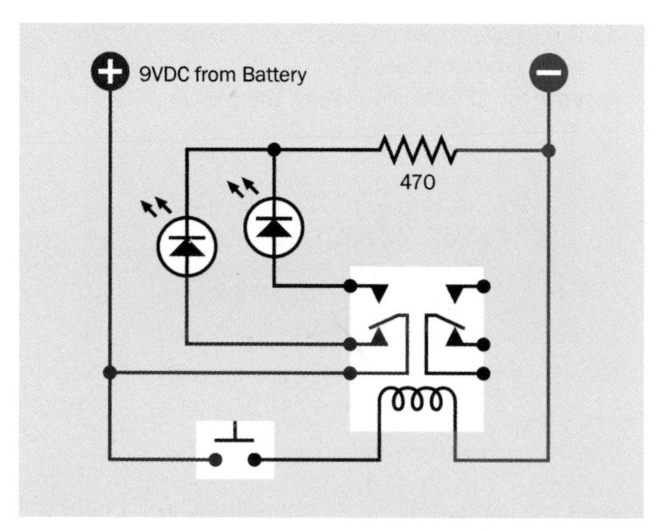

Figure 2-68 *A schematic that corresponds with the breadboard connections already shown.*

If you're wondering why there is only one 470-ohm resistor to protect two LEDs, it's because the LEDs only light up one at a time.

Making It Buzz

The next step is to modify your circuit, to make it more interesting. Look at the new schematic in Figure 2-69. Compare it with the previous version in Figure 2-68. Can you spot the difference? In the old version, the pushbutton that energized the coil received its power directly from the 9V supply. In the new version, the pushbutton gets its power through the lower contact of the relay. What effect will this have?

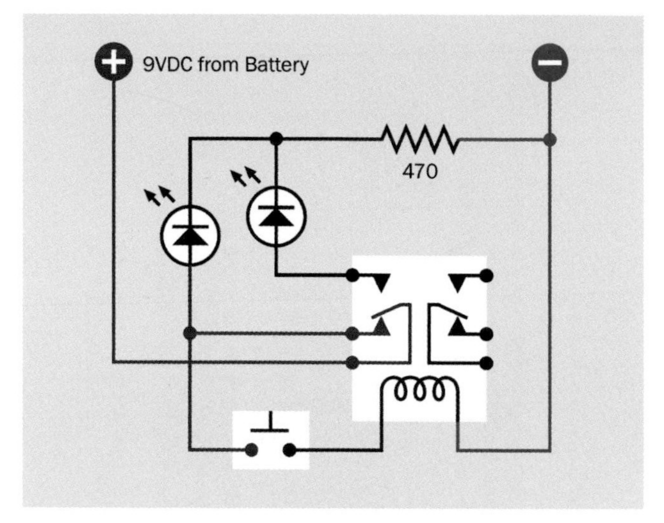

Figure 2-69 *A revised version of the previous schematic now delivers power to the pushbutton through the lower contact of the relay.*

Figure 2-70 shows how you can adapt your previous breadboarded circuit to match the new schematic. All you have to do is rotate the pushbutton by 90 degrees, and use an extra jumper wire (colored green in the figure) to link it with the same relay pin that energizes the lefthand LED.

Push the button—briefly!—and what happens? The relay makes a buzzing sound. (If your hearing is not good, touch the relay to feel it vibrating.)

Can you see what's happening, here? In its relaxed state, the switch inside the relay rests against the lower contact. This feeds positive voltage to the lefthand LED, and also to the pushbutton. Consequently, when you press the button, the power connects with the relay coil. The coil pushes the internal switch upward—but as soon as it does that, it breaks the connection feeding voltage to the coil. So, the switch falls back to its relaxed position. But this energizes the coil again, so the cycle repeats.

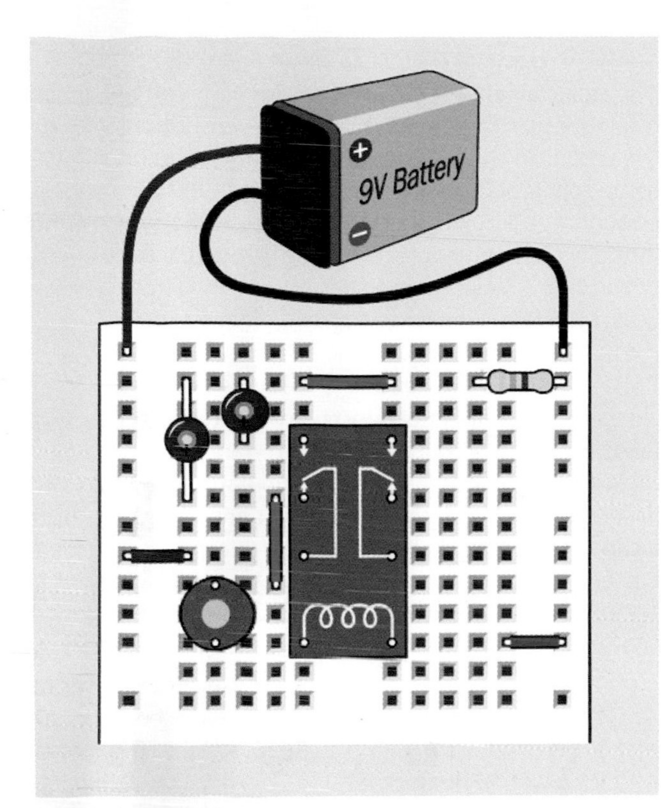

Figure 2-70 *The previous breadboarded circuit has been modified to match the revised schematic.*

The relay is *oscillating* between its two states.

Because you're using a small relay, it switches on and off quite fast. In fact, it oscillates perhaps 20 times per second (too fast for the LEDs to show what's really happening).

- When you force a relay to behave like this, it's liable to burn itself out or destroy its contacts. You are also controlling a bit more current than your tactile switch is designed to handle. So don't hold down the button for long! To make the circuit less self-destructive, we need to make everything happen more slowly. I'm going to achieve this by using a capacitor.

Adding Capacitance

Add a 1,000µF electrolytic capacitor in parallel with the coil of the relay as shown in the diagram in Figure 2-71, making sure that the capacitor's *short* wire is connected to the *negative* side of the circuit; otherwise, it won't work. In addition to the short wire, you should find a minus sign on the body of the capacitor, which is there to remind you which side should be more negative. In the diagram, I used a plus symbol, because it's more obvious than a minus symbol, and I wanted to be consistent with the style I used for LEDs.

- Electrolytic capacitors react very badly to being connected the wrong way around. They may self-destruct. Double-check the polarity.

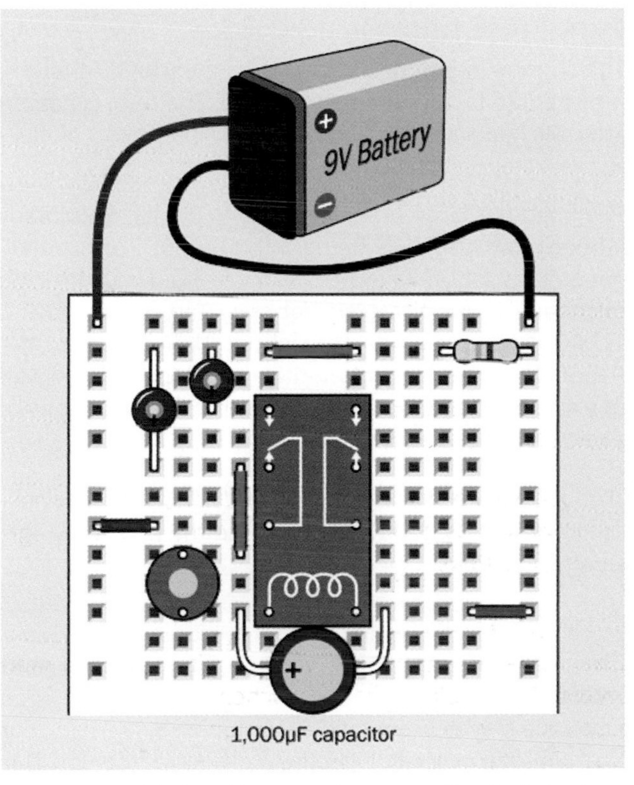

Figure 2-71 *With the addition of a large capacitor, the behavior of the circuit slows down.*

When you press the button now, the relay should click intermittently instead of buzzing.

A capacitor is like a tiny rechargeable battery. It's so small that it charges in a fraction of a second, before the relay has time to open its lower pair of contacts. Then, when the contacts are open, the capacitor releases its power to the relay (and to the lefthand LED). This keeps the coil of the relay energized for a moment. After the capacitor exhausts its power reserve, the relay relaxes and the process repeats.

During this process, the capacitor is *charging and discharging*.

Disconnect the righthand LED, and you should find that the lefthand LED pulses in a pleasing manner, gradually fading as the voltage from the capacitor diminishes.

Because the capacitor takes a big surge of current when it charges, your tactile switch may overheat if you hold it down for too long during this experiment.

Fundamentals: Farad Basics

The storage ability of a capacitor is measured in *farads*, represented by an uppercase letter F. The term is named after Michael Faraday, another in the pantheon of electrical pioneers.

The farad is a large unit, and is divided into microfarads (abbreviated µF, each being 1/1,000,000th of a farad), nanofarads (abbreviated nF, each being 1/1,000th of a microfarad), and picofarads (abbreviated pF, each being 1/1,000th of a nanofarad). In the United States, the nanofarad is used less often than in Europe. Instead, values may be expressed using picofarads and fractions of a microfarad.

A conversion table for picofarads, nanofarads, microfarads, and farads is shown in Figure 2-72.

Picofarads	Nanofarads	Microfarads	Farads
1pF	0.001nF	0.000001µF	
10pF	0.01nF	0.00001µF	
100pF	0.1nF	0.0001µF	
1,000pF	1nF	0.001µF	
10,000pF	10nF	0.01µF	
100,000pF	100nF	0.1µF	
1,000,000pF	1,000nF	1µF	0.000001F
		10µF	0.00001F
		100µF	0.0001F
		1,000µF	0.001F
		10,000µF	0.01F
		100,000µF	0.1F
		1,000,000µF	1F

Figure 2-72 *Conversion table for fractions of farads.*

Caution: Getting Zapped by Capacitors

If a large capacitor is charged with a high voltage, it can retain that voltage for minutes or even hours. Because the circuits in this book use low voltages, you don't have to be concerned about this here, but if you are reckless enough to open an old TV set and start digging around inside (which I do not recommend), you may have a nasty surprise. A large charged capacitor can kill you as easily as if you stick your finger into an electrical outlet.

Fundamentals: Capacitor Basics

No electrical connection exists inside a capacitor. Its two leads are connected internally with *plates* that are a tiny distance apart, separated by an insulator known as the *dielectric*. Consequently, DC current cannot flow through a capacitor. However, if you connect a capacitor across a battery, it will charge itself as suggested in Figure 2-73, because the charge on one plate attracts an opposite charge on the other plate.

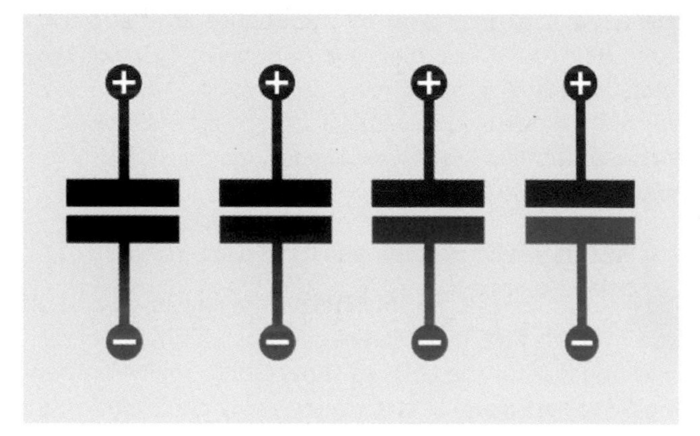

Figure 2-73 *A capacitor connected across a battery will accumulate equal and opposite charges, as suggested here.*

In modern capacitors, the plates have been reduced to strips of very thin, flexible, metallic film.

The two most common varieties of capacitors are ceramic (usually small, to store a relatively small charge) and electrolytic (which can be much larger). Electrolytics are usually shaped like miniature tin cans and may be any color, although black is most common. Older ceramics are often disc-shaped, while newer ones are little rounded blobs.

Ceramic capacitors have no polarity, meaning that you don't have to worry about which way around they are in a circuit. Electrolytics do have polarity, and won't work unless you connect them the right way around.

The schematic symbol for a capacitor contains two lines representing the two plates inside it. If both of the lines are straight, the capacitor does not have a polarity—it can be used either way around. If one of the lines is curved, that side of the capacitor must be more negative than the other. A + sign may be included to remind you of the polarity. The variants are shown in Figure 2-74.

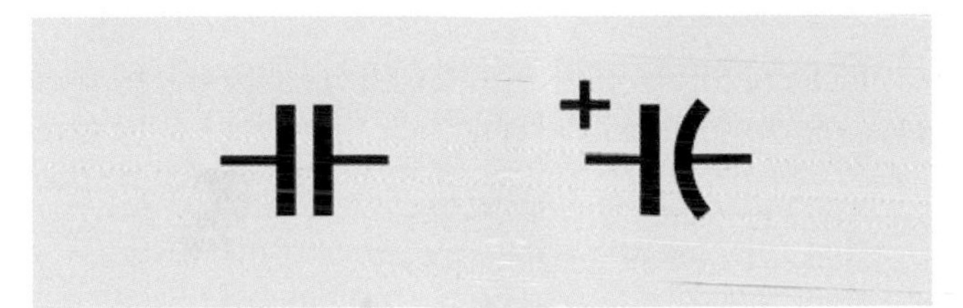

Figure 2-74 *Two variants of the symbol representing a capacitor. See text for details.*

The symbol with the curved plate is not used so much anymore. People assume that if you have an electrolytic capacitor, you'll be smart enough to connect it the right way around. Also, multilayer ceramic capacitors have become available in higher values, and may be substituted for electrolytics.

- My schematic diagrams will only use the nonpolarized capacitor symbol. Whether you use an electrolytic capacitor or a ceramic capacitor is your choice.
- My breadboard diagrams will show electrolytic capacitors where I think you are most likely to use them. Still, you can substitute ceramic capacitors if you wish.

Caution: Observe Capacitor Polarity!

The most common type of electrolytic capacitor uses aluminum plates. Two other types use tantalum and niobium, respectively. All of these capacitors are fussy about polarity. In Figure 2-75, a tantalum capacitor was plugged into this breadboard, connected the wrong way to a power source capable of delivering a lot of current. After a minute or so of this abuse, the capacitor rebelled by popping open and scattering small flaming pieces, which burned their way into the breadboard. Lesson learned: observe polarity!

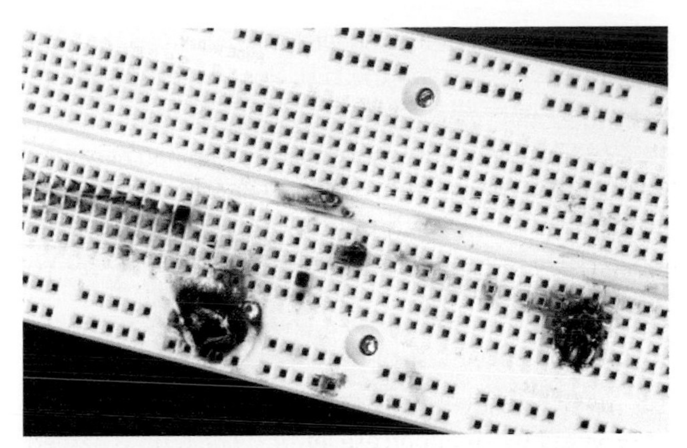

Figure 2-75 *The aftermath of an error in which a polarized tantalum capacitor was connected the wrong way around to a power source capable of supplying substantial current.*

Fundamentals: Fault Tracing

As you build more circuits on breadboards, and they become more complex, errors become more likely. No one is exempt from this unfortunate fact.

One frequent breadboarding error is to plug a wire into the wrong row on a breadboard. This is especially easy when you have a component such as a relay, where the pins are hidden. Often I pull that component out, take another look, and push it back in, just to make sure.

A more subtle error occurs when you forget the connections made by the strips hidden inside the board. Take a look at Figure 2-76. What could be simpler? Clearly, the positive source of power feeds current through the LED, through a jumper, and then through a resistor to the negative bus. But if you assemble the components as I have shown them, I absolutely, positively guarantee that they won't work.

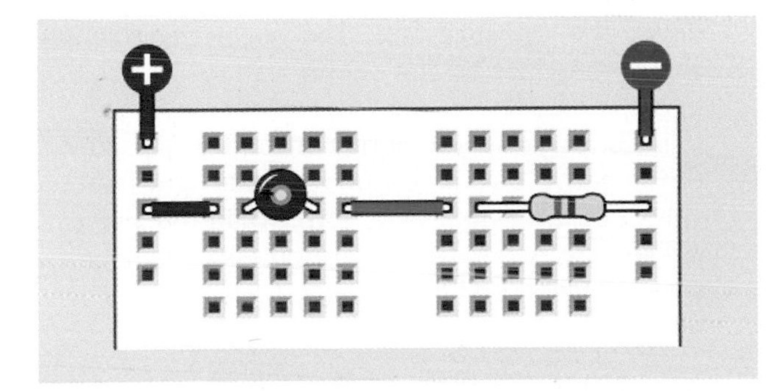

Figure 2-76 *This breadboarded circuit won't work. Can you see why?*

The situation will get much worse if you swap the positions of the resistor and the LED. Now the circuit will immediately burn out the LED.

The explanation becomes apparent when you check the x-ray view in Figure 2-77. The problem is that both leads of the LED are plugged into the same strip inside the board. The electricity has the option to run through the LED, or run through the copper strip instead—and since the resistance of the strip is a tiny fraction of the resistance of the LED, most of the electrons prefer the strip, and the LED remains dark.

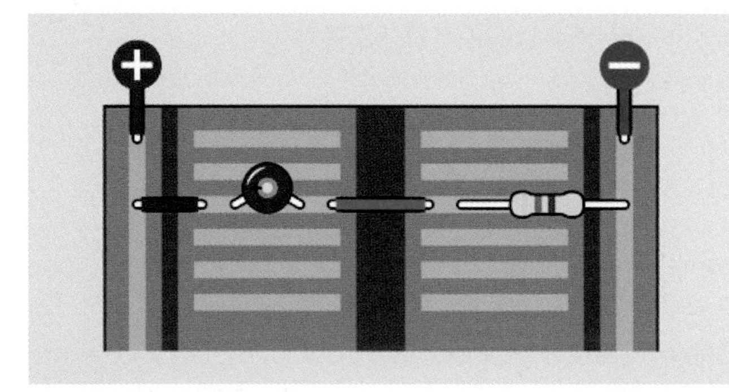

Figure 2-77 *An x-ray view helps to explain why this won't work.*

Many other kinds of errors are possible. How can you find them most quickly and efficiently? You simply have to be methodical. Try to follow these steps:

1. **Check voltages**. Attach the red lead from your meter to a connection near the top of the positive bus on your breadboard. Set the meter to measure volts (DC volts, unless an experiment suggests otherwise). Make sure the power to your circuit is switched on. Now touch the black probe from your meter to various locations down the negative bus. The meter reading should be close to the supply voltage. If you find a value close to zero, you probably forgot to include a jumper to bridge one of the gaps in the negative bus. If you find a value of a few volts, but it is significantly lower than the supply voltage, you may have a short circuit somewhere, pulling down the voltage from the battery (if you are using a battery).

   Now secure the black probe to a connection near the top of the negative bus, and check the positive bus from top to bottom.

   Finally, with the black probe still attached near the top of the negative bus, use the red probe to check voltages at random locations in your circuit. If you find a voltage close to zero, probably a connection is missing somewhere, or a component or wire is not making contact inside the breadboard.

2. **Check placement**. Make sure that all the jumper wires and component leads are exactly where they should be on the breadboard.

3. **Check component orientation.** Diodes, transistors, and capacitors that have polarity must be the right way around. When you start to use integrated circuit chips, later in this book, check that they are the right way up, and make sure that none of the pins on a chip has been bent so that it is hidden underneath the chip.

4. **Check connections.** Sometimes (seldom, but it can happen) a component may make a bad connection inside the breadboard. If you have an inexplicable intermittent fault or zero voltage, try relocating some of the components. In my experience this problem is more likely to occur if you buy very cheap breadboards. It is also more likely if you use wire that has a smaller diameter than 22 gauge. (Remember, a higher gauge number means a thinner wire.)

5. **Check component values.** Verify that all the resistor and capacitor values are correct. My standard procedure is to check each resistor with a meter before I plug it in. This is time-consuming but can actually save time in the long run.

6. **Check for damage.** Integrated circuits and transistors can be damaged by incorrect voltages, wrong polarity, or static electricity. Keep spares on hand so that you can make substitutions.

7. **Check yourself!** When all else fails, take a break. Working obsessively for long periods can create tunnel vision, which prevents you from seeing what's wrong. If you move your attention to something else for a while, then come back to your problem, the answer may suddenly appear obvious.

You may want to bookmark this list of fault-tracing procedures and come back to it later if something doesn't work.

Background: Michael Faraday and Capacitors

As previously noted, the farad is named after Michael Faraday. He was an English chemist and physicist who lived from 1791 to 1867. See Figure 2-78.

Figure 2-78 *Michael Faraday, after whom the farad is named.*

Although Faraday was relatively uneducated and had little knowledge of mathematics, he had an opportunity to read a wide variety of books while working for seven years as a bookbinder's apprentice, and thus was able to educate himself. Also, he lived at a time when relatively simple experiments could reveal fundamental properties of electricity. He made major discoveries including electromagnetic induction, which led to the development of electric motors. He also discovered that magnetism could affect rays of light.

His work earned him numerous honors, and his picture was printed on English bank notes denominated in 20 pounds sterling, from 1991 through 2001.

Experiment 9: Time and Capacitors

Electrons travel almost at the speed of light, yet we can use them to measure time in seconds, minutes, or even hours. This experiment will show you how.

What You Will Need

- Breadboard, hookup wire, wire cutters, wire strippers, test leads, multimeter
- 9-volt battery and connector (1)
- Tactile switches (2)
- Generic LED (1)
- Resistors: 470 ohms, 1K, 10K (one of each)
- Capacitors: 0.1µF, 1µF, 10µF, 100µF, 1,000µF (one of each)

Charging a Capacitor

First set your meter to measure volts DC, and measure the voltage of a 9-volt battery with your meter. If it's less than 9.2V, you need a newer battery for this particular experiment.

Install two tactile switches, a 1K resistor, and a 1,000µF capacitor on your breadboard as shown in Figure 2-79. Use a couple of test leads to connect your meter so that you can measure the voltage across the leads of the capacitor while keeping your hands free.

Snap a connector on your battery, and plug the wires into the breadboard to provide the 9VDC supply on the two buses of the breadboard, with positive on the left, as shown in the figure.

If the meter measures more than 0.1V, discharge the capacitor by pressing button B, which shorts together the two sides of the capacitor.

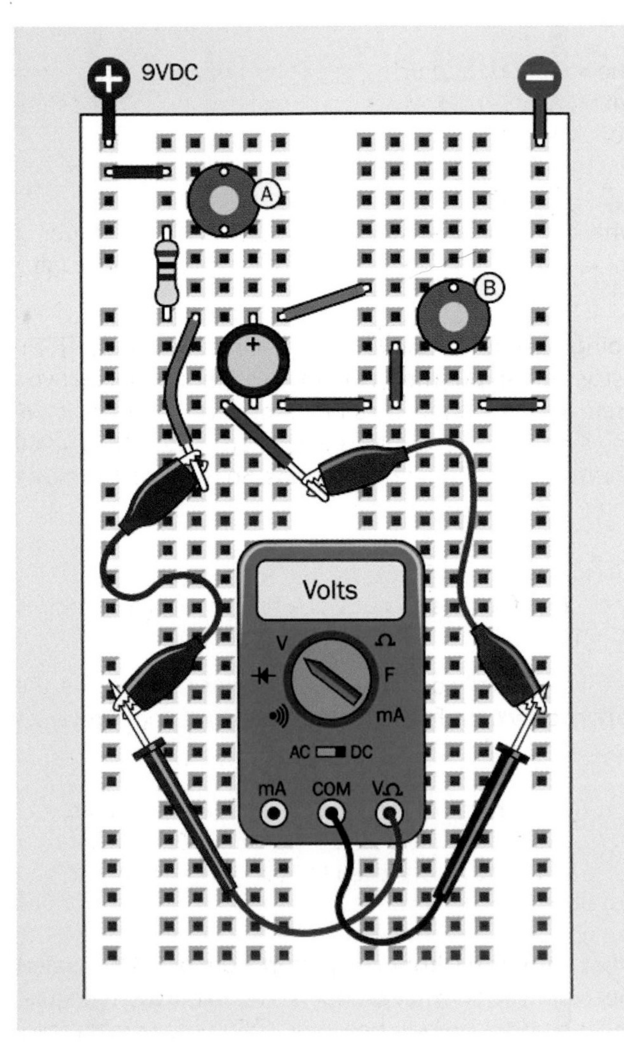

Figure 2-79 *A simple setup for timing the charging of a capacitor. The capacitor is 1,000μF and the resistor is 1K.*

A schematic in Figure 2-80 shows the same circuit, and may help you to see what's going on.

Now hold down button A while you use a watch, a clock, or a smartphone to count how many seconds the capacitor takes to charge to 9.0V. If you have an autoranging meter, it should automatically switch from measuring millivolts, at first, to measuring volts as the charge increases. When I performed this experiment, the meter took just over three seconds.

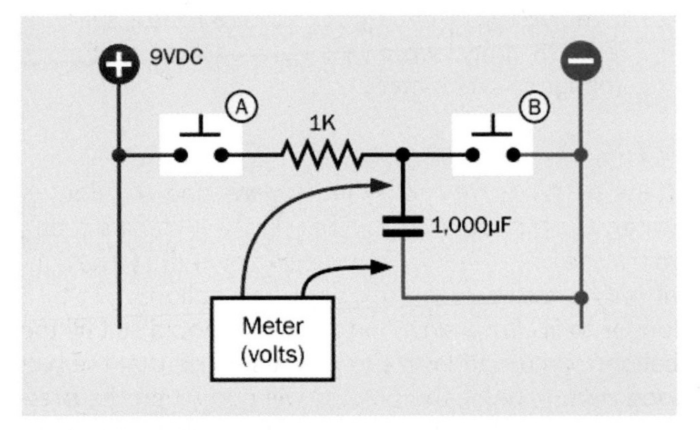

Figure 2-80 *This schematic shows the same circuit as the preceding figure, which shows the breadboarded version.*

The positive side of the capacitor has become "more positive" and the negative side has become "more negative" as electron-holes and electrons have been attracted to each other on the plates. The potential difference between the leads of the capacitor has increased, while current has not passed through it. One of the first statements you will find, when you read any introductory electronics text, is:

- A capacitor blocks DC (direct current).

So long as you apply a steady electrical potential to the capacitor, this is true.

An RC Network

Remove the 1K resistor and substitute a 10K resistor. If the meter shows that there is still some voltage across the capacitor, discharge it by pressing button B.

Now repeat the test. How long does the capacitor take to reach 9.0V, charging through a 10K resistor?

This simple combination of a capacitor and a resistor is known as an *RC network* (R for resistor, C for capacitor). It's a very important concept in electronics. Before I explain what it does, here are some questions to consider:

- Did the capacitor take exactly 10 times as long to reach 9 volts when you used the 10K resistor instead of the 1K resistor?
- Did the voltage across the capacitor rise at a steady rate, or did it increase faster at the beginning of the experiment—or toward the end?

- If you wait long enough, will the capacitor ever reach the initial value that you measured as the voltage of your battery?

Voltage, Resistance, and Capacitance

Think of the resistor as a faucet restricting the flow of water, and the capacitor as a balloon that you are trying to fill (see Figure 2-81). If you screw down the faucet until only a trickle comes through, the balloon will take longer to fill. But a slow flow of water should still fill the balloon if you wait long enough. Assuming that the balloon doesn't burst, the process will end when the pressure inside the balloon is equal to the water pressure in the pipe supplying the faucet.

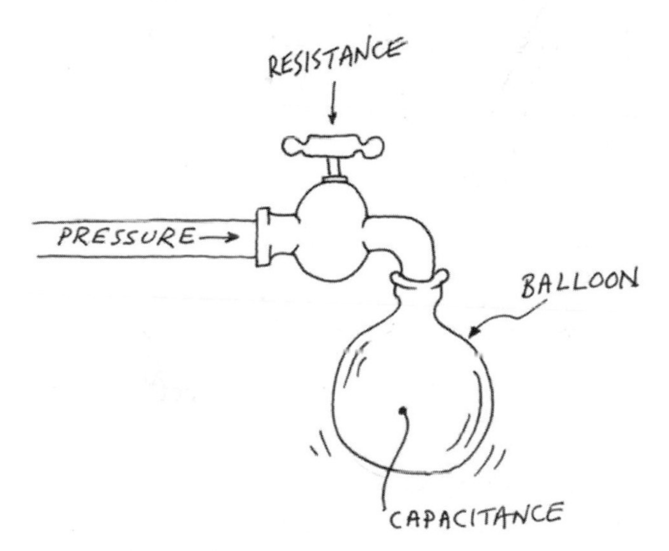

Figure 2-81 *Water flowing into a balloon can be compared with electrons flowing into a capacitor.*

But this description leaves out an important factor. As the balloon starts to fill, it stretches, exerting more pressure on its contents. As the pressure inside the balloon increases, it pushes back against the incoming flow of water. Consequently we may expect the water to flow in more slowly as the process continues.

How does this compare with electrons flowing into a capacitor? The concept is similar. Initially, the electrons go rushing in, but as they occupy more of the electron-holes, the newcomers take longer to find a resting place. The charging process gets slower, and slower. In fact, theoretically, the charge on the capacitor never quite reaches the voltage being applied to it.

Background: The Time Constant

The speed with which a capacitor charges is measured with a function known as the "time constant." The definition is very simple:

```
T = R × C
```

where T is the time constant, in seconds, if a capacitor of value C (measured in farads) is being charged through a resistor of R ohms.

Going back to the circuit you first tested, using a 1K resistor, we can put the values of the components that you used into the time-constant formula—but only if we convert the units to ohms and farads. Well, 1K is 1,000 ohms, and 1,000µF is 0.001 farads. So the calculation becomes very easy:

```
TC = 1,000 × 0.001
```

Therefore, for those values of a resistor and a capacitor, TC = 1.

But what exactly does this mean? Does it mean that the capacitor will be fully charged in one second? Sorry, it's not that simple.

- T, the time constant, is the number of seconds required for a capacitor to acquire 63% of the voltage being supplied to it, if it starts with zero volts.

What if the capacitor doesn't start from zero? If we start measuring after the capacitor has already acquired some voltage, the definition becomes a bit more complicated. If V_{DIF} is the difference between the voltage on the capacitor and the supply voltage, T is the number of seconds required for a capacitor to add 63% of V_{DIF} to its current charge.

(Why 63%? Why not 62%, or 64%, or 50%? The answer to that question is too complicated for this book, and you'll have to read about time constants elsewhere if you want to know more. Be prepared for differential equations.)

A comparison may be helpful. Figure 2-82 shows a greedy guy who is ready to eat some cake. At first he's ravenously hungry, so he takes 63% of the cake and eats it in one second, which is his time constant for eating cake. In his second bite, he takes another 63% of the cake that is left—and because he's not feeling so hungry anymore, he requires one more second (remember, this is his time constant). In his third bite, he takes 63% of what still remains, and still ingests it in one second. And so on. He is gradually filling up with cake, like a capacitor filling up with electrons. But he never quite eats all the cake, because he only takes 63% of whatever is left.

Figure 2-82 *If our gourmet always eats just 63% of the cake still on the plate, he "charges up" his stomach in the same way that a capacitor charges itself. No matter how long he keeps at it, the cake is never quite gone and his stomach is never completely filled.*

Figure 2-83 shows the process another way. After each time constant (which is one second, if we have a 1,000µF capacitor and a 1K resistor), the capacitor acquires another 63% of the difference between the charge it had and the voltage being applied.

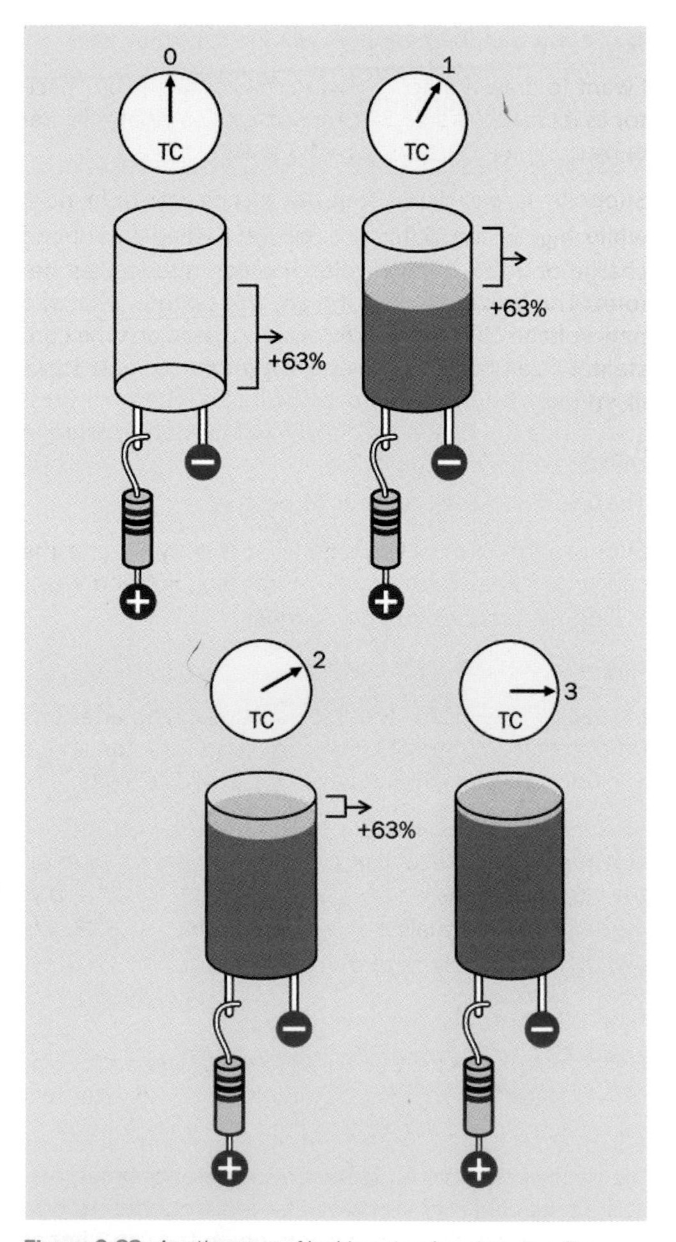

Figure 2-83 *Another way of looking at a charging capacitor.*

In a perfect world of perfect components, the charging process for a capacitor would continue for an infinite time. In the real world, we say rather arbitrarily:

- After five time constants, the charge on the capacitor will be so close to 100 percent, we can think of the process as being complete.

Background: Graphing It

I want to draw a graph showing the voltage in a capacitor as it charges. To do this, I'm going to calculate the data by using the time-constant formula.

Suppose V_{CAP} is the voltage on a capacitor right now, while V_{DIF} is the difference between that amount of charge and the battery voltage being applied (as before). The formula shown below will tell me what the new voltage on the capacitor will be after one time constant. I'll call the new voltage V_{NEW}. The formula looks like this:

$$V_{NEW} = V_{CAP} + (0.63 \times V_{DIF})$$

The 0.63 value is the same as 63 percent.

Suppose the battery was supplying exactly 9V and the capacitor started with exactly 0V. So, $V_{CAP} = 0$ and $V_{DIF} = 9$. Plug those values into the formula:

$$V_{NEW} = 0 + (0.63 \times 9)$$

My calculator tells me that $0.63 \times 9 = 5.67$. So after one time constant (one second, with a 1K resistor and a 1,000µF capacitor) the capacitor acquired 5.67 volts.

What about the next second? We have to repeat the calculation, using the new values. The current voltage on the capacitor, V_{CAP}, is now 5.67. The battery is still applying 9V, so V_{DIF} equals 9 minus 5.67, which is 3.33. We take those values back to the same formula:

$$V_{NEW} = 5.67 + (0.63 \times 3.33)$$

My calculator tells me that 0.63 times 3.33 equals 2.1, approximately. And 2.1 plus 5.67 equals 7.77. So, after two seconds, the capacitor has acquired 7.77 volts.

We can repeat this calculation any number of times, creating a sequence of numbers like this (rounded to two decimal places), showing the voltage on the capacitor at the end of each second, assuming a power supply of 9V:

```
After 1 second: 5.67 volts
After 2 seconds: 7.77 volts
After 3 seconds: 8.54 volts
After 4 seconds: 8.83 volts
After 5 seconds: 8.94 volts
After 6 seconds: 8.98 volts
```

The graph in Figure 2-84 was created by drawing a smooth curve through those values. I didn't bother to go beyond six seconds, because the values climb so close to 9V.

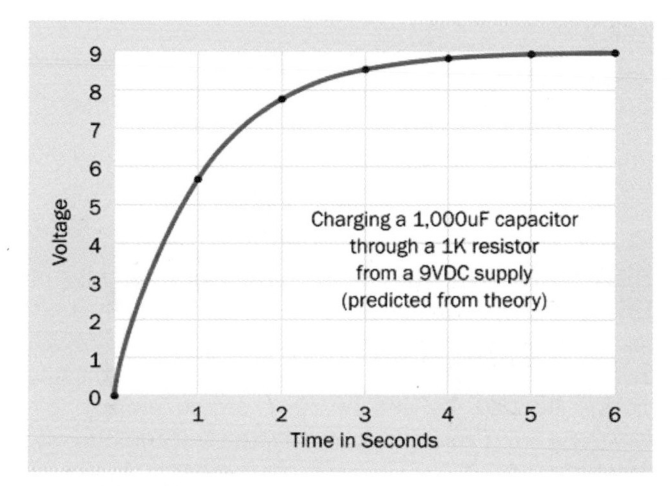

Figure 2-84 *A graph can help to show how charge accumulates on a capacitor over a period of time.*

Experimental Verification

I've told you how to calculate the charge on a capacitor in an RC network. But—how do you know I'm right? Should you just take my word for it?

Maybe you should test it yourself. In other words, you could do some *experimental verification*, which is a big part of Learning by Discovery.

Go back to the circuit that you used before, and make sure you have the 10K resistor in it, not the 1K resistor. Ask a friend to sit beside you, keeping track of the time, while you watch the display of volts on your multimeter. Every 10 seconds, your friend says "Now!" and you write down the voltage on the meter at that moment. You follow this procedure for one minute.

Because you're using a 10K resistor, not a 1K resistor, the time constant is now 10 seconds instead of 1 second. So, your readings should look like the series of voltages that I tabulated above at 1-second intervals, except that yours will be at 10-second intervals.

The voltages that you measure should be close to mine, but they won't match precisely. Why? I can think of many reasons.

- Your battery doesn't provide exactly the same voltage as mine.
- Your resistor isn't precisely 10,000 ohms.
- Your capacitor is not exactly 1,000 microfarads.
- Your meter is not totally accurate.
- You will take a few microseconds to read the meter.
- Your friend may not prompt you precisely every 10 seconds.

There are two other factors that you might not think of. First, capacitors don't store electricity perfectly. They suffer from *leakage*, in which the charge gradually leaks away. This happens even while the capacitor is gaining charge. Toward the end of the charging process, the electrons are trickling in so slowly, the leakage (the rate at which they are trickling out) becomes significant by comparison.

Additionally, your meter has an internal resistance. It's very high, but still, it passes a small amount of current. This means that the meter is stealing a little bit of the charge from your capacitor while you are measuring the voltage. Yes, the process of making a measurement changes the value that you are trying to measure! This is actually a common problem in physics and engineering.

I can think of ways to minimize all of these factors, but I can't imagine a way to eliminate them completely. There will always be some experimental error. This is the challenge when you use an experiment to verify a theory. Verification can be a very lengthy process, requiring a lot of patience—which is why theorists tend to have different personalities from experimentalists.

Capacitive Coupling

Now that I've told you how capacitors become charged and discharged, I have to go back to the statement I made before:

- A capacitor blocks DC (direct current).

You may remember, I also said, "So long as you apply a steady electrical potential to the capacitor, this is true."

But what if there isn't a steady electrical potential? What happens at the moment when a capacitor goes from having no charge on it to connecting suddenly to a source of voltage?

Ahhh, that's a different matter. Under those circumstances, the capacitor *allows a signal to pass through*.

How can this be? The plates inside a capacitor don't touch each other, so, how can a pulse of electricity jump from one to the other?

I'll deal with the "how" and "why" in a moment. First you need to make sure that what I'm talking about actually happens.

Take a look at the components on the breadboard in Figure 2-85. The layout is similar to the circuit in Figure 2-79, but the 10K resistor has moved from the left side to the right side, and an LED and a 470-ohm resistor have been added.

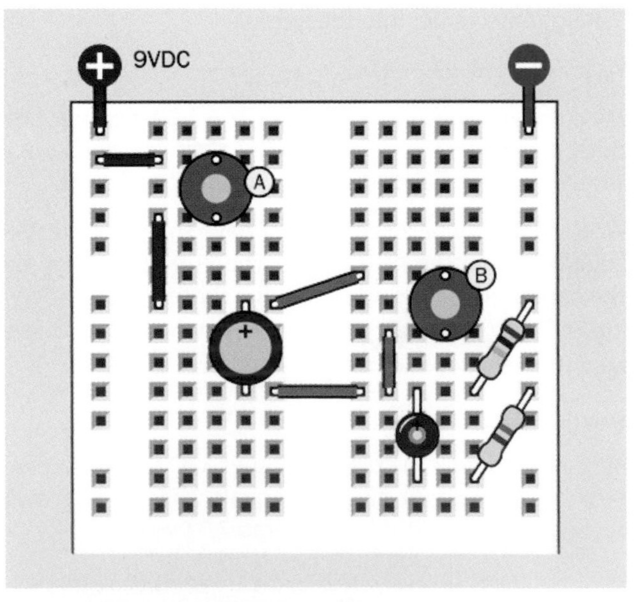

Figure 2-85 *The flashing of a red LED shows how the behavior of a capacitor changes when there is a quick change in voltage.*

The breadboarded circuit is redrawn as a schematic in Figure 2-86, which may help to clarify it.

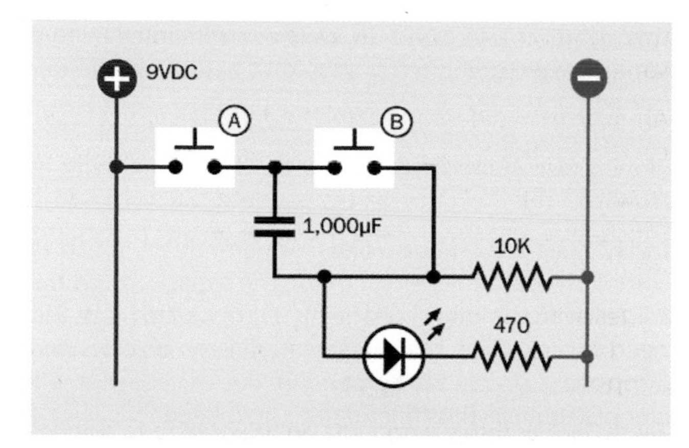

Figure 2-86 *This schematic shows the same circuit as in the preceding figure depicting a breadboarded version.*

And just to make sure there is no confusion, I'm showing the component values in Figure 2-87.

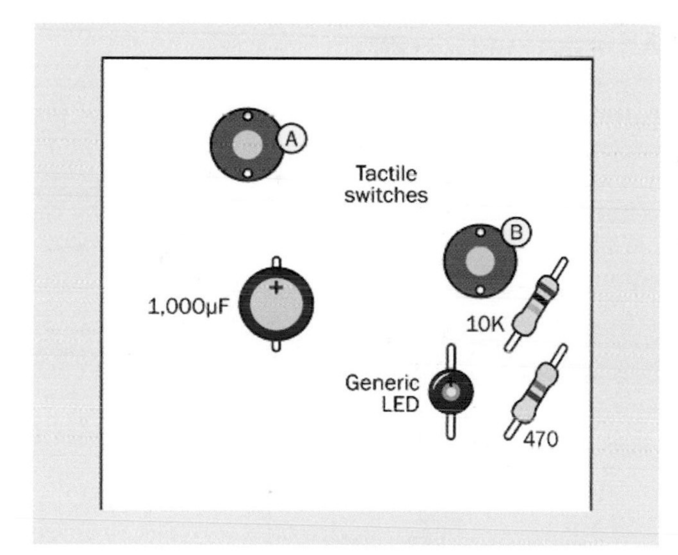

Figure 2-87 *Component values for the breadboarded circuit.*

After you assemble the circuit, first remember to press button B to discharge the capacitor. Now press button A, and—why did the LED just flash and slowly fade away?

Press button A again. This time, little or nothing happens. Evidently, the capacitor has to begin in a discharged state for this to work. So, press button B to discharge it. Now press button A again, and the LED flashes again.

We know that the capacitor started with almost no positive voltage on its lower pin, because it was connected to negative ground through the 10K resistor. We also know that the capacitor started with almost no positive voltage on its upper pin, because button B shorted both sides of the capacitor together. (That was why I asked you to discharge it.)

Then you pressed button A, which applied a sudden positive pulse, and the LED lit up on the other side of the capacitor. The current through the LED had to come from somewhere, and the only explanation is that it came through the capacitor.

Displacement Current

Let's try this again, using your meter instead of the LED and its series resistor. Figure 2-89 shows the breadboard layout, while Figure 2-88 shows the schematic. Discharge the capacitor by pressing button B, and then check the reading on your meter. It should be near 0V.

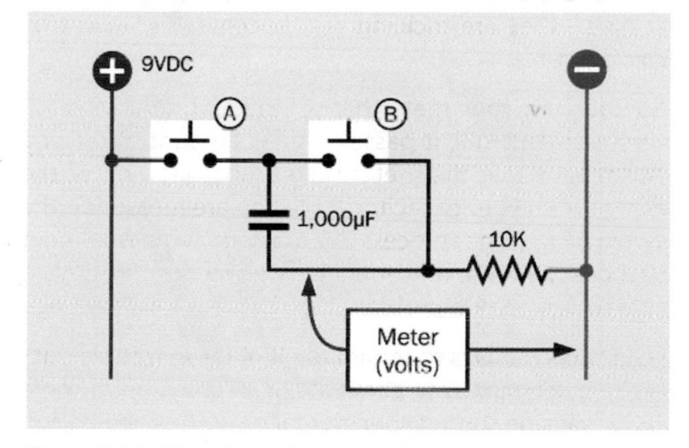

Figure 2-88 *The schematic version of the preceding breadboarded circuit.*

Watch your meter very carefully while you press button A. A digital meter does not respond very quickly, but still you will see a sudden rise in the voltage, after which the voltage gradually diminishes.

When I connected this circuit with an oscilloscope, which can measure and display very rapid changes in voltage, the trace looked look like the curve that I added at the bottom of Figure 2-89. The rise in voltage was so fast, it appeared to be instant.

The way that a capacitor allows a sudden fluctuation in voltage to pass through is well known, and is often used in electronics. But how can it happen?

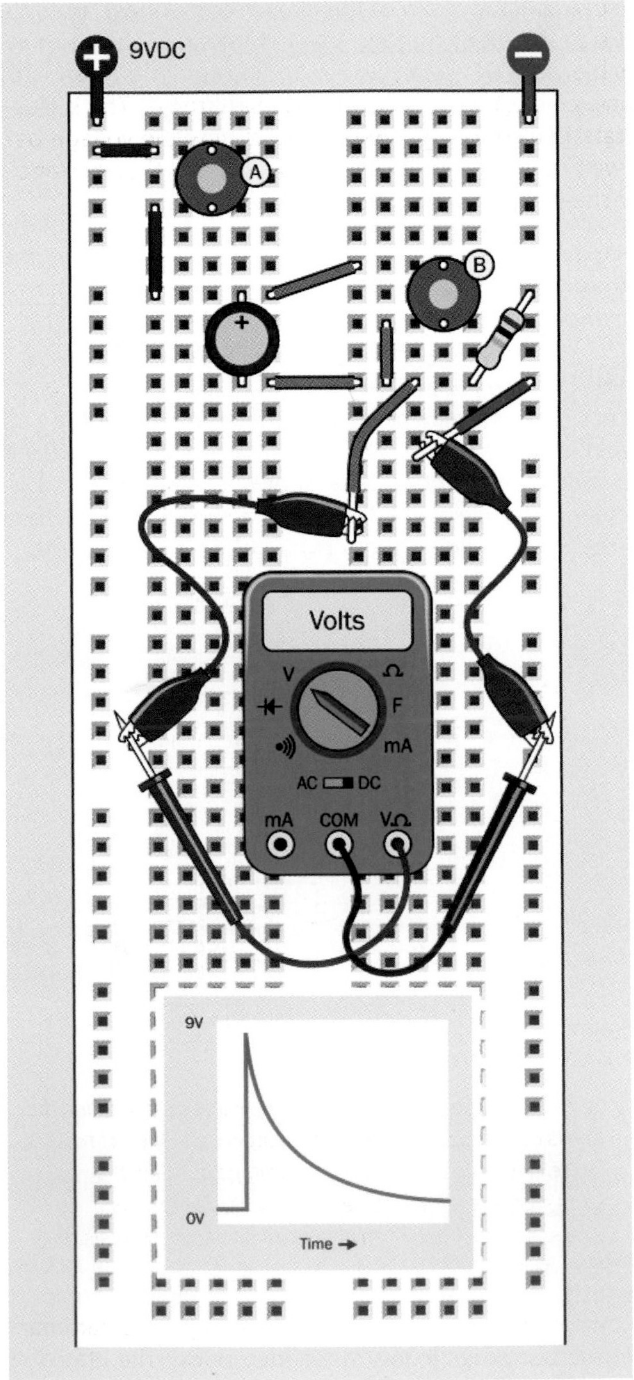

Figure 2-89 *A meter has been substituted for the LED and 470-ohm series resistor that were used in the previous version of this circuit.*

This question interested an early experimenter named James Maxwell, who felt it shouldn't happen; so he developed a theory and invented a phrase to describe what he saw. He called it *displacement current*. This conformed with some theories that he was developing at the time.

Today, there are other theories. Apparently an inrush of current creates a field effect inside the capacitor, and the field effect can induce voltage on the opposite plate. But this concept becomes complicated very quickly, and most books simply say something like, "a capacitor will block DC but will pass fluctuations in voltage."

If you substitute a smaller capacitor, you'll find that it passes a briefer pulse. Remove the meter, put the LED and its 470-ohm resistor back into the circuit, and try 100µF, 10µF, 1µF, and 0.1µF capacitors. By the end of the series, the LED barely flickers.

Alternating Current

If you reverse the circuit, it still works, although current flows in the opposite direction. Figure 2-91 shows the circuit, with the 10K resistor moved to the left, and button A moved to the right. The meter still measures the voltage from the point between the resistor and the capacitor. The schematic in Figure 2-90 shows the same revision.

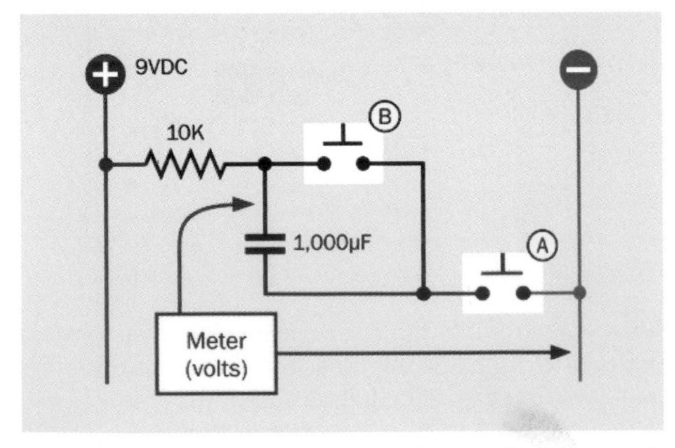

Figure 2-90 *A schematic of the preceding breadboarded circuit.*

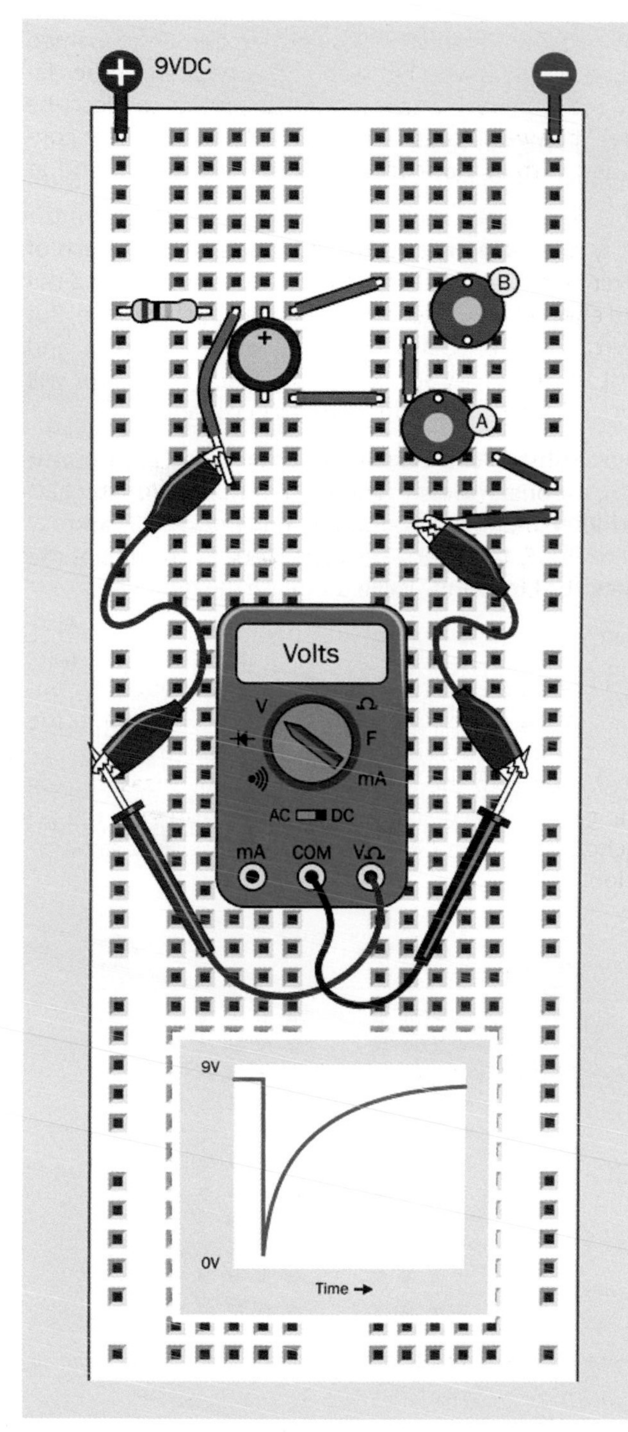

Figure 2-91 *The previous circuit has been modified, with the voltages reversed.*

After you press and release button B to discharge the capacitor, the meter measures approximately 9VDC, because the upper pin of the capacitor is connected with the positive bus through the 10K resistor. The capacitor is blocking DC, so it appears to have infinite resistance, and the positive charge has "nowhere to go." This is illustrated in Figure 2-92, which shows how the voltage between a pair of resistors increases when the resistance increases between that point and ground.

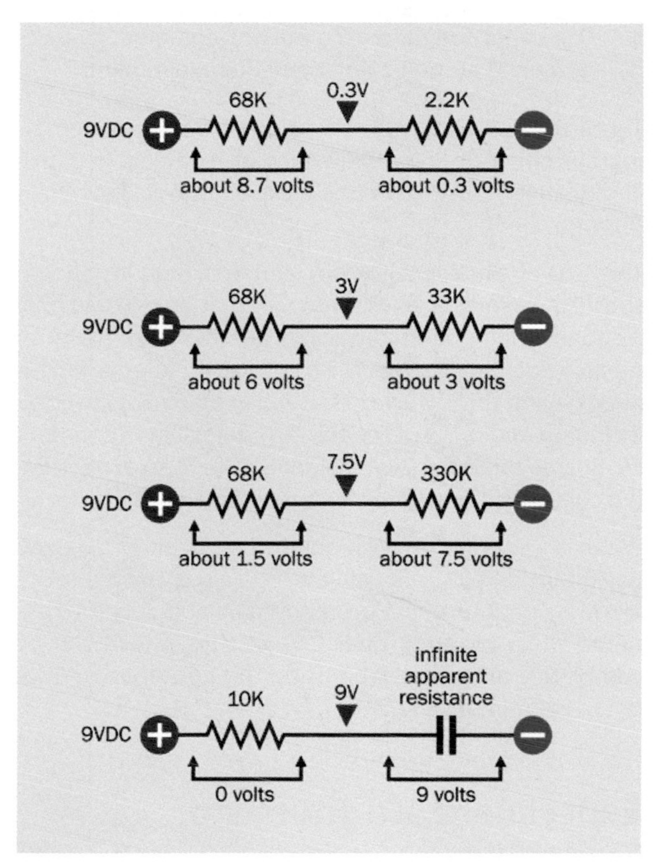

Figure 2-92 *When you have a pair of resistors in series, and the left one is connected with the power supply while the right one is connected with negative ground, the voltage between the resistors increases as the value of the righthand resistor increases. A capacitor has an almost infinite effective resistance to DC current.*

However, when you press button A in your breadboarded circuit, this creates a negative pulse. The effective resistance of the capacitor disappears momentarily as the pulse passes through, causing your meter reading to fall. Then the capacitor slowly recharges, just as it did in your very first test in this experiment.

The graph in Figure 2-91 gives a rough idea of how the charge on the capacitor changes.

- A capacitor does block direct current (DC).
- The same capacitor will allow a brief fluctuation to pass through, regardless of which way the current is flowing.
- The capacitor then accumulates a charge, as I described at the beginning of this experiment.

This leads to an important conclusion. Because alternating current (AC) is a rapid series of relatively negative and relatively positive pulses, a capacitor will allow them through.

The size of the capacitor will be important. When you substituted smaller values, you saw that they would only respond briefly. A smaller capacitor will pass high-frequency fluctuations, but will block low-frequency fluctuations—and this behavior is useful in many applications, including audio. You'll see this for yourself in Experiment 29. Bear in mind that audio signals are a form of alternating current, because they fluctuate rapidly.

When a capacitor is positioned in a circuit to pass AC while blocking DC, we call it a *coupling capacitor*. It can allow a signal to travel from one part of a circuit to another, while blocking their DC voltages, which can be completely different. I'll be using this concept when we get to Experiment 11.

Experiment 10: Transistor Switching

Now that you've seen the behavior of capacitors, I'm going to move on to another fundamental component: the transistor. After you learn how that works, you'll see how capacitors and transistors can be used together.

What You Will Need

- Breadboard, hookup wire, wire cutters, wire strippers, multimeter
- Transistor, 2N2222 (1)
- 9-volt battery and connector (1)
- Resistors: 470 ohms (2), 1M (1)
- Trimmer potentiometer, 500K (1)
- Generic LED (1)

The Finger Test

I'm going to be using the 2N2222 transistor, which is the most widely used semiconductor of all time (it was introduced by Motorola in 1962 and has been in production ever since, in one form or another).

Because Motorola's patents on the 2N2222 ran out long ago, any company can manufacture their own version of it. Some versions are packaged in a little piece of black plastic, while others are enclosed in a little metal "can." I showed these two versions in Figure 2-23. For our purposes, either will do. However, note the warning that I included previously about part numbers (see "Essential: Transistors" on page 49). Some 2N2222s are not the same as others, and you need to use the right type.

Plug your transistor into your breadboard with an LED and a 470-ohm resistor, as shown in Figure 2-93. Make sure that the longer lead on your LED is facing to the left, as indicated by the + sign. Also, check that the transistor has its flat side facing to the right. In the unlikely event that you are using a transistor in a metal can, the tab sticking out from the can should point down and to the left.

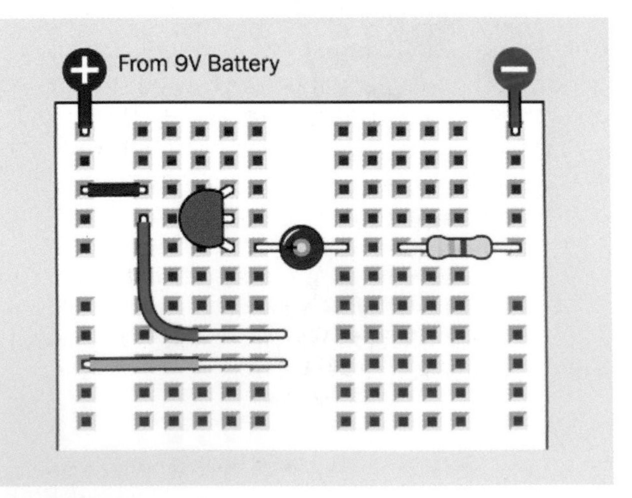

Figure 2-93 *Breadboard setup for your first transistor test.*

Notice that the wires that I have shown as green and orange have had some extra insulation removed. If you are using precut jumpers, you'll have to bend out one leg of each, so that they rest flat on the breadboard.

Now for the fun part. Press your finger to the exposed metal of the green and orange jumpers as shown in Figure 2-94, while you watch the LED. If nothing happens, moisten your finger slightly and try again. The harder you press, the brighter the LED becomes. The transistor is amplifying the tiny amount of current flowing through your finger.

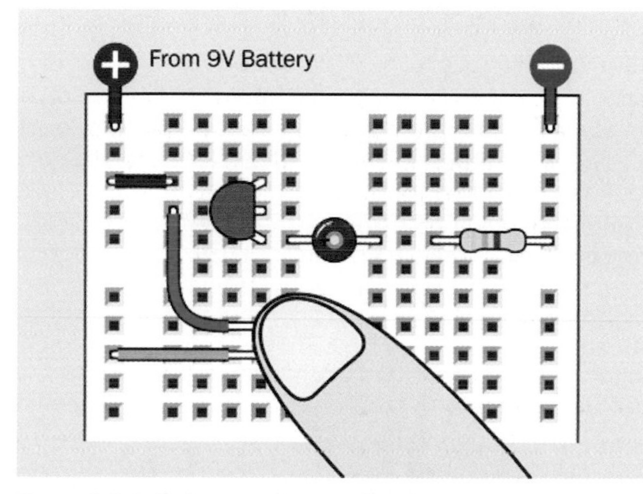

Figure 2-94 *Adding your finger makes the experiment work.*

Caution: Never Use Two Hands

The fingertip switching demo is safe if the electricity only passes through your finger. You won't even feel it, because it's 9 volts DC from a small battery. But it's not a good idea to put the finger of one hand on one wire, and the finger of your other hand on the other wire. This would allow the electricity to pass through your body. Although there is no chance of hurting yourself in this circuit because the current is so small, you should *never get into the habit of allowing electricity to run through you from one hand to the other*. Also, when touching the wires, *don't allow them to penetrate your skin*. This also means that you shouldn't apply voltage to any body ornaments that already pierce your skin.

Inside the Finger Test

Take a look at Figure 2-95, which reveals the connectors inside the breadboard, omitting the ones that are not connected in this experiment. Notice that the bottom lead of the transistor is connected through the breadboard to the LED, and then through the 470-ohm resistor to the negative bus. So, enough current flowed out of the transistor to light the LED.

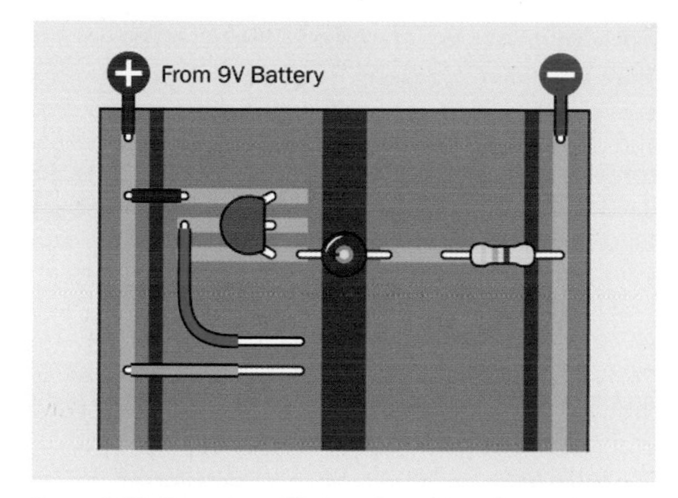

Figure 2-95 *X-ray view of the breadboard from the previous figure.*

Where did this current come from? Well, some electricity flowed in through the skin of your finger, to the center lead of the transistor. But this wasn't enough to light an LED.

There's only one other explanation. The transistor has a third lead, at the top, which is connected to the positive bus. Electricity entered the transistor through this lead. And then, somehow, this flow of current was controlled by the smaller amount of current that flowed through your finger into the center lead of the transistor.

This principle is illustrated in Figure 2-96.

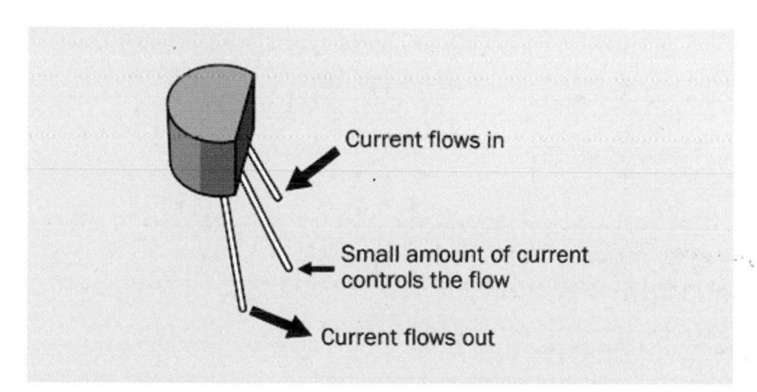

Figure 2-96 *The basic function of an NPN transistor.*

Incidentally, this phenomenon is very different from the behavior of the capacitor that you saw in the previous experiment. A capacitor just passed a quick pulse of electricity. A transistor controls a steady flow.

Fundamentals: Transistor Variants

The component that you're using in this experiment is a *bipolar transistor*. It comes in two variants named *NPN* and *PNP*. The NPN type, which you have been using, contains three layers of silicon, of which two "N" layers have a surplus of negative charge carriers. A third layer, sandwiched between the other two, is the "P" layer, with a surplus of positive charge carriers. I won't be going into a lot of detail about how the transistor works on an atomic level, because in this book I'm more interested in what a transistor does than in the theory that explains how it does it. You can find that information in any technical book, or in many sources online.

The three leads on an NPN bipolar transistor are named the collector, base, and emitter, as shown in Figure 2-97.

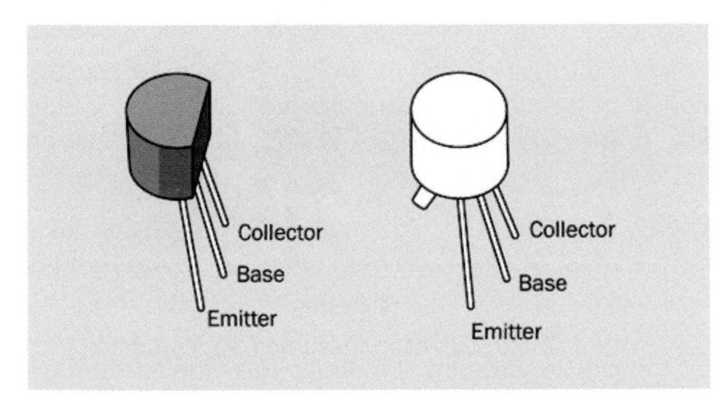

Figure 2-97 *The names of the three leads of an NPN bipolar transistor, supplied in a plastic body (left) and metal can (right).*

- When the base of an NPN transistor is slightly more positive than its emitter, the transistor allows positive current to flow in through the collector and out through the emitter.
- In this way, a very small current entering through the base of the transistor can control a larger current entering through the collector.

A PNP transistor functions oppositely to an NPN transistor. It allows negative current to flow in through the collector and out through the emitter when the base is slightly more negative than the emitter. PNP transistors are sometimes more convenient in a circuit, but are less commonly used. I will not be using them in *Make: Electronics.*

Four variants of the schematic symbol for an NPN transistor are shown in Figure 2-98. They all function the same way. The letters C, B, and E remind you that these connections are for the collector, base, and emitter.

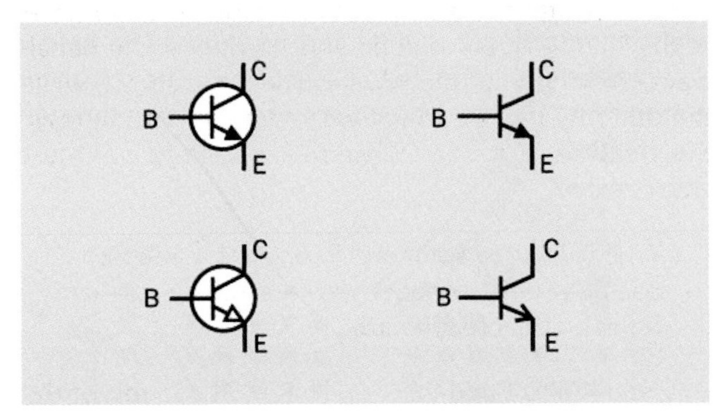

Figure 2-98 *Any of these symbols may be used to represent an NPN transistor.*

Four variants of the schematic symbol for a PNP transistor are shown in Figure 2-99. They all function the same way.

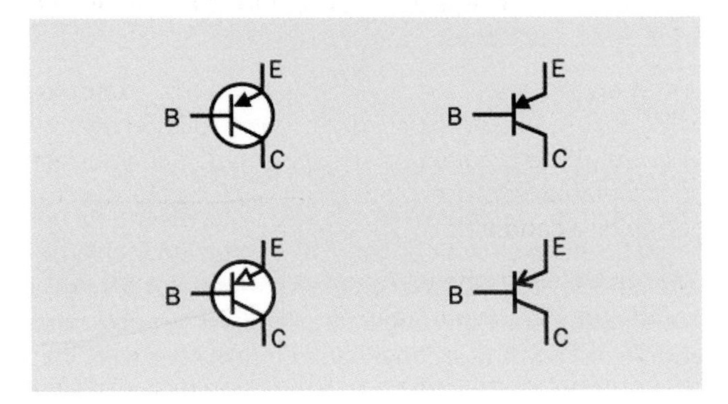

Figure 2-99 *Any of these symbols may be used to represent a PNP transistor.*

It's easy to get the symbols for PNP and NPN transistors mixed up, but there's a simple way to remind yourself which is which. The arrow in the NPN symbol points outward, never in. So, think of "NPN" as being short for "never pointing in."

Adding a Potentiometer

To learn more about the way in which a transistor works, we need a component that is a little more controllable than the tip of your finger. I think a potentiometer will do the job, but not the large type that you used previously. A *trimmer potentiometer* is what I have in mind, as pictured in Figure 2-22.

Even though they are dissimilar in shape and size, they all have three pins. These pins are functionally the same as the three tags that stick out of a large potentiometer of the kind that you used previously. The middle pin always connects with the wiper inside the trimmer, while the other two pins connect with each end of the resistive track inside it. Here is the basic rule that you must always follow:

- When you plug a trimmer into your breadboard, each pin must connect with a separate row of holes on the breadboard.

This rule is illustrated in Figure 2-100. At the top of the figure I have drawn a plan view of three types of trimmers, including the multi-turn type, because even though I am not recommending it, you may find yourself using one, some day. The pins are not visible from above, but I have shown their positions as if you can see them through the component. The positioning varies, but there are always three pins, and they should be 1/10" apart vertically.

In the lower part of the figure, the two "Yes" examples will work because each pin connects with a different row of holes in the breadboard. The two "No" examples are unacceptable, because a pair of pins will be shorted together by a conductor inside the board.

Having dealt with the trimmer basics, I want you to add a 500K trimmer to your transistor circuit. This is shown in Figure 2-101. Connect the power and use a small screwdriver to rotate your trimmer all the way clockwise, and all the way counterclockwise. Notice that if you start with the LED completely dark, you have to turn the screw on the trimmer a little way before the LED begins to glow.

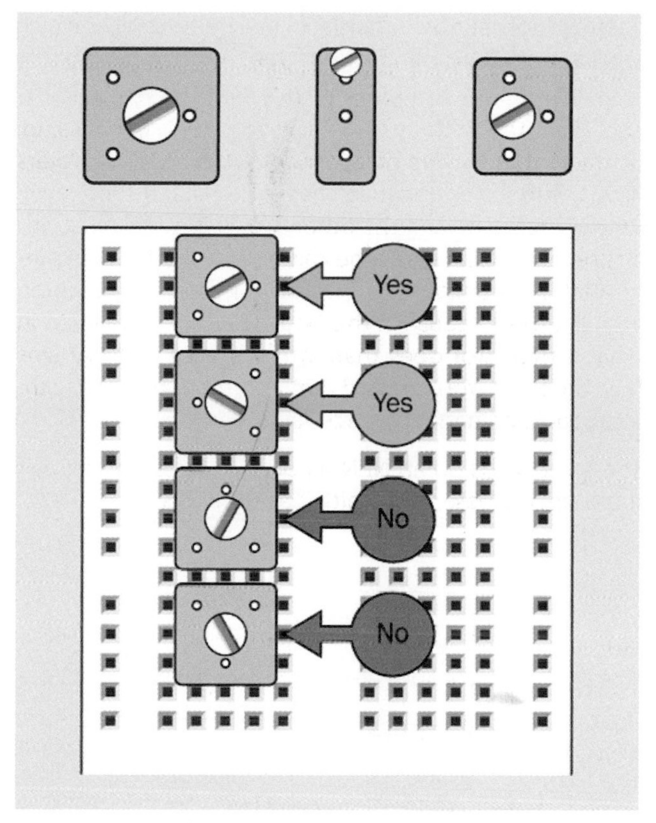

Figure 2-100 *Three types of trimmers, and the correct orientation of their pins.*

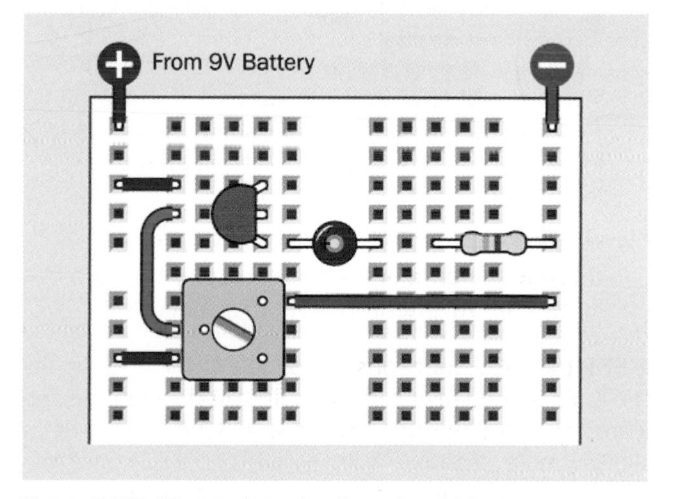

Figure 2-101 *The previous circuit now has a trimmer potentiometer so that you can control the transistor more precisely than was possible with your finger.*

Take a look at the schematic in Figure 2-102, which shows the same connections that you made on the breadboard, but in a format that is easier to understand.

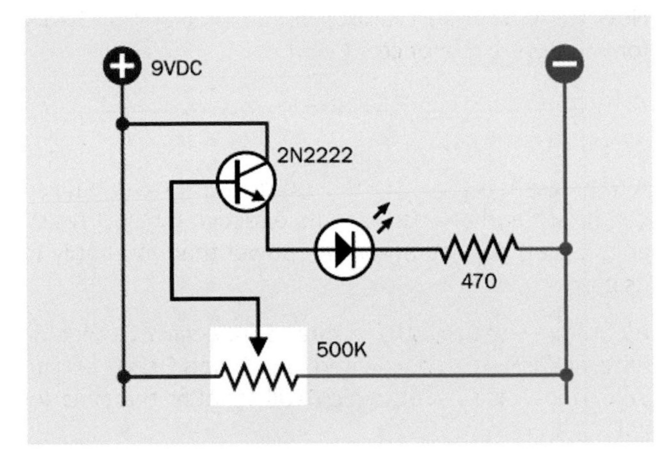

Figure 2-102 *This schematic shows the same breadboard circuit that you built using a trimmer.*

The component values are also shown in Figure 2-103.

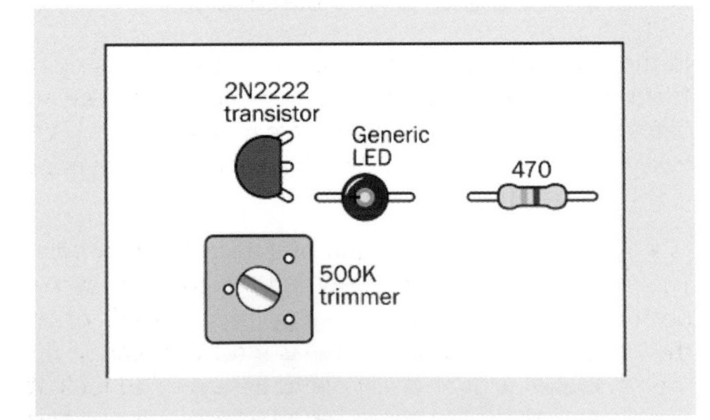

Figure 2-103 *Values for the breadboarded components.*

The potentiometer is connected between the positive bus and the negative bus. In this orientation we call it a *voltage divider*. When the wiper is at one end of the track, it connects directly with the positive side of the power supply. At the other end of the track, it connects directly with negative ground. At positions in between, it divides the voltage. Potentiometers are often used in this way, to provide a full range of voltages.

I mentioned that the LED didn't light up when you first started moving the wiper of the potentiometer from negative to positive. Is this just because the LED isn't getting enough power? Not exactly. The bipolar transistor deducts some of the power as its fee for providing a service. It won't respond unless the voltage at its base is higher than the voltage at its emitter (usually by around 0.7V). In this mode, the transistor is *positively biased*.

The general concept is illustrated in Figure 2-104.

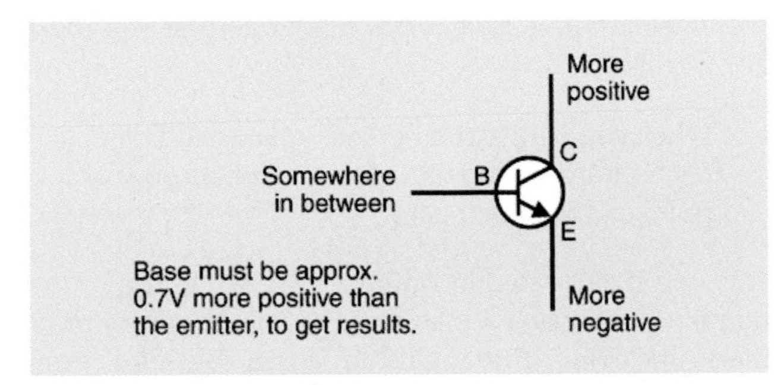

Figure 2-104 *Rule of thumb for using an NPN transistor.*

Voltage and Current

You've seen that the voltage at the base of a bipolar transistor controls the output from the transistor. Does that mean the transistor is amplifying the voltage?

You can discover this for yourself. Take your meter, set it to measure volts, and use a test lead to anchor the negative probe to the negative bus on the breadboard, as shown in Figure 2-105. Touch the red probe to the emitter pin of the transistor, make a note of the voltage, and move the probe to the base pin. I guarantee that the voltage at the emitter will be lower than the voltage on the base.

Readjust the trimmer to a different position, and try again. Regardless of how you change the voltage on the base pin, the emitter pin will always be lower.

Could this be because the 470-ohm resistor doesn't provide much resistance between the emitter of the transistor and the negative bus? Could it be pulling the voltage down?

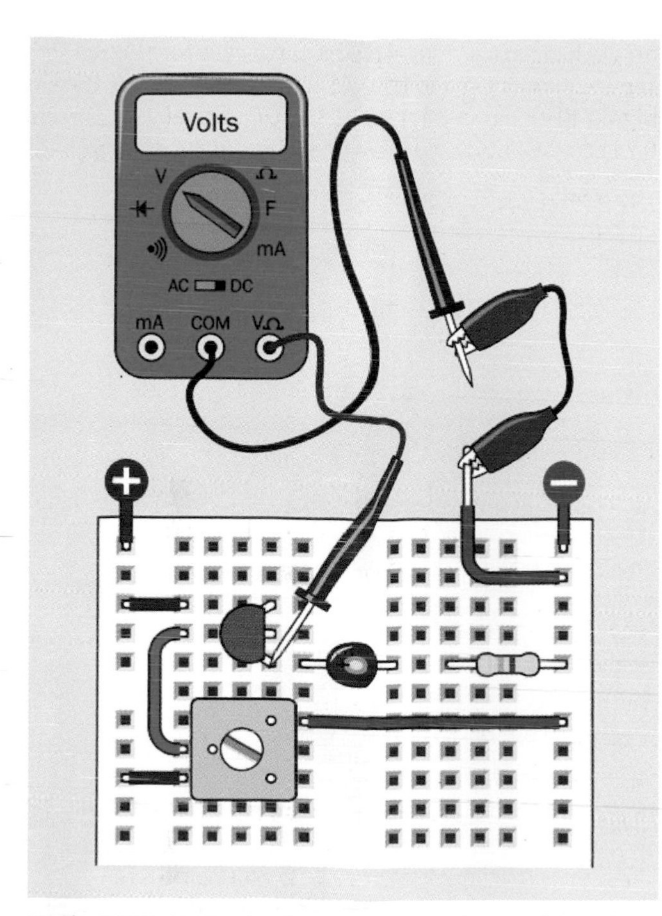

Figure 2-105 *Testing to discover if a transistor amplifies voltage.*

Let's find out. Remove the LED and the 470-ohm resistor, and substitute a 1M resistor between the emitter of the transistor and negative ground. It won't make much difference. The voltage on the emitter will still be lower than the voltage on the base.

If you have the patience to test the *current* into the base and out of the emitter, you would find something very different. This would entail setting your meter to measure milliamps and inserting it into the circuit in each location. Remember, current must pass *through* the meter to be measurable.

But I'll tell you what you would find. This particular transistor amplifies the current entering its base by a factor of more than 200:1. This is called the *beta value* of a transistor, and leads me to a fundamental fact:

- A bipolar transistor amplifies current, not voltage.

In *Make: More Electronics* you can find a lot more on this topic. I'm keeping it brief, here, because this is an introductory book.

Now I'll summarize the facts about bipolar transistors, for your future reference.

Fundamentals: All About NPN and PNP Transistors

A transistor is a *semiconductor*, midway between being a conductor and an insulator. Its effective internal resistance varies, depending on the power that you apply to its base.

All bipolar transistors have three connections: Collector, Base, and Emitter, abbreviated as C, B, and E on the manufacturer's data sheet, which will identify the pins for you.

- NPN transistors are activated by *positive* voltage on the base relative to the emitter.
- PNP transistors are activated by *negative* voltage on the base relative to the emitter.

In their passive state, both types block the flow of electricity between the collector and emitter, just like an SPST relay in which the contacts are normally open. (Actually a transistor allows a tiny bit of current known as *leakage*.)

In a schematic, the orientation of a transistor may vary. The emitter may be at the top and the collector at the bottom, or vice versa. The base may be on the left, or on the right, depending on what was most convenient for the person drawing the schematic. Be careful to look at the arrow in the transistor to see which way up it is, and whether the component is NPN type or PNP type. You can damage a transistor by connecting it incorrectly.

Transistors come in many different sizes and configurations. In a lot of them, there is no way to tell which wires connect to the emitter, the collector, or the base. You may need to check the manufacturer's datasheet to find out.

If you forget which lead is which, many multimeters have a function that will identify the emitter, collector, and base for you. Typically there are four holes labeled E, B, C, and E. When you have the emitter lead of your transistor inserted in either of the holes marked E, the base lead in B, and the collector lead in C, the meter will display the beta value of the transistor. In any other orientation, the meter reading will be unstable, or blank, or zero, or much lower than it should be (almost always below 50, and usually below 5).

Caution: Fragile Component!

Transistors are easily damaged, and the damage will be permanent.

- Never apply a power supply directly between any two pins of a transistor. You can burn it out with too much current.
- Always limit the current flowing between the collector and the emitter of a transistor by using another component such as a resistor, in the same way you would protect an LED.
- Don't apply voltage in reverse orientation. The collector of an NPN transistor should always be more positive than the base, which should be more positive than the emitter.

Background: Transistor Origins

Although some historians trace the origins of the transistor back to the invention of diodes (which allow electricity to flow in one direction while preventing reversal of the flow), the first practical and fully functional transistor was developed at Bell Laboratories in 1948 by John Bardeen, William Shockley, and Walter Brattain.

Shockley was the leader of the team, and had the foresight to see how potentially important a solid-state switch could be. Bardeen was the theorist, and Brattain actually made it work. This was a hugely productive collaboration—until it succeeded. At that point, Shockley started maneuvering to have the transistor patented exclusively under his own name. When he notified his collaborators, they were naturally unhappy about this.

A widely circulated publicity photograph didn't help, in that it showed Shockley sitting at the center in front of a microscope, as if he had done the hands-on work, while the other two stood behind him, implying that they had played a lesser role. A copy of this picture appeared on the cover of *Electronics* magazine (see Figure 2-106). In fact Shockley, as the supervisor, was seldom present in the laboratory where the real work was done.

Figure 2-106 *Front, William Shockley. Rear, John Bardeen. Right, Walter Brattain. For their collaboration in development of the world's first working transistor in 1948, they shared a Nobel prize in 1956.*

The productive collaboration quickly disintegrated. Brattain asked to be transferred to a different lab at AT&T. Bardeen moved to the University of Illinois to pursue theoretical physics. Shockley eventually left Bell Labs and founded Shockley Semiconductor in what was later to become Silicon Valley, but his ambitions outstripped the capabilities of the technology of his time. His company never manufactured a profitable product.

Eight of Shockley's coworkers in his company eventually betrayed him by quitting and establishing their own business, Fairchild Semiconductor, which became hugely successful as a manufacturer of transistors and, later, integrated circuit chips.

Fundamentals: Transistors and Relays

One limitation of NPN and PNP transistors is that they require power to perform their function, unlike a relay, which can be passively on or off without any power input at all.

Relays also offer more switching options. Different versions can be normally open, normally closed, or can latch in either position. A relay can contain a double-throw switch, which gives you a choice of two "on" positions. It can also contain a double-pole switch, which makes (or breaks) two entirely separate connections. Single-transistor devices cannot provide the double-throw or double-pole features, although you can design more complex circuits that emulate this behavior.

A table comparing the attributes of transistors and relays appears in Figure 2-107.

	Transistor	Relay
Long-term reliability	Excellent	Limited
Can switch in DP or DT mode	No	Yes
Can switch large currents	Limited	Yes
Can switch alternating current	Usually not	Yes
Triggerable by alternating current	Usually not	Optional
Suitable for miniaturization	Excellent	Very limited
Able to switch at very high speed	Yes	No
Price advantage for high voltage/current	No	Yes
Price advantage for low voltage/current	Yes	No
Current leakage when nonconducting	Yes	No

Figure 2-107 *Principal attributes of relays and transistors, compared.*

The decision to use a relay or a transistor will depend on each particular application.

So much for the theory. Now what can we do with a transistor that's fun, or useful, or both? We can do Experiment 11!

Experiment 11: Light and Sound

It is time now for your first project that has a function and a purpose. It will culminate with an ultra-simple sound synthesizer.

What You Will Need

- Breadboard, hookup wire, wire cutters, wire strippers, multimeter
- 9-volt battery and connector (1)
- Resistors: 470 (2), 1K (1), 4.7K (4), 100K (2), 220K (2), 470K (4)
- Capacitors: 0.01µF (2), 0.1µF (2), 0.33µF (2), 1µF (1), 3.3µF (2), 33µF (1), 100µF (1), 220µF (1)
- Transistors: 2N2222 (6)
- Generic LED (1)
- Loudspeaker, 8-ohm, 1" or (preferably) 2" (1)

Fluctuations

Figure 2-108 shows the new circuit that I'd like you to create on your breadboard. There's not a lot of room among the components, so you may find that it's easier to install them using your pliers instead of your fingers. Count the holes in the board carefully, and double-check to make sure that everything is in exactly the right place.

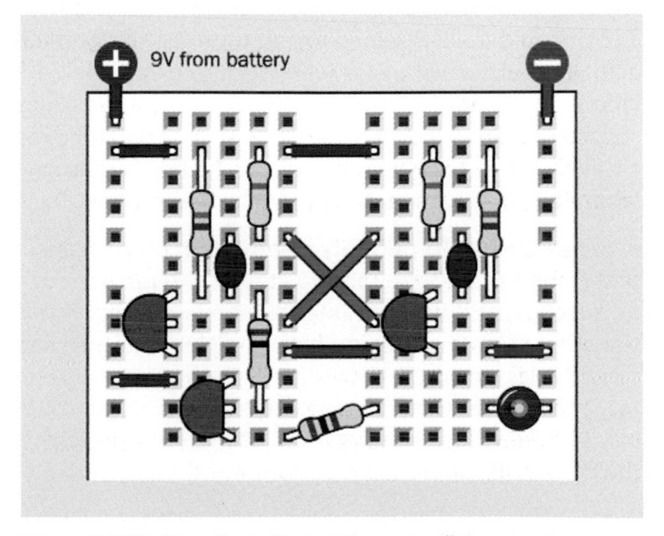

Figure 2-108 *Breadboard layout for an oscillator circuit.*

The component values are shown in Figure 2-109.

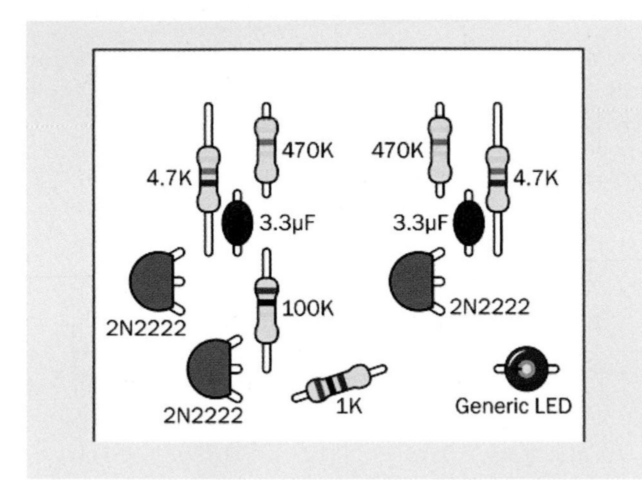

Figure 2-109 *Component values in the breadboarded circuit.*

Apply power, and the LED should flash on for approximately one second, and off for approximately one second.

Is that all? No, we've just begun. First, however, you have to understand how it works. Check the layout in Figure 2-110 if you have difficulty visualizing how the components are connected inside the board. Then look at the schematic in Figure 2-111, and you'll see that the connections between components are the same. I'm going to be using the schematic to explain what's going on.

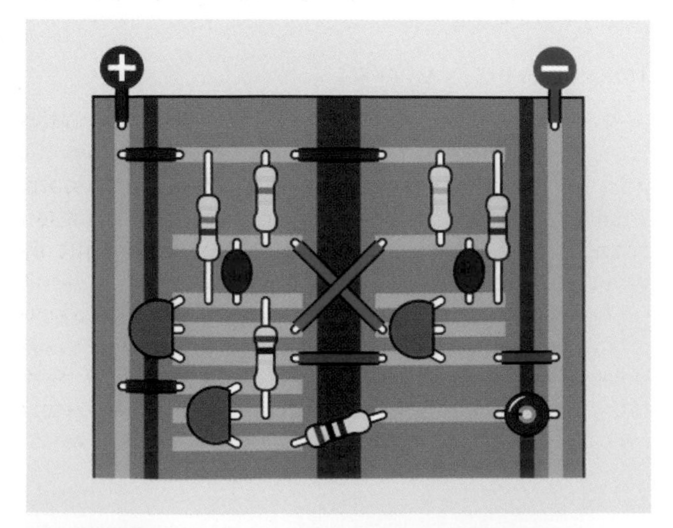

Figure 2-110 *This x-ray view may help to elucidate the interconnections between components.*

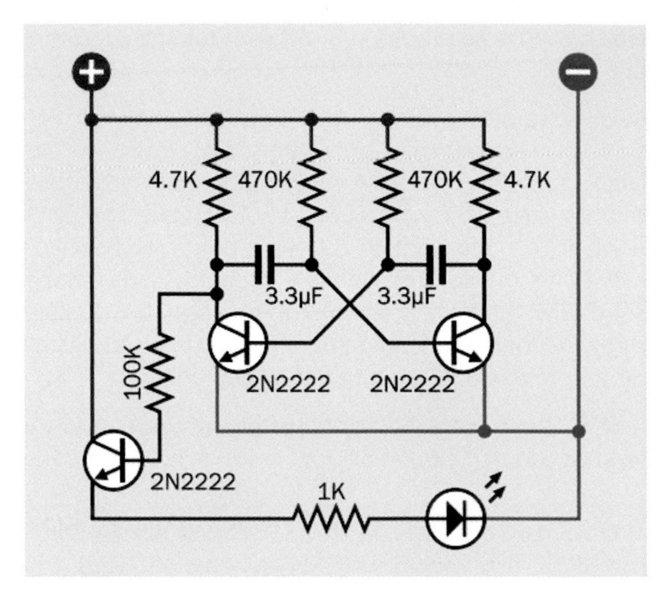

Figure 2-111 *The components in this schematic are in similar locations in the breadboard layout.*

The first thing you'll notice is that it is somewhat symmetrical. Does that mean that the left half and the right half are both doing the same thing? Yes, but not at the same time. In effect, one half turns the LED on, and the other half turns it off.

Understanding this in detail is tricky, because the voltages are constantly fluctuating, and more than one thing is happening at any given moment. However, I have created four snapshots showing the inner workings at intervals, and I'm hoping they will make everything clear.

I have omitted the third transistor and the LED from each snapshot, as they have no role in creating the oscillation in the circuit.

The first snapshot is in Figure 2-112. I have color-coded the wires like this:

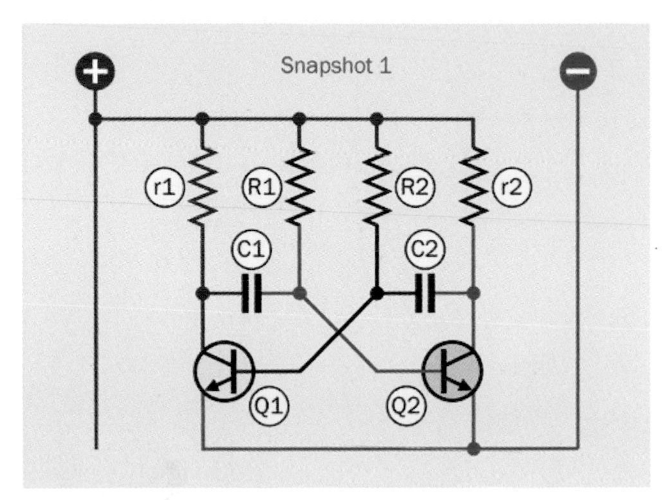

Figure 2-112 *The first of four snapshots showing voltages in the LED-flasher circuit. See text for details.*

- Black conductors and components have an unknown or indeterminate voltage.
- Blue conductors are at a near-zero voltage.
- Red conductors are rising toward positive supply voltage.
- White conductors are briefly pulled down to a more negative voltage (below negative ground), for reasons I'll get to in a moment.

As for the transistors:

- A gray transistor is not conducting electricity from collector to emitter. You can think of it as "switched off."
- A pink transistor is conducting.

The transistors are labeled Q1 and Q2, because this is a common way to identify semiconductors. The little tab sticking out of an old-fashioned metal-can transistor made it look like a letter Q when seen from above, and people got into the habit of identifying transistors this way.

To distinguish between the two sides of the circuit, r1 and R1 are on the left, and r2 and R2 are on the right. Lowercase letters identify the lower-value resistors.

One last note before I begin the explanation. Bear in mind the fundamental behavior of a transistor:

- When current into its base switches it "on," its effective internal resistance drops very low. Consequently, if its emitter is grounded at approximately 0 volts, its collector will also be near 0 volts, and so will anything connected directly to the collector. The base can also be at a relatively low voltage, so long as it is not quite as low as the emitter. You can see this happening to Q2 in Snapshot 1.
- When it switches "off," the effective internal resistance of the transistor rises to at least 5K. Consequently, any component attached to its collector isn't grounded through the transistor anymore, and can accumulate a positive charge.

Step by Step

I'm going to begin at an arbitrary moment while the circuit is already running. After I proceed through the sequence of events, I'll get to the issue of how it begins to oscillate initially.

In Snapshot 1, let's suppose that Q1 has just switched off and Q2 has just switched on. The bottom end of r1 was being grounded through Q1, but now that Q1 is off, the voltage on its collector starts to rise, and this increases the voltage on the left side of C1. The voltage on the base of Q1 also starts to rise, but not as rapidly, because R2 has a higher value. Meanwhile, because Q2 is on, it is sinking current from r2, pulling the voltage down. The base of Q2 is also sinking current through the transistor to negative ground.

That's the setup. What next?

In Snapshot 2, shown in Figure 2-113, the voltage on the base of Q1 increased enough to start it conducting. It is now sinking current from C1 and also through its base, which is why those wires are now blue. The sudden change in the voltage on the left side of C1 momentarily induces an equal drop on its righthand side, as a result of the field effect that can also be described as displacement current, described in Experiment 9. This actually pulls the voltage on the righthand side of C1 below zero, represented by the white wire. This immediately switches off Q2 by inflicting negative bias on its base.

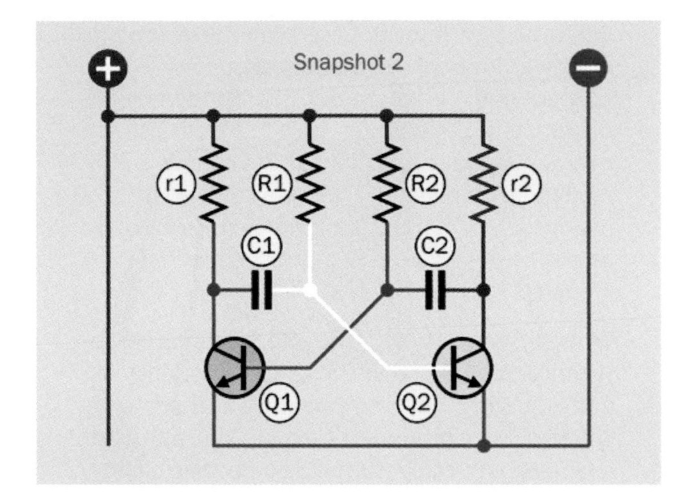

Figure 2-113 *The second snapshot.*

In Snapshot 3, shown in Figure 2-114, Q1 is still switched on while Q2 is still off. This is the mirror image condition of Snapshot 1. C1 has started to charge in the opposite direction, through R1. Gradually this is increasing the voltage on the base of Q2.

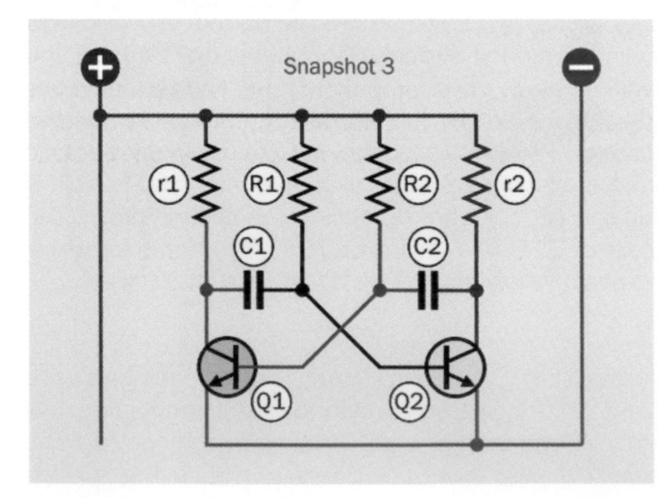

Figure 2-114 *The third snapshot.*

In Snapshot 4, shown in Figure 2-115, Q2 has started conducting, grounding the right side of C2. This transition pulls down the voltage of the left side of C2 below zero, and switches off Q1 by grounding its base. This is the mirror image of Snapshot 2.

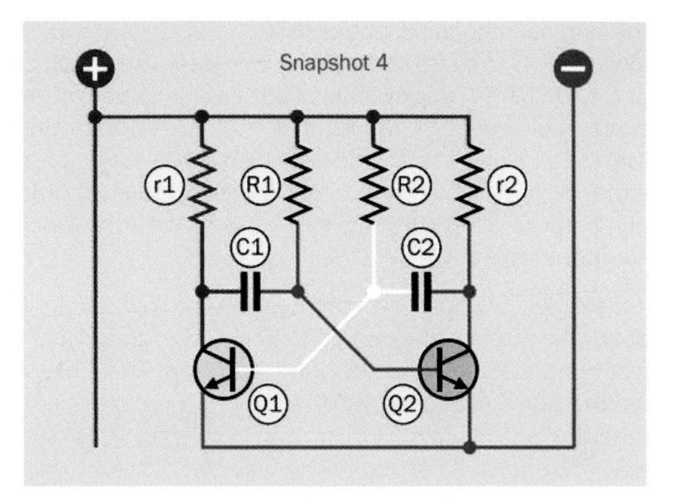

Figure 2-115 *The fourth snapshot. After this, the sequence repeats.*

After the fourth snapshot, the sequence repeats from Snapshot 1. If the extra transistor and LED are connected as in Figure 2-111, the LED should light up during snapshots 1 and 4.

Coupling Capacitor

As you can see, oscillators can be difficult to understand. This particular circuit is extremely common. In fact, if you search Google Images for "oscillator," this is probably what you'll find. Still, many people have difficulty with it.

The key is that in Snapshots 2 and 4, a sudden voltage drop on one side of a capacitor creates an equal voltage drop on the other side—the coupling effect that you saw for yourself in Experiment 9.

But How Does It Start?

Bear in mind that the circuit is basically symmetrical. When you first power it up, why shouldn't both of the transistors be on, or both of them off?

In a perfect world, where two transistors or two resistors can be absolutely identical, the circuit would initialize itself symmetrically. But in reality, there's always a tiny manufacturing difference among resistors and capacitors, causing one transistor to start conducting before the other. As soon as that happens, the circuit is out of balance, and it starts to oscillate as I described above.

Another issue to explain would be, how did I decide where to take an output from the oscillator circuit? In the original schematic, notice that r1 and r2 have much lower values than R1 and R2. This will allow the left side of C1 to charge quickly, almost to the full value of the supply voltage—and the right side of C2 behaves the same way. So, we can get a nice wide voltage range from either of those locations. I chose the one on the left, simply because it enabled an easier layout of additional components in the schematic.

If the circuit is tapped for too much current, this will slow down the process of charging the capacitor, and will affect the timing and balance of the oscillator. Therefore, I fed the signal through a 100K resistor to the base of another transistor. That transistor will draw very little current through its base but will amplify the signal so that you can do something useful with it.

Why So Complicated?

In the first edition of this book, I suggested a flashing-LED project using a component called a Programmable Unijunction Transistor (PUT). Its behavior is much easier to understand, and you only need one of them to get a result. But PUTs are not used much anymore, and some readers complained that you can't buy them easily, while others said that using a PUT was just too old-school.

Really you can still buy PUTs, but they are almost obsolete. Bipolar transistors are still widely used, so I listened to my reader feedback and abandoned the PUT. I considered various alternative oscillator schematics before settling on this one, mainly because this circuit is more common than any of the others. Also, I think all oscillator circuits can be somewhat difficult to understand.

A Processed Pulse

You've learned that two transistors can create a signal that pulses on and off, and a third transistor can amplify it to power an LED. Think back for a moment to the previous experiments. What have you learned from them that you might be able to apply to this one?

We have an output that is fluctuating quite slowly. So, we can make it more interesting by adding an RC network. (See "An RC Network" on page 76 if you need to refresh your memory regarding this concept.)

Take a look at Figure 2-116. The new RC section is at the bottom.

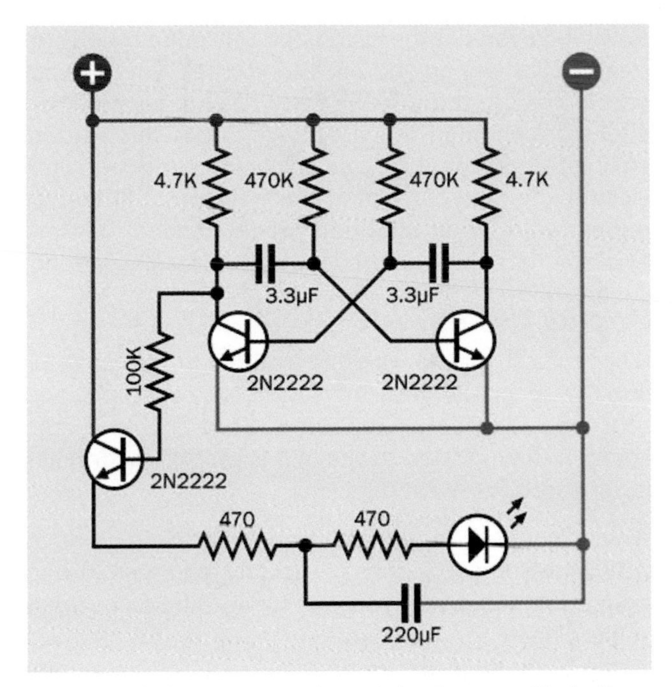

Figure 2-116 *The previous schematic has been modified with one additional resistor and a 220μF capacitor at the bottom, making an RC network.*

In Figure 2-117 the components that have been added or repositioned are in color at bottom-right, while the components that are unchanged have been grayed out.

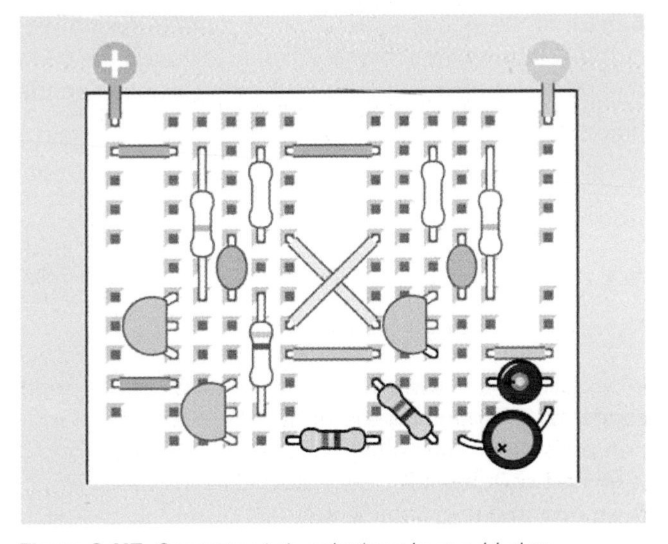

Figure 2-117 *Components in color have been added or repositioned. The new capacitor is a 220μF electrolytic.*

Now when you run the circuit, the LED pulses gently instead of flashing on and off. Can you see why? The capacitor charges through the first 470-ohm resistor, then discharges through the other. Why does this matter? Well, let's suppose you were thinking of making some electronic jewelry. Adjusting the way that it flickers or pulses could be an important aesthetic factor. On old Apple laptops, the logo used to pulse instead of flashing.

Upping the Speed

What else can you do with this circuit? You can easily adjust the speed. Remove the two 3.3µF capacitors, and substitute two 0.33µF capacitors. They should charge about 10 times faster, so the LED should flash 10 times faster, too. Is that what happens?

What if you reduce the capacitor values even more, to 0.01µF? Beyond 50 flashes per second, you have moved from a frequency that you can see to a frequency that you can hear.

How can you change the output of this circuit to be audible instead of visible? Easily! Remove the LED, the 470-ohm resistors, and the 220µF capacitor, and substitute a little loudspeaker, a 100µF coupling capacitor, and a 1K resistor as shown in Figure 2-118. The resistor grounds the emitter of the transistor, because a transistor will only work if its emitter has a defined voltage lower than the voltage at its base. The capacitor blocks the DC component of the signal, while allowing the alternating current through. In the schematic, I only included the parts that have changed. How should they be installed on the breadboard? I think you can figure that out.

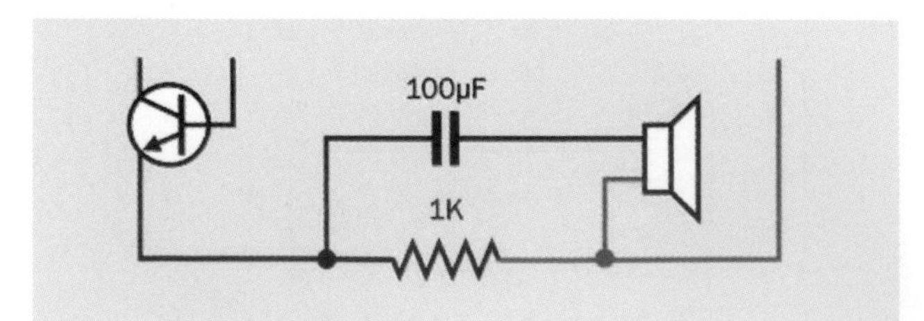

Figure 2-118 *Modifying the circuit to generate audio.*

Still More Mods

Now that you have sound, how about raising the pitch of the sound? Just substitute smaller resistors or capacitors in the oscillator circuit. You could remove the 470K resistors and substitute 220K resistors (or a value in between). Transistors can switch a signal more than a million times per second, so you certainly won't push their limits by making your oscillator run faster. A signal that oscillates 10,000 times per second sounds extremely high-pitched. If you push it to 20,000 times per second, it's above the hearing range of almost all human ears.

How about changing the character of the sound?

In the upper half of Figure 2-119 I've substituted a 1µF coupling capacitor in series with the loudspeaker, instead of the 100µF capacitor that was there previously. The lower value for the capacitor will only pass high frequencies (short pulses), and will deprive the sound of some of its lower resonance.

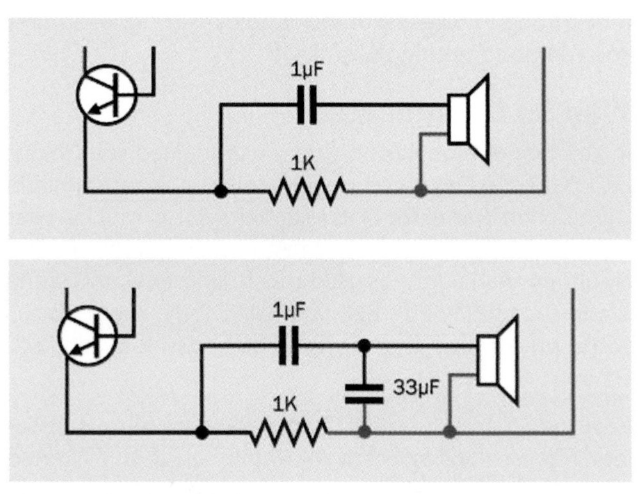

Figure 2-119 *Substituting a smaller value for the coupling capacitor blocks the lower audio frequencies, so that you only hear the higher frequencies. Placing a capacitor so that it bypasses the speaker reroutes the higher frequencies to negative ground, so that you only hear the lower frequencies.*

What if you put a capacitor across the loudspeaker, as shown in the lower part of Figure 2-119? Now the opposite effect occurs, because the capacitor is still passing higher frequencies, but routes them past the loudspeaker. In this orientation, you have a *bypass* capacitor.

These are all simple ways to modify a circuit. If you feel a little more ambitious, you could duplicate the circuit and use one section to control the other.

Restore the original component values from Figure 2-109, so that it runs at the slow original speed. Then use its output to power the duplicate section below it on the breadboard with 0.01µF capacitors to generate an audio frequency. This is shown in Figure 2-120, where the part that you built originally in this project has been grayed out, and the audio section is at the bottom.

The red piece of wire labeled A has been repositioned so that the lower section of the circuit now gets its power from the output of the upper section. The red and blue wire segments labeled B have been added to bridge the gaps that you may have in the buses of your breadboard.

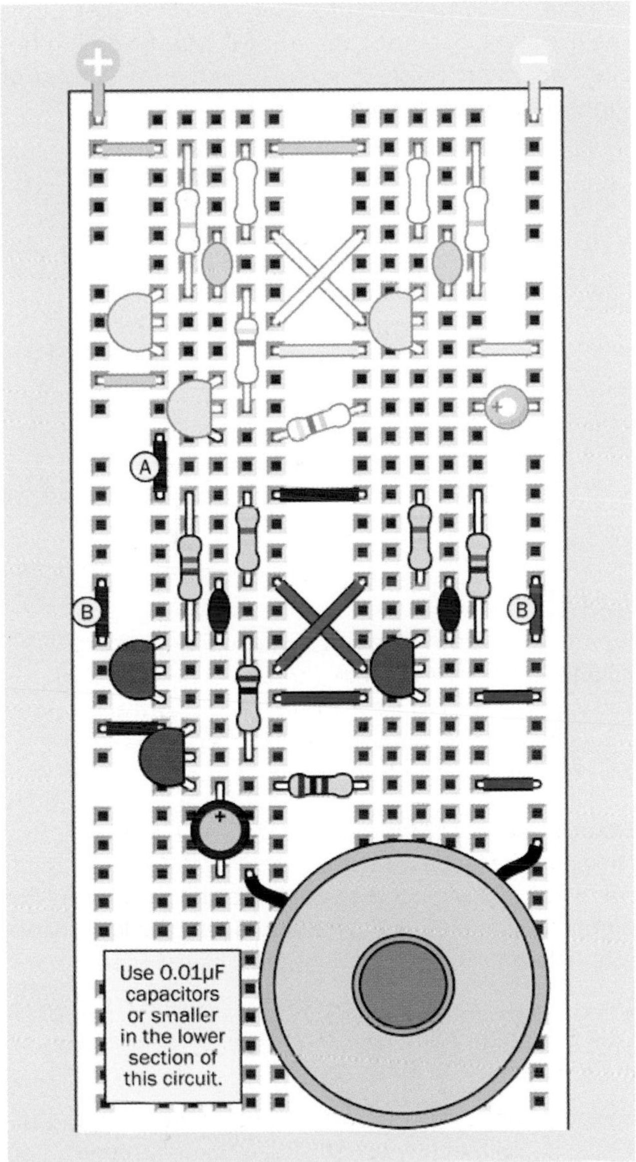

Figure 2-120 *Driving the audio section of the circuit by using a slower-running duplicate to provide fluctuating power.*

What happens if you change the capacitor or resistor values in the top half of the circuit, to switch the bottom half faster?

What would happen if you take a 220µF capacitor and apply it between various points (either in the top half of the circuit, or the bottom half) and negative ground? You won't damage any components, so feel free to experiment.

Another option is to go back to the "processed pulse" of light created in Figure 2-116, and change the way in which the components are physically joined together. You can remove them from the breadboard and rebuild them into a small wearable object.

I'm going to show you how to do that in Experiment 14. Of course, this will entail some soldering, but learning to solder components together is what I want you to do in Experiment 12, which is next.

Background: Mounting a Loudspeaker

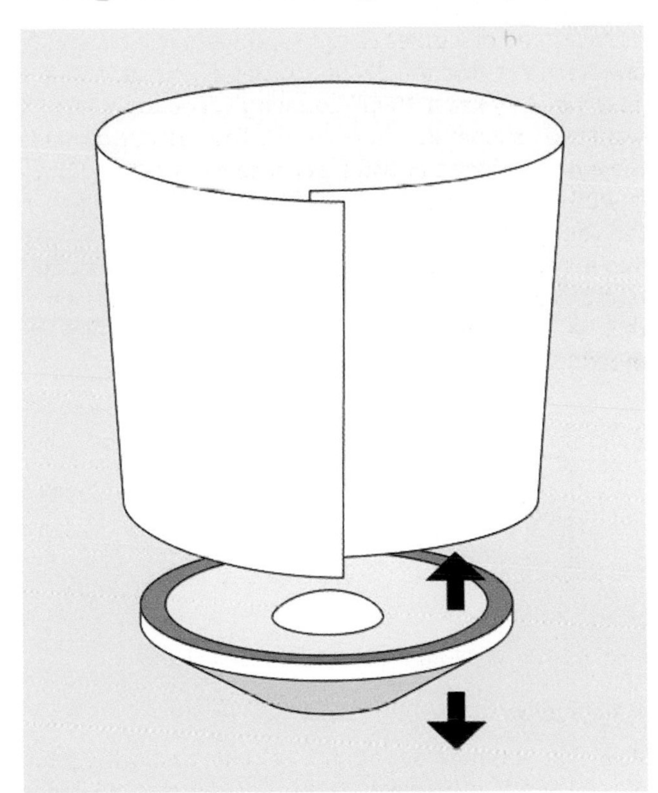

Figure 2-121 *A paper or cardboard tube increases the perceived volume from a loudspeaker.*

The *diaphragm* of a loudspeaker, also known as its *cone*, is designed to radiate sound. However, as it oscillates up and down, it emits sound from its back side as well as its front side. Because the sounds are opposite in phase, they tend to cancel each other out.

The perceived output from a loudspeaker can increase dramatically if you add a horn around it in the form of a tube that focuses the output from the front. For a miniature 1-inch loudspeaker, you can bend and tape a large file card around it. See Figure 2-121.

Better still, mount it in a box with holes drilled in it to allow sound from the front of the speaker to radiate, while the closed back side of the box absorbs sound from the rear of the speaker.

3 Getting Somewhat More Serious

This third chapter of the book applies what you have learned so far to some finished projects. I'll show you how to make a wearable version of the "pulsing glow" project from Experiment 11, and will take you through the initial development process of an intrusion alarm. Looking further ahead, in Chapter Four you'll move into the world of integrated circuit chips.

The tools, equipment, components, and supplies described below will be useful in Experiments 12 through 15, in addition to items that have been recommended previously.

Necessary Items for Chapter Three

As before, when buying tools and equipment, see "Buying Tools and Equipment" on page 324 for a shopping list. If you want kits containing components and supplies, see "Kits" on page 311. If you prefer to buy your own components from online sources, see "Components" on page 317. For supplies, see "Supplies" on page 316.

Essential: Power Supply

You could continue to use 9-volt batteries for all the projects in the book, but I am now classifying an *AC adapter* as essential, because it's so much more convenient. Also, I think it should turn out to be cheaper than buying batteries, as you start to build circuits that take more power.

To convert AC current from an outlet in your home, you have three options:

A *universal adapter* such as the one in Figure 3-1 is the most versatile option, providing a switchable range of outputs. Typically these will include 3V, 4.5V or 5V, 6V, 9V, and 12V. Universal adapters are intended to power small devices such as voice recorders, phones, and media players. They may not deliver a perfectly smooth or accurate DC output, but you should be able to smooth the power yourself with a couple of capacitors, as I will illustrate when we get to a project that uses the adapter.

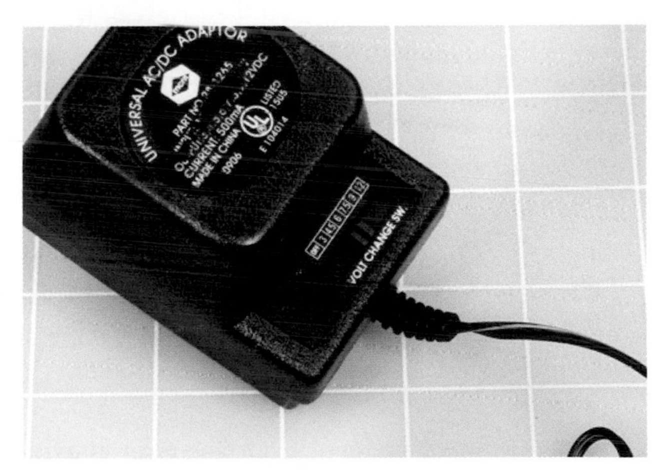

Figure 3-1 *This AC adapter plugs into the wall and allows you a choice of DC voltages by moving a small switch.*

Alternatively you can buy a *single voltage* AC adapter to deliver 9 volts DC, as shown in Figure 3-2. When you start using digital logic chips that require 5 volts, you can convert the 9 volts with a small, cheap component called a *voltage regulator*. (The regulator will also work with power from a 9-volt battery.)

Figure 3-2 *An AC adapter that delivers a fixed 9-volt DC output.*

Your third option is to spend much more money on a proper *benchtop power supply* that provides variable DC outputs of 0V to +15V and 0V to −15V, plus a fixed 5V output. It should also have several breadboards mounted conveniently on top of its box. There is no doubt that this will be very useful if you continue into the field of electronics, but you may not feel sure about that yet.

If you decide to buy a universal adapter, see "Other Components" on page 319 and scroll down to the part referring to Chapter Three of the book, for search instructions.

Whichever kind of adapter you buy, it must have these attributes:

- The output must be DC, not AC. Almost all AC adapters give you a DC output, but there are a few exceptions.
- Outputs must be rated for at least 500mA (which may be written as 0.5A).
- It doesn't matter what kind of plug is on the end of the DC power-output wire, because you'll be cutting it off anyway.
- For the same reason, if a universal adapter offers you a range of output plug adapters, don't worry about what type they are, as you won't be using them.
- Very cheap AC adapters may become unreliable if you draw current up to their rated limit. In the United States, look for the UL symbol indicating certification from Underwriters Laboratories.

Essential: Low-Power Soldering Iron

While a breadboard is indispensable for putting together a circuit quickly to understand how it works, a *soldering iron* is necessary for making permanent electrical connections in a circuit that you want to keep. The soldering iron works by melting a thin wire composed of an alloy called *solder* until it forms a blob around the copper wires or components that you wish to join. When the solder cools, the joint becomes durable.

You don't absolutely have to have a soldering iron. You can complete all the projects in this book just by breadboarding them. However, there is a special pleasure in building something durable, and soldering is a useful skill. For these reasons, I have categorized a soldering iron as "essential."

Personally I like to have a low-power soldering iron for small parts that are vulnerable to excessive heat, and a separate general-duty soldering iron for heavier tasks (described immediately below). Some people prefer to use just one thermostatically controlled soldering iron for everything, but if it's small, I don't think it delivers the heat capacity that is sometimes required, and if it's medium-sized, it isn't so easy to use for delicate work. Also, a thermostatically controlled unit can be expensive.

A low-power soldering iron should be rated at 15W, and the smaller it is, the easier it is to handle. The tip should taper to a slender but rounded point, like a newly sharpened pencil. A plated tip is preferable, although the manufacturer may not state that the tip is plated. A well-used 15W soldering iron is shown in Figure 3-3. The discoloration is a normal consequence of heat, and does not degrade function.

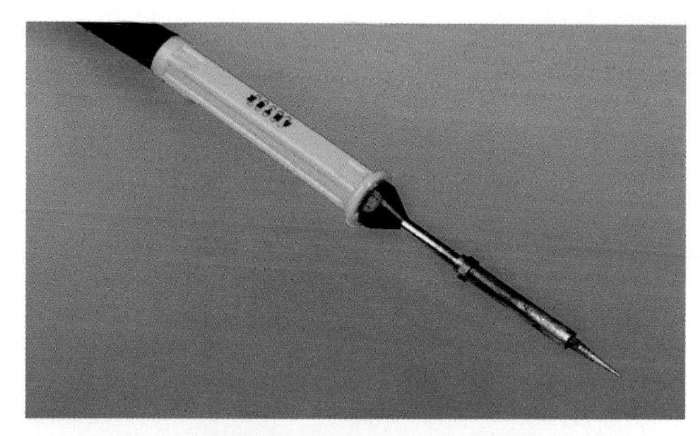

Figure 3-3 *A low-wattage soldering iron designed for precise electronics work.*

Essential: General Duty Soldering Iron

The limited heat capacity of a 15W soldering iron will be insufficient if you need to connect thicker wires, especially to components such as heavy-duty switches with terminals designed to handle substantial current. The terminals may sink heat so rapidly, the low-wattage iron will be unable to create a temperature high enough to melt solder. You may run into a similar situation when you try to attach a wire to a solder tab on a full-size potentiometer.

For these situations, you will need a soldering iron rated for 30 to 40 watts. While it is not required for most of the projects in this book, I do recommend it when you are creating your first solder joints, because the greater heat capacity will make the joints easier. A 30W iron is usually cheaper than a 15W iron, and represents a relatively small additional expense. I think a chisel-shaped tip enables better heat transfer, and since you will not be using this iron for delicate work, you don't really need a pointed tip.

Soldering Iron Terminology

Some soldering irons have a *desoldering pump* built in, to help you undo a solder joint. This is a plunger which you pull with your fingers to suck a little air in through the tip of the iron. I don't think it works very well. In any case, I have only seen it on 30W irons, which are too powerful to be used in many electronics applications.

A soldering iron may have the term "welding" in its product description. You can ignore the inaccurate use of this term, because soldering irons do not do welding in the usual sense.

A few soldering irons are packaged with a *Helping Hand* device that can hold small parts while you are working on them. This combination is worth considering, as it should cost less than if you buy the items separately. The Helping Hand device is described below.

If a soldering iron is sold with some solder included, don't use it unless it is described as electrical solder that has a *rosin core*.

Many soldering irons are described as *pencil type*. The term is not very informative, because it can be applied either to a 15W or a 30W iron.

However, a pencil-type soldering iron is different in appearance from a *pistol-grip* soldering iron, such as the Weller Therma-Boost, shown in Figure 3-4. Some people prefer the ergonomics of this kind of grip, and the Therma-Boost has a nice quick-start feature that enables it to reach working temperature in less than a minute, making it ideal for someone who tends to be impatient. However, pistol-grip irons are all rated at 30W or higher, and tend to be more expensive than the ordinary pencil-type.

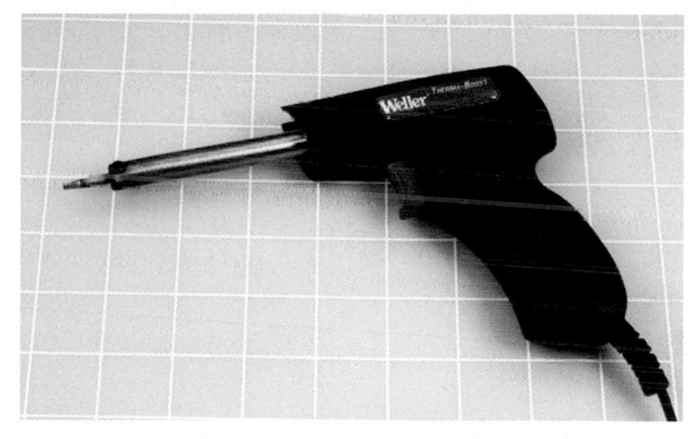

Figure 3-4 *A Weller Therma-Boost 30W soldering iron can be useful when working with heavier wire and larger components.*

Essential: Helping Hand

The so-called *Helping Hand* (sometimes known as a *third hand*) has two alligator clips that can hold components or pieces of wire precisely in position while you join them with solder. Some versions of the Helping Hand also feature a magnifying lens, a wire spiral in which you can rest your soldering iron, and a little sponge that you may use to clean the tip of your iron when it becomes dirty. These additional features are nice but not essential. See Figure 3-5.

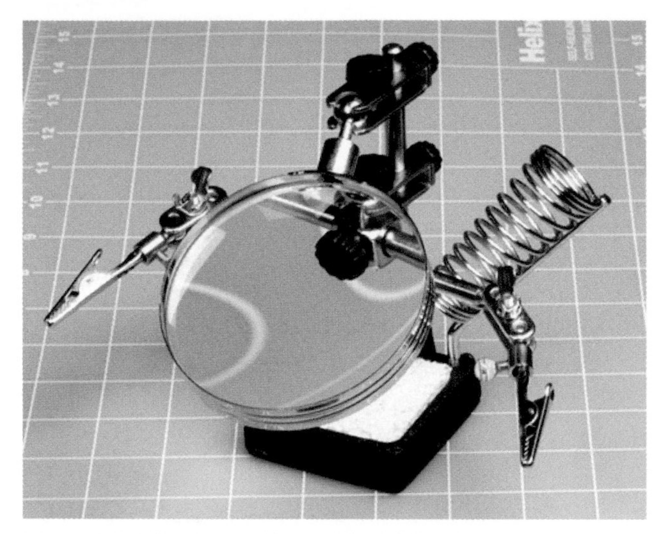

Figure 3-5 *The Helping Hand, with additional accessories.*

Essential: Magnifying Lens

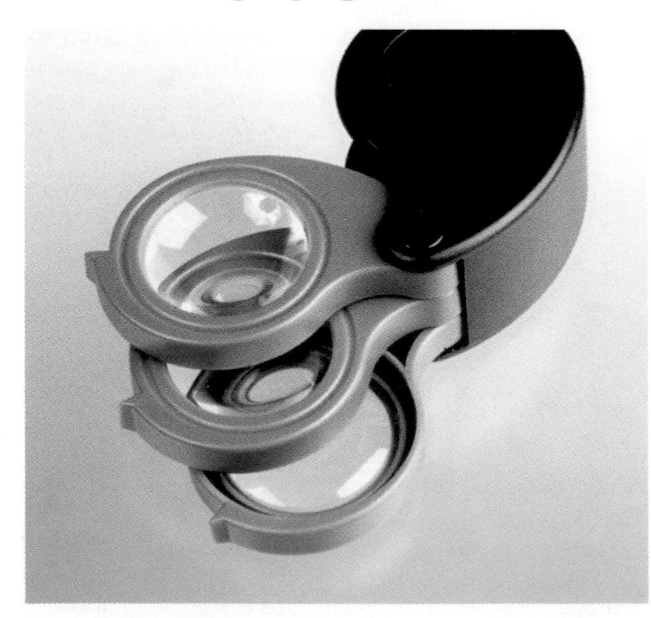

Figure 3-6 *Handheld magnification is essential for inspecting solder joints.*

No matter how good your eyes are, a small, handheld, powerful magnifying lens is essential when you are checking solder joints on perforated board. The three-lens set in Figure 3-6 is designed to be held close to your eye, and is more powerful than the large lens on a Helping Hand, which I do not find very useful. The folding lens in Figure 3-7 stands on your workbench for hands-free operation. Both are available from hobby stores or sources such as eBay and Amazon. Plastic lenses are quite acceptable if you are gentle with them.

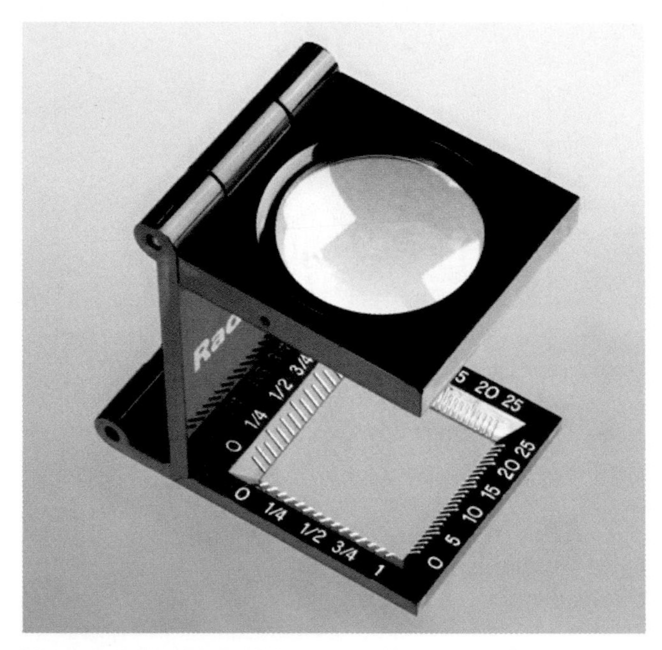

Figure 3-7 *This kind of folding magnifier can stand on your desktop.*

Optional: Clip-on Meter Test Leads

In previous experiments I have suggested that you can grab one of your meter probes with the alligator clip on a test lead, and use the alligator at the other end of the lead to grip a wire or a component.

A more elegant alternative is to buy a pair of "minigrabber" probes with little spring-loaded clips at the end. The Pomona model 6244-48-0 (shown in Figure 3-8) will do the job. However, this is a relatively expensive choice. You may prefer to look for meter leads that terminate in small alligator clips, such as those in Figure 3-9. These are usually the cheapest option. Or, you can just continue to use test leads in the manner that I suggested previously.

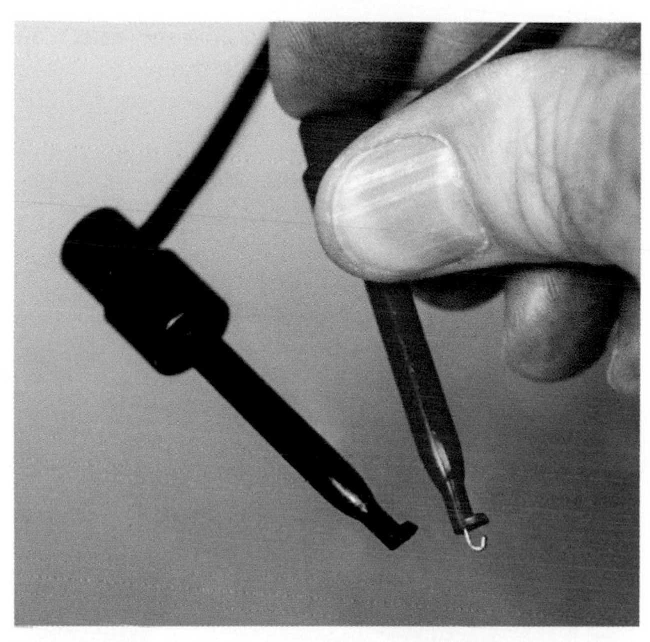

Figure 3-8 *These "minigrabbers" on meter leads will grip wires or component leads.*

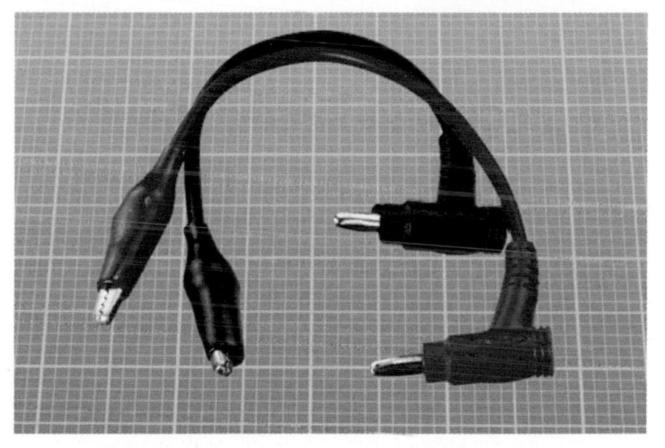

Figure 3-9 *Meter leads that terminate in miniature alligator clips.*

Optional: Heat Gun

If you join two wires with solder, you often need to insulate them. You can use electrical tape (sometimes called insulating tape), but it tends to come unstuck. A better option is heat-shrink tubing, which forms a safe, permanent sheath around a bare-metal joint. To make the tubing shrink, you use a heat gun, which is like a very powerful hair dryer. See Figure 3-10. They're available from any hardware supply source, and I suggest you buy the cheapest one you can find.

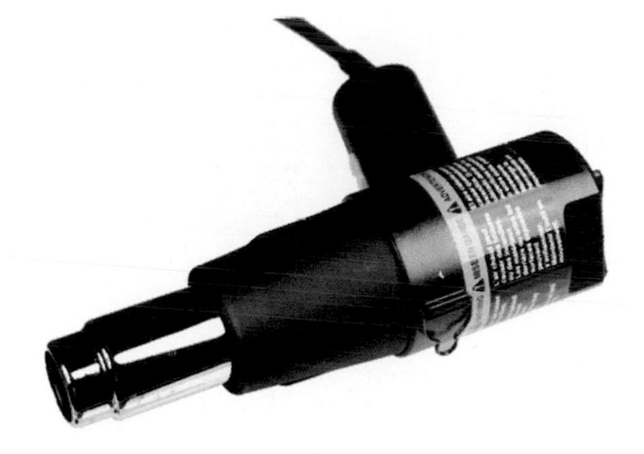

Figure 3-10 *When a heat gun is applied to heat-shrink tubing, the tubing can create a snug, insulated sheath around bare wire.*

For precise work, you may prefer to use a miniature heat gun, such as the one shown in Figure 3-11.

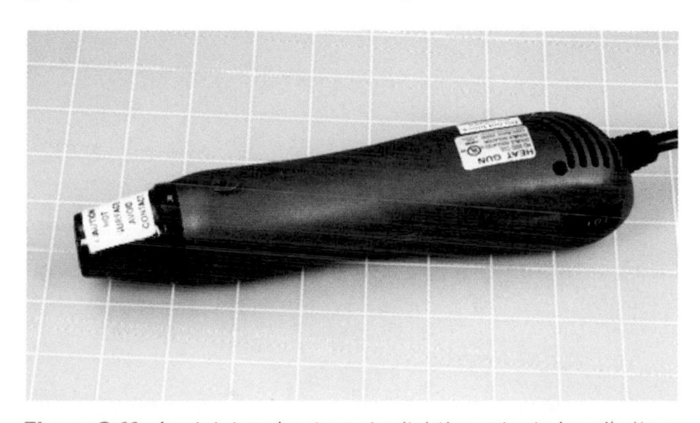

Figure 3-11 *A miniature heat gun is slightly easier to handle than the full-sized type.*

Optional: Desoldering Equipment

A *desoldering pump* is supposed to suck up hot, melted solder when you are trying to remove a solder joint that you made in the wrong place. See Figure 3-12. Some of my readers insist that this should be essential, not optional, but it's a matter of taste. Personally if I make a soldering error, I prefer to snip it out and do it over.

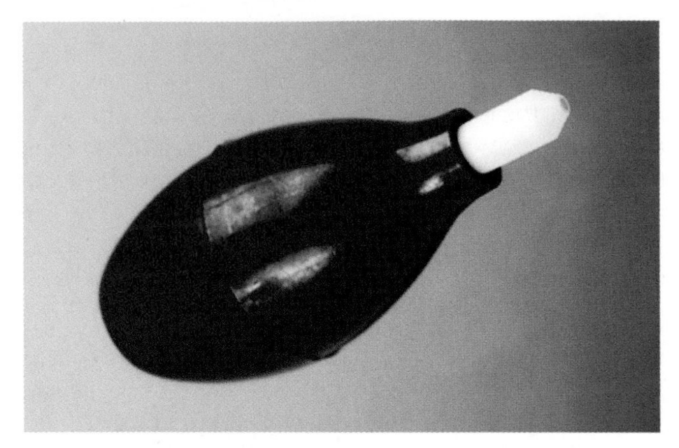

Figure 3-12 *To remove a solder joint, you can suck melted solder into this squeezable rubber bulb.*

Desoldering wick, also known as *desoldering braid*, is intended to soak up solder in conjunction with the solder pump. See Figure 3-13.

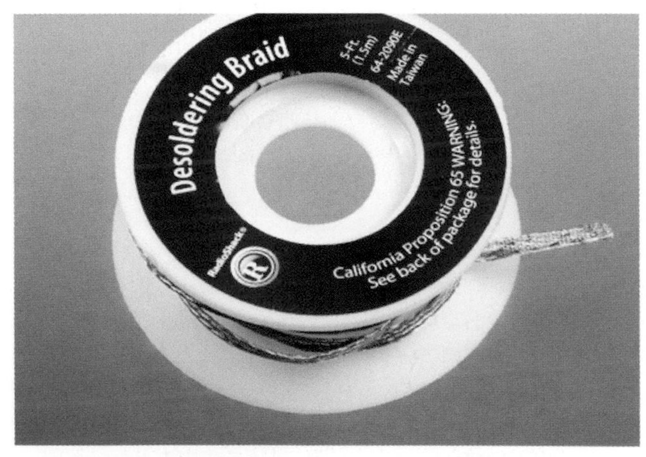

Figure 3-13 *An additional option for removing liquid solder is to soak it up in this copper braid.*

Optional: Soldering Stand

You place your hot soldering iron into a stand when you are not using it, in the same way that you would put a kitchen knife into a rack. See Figure 3-14. If you don't want to spend the money on a stand, you can improvise a substitute, such as a piece of steel electrical conduit or even an old tin can nailed to a piece of wood. Or, you can rest the soldering iron on the edge of your work bench, and promise yourself to be veeery careful not to dislodge it. (Been there, done that.) When—not if—the soldering iron falls onto the floor, it will melt synthetic carpet or plastic floor tiles. Knowing this, you may attempt to catch it when you see it fall. If you grab it by the hot end, you will let go of it, so you might as well let it fall on the floor without the intermediate step of burning yourself.

Perhaps a soldering stand should be considered essential.

Figure 3-14 *A safe and simple stand for a hot soldering iron. The yellow sponge can be wetted with water to clean the tip of the iron.*

Optional: Miniature Hand Saw

Sooner or later, you may want to mount a finished electronics project in a decent-looking enclosure. For this purpose, you are likely to need tools to cut, shape, and trim thin plastic. For example, you may want to cut a square hole so that you can mount a square power switch in it.

Power tools are overkill for this kind of delicate work. A miniature handsaw (a.k.a. a "hobby saw") is ideal for trimming things to fit. X-Acto makes a range of tiny saw blades. See Figure 3-15.

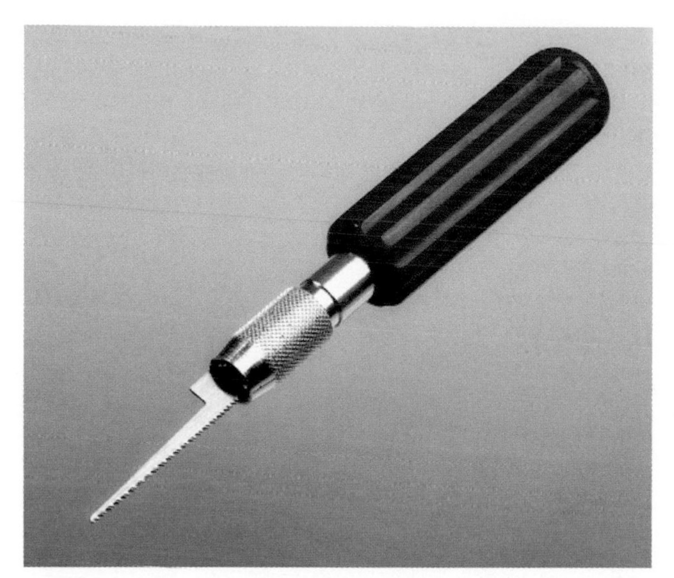

Figure 3-15 *Useful for cutting small holes to mount components in plastic boxes.*

Optional: Deburring Tool

A deburring tool instantly smooths and bevels any rough-sawn plastic or aluminum edge, and also can enlarge holes slightly. This may be necessary because some components are manufactured to metric sizes, which don't fit in the holes that you drill with American bits. See Figure 3-16.

Figure 3-16 *A deburring tool.*

Optional: Calipers

These may seem like a luxury, but are very useful for measuring the external diameter of a round object (such as the screw thread on a switch or a potentiometer) or the internal diameter of a hole (into which you may want a switch or potentiometer to fit). See Figure 3-17. If you choose one that has a digital output powered by a button battery, it will be switchable between metric and inches.

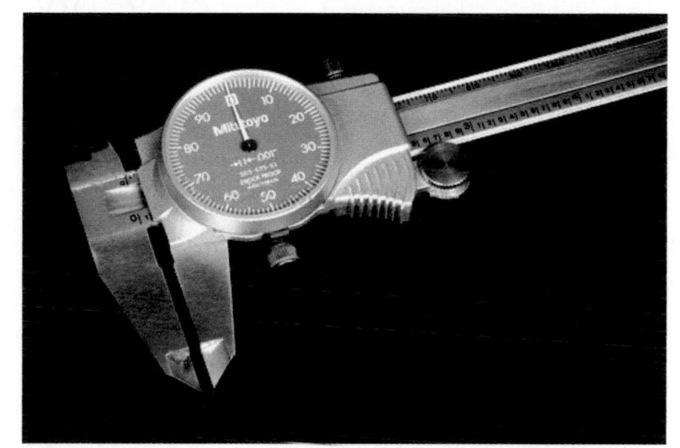

Figure 3-17 *Calipers can measure internal and external diameters.*

Supplies

While many of the tools that I listed are optional, supplies tend to be essential—unless you are absolutely positive that you never want to build a permanent version of a device. The tools and materials to build permanent projects will cost about the same as a month of cable TV. I think they're a worthwhile investment.

Essential: Solder

This is the stuff that you will melt to join components together on a permanent (we hope) basis. It's good to have very thin solder, 0.02" to 0.04" (0.5mm to 1mm) in diameter, for very small components. A range of solder thicknesses is shown in Figure 3-18. For projects in this book, a minimal amount of solder (half an ounce, or maybe three feet) will be sufficient.

Avoid buying solder that is intended for plumbers, or for craft purposes such as creating jewelry. The word "electronics" should appear in the manufacturer's description of suitable purposes for the solder.

Figure 3-18 *Spools of solder in various thicknesses.*

There is some controversy about using solder that has lead in it. An experienced machinist assures me that this older type of solder makes better, easier joints at a lower temperature, and entails minimal health risk if used sparingly. He points out that lead-free solder has its own problems, as it contains more rosin, which creates more fumes. The issue has roused significant debate online, as you can see if you try this search:

```
lead tin solder safety
```

Personally, I lack the specialized knowledge to make a judgment call. I do know that if you live in the European Union, you're not supposed to use solder with lead in it, for environmental reasons.

What you do definitely need is rosin-core solder intended for electronics use. Whether it contains lead is your choice.

Optional: Heat-Shrink Tubing

For use in conjunction with your heat gun, described previously. It's good to have a range of sizes in any colors of your choice. See Figure 3-19. You slide heat-shrink tubing over a solder joint and then apply heat from a heat gun. The tube shrinks around the joint, insulating it. The diameter after shrinking is typically half of the original diameter, but some tubing has a higher shrink ratio. Different materials offer varying characteristics relating to insulation, abrasion resistance, and other factors. An amazing range of heat-shrink tubing will be found at McMaster-Carr (*http://bit.ly/mm-hst*), with details about the various properties. For our purposes, the cheapest tubing should be satisfactory, so long as it is rated for 240 volts (or higher). One bag or box containing an assortment of five or six diameters will be sufficient. You are more likely to use the small sizes than the large sizes.

Figure 3-19 *A variety of heat-shrink tubing.*

Essential: Copper Alligator Clips

These absorb heat when you are soldering delicate components. Don't be deceived by steel clips that are copper-plated; you should really get the solid copper type. Buy as few as possible, as you can reuse them indefinitely. Just two will be enough.

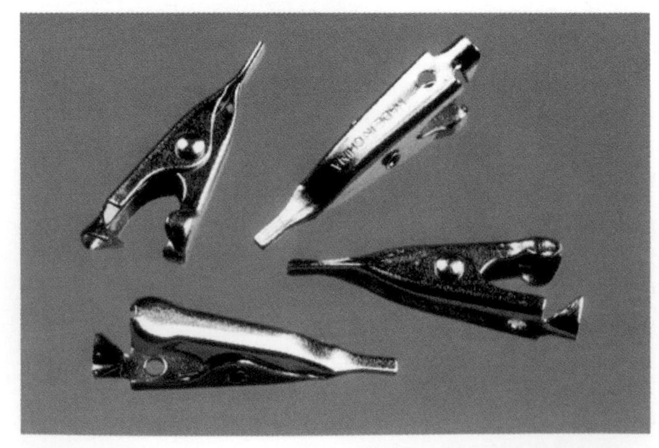

Figure 3-20 *These small copper clips absorb heat to protect components when you're soldering them.*

Optional: Perforated Board

When you're ready to move your circuit from a breadboard to a more permanent location, you'll want to solder it to a piece of perforated board, often known as "prototyping board" and also called "perf board."

The easiest type to use is plated with copper strips on the back in exactly the same layout as the conductors hidden inside a breadboard. This enables you to minimize errors by keeping the same layout of your components when they migrate to the perforated board. See Figure 3-21. Just buy one board, initially.

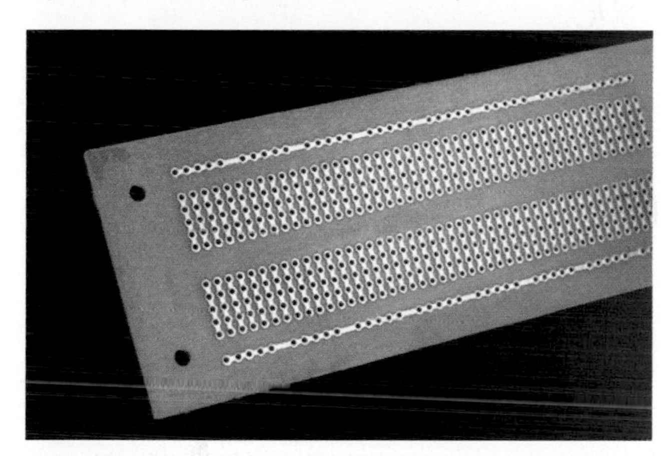

Figure 3-21 *This perforated board has a pattern of copper traces identical to the pattern of conductors inside a breadboard.*

The disadvantage of using the breadboard component layout is that it is not very space-efficient. To compress a circuit to minimal size, you can try point-to-point wiring on a plain perforated board—and I'll show you how this is done in Experiment 14. You may only require a small piece of board, but you can buy a larger piece and cut as much as you need. See Figure 3-22.

Another possibility is to use perforated board with a different pattern of copper traces. *Cut-board*, for instance, features parallel traces that you can cut with a knife where you want to break a connection. Everyone who does much soldering seems to have a favorite type of board configuration, but I think you'll need to get acquainted with the soldering process before you start exploring the options.

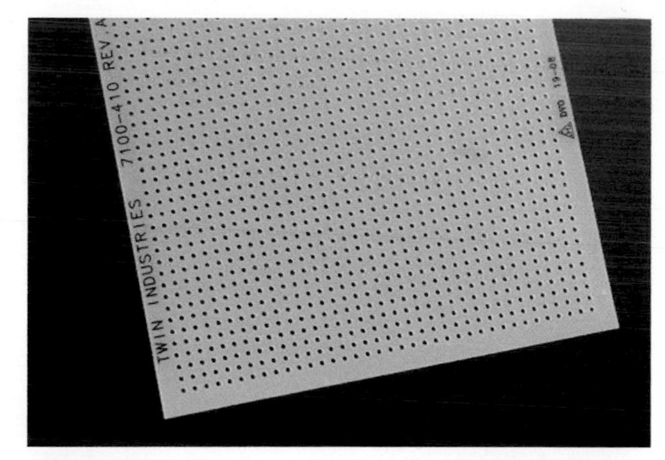

Figure 3-22 *Plain perforated board (with no copper traces) for point-to-point wiring.*

Optional: Plywood

When you use a soldering iron, hot drops of solder tend to fall onto your table or workbench. The solder solidifies almost instantly, can be difficult to remove, and will leave a scar. Consider using a two-foot square of half-inch plywood to provide disposable protection. You can buy it precut at any big-box hardware store.

Optional: Machine Screws

To mount components behind a panel, you need small *machine screws* (more commonly referred to as "bolts"). They look nice if they have flat heads that fit flush against the panel, provided you coutersink the holes. I suggest stainless-steel machine screws, #4 size, in lengths of 3/8" and 1/2", with #4 locknuts of the type that have nylon inserts, so that they won't work loose.

Essential: Project Boxes

A project box is just a small box (usually plastic) with a removable lid. As its name implies, its purpose is to hold one of your electronic projects. You mount your switches, potentiometers, and LEDs in holes that you drill in the lid, and you attach your circuit on a perforated board that goes inside the box. You can also use a project box to contain a small loudspeaker.

For the intrusion alarm project in Experiment 15, you can use a box measuring approximately 6 inches long, 3 inches wide, and 2 inches high.

Essential: Power Connectors

After you finish a project and put it in a box, you'll need a convenient way to supply it with power. You can buy the kind of low-voltage DC plug-and-socket pair pictured in Figure 3-23. They are properly known as "barrel plugs" and "barrel sockets" but you may also find them described as "6VDC plug" and "6VDC socket." They are available in various sizes, but this is unimportant, so long as your connectors are the same size as each other.

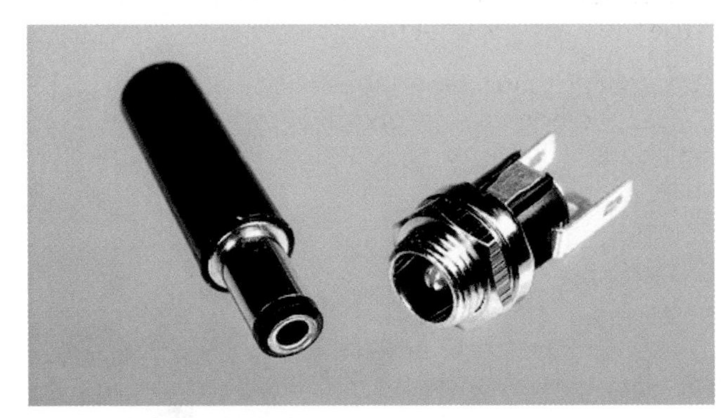

Figure 3-23 *The socket on the right can be mounted on a project box to receive power from the plug on the left, which you can attach to the wire from an AC adapter.*

Optional: Headers

After you have soldered a circuit to perforated board, you will need to connect it with separate switches or pushbuttons. An unpluggable connection is preferable, in case you ever need to fix a fault on the board.

Sometimes known as *single inline sockets and headers*, but also known as *boardmount sockets and pinstrip headers*, small connectors come in strips of 36 or more, and you can snap off as many as you need.

Figure 3-24 shows headers before and after being snapped into small sections. Make sure that the interconnects have a terminal spacing of 0.1 inch, to match your perforated board. (Some interconnects have metric spacing.)

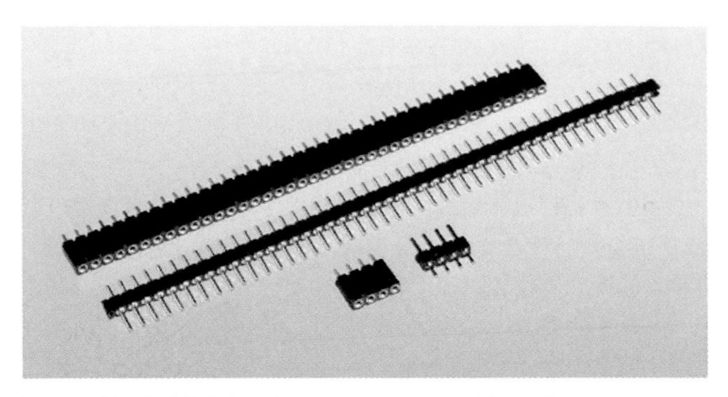

Figure 3-24 *Miniature interconnects are often referred to as "headers."*

Components

Again I must remind you that component kits are available. See "Kits" on page 311. If you prefer to buy your own components from online sources, see "Components" on page 317. In addition to the components that I described at the beginning of Chapter Two (see "Components" on page 47), you're going to need the following.

Diodes

A diode passes electric current in one direction, while blocking it in the opposite direction. The end of the diode that should be more negative is known as the *cathode*. It is marked with a line, as shown in Figure 3-25. The diode on the right, in this photograph, is a 1N4001, which is rated for slightly more current than the 1N4148 on the left. They're cheap, and likely to be useful in the future, so buy 10 of each. The manufacturer is irrelevant.

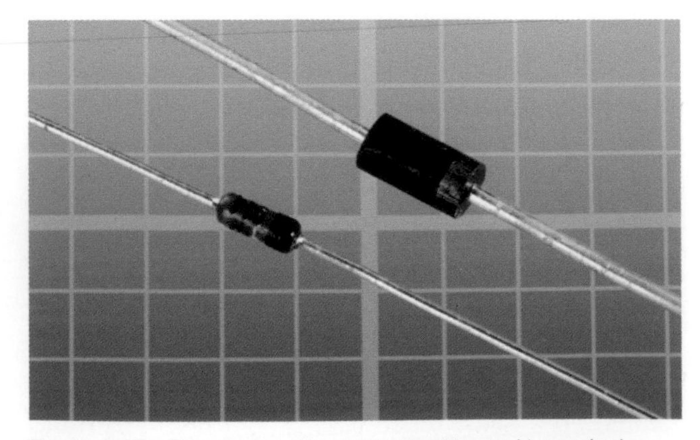

Figure 3-25 *Two diodes. The more-negative end is marked.*

Experiment 12: Joining Two Wires Together

Now you're ready to begin using everything that I have just described, and the first item has to be a soldering iron.

Your adventure into soldering begins with the prosaic task of joining one wire to another, but will lead quickly to creating a full electronic circuit on perforated board. So let's get started!

What You Will Need

- Hookup wire, wire cutters, wire strippers
- 30-watt or 40-watt soldering iron
- 15-watt soldering iron
- Thin solder
- Optional: medium solder
- "Helping Hand" to hold your work
- Optional: shrink-wrap tubing, assorted
- Optional: heat gun
- Optional: a piece of heavy cardboard or plywood to protect your work area from drops of solder

Caution: Soldering Irons Do Get Hot!

Please take these basic precautions:

Use a proper stand (such as the one incorporated in a Helping Hand) to hold your soldering iron. Don't leave the iron lying on a workbench.

If you have infants or pets, remember that they may play with, grab, or snag the wire to your soldering iron. They could injure themselves (or you).

Be careful never to rest the hot tip of the iron on the power cord that supplies electricity to the iron. It can melt the plastic in seconds and cause a dramatic short circuit.

If you drop a soldering iron, don't be a hero and try to catch it.

Most soldering irons have no warning lights to tell you that they're plugged in. As a general rule, always assume that a soldering iron is hot, even if it's unplugged. It may retain sufficient heat to burn you for longer than you expect.

Your First Solder Joint

We'll start with your general-duty soldering iron—the one rated for 30 or 40 watts. Plug it in, leave it safely in its holder, and find something else to do for five minutes. If you try to use a soldering iron without giving it time to get fully hot, you will not make good joints, because the solder may not melt completely.

Strip the insulation from the ends of two pieces of 22-gauge solid hookup wire and clamp them in your Helping Hand so that they cross each other and touch each other, as shown in Figure 3-26.

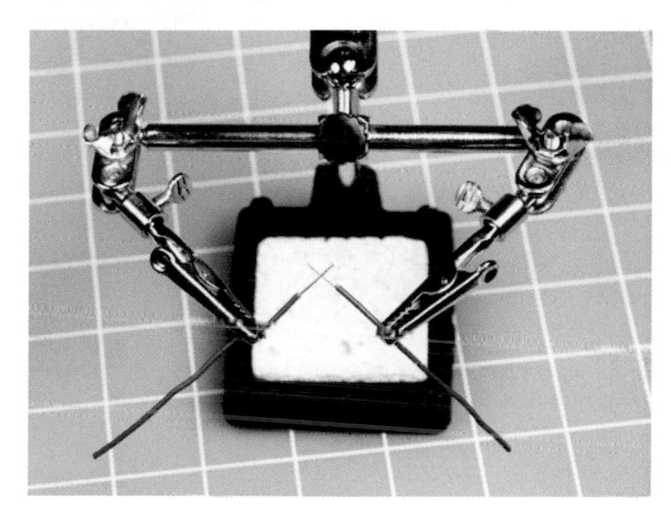

Figure 3-26 *Ready for your first soldering adventure.*

To make sure that the iron is ready, touch a thin piece of solder on the tip of it. The solder should melt instantly. If it melts slowly, the iron isn't hot enough yet.

If the tip of your soldering iron is dirty, you should clean it. The usual procedure is to wet the sponge that is built into your soldering iron stand, and rub the tip of the iron in the sponge. Personally I prefer not to do this, as I believe that getting moisture on the tip causes thermal expansion and contraction, which may open small cracks in the plating of the tip. I rub a piece of crumpled paper over the hot tip of the soldering iron, quickly enough to avoid charring the paper. Then I apply a tiny amount of solder and rub the tip again, until it is uniformly shiny.

When the tip of your soldering iron is clean. follow the steps shown in Figure 3-27, Figure 3-28, Figure 3-29, Figure 3-30, and Figure 3-31.

Step 1. Touch the hot tip of the iron against the intersection of the wires steadily for three seconds to heat them.

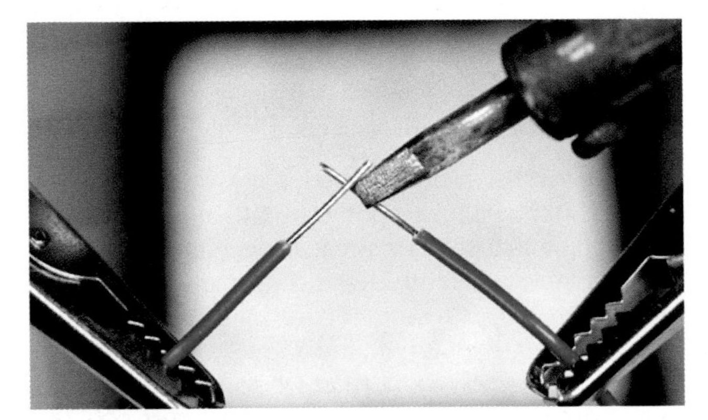

Figure 3-27 *Step 1.*

Step 2. While maintaining the iron in this position, feed a little solder to the intersection of the wires, also touching the tip of the soldering iron. Thus, the two wires, the solder, and the tip of the iron should all come together at one point.

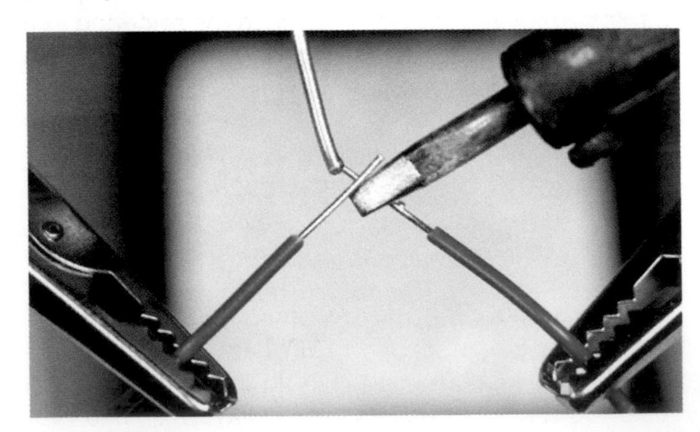

Figure 3-28 *Step 2.*

Step 3. At first, the solder may melt slowly. Be patient.

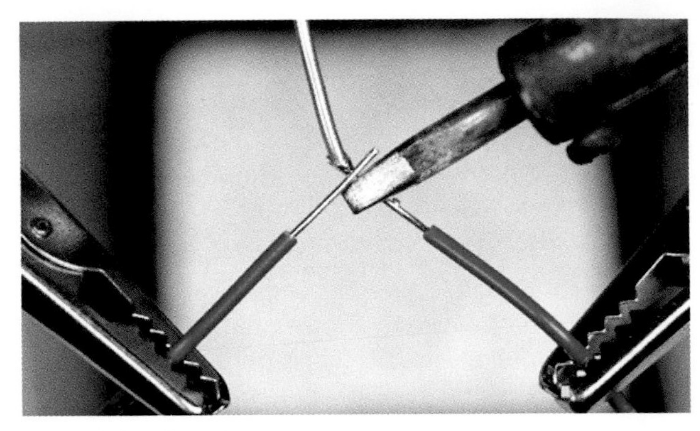

Figure 3-29 *Step 3.*

Step 4. Now you see the solder forming a nice round blob.

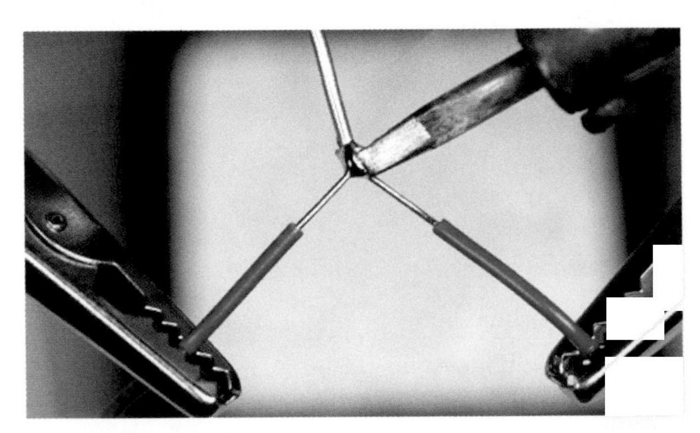

Figure 3-30 *Step 4.*

Step 5. Remove the iron and the solder. Blow on the joint to cool it. Within 10 seconds, it should be cool enough to touch. The completed joint should be shiny, uniform, and rounded in shape

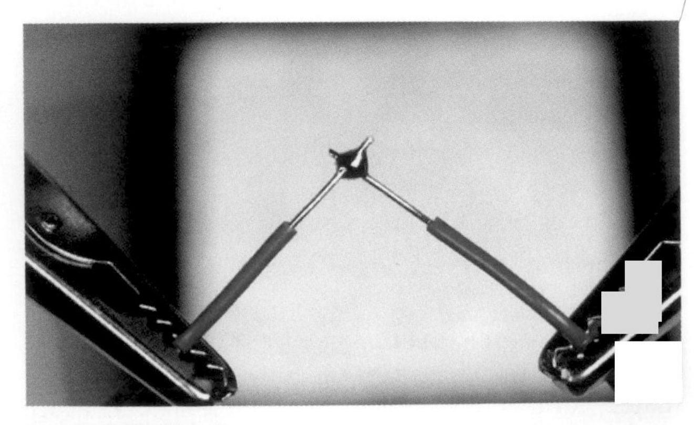

Figure 3-31 *Step 5.*

When the joint has cooled, unclamp the wires and try to tug them apart. Tug hard! If they defeat your best attempts to separate them, the wires are electrically joined and should stay joined. If you didn't make a good joint, you will be able to separate the wires relatively easily, probably because you didn't apply enough heat or enough solder. Figure 3-32 gives you the general idea.

Figure 3-32 *Telling the difference between a bad solder joint (left) and a good one (right) is really not very difficult.*

The reason I asked you to begin by using the higher-powered soldering iron is that it delivers more heat, which makes it easier to use.

The soldering steps can be summarized like this: apply heat to the wires, bring in the solder while maintaining the heat, wait for the solder to start to melt, wait a moment longer for it to form a completely molten bead, and then remove the heat. The whole process should take between four and six seconds.

Background: Soldering Myths

Myth #1: Soldering is very difficult. Millions of people have learned how to do it, and they can't all be more competent than you. I have a lifelong problem with a tremor in my hands that makes it difficult for me to hold small things steadily. I also get impatient with repetitive detail work. If I can solder components, almost anyone should be able to do it.

Myth #2: Soldering endangers your health with poisonous chemicals. You should avoid inhaling the fumes, but that also applies to everyday products such as bleach and paint. You should wash your hands after handling solder, and to do a really thorough job, use a nail brush. But if soldering was a significant health hazard, we should have seen a high death rate among electronics hobbyists decades ago.

Myth #3: Soldering irons are dangerous. A soldering iron is less hazardous than the kind of iron that you might use to iron a shirt, because it delivers less heat. In fact, in my experience, soldering is safer than most activities in a typical home or basement workshop. Of course, that doesn't mean you can be careless. The iron is hot enough to burn your skin if you touch it.

Fundamentals: Eight Soldering Errors

Not enough heat. The joint looks OK, but because you didn't apply quite enough heat, the solder didn't melt sufficiently to realign its internal molecular structure. It remained granular instead of becoming a solid, uniform blob, and you end up with a *dry joint*, also known as a *cold joint*, which will come apart when you pull the wires away from each other. Reheat the joint thoroughly and apply new solder.

Carrying solder to the joint. A leading cause of underheated solder is the temptation to melt some solder onto the iron alone, and then carry the solder to the location where you want to apply it. This means that the wires will be cold when you try to make the solder stick to them. What you should do is touch the soldering iron to heat the wires first, and then apply the solder. This way, the wires are hot and help to melt the solder.

- Because this is such a universal problem, I'll repeat myself. You don't want to put hot solder on cold wires. You want to put cold solder on hot wires.

Too much heat. This may not hurt the joint, but can damage everything around it. Vinyl insulation will melt, exposing too much wire and raising the risk of short circuits. You can easily damage semiconductors, and may even melt the internal plastic parts of switches and connectors. Damaged components must be desoldered and replaced, which will take time and is a big hassle. If your attempt at soldering isn't working for some reason, pull back, pause, and allow everything to cool down a little before you try again.

Not enough solder. A thin connection between two conductors may not be strong enough. When joining two wires, always check the underside of the joint to see whether the solder penetrated completely.

Moving the joint before the solder solidifies. This may create a fracture that you won't necessarily see. It may not stop your circuit from working, but at some point in the future, as a result of vibration or thermal stresses, the fracture can separate just enough to break electrical contact. Tracking it down will then be a chore. If you clamp components before you join them, or use perforated board to hold the components steady, you can avoid this problem.

Dirt or grease. Electrical solder contains rosin that cleans the metal that you're working with, but contaminants can still prevent solder from sticking. If any component looks dirty, clean it with fine sandpaper before joining it.

Carbon on the tip of your soldering iron. The iron gradually accumulates flecks of black carbon during use, and they can act as a barrier to heat transfer. Clean the tip, as described previously.

Inappropriate materials. Electrical solder is designed for electronic components. It will not work with aluminum, stainless steel, or various other metals. You may be able to make it stick to chrome-plated items, but only with difficulty.

Failure to test the joint. Don't just assume that it's OK. Always test it by applying manual force. If you can't get a grip on the joint, slip a screwdriver blade under it and flex it just a little, or use small pliers to try to pull it apart. Don't be concerned about ruining your work. If your joint doesn't survive rough treatment, it wasn't a good joint.

Of the eight errors, dry/cold joints are by far the worst, because they are easy to make and can look OK.

Background: Soldering Alternatives

As recently as the 1950s, connections inside electronic appliances such as radio sets were still being hand-soldered by workers on production lines. But the growth of telephone exchanges created a need for a faster way to make large numbers of rapid, reliable point-to-point wiring connections, and *wire wrap* became a viable alternative.

In a wire-wrapped electronics project, components are mounted on a circuit board that has long, gold-plated, sharp-cornered square pins sticking out of the rear. Special silver-plated wire is used, with an inch of insulation stripped from its ends. A manual or power-driven wire-wrap tool twirls the end of a wire around one of the pins, applying sufficient tension to "cold-weld" the soft silver plating of the wire to the pin. The wrapping process exerts sufficient pressure to make a very reliable joint, especially as seven to nine turns of wire are applied, each turn touching all four corners of the pin.

During the 1970s and 1980s, this system was adopted by hobbyists who built their own home computers. A wire-wrapped circuit board from a hand-built computer is shown in Figure 3-33. The technique was used by NASA to wire the computer in the Apollo spacecraft that went to the Moon, but today, wire-wrapping has few commercial applications.

Figure 3-33 *This picture shows some of the wire-wrapping in Steve Chamberlin's custom-built, retro 8-bit CPU and computer. Connecting such a network of wires with hand soldering would have been unduly time-consuming and prone to faults. Photo credit: Steve Chamberlin.*

The widespread industrial use of "through-hole" components, such as the chips on early desktop computers, encouraged development of wave soldering, in which a wave or waterfall of molten solder is applied mechanically to the underside of a preheated circuit board where chips have been inserted. A masking technique prevents the solder from sticking where it isn't wanted.

Today, surface-mount components (which are significantly smaller than their through-hole counterparts) are glued to a circuit board with a solder paste, and the entire assembly is then heated, melting the paste to create a permanent connection.

Your Second Solder Joint

Time now to try your 15W soldering iron. Once again, you must leave it plugged in for a good five minutes to make sure it's hot enough. In the meantime, don't forget to unplug your other soldering iron, and put it somewhere safe while it cools.

Use thin solder to make this joint. It will take less heat from the less-powerful soldering iron.

This time I'd like you to align the wires parallel with each other. Joining them this way is a little more difficult than joining them when they cross each other, but it's a necessary skill. Otherwise, you won't be able to slide heat-shrink tubing over the finished joint to insulate it.

Five steps to create this joint are shown in Figure 3-34, Figure 3-35, Figure 3-36, Figure 3-37, and Figure 3-38. At the beginning, the two wires do not have to make perfect contact with each other; the solder will fill any small gaps. But as before, the wires must be hot enough for the solder to flow, and this can take an extra few seconds when you use the low-wattage iron.

Be sure to feed the solder in as shown in the pictures. Remember: don't try to carry the solder to the joint on the tip of the iron. Heat the wires first, and then touch the solder to the wires and the tip of the iron, while keeping it in contact with the wires. Wait until the solder liquifies, and you will see it running eagerly into the joint. If this doesn't happen, be more patient and apply the heat for a little longer.

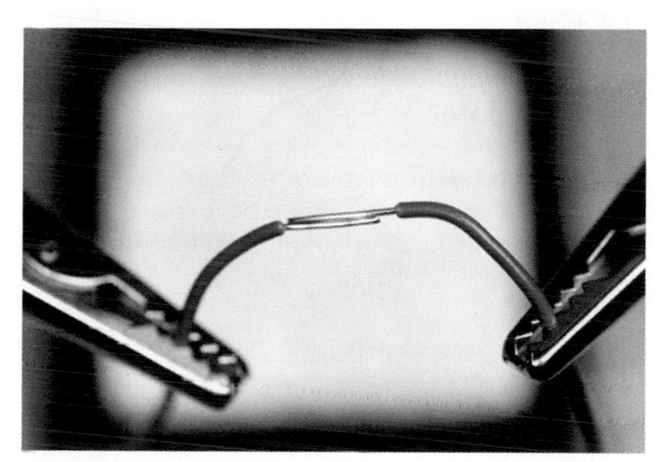

Figure 3-34 *Step 1: Align the wires.*

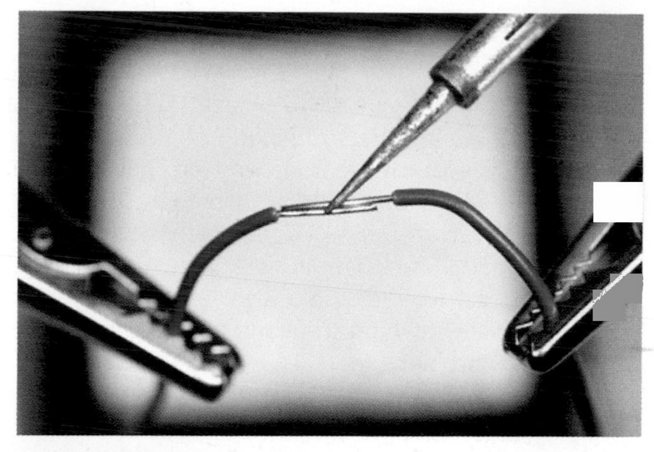

Figure 3-35 *Step 2: Heat the wires.*

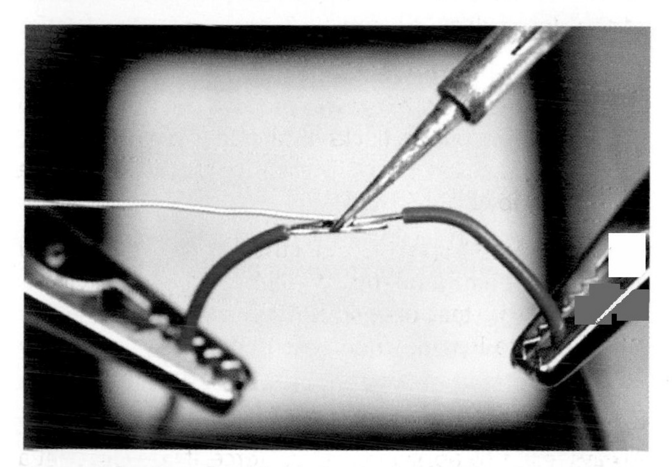

Figure 3-36 *Step 3: While continuing to heat the wires, apply the solder. Wait for it to start melting. Be patient.*

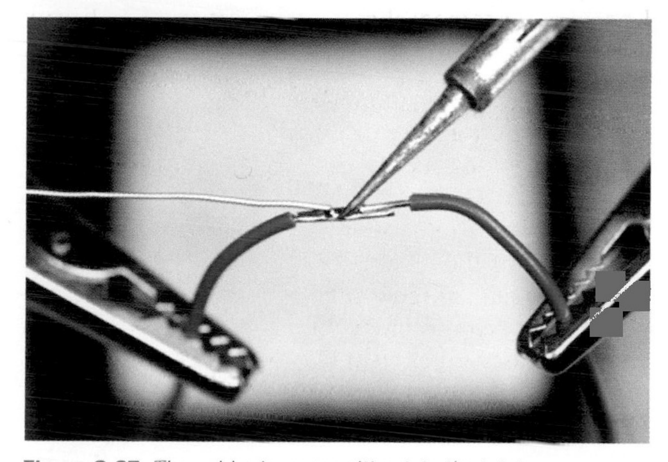

Figure 3-37 *The solder is now melting into the joint.*

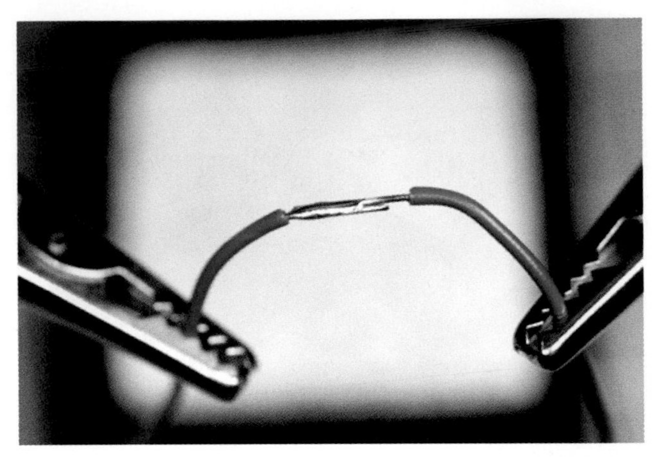

Figure 3-38 *The finished joint is shiny, and the solder has spread across the copper wires.*

The finished joint has enough solder for strength, but not so much solder that it will prevent heat-shrink tubing from sliding over it. I'll get to that in a moment.

Theory: Heat Transfer

The better you understand the process of soldering, the easier it should be for you to make good solder joints.

The tip of the soldering iron is hot, and you want to transfer that heat into the joint that you are trying to make. For this reason, you should adjust the angle of the soldering iron so that it makes the widest possible contact. See Figure 3-39 and Figure 3-40.

Figure 3-39 *A small contact area between the iron and the working surface allows insufficient heat transfer.*

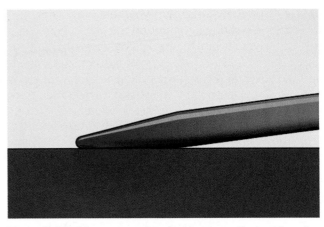

Figure 3-40 *A larger area of contact increases the heat transfer.*

Once the solder starts to melt, it broadens the area of contact, which helps to transfer more heat, so the process accelerates naturally. Initiating it is the tricky part.

The other aspect of heat flow that you should consider is that it can suck heat away from the places where you want it, and deliver it to places where you don't want it. If you're trying to solder a very heavy piece of copper wire, the joint may never get hot enough to melt the solder, because the heavy wire conducts heat away from the joint. You may find that even a 40-watt iron isn't powerful enough to overcome this problem. At the same time, while the copper doesn't get hot enough to melt the solder, it can be quite hot enough to melt insulation off the wire.

As a general rule, if you can't complete a solder joint in 10 seconds, you aren't applying enough heat.

Insulating the Solder Joint

After you've succeeded in making a good inline solder connection between two wires, it's time for the easy part. Choose some heat-shrink tubing that is just big enough to slide over the joint with a little bit of room to spare.

Of course, you do need to plan ahead. Usually you need to slip some tubing over one of the wires *before* you join them together. You'll see how this works when I get to the step-by-step procedure, below.

Assuming you have heat-shrink tubing on one of the wires, slide it along until the joint is centered under it. Hold it in front of your heat gun, and switch on the gun (keeping your fingers away from the blast of superheated air). Turn the wire so that you heat both sides. The tubing should shrink tight around the joint within half a minute. If you overheat the tubing, it may shrink so much that it splits, at which point you must remove it and start over. As soon as the tubing is tight, your job is done, and there's no point in making it any hotter. Note that while tubing mostly shrinks at right angles to its length, a little shrinkage also occurs along its length.

Figure 3-41, Figure 3-42, and Figure 3-43 show the desired result. I used white tubing because it shows up well in photographs. Different colors of heat-shrink tubing all perform the same way.

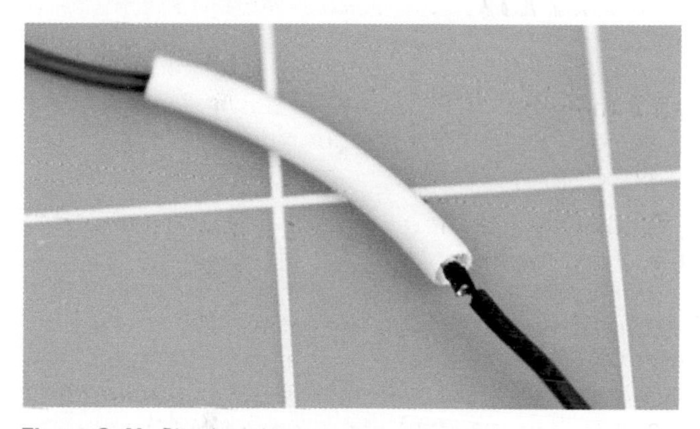

Figure 3-41 *Slip the tubing over your wire joint.*

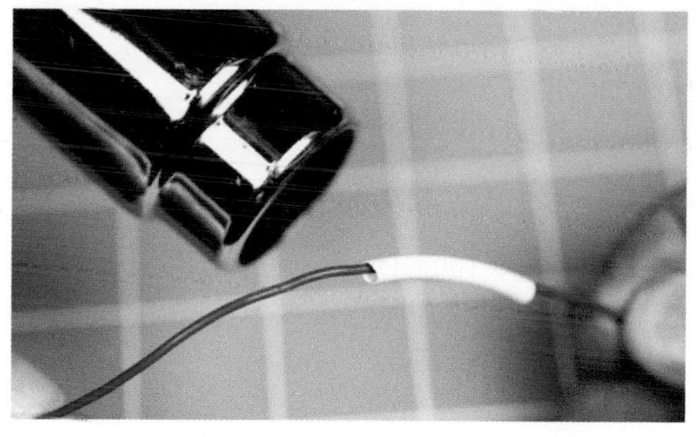

Figure 3-42 *Apply heat to the tubing.*

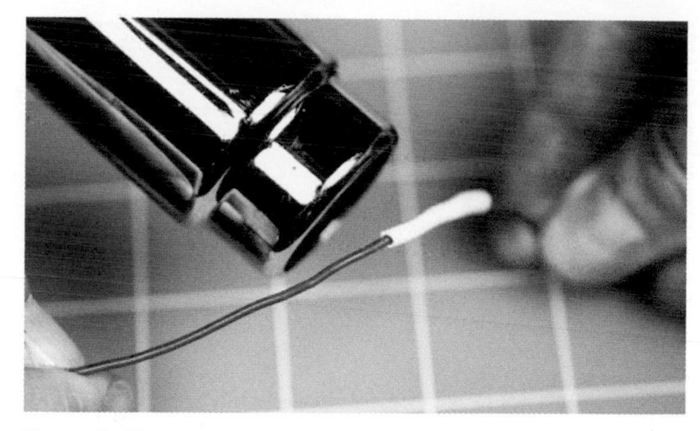

Figure 3-43 *Leave the heat on the tubing until it shrinks to firmly cover the joint.*

Caution: Heat Guns Do Get Hot!

Notice the chromed steel tube at the business end of a full-sized heat gun. Steel costs more than plastic, so the manufacturer must have put it there for a good reason—and the reason is that the air flowing through it becomes so hot that it would melt a plastic tube.

The metal tube stays hot enough to burn you for several minutes after you've used it. And, as in the case of soldering irons, other people (and pets) are vulnerable, because they won't necessarily know that the heat gun is a hazard. Most of all, make sure that no one in your home ever makes the mistake of using a heat gun as a hair dryer (see Figure 3-44).

Figure 3-44 *Members of your family should understand that a heat gun isn't quite the same as a hair dryer.*

This tool is just a little more hazardous than it appears. A mini heat gun may entail slightly less risk, but still should be treated with respect.

Wiring Your Power Source

The next application for your soldering skills will be more practical. You can add color-coded, solid-core wires to your AC adapter—or if you don't have an adapter, you can add extensions to the thin wires from a 9V battery connector. Either way, your extended 22-gauge wires will plug conveniently into a breadboard.

You can use your larger soldering iron for this, as no heat-sensitive components are involved.

If you have acquired an AC adapter, I'm assuming it is a little plastic module that plugs straight into the wall. A pair of wires emerges from it, carrying the low DC voltage that you need, and they terminate in some kind of miniature plug. The plug is suitable for a device such as a media player or a phone that has a matching socket, but for the purposes of this book, the plug is not useful, because you will want to apply the power to a breadboard.

How can this be done? I will show you.

Step One: Cut and Measure

First, let's make sure your AC adapter is doing what it's supposed to do.

Do not plug the adapter into the wall just yet. Begin by cutting the little plug off the end of the low-voltage wire, as shown in Figure 3-45. (You may notice that this photograph is of a RadioShack adapter. Ah, memories.)

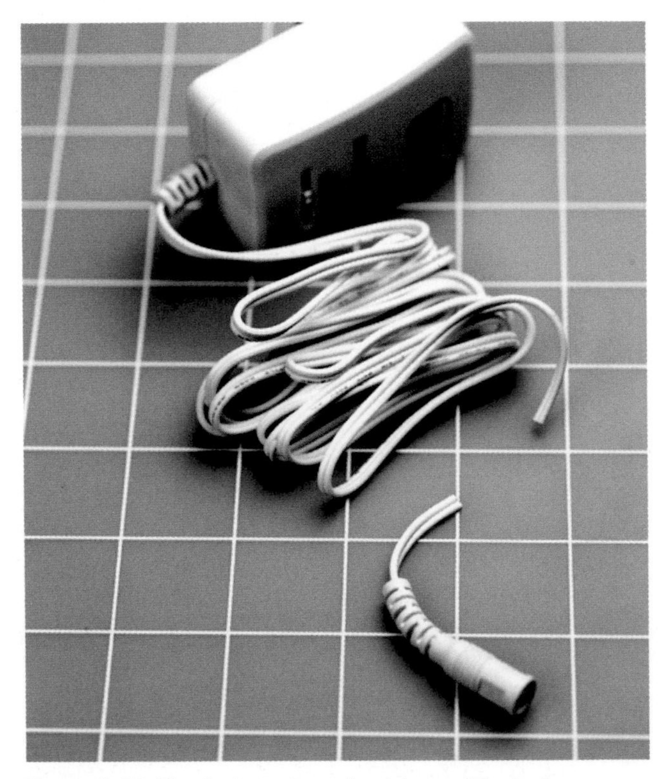

Figure 3-45 *The first step in customizing an AC adapter.*

Separate the two conductors, using wire cutters or a utility knife, and strip about 1/4" of insulation off each one. See Figure 3-46. The wires should be of unequal lengths, to reduce the risk of them touching.

Figure 3-46 *The stripped wires.*

If the bare ends do touch each other while the adapter is plugged in, they may overload it or blow a fuse inside it. You may also get a spark that will be disconcerting, although I doubt it will hurt you. Not a big deal, but inconvenient.

Now set your meter to measure volts DC and attach it to the two wires from the AC adapter, preferably using alligator test leads to keep everything under control. After double-checking to make sure that your red meter lead is in the volts socket on the meter, not the mA socket, plug the AC adapter into the wall and measure how much voltage it is providing.

If you get a confusingly high reading, it could be because the voltage delivered by AC adapters is often greater when the adapter is not powering anything. The internal resistance of your meter is so high, the adapter will behave as if it isn't being loaded at all.

For a more meaningful test, select a resistor with a value of 680 ohms and clip it across the output from the adapter, in parallel with your meter. This will pull the voltage from the adapter down to a more appropriate level. Now you should get a value that makes sense.

It's not a good idea to use a resistor of much less than 680 ohms because the resistors on your shopping list are only rated at 1/4 watt, and if you try to push more power through them than that, they will overheat. If the 680-ohm resistor is attached to a 9-volt supply, Ohm's Law tells us that the current flowing through it will be about 13mA, and therefore the power dissipation will be about 120mW, or 0.12W, which is well within the rated maximum of a 0.25W resistor.

If you want to see how the voltage output from your AC adapter varies with a load of lower resistance, you can clip several 680-ohm resistors together in parallel. This could be an interesting test—but let's get back to the primary purpose here, which is to obtain power for your breadboard.

Step Two: Soldering

Use your meter to make absolutely sure there is no minus sign in front of the voltage that it measures on the wires from your AC adapter. If there is a minus sign, reverse the leads from your meter.

If the reading on your meter is positive, not negative, you know that the red wire from your meter is clipped to the positive side of your AC adapter. This is important, as you don't want your adapter to destroy components in your circuits by applying power to them the wrong way around.

The next steps will be the same regardless of whether you are adding solid 22-gauge wires to an AC adapter or a 9V battery connector.

Cut two pieces of solid-conductor 22-gauge wire—one of them red, the other black or blue. Each should be about 2 inches long. Strip 1/4" of insulation from both ends of each piece of wire.

Solder the 22-gauge wires to the wires from the AC adapter or the battery connector, using the technique that you practiced previously. Naturally, you attach the red wire to the positive side of your power supply.

If you have heat-shrink tubing and a heat gun, use it as you did in the practice session. The result should look like Figure 3-47. Once again the wires should be of unequal lengths, to reduce the risk of them touching each other. When the job is done, you can plug the ends of the 22-gauge wires into your breadboard.

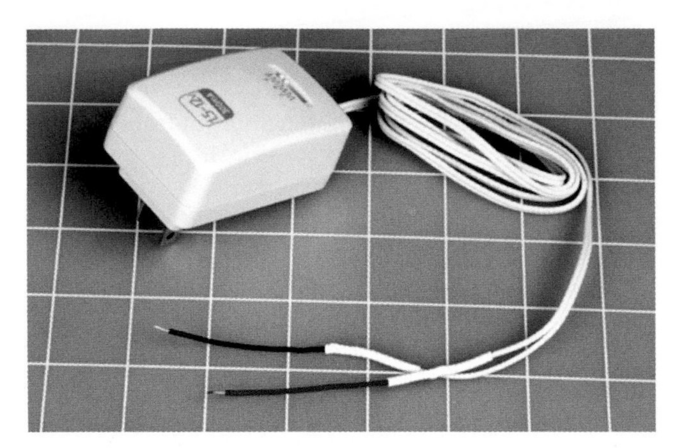

Figure 3-47 *The 22-gauge wires can be plugged into a breadboard, to supply it with power.*

Pruning a Power Cord

What else can you do with your newly acquired soldering skills? Here's a suggestion. Those of us who don't use Apple products may find ourselves with a laptop power supply that has a detachable AC power cord for plugging into the wall, in addition to the low-voltage DC wire that plugs into the computer. A typical power cord is shown in Figure 3-48.

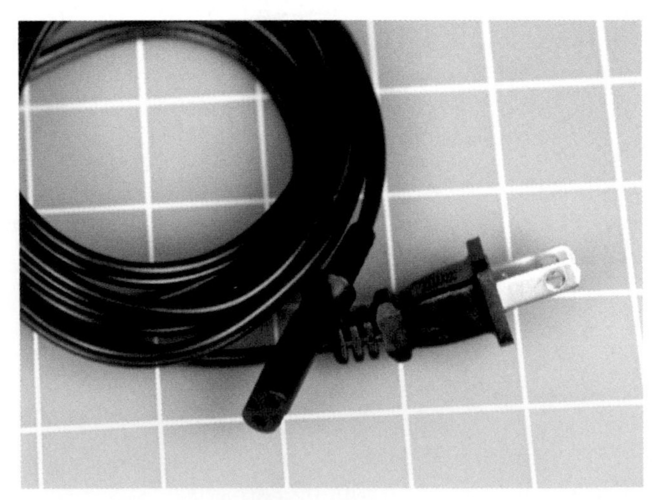

Figure 3-48 *A detachable AC power cord for a non-Apple laptop computer.*

What if you are an Apple fan? Maybe you have detachable power cords for other devices, such as a printer or a scanner. The purpose of this exercise is to reduce the length of a power cord so that it is exactly what you want, instead of lying around in a tangle. And if, like me, you have a laptop power cord that is longer than it needs to be, and you like to travel as light as possible, this exercise could be useful.

Twelve Steps to a Shorter Cord

In Figure 3-49 we see the first step, in which you boldly apply your wire cutters to chop a power cord. Needless to say, all of these steps require that the power cord should *not be plugged in* while you are working on it.

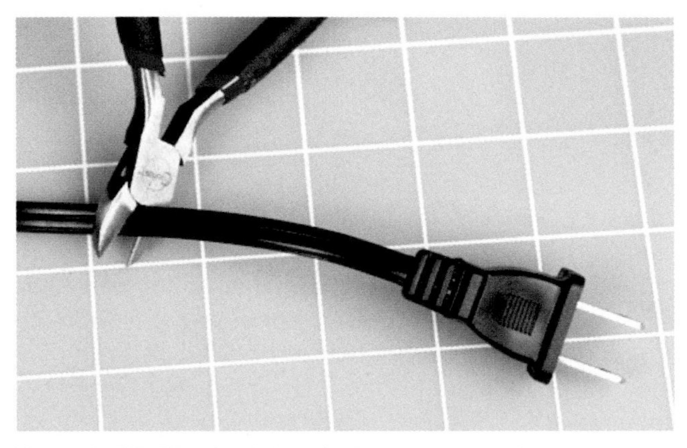

Figure 3-49 *Step 1 of 12, to shorten a power cord.*

Figure 3-50 shows the ends that you want to keep. You can store the rejected middle section of the power cord for some other purpose in the future.

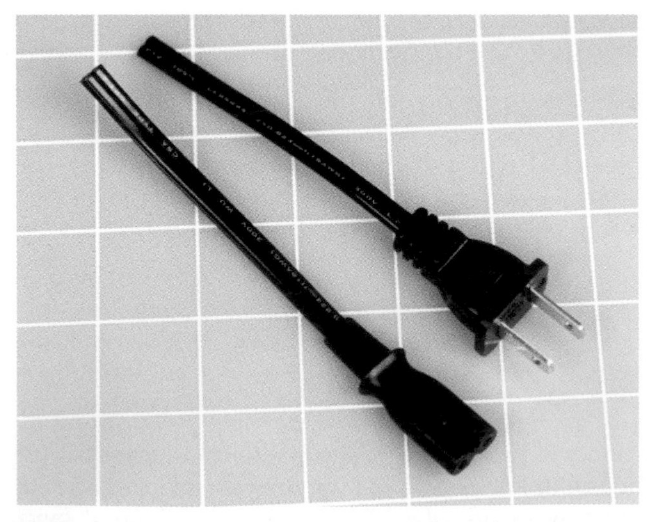

Figure 3-50 *Step 2 of 12, to shorten a power cord.*

In Figure 3-51, a utility knife on a cutting mat is an easy way to separate the two conductors in each segment of the power cord.

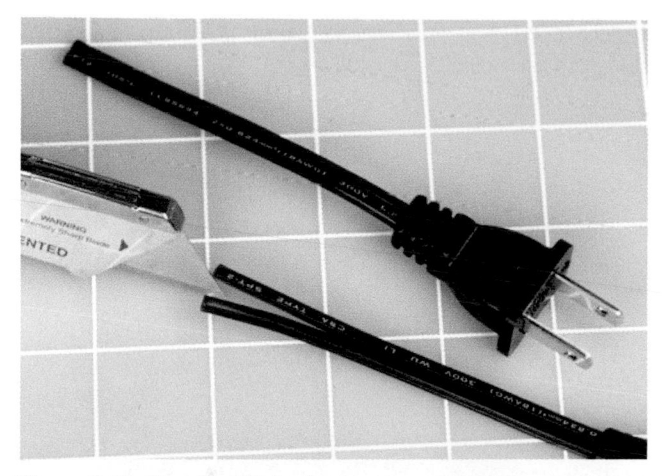

Figure 3-51 *Step 3 of 12, to shorten a power cord.*

In Figure 3-52, the segments of the power cord have been trimmed so that the conductors are of matching but unequal lengths. This way they will take up less room when you rejoin them, and there will be less risk of a short circuit if one of the joints fails for some reason.

Note that one conductor will be always marked, either with print or with molded ridges. Be sure that the marked conductors match each other when you join them.

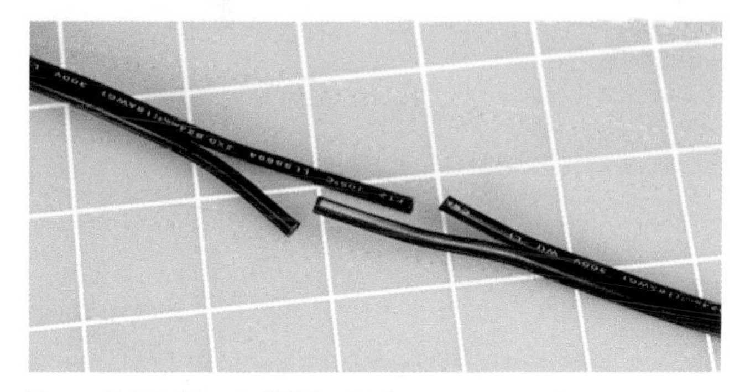

Figure 3-52 *Step 4 of 12, to shorten a power cord.*

Strip off a minimal amount of insulation. Just 1/8" (3mm) will be enough. Then cut some pieces of heat-shrink tubing, if you have it. Each smaller piece must be just large enough to slide over each individual conductor in the power cord, while the bigger piece, about 2" long, will cover the whole joint. See Figure 3-53.

Note that some heat-shrink tubing is rated only for low voltages. It should not be used for this project.

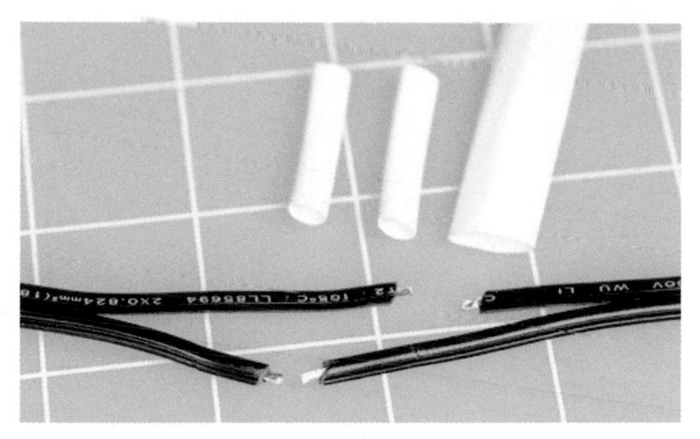

Figure 3-53 *Step 5 of 12, to shorten a power cord.*

Now for the most difficult part: activating your human memory. You have to remember to slide the tubing onto the wire *before* you make your solder joint, because the plugs on the ends of the wires will prevent you from adding any heat-shrink tubing later. If you're as impatient as I am, it's very difficult to remember to do this every time. See Figure 3-54.

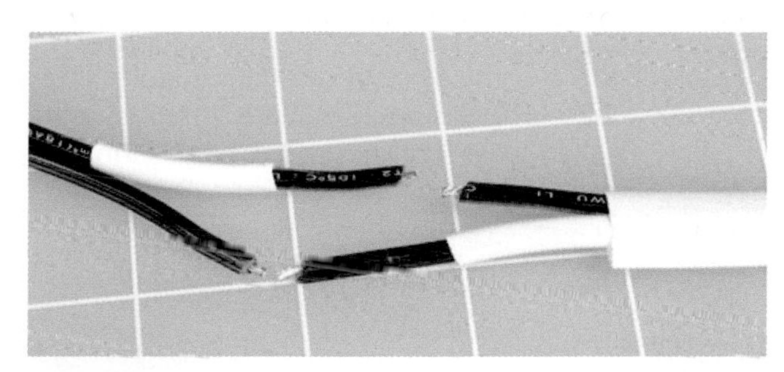

Figure 3-54 *Step 6 of 12, to shorten a power cord.*

Use your Helping Hand to align the first joint. Push the two pieces of wire together so that the strands intermingle, and then squeeze them tight between finger and thumb, so that there are no little bits sticking out. A stray strand of wire can puncture heat-shrink tubing when the tubing is hot and soft and is shrinking around the joint. See Figure 3-55.

Figure 3-55 *Step 7 of 12, to shorten a power cord.*

The wire that you're joining is much heavier than the 22-gauge wire that you worked with previously, so it will suck up more heat, and you must touch the soldering iron to it for a longer time. Make sure that the solder flows all the way into the joint, and check the underside after the joint is cool. Most likely you'll find some bare copper strands there. The joint should become a nice solid, rounded, shiny blob. See Figure 3-56.

Be careful to keep the heat-shrink tubing as far away from the joint as possible while you're using the soldering iron, so that heat from the iron doesn't shrink the tubing prematurely, preventing you from sliding it over the joint later.

Figure 3-56 *Step 8 of 12, to shorten a power cord.*

Slide a section of heat-shrink tubing over the solder joint, and apply the heat gun, as shown in Figure 3-57. Don't allow the other pieces of heat-shrink tubing to pick up stray heat.

Figure 3-57 *Step 9 of 12, to shorten a power cord.*

In Figure 3-58, the heat-shrink tubing has now shrunk.

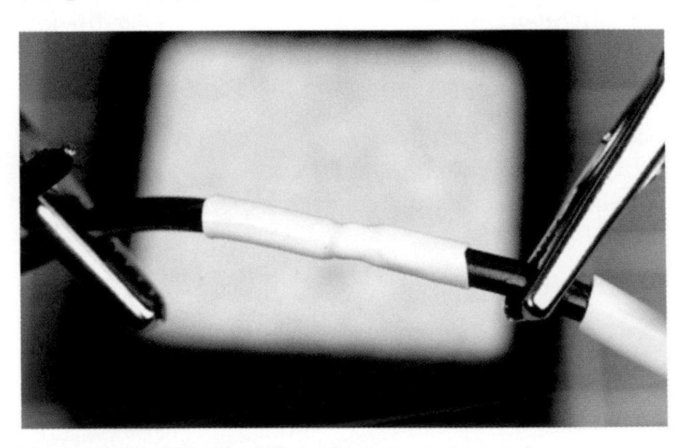

Figure 3-58 *Step 10 of 12, to shorten a power cord.*

Get ready to solder the other pair of conductors. See Figure 3-59.

Figure 3-60 shows that the second joint has been made. After you protect it with its own piece of heat-shrink tubing, you'll be ready to slide the larger section of tubing over the whole assembly. Er—you did remember to put the large tubing onto the wire at the beginning, didn't you?

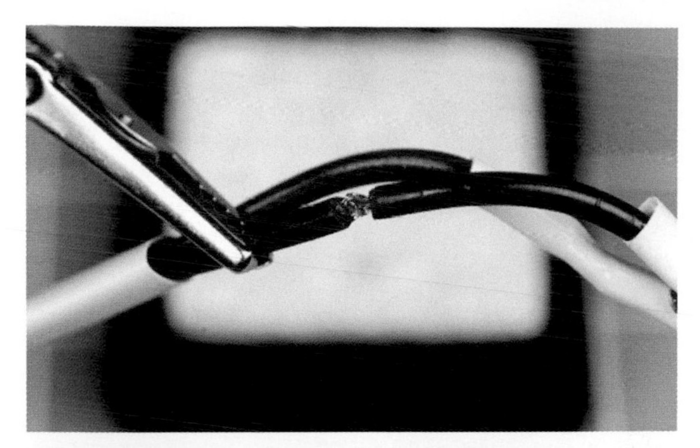

Figure 3-59 *Step 11 of 12, to shorten a power cord.*

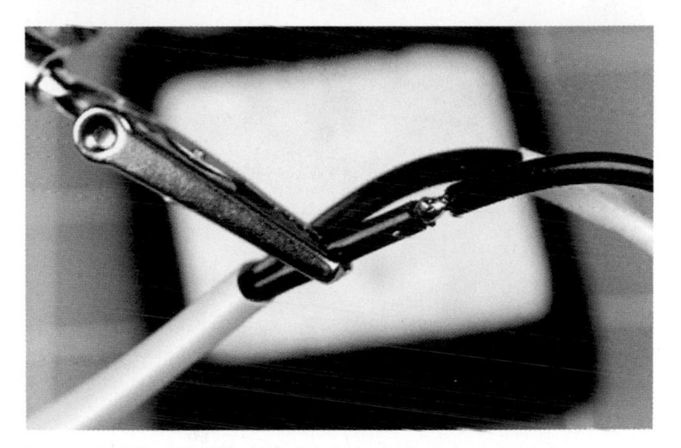

Figure 3-60 *Step 12 of 12, to shorten a power cord.*

Figure 3-60 shows the finished shortened power cord.

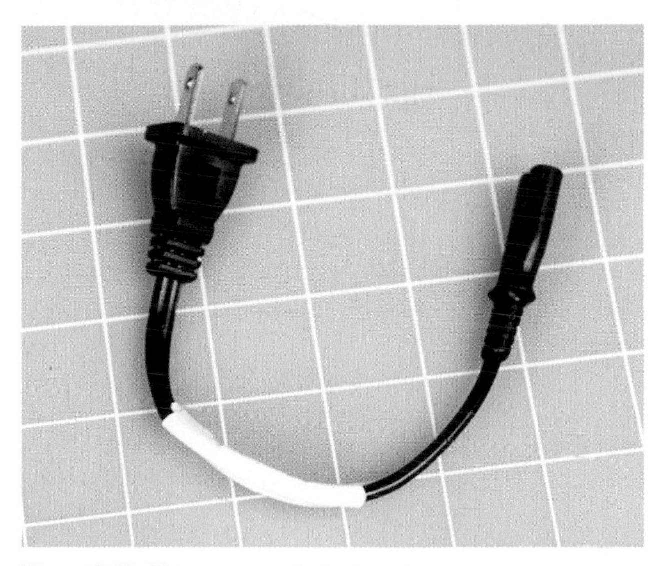

Figure 3-61 *The power cord, shortened.*

What's Next?

If you have completed the soldering exercises so far, you have sufficient basic skills to build your first soldered electronic circuit. Although—maybe first we should just run through a quick demonstration of the consequences of unintentionally excessive heat. I'd hate for you to take a lot of trouble soldering things together, only to discover that you melted a transistor or an LED. Unsoldering damaged parts is much less fun than soldering them.

Experiment 13: Roasting an LED

In Experiment 4, you discovered how easy it is to burn out an LED. What really happened, in that little adventure, is that excessive current passing through the LED created excessive heat, and the heat killed the component.

If heat caused by electricity can destroy an LED, do you think heat from a soldering iron can do the same thing? It sounds plausible, but there's only one way to make absolutely sure.

What You Will Need

- 9V battery and connector, or 9V AC-DC adapter
- Long-nosed or sharp-nosed pliers
- 30-watt or 40-watt soldering iron
- 15-watt soldering iron
- LEDs, generic (2)
- 470-ohm resistor (1)
- Helping Hand to hold your work
- One large or two small pure-copper alligator clips

The purpose of the experiment is to study the effects of heat. This means we need to know where the heat is going.

For this reason, you're not going to use a breadboard. The contacts inside the board would absorb an unknown amount of heat. I don't want you to use test leads, either, because they too will absorb heat.

Instead, please use some sharp-nosed pliers to bend each of the leads from an LED into little hooks, and do the same thing with the wires on a 470-ohm resistor. See Figure 3-62, where you will see that the wires from a 9-volt battery have been bent the same way. To make them maintain the hook shape, you may need to strip off a little extra insulation and apply a little solder.

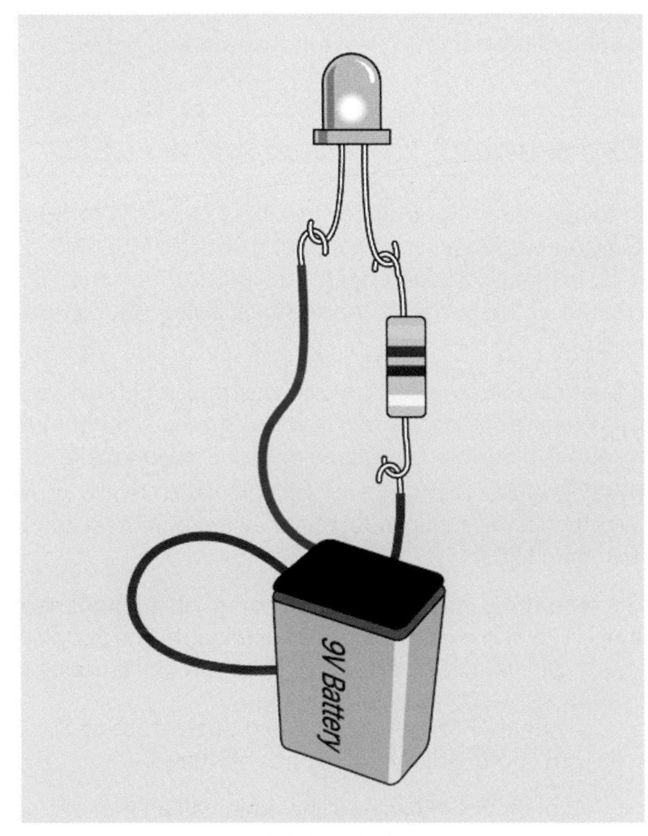

Figure 3-62 *Measuring the heat tolerance of an LED. An AC adapter can substitute for the 9-volt battery.*

To minimize heat loss through conduction, the resistor dangles from one of the leads on the LED, and the power-supply wire hangs from that, a little farther down. Gravity should be sufficient to make this work.

Grip the plastic body of the LED in your Helping Hand. Plastic is not a good thermal conductor, so the lens of the LED should not allow much heat to be conducted away through the Helping Hand.

Apply the 9 volts, and your LED should be shining brightly. I used a white LED in this experiment, because it's easier to photograph.

You'll need your low-powered 15W soldering iron, and the higher-powered one, too. Plug them in and wait at least five minutes, to make sure they are really hot. Now hold the tip of the 15W iron firmly against one of the leads on your glowing LED, while you check the time with a watch. Figure 3-63 shows the setup.

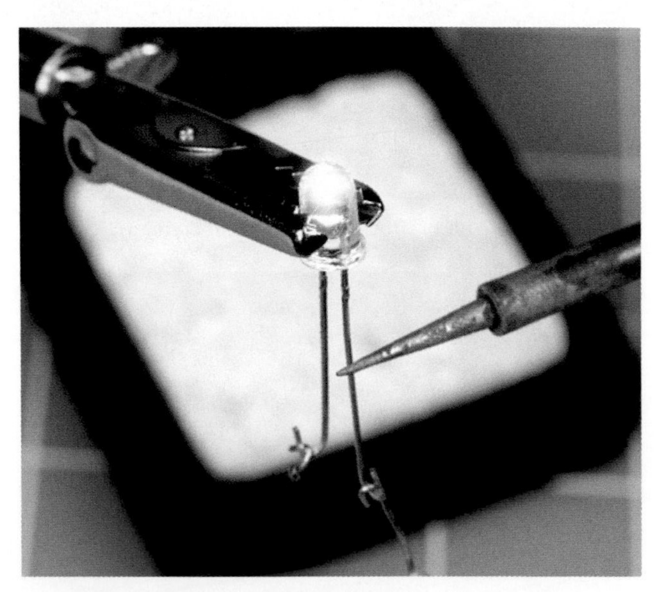

Figure 3-63 *Applying heat with a 15-watt soldering iron.*

I'm betting that you can sustain this contact for a full three minutes without burning out the LED. Now you know why a 15-watt soldering iron is recommended for delicate electronics work.

Allow your LED wire to cool, and then apply your more powerful soldering iron in the same location as before. I think you'll find that the LED will go dark after as little as 10 seconds (note that some LEDs can survive higher temperatures than others). This is why you *don't* use a 30-watt soldering iron for delicate electronics work.

The large iron doesn't necessarily reach a higher temperature than the small one. It just has a larger heat capacity. In other words, a greater quantity of heat can flow out of it, at a faster rate.

Your LED has been sacrificed to satisfy the need for knowledge. It was an honorable death. Lay it to rest in your trash, and substitute a new LED, which we will try to treat more kindly. Connect it as before, but this time add a full-size copper alligator clip (or two small clips) to one of the leads near the body of the LED, as shown in Figure 3-64. Press the tip of your 30-watt or 40-watt soldering iron against the lead just *below* the alligator clip. This time, you should be able to hold the powerful soldering iron in place for a full two minutes without burning out the LED.

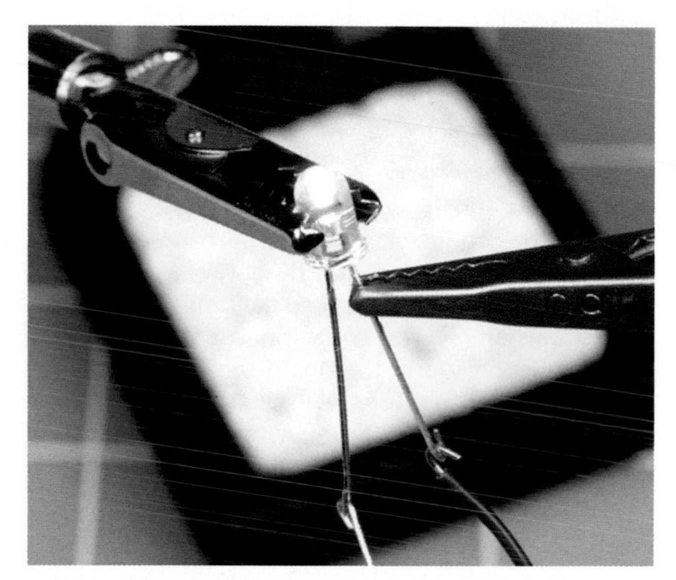

Figure 3-64 *Using a copper alligator clip as a heat sink to protect an LED.*

Where the Heat Goes

At the end of your experiment, if you touch the clip, you'll find that it's relatively hot, while the LED remains not so hot. Imagine the heat flowing out through the tip of your soldering iron, into the wire that leads to the LED —except that the heat meets the alligator clip along the way, as shown in Figure 3-65. The clip is like an empty container waiting to be filled. The heat prefers to flow into the copper clip, leaving the LED unharmed.

The alligator clip functioned as a *heat sink*. It works better than an everyday nickel-plated steel alligator clip, because copper is such a good conductor of heat.

Going back to the first part of this experiment, you saw that a 15-watt soldering iron failed to harm the LED, with no heat sink needed. Does this mean that a 15-watt iron is completely safe?

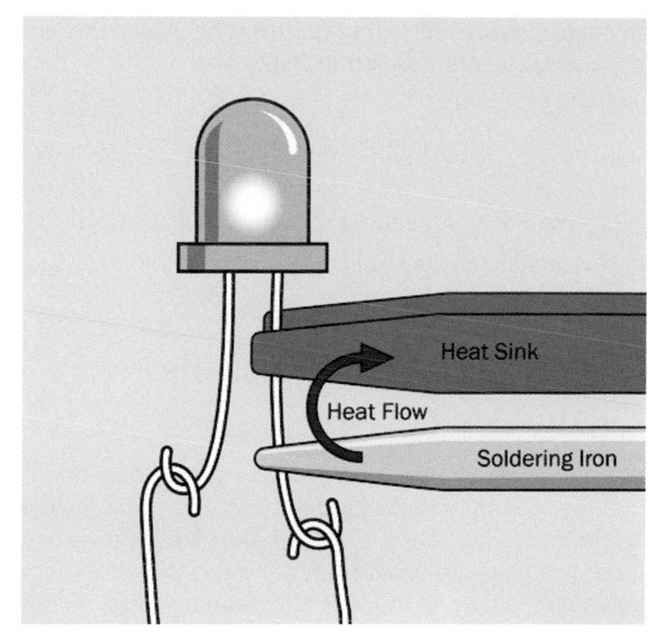

Figure 3-65 *A copper alligator clip conducts heat away from an LED.*

Well, maybe. The problem is, you don't really know whether some semiconductors may be more heat-sensitive than LEDs.

Because the consequences of burning out a component can be so annoying, I suggest you should consider playing it safe and use a heat sink in these circumstances:

- If you apply a 15-watt iron extremely close to a semiconductor for 20 seconds or more.
- If you apply a 30-watt iron within half an inch of resistors or capacitors for 10 seconds or more. (Never use it near semiconductors.)
- If you apply a 30-watt iron near anything meltable for 20 seconds or more. Meltable items include insulation on wires, plastic connectors, and plastic components inside switches.

Rules for Heat Sinking

- Full-size copper alligator clips work best, but may not fit into tight corners. Ideally you should have small ones available too.
- Clamp the alligator clip as close as possible to the component and as far as possible from the solder joint that you are trying to make. The

joint does need to get hot. Divert heat from the component, not from the joint.

- Make sure there is a metal-to-metal connection between the alligator clip and the wire to promote good heat transfer.

If you keep these points in mind, we can proceed with the fascinating challenge of point-to-point wiring.

Experiment 14: A Wearable Pulsing Glow

Until now, I've encouraged you to start putting things together without much of a theory or a plan. That's what Learning by Discovery tends to be like. Sometimes, though, a plan can be essential, and this is one of those times. I'm going to outline the requirements of this project, and then I will take you step by step through the process of building it.

What You Will Need

- 9V battery and connector, or 9V AC-DC adapter
- Hookup wire, wire cutters, wire strippers, multimeter
- 15-watt soldering iron
- Thin solder (0.022 inches)
- Plain perforated board (no copper plating necessary)
- Helping Hand
- Resistors: 470 (2), 100K (1), 4.7K (2), 470K (2)
- Capacitors: 3.3µF (2), 220µF (1)
- Transistors: 2N2222 (3)
- Generic LED (1)

Fluctuations Revisited

Please turn back to Figure 2-116 to refresh your memory of that circuit. The task, now, is to make it as small as possible, so that someone could wear it.

Imagine that the leads of the components are interconnected with rubber bands, allowing you to shuffle them around on a tabletop without them losing their connections with each other. When the rubber bands are stretched as little as possible, the circuit is as small as possible, and you can link the parts with bare wires, supporting everything with a piece of perforated board.

The only problem is that bare wires under the board cannot cross each other. The idea is that after you have verified the function of your circuit, you could send the specification to a service that etches circuit boards.

Of course, modern printed circuits are double-sided, at the very least, and many have intermediate layers, allowing multiple conductors to cross each other without making electrical contact. But it's always good to start on a simple, traditional basis, and the simplest board has components on one side and connections on the other. Components on top of the board can bridge the conductors beneath, because the insulating material of the board separates them. But conductors cannot cross each other.

My best attempt to reduce the size of this circuit is shown in Figure 3-66, on a piece of plain perforated board measuring 0.9" x 1.3". If you can come up with a design that is significantly smaller, I'd love to see it. Here are some ideas:

- Use smaller resistors rated for 1/8 watt instead of 1/4 watt.
- Mount resistors vertically.
- Thread two leads through one hole, if the holes in your board are big enough.

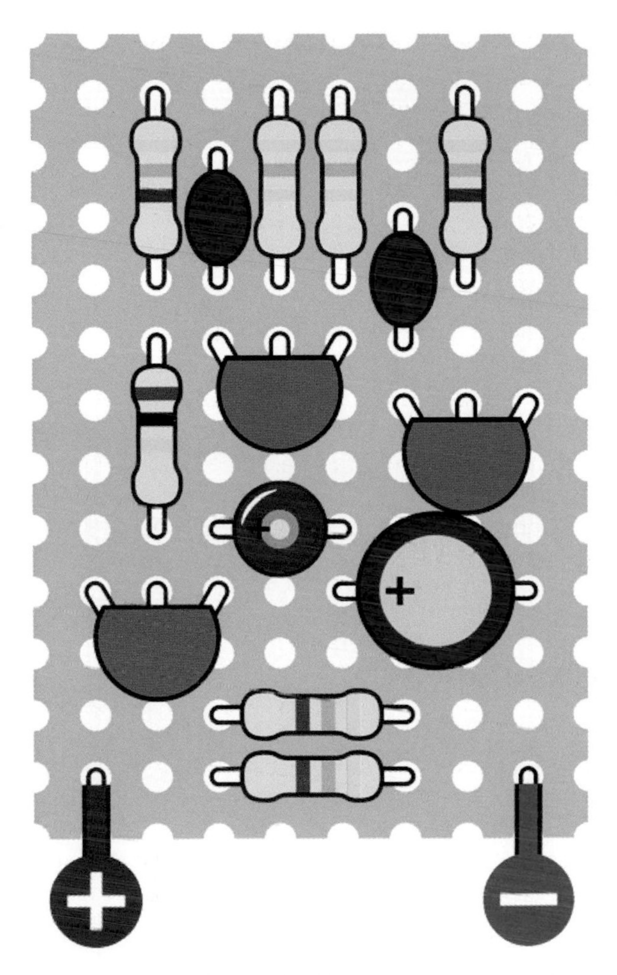

Figure 3-66 *The oscillator circuit, reduced to occupy minimal space on perforated board.*

Where are the connections between the components? They're under the board. In Figure 3-67 I grayed out the components and disappeared the board, so that you can see the wiring.

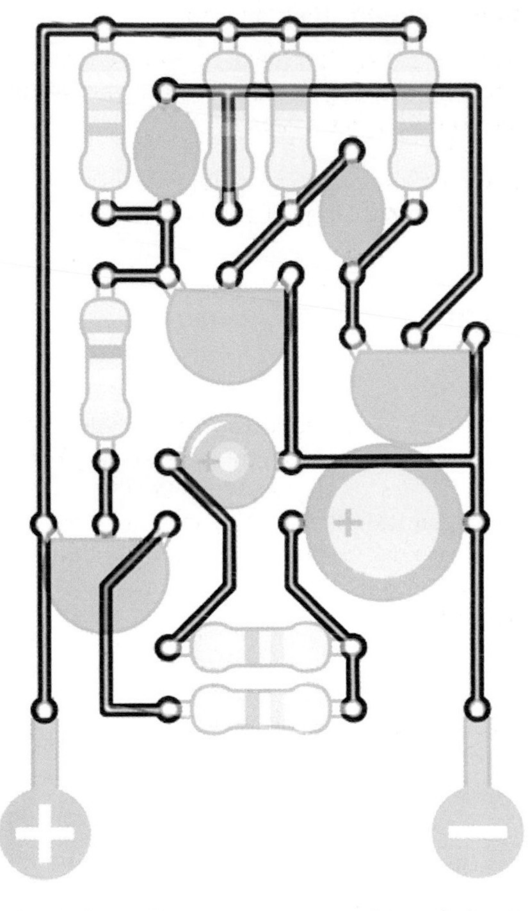

Figure 3-67 *Connections in black are wires beneath the circuit board, which is transparent in this view.*

If you compare this diagram very carefully with the schematic in Figure 2-116, you should find that the connections between components are the same—unless I made an error. (I sure hope not. I don't want to have to redraw everything.)

Figure 3-68 shows yet another view, this time omitting the components while including the board, so that you can see how the connections fit the 0.1″ x 0.1″ grid of holes in the board.

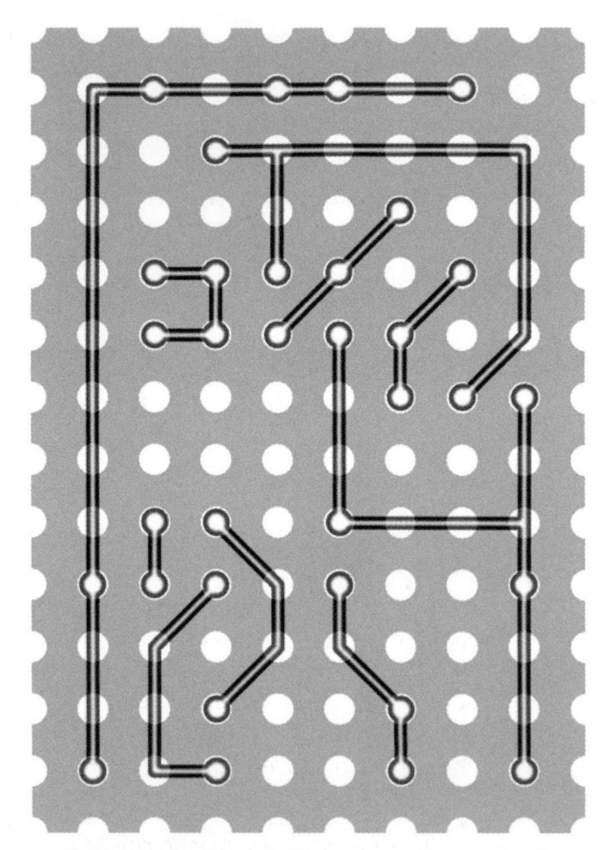

Figure 3-68 *In this view, only the board and the connections are shown. Every circular dot indicates a connection that comes through a hole in the board.*

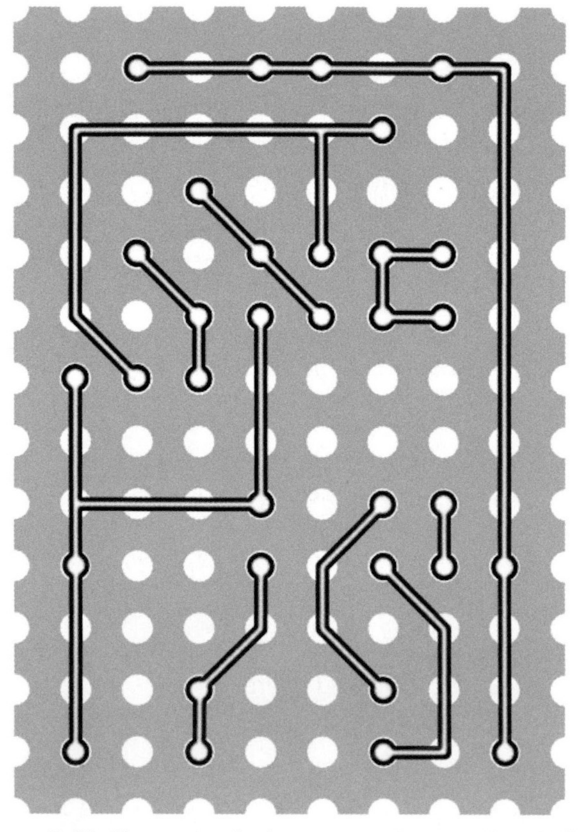

Figure 3-69 *The previously shown connections are flipped left to right, so that this is the way the board should look from the underside.*

Finally, Figure 3-69 shows the board flipped over, left to right, so that you are viewing it from behind. This will help you to make the connections when you're putting the components together. You are going to give this a try, aren't you?

Bend Wires, Add Solder

Now that you've seen the plan for this project, how are you supposed to make all those connections?

It's not so difficult. Resistors, capacitors, and transistors have wire leads that are usually at least 1/2" long. So, you can push them through the holes in the perforated board, and then bend the leads over to touch each other. While they are touching, you solder them together. Snip off any surplus, add a battery connection, and you're done.

There are three main issues to watch out for.

- Holding the board steady while you work on it requires some care and patience. Your Helping Hand will be necessary.
- The components and the solder joints that you make will be very close together. Use your copper alligator clips to provide heat protection.
- Flipping between the top side of the board and the bottom side is confusing. You can easily put a wire in the wrong place. I think this is actually the hardest part.

Perhaps you have seen perforated board on which a little circle of copper has been added around each hole. Would that be suitable for this project? The copper circles have the advantage of anchoring components securely, but they can also create a short circuit between wires that are close together. I think bare board is easier for a small project like this. A sample was shown in Figure 3-22. Some perforated board has holes that are larger, but this does not make a significant difference.

Step by Step

Here's the specific procedure for building the circuit:

Cut a piece of perfboard measuring 0.9" by 1.3" out of a plain sheet. (You don't need a ruler calibrated in tenths of an inch. Just count the rows of holes in the board.) You can use a miniature hobby saw, or you may be able to snap the board along its lines of holes, if you're careful. A hacksaw will also work. I suggest you should not use a good wood saw, because perforated board often has glass fibers in it, and they can blunt a saw.

Gather all the components and carefully insert three or four of them through holes in the board, counting the holes to make sure everything is in the right place. Flip the board over and bend the wires from the components to anchor them to the board and create the connections shown in Figure 3-69. If any of the wires isn't long enough, you'll have to supplement it with an extra piece of 22-gauge wire from your supply. Remove the insulation, which just gets in the way.

Trim the wires approximately with your wire cutters.

Make the joints with your soldering iron.

Now, the important part: check each joint using a close-up magnifying glass, and wiggle the wires with pointed-nosed pliers. If there isn't enough solder for a really secure joint, reheat it and add more. If solder has created a connection that shouldn't be there, use a utility knife to make two parallel cuts in the solder, and scrape away the little section between them.

Generally, I deal with just three or four components at a time, because I get confused if there are more than that. If I solder one component in the wrong location, undoing the error is not too difficult—unless I have already added more components to it by the time I discover the error.

Caution: Flying Wire Segments

The jaws of your wire cutters exert a powerful force that peaks and then is suddenly released when they cut through wire. This force can be translated into sudden motion of the snipped wire segment. Some wires are relatively soft, and don't pose a risk, but transistors and LEDs tend to have harder wires. Little segments can fly in unpredictable directions at high speed, creating a real hazard for your eyes when you are doing close-up work.

Everyday eyeglasses can protect you when trimming wires. If you don't use eyeglasses, plastic safety glasses are really a good idea.

Finishing the Job

I always use bright illumination. This is not a luxury; it is a necessity. Buy a desk lamp, if you don't already have one. It doesn't have to be expensive; a thrift-store item is okay.

I use a daylight-spectrum LED desk lamp, because it helps me to identify the colored bands on resistors more reliably. I stopped using a fluorescent desk lamp when I discovered that any small imperfection in the coating inside the tube can allow ultraviolet light to escape. This constitutes a hazard when you are working so close to the light.

No matter how good your vision is, you need to examine each joint with a magnifier. You'll be surprised how imperfect some of them are. Hold the magnifier as close as possible to your eye, then pick up the board and bring it closer until the joint that you are inspecting comes into focus.

Finally, you should end up with a working circuit that pulses like a heartbeat. Or does it? If you have difficulty making it work, retrace every connection and compare it with the schematic. If you don't find an error, apply power to the circuit, attach the black lead from your meter to the negative side, and then go around the circuit with the red lead, checking the presence of voltage. Every part of this circuit should show at least some voltage while it's working. If you find a dead connection, you may have made a bad solder joint, or missed one entirely.

When you're done, now what? Well, now you can stop being an electronics hobbyist and become a crafts hobbyist. You can try to figure out a way to make this thing wearable.

First you have to consider the power supply. Because of the components that I used, you really need 9 volts to make this work well. How are you going to make this 9-volt circuit wearable, with a bulky 9-volt battery?

I can think of three answers:

- You can put the battery in a pocket, and mount the flasher on the outside of the pocket, with a thin wire penetrating the fabric.
- You could mount the battery inside the crown of a baseball cap, with the flasher on the front.
- You can put three 3-volt button batteries in a stack, held in some kind of plastic clip. I'm not sure how long they'll last, though.

I have to note that the 2N2222 transistors in this project are not ideal, because they tend to use more power than field-effect transistors, also known as MOSFETS. However, I made a decision in this book that I only had space for one transistor family, and bipolar NPNs are the most fundamental type.

Regarding your choice of LED, those with a clear lens create a defined beam of light, which may not be appropriate for this project. Those that are diffused create a more pleasing glow. You can diffuse the light more by embedding the LED in a piece of transparent acrylic plastic, at least 1/4 inch thick, as shown in Figure 3-70. Roughen the front of the acrylic with fine sandpaper, ideally using an orbital sander that won't make an obvious pattern. This will make the acrylic translucent rather than transparent.

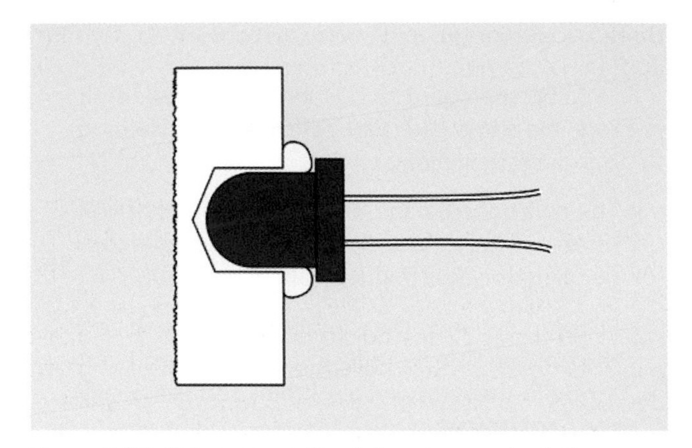

Figure 3-70 *This cross-sectional view shows a sheet of transparent acrylic in which a hole has been drilled part of the way from the back toward the front. Because a drill bit creates a hole with a conical shape at the bottom, and because the LED has rounded contours, transparent epoxy or silicone caulking can be injected into the hole before mounting the LED.*

Drill a hole slightly larger than the LED in the back of the acrylic. Don't drill all the way through the plastic. Remove all fragments and dust from the hole by blasting some compressed air into it, or by washing it if you don't have an air compressor. After the cavity is completely dry, get some transparent silicone caulking or mix some clear five-minute epoxy and put a drop in the bottom of the hole. Then insert the LED, pushing it in so that it forces the epoxy to ooze around it, making a tight seal.

Try illuminating the LED, and sand the acrylic some more if necessary. Finally, you can decide whether to mount the circuit on the back of the acrylic, or whether you want to run a wire to it elsewhere.

You can choose the resistors in the oscillator circuit to make the LED flash at about the speed of a human heart while the person is resting. This may make it look as if it's measuring your pulse, especially if you mount it on the center of your chest or in a strap around your wrist. If you enjoy hoaxing people, you can suggest that you're in such amazingly good shape, your pulse rate remains constant even when you're taking strenuous exercise.

To make a good-looking enclosure for the circuit, I can think of options ranging from embedding the whole thing in clear epoxy to finding a Victorian-style locket. I'll leave you to consider alternatives, because this is a book about electronics rather than craft projects. However, there is one craft-related issue that I want to mention, and this is a good moment to do so.

Background: Maddened by Measurement

Throughout this book, I've mostly used measurements in inches. Sometimes, though, I've ventured into the metric system, as when referring to "5mm LEDs." This isn't inconsistency on my part; it reflects the conflicted state of the electronics industry, where you'll find inches and millimeters both in daily use, often in the very same datasheet. For instance, the pin spacing of surface-mount chips tends to be measured in millimeters, but through-hole chips still have pins 0.1" apart, and probably always will.

To complicate matters further, where inches are used, there are two different systems for dividing them into fractional amounts. Drill bits, for instance, are measured in multiples of 1/64ths of an inch. Metal shims are graded in 1/1,000ths of an inch (0.001", 0.002", and so on). To make things even more confusing, the thickness of sheet metal is often measured by "gauge," as in 16-gauge steel, which happens to be about 1/16" thick.

Why doesn't the US move to the metric system, since it's so much more rational?

We can debate whether it really is rational. When it was formally introduced in 1875, the meter was defined as being 1/10,000,000 of the distance between the North Pole and the equator, along a line passing through Paris. Why Paris? Because the French came up with the idea. Since then, the meter has been redefined three times, in a series of efforts to achieve greater accuracy in scientific applications.

As for the usefulness of a 10-based system, moving a decimal point is certainly simpler than doing calculations in 1/64ths of an inch, but the only reason we count in tens is because we happen to have evolved with that number of digits on our hands. A 12-based system would really be more convenient, as numbers would be evenly divisible by 2 and 3.

This is all very hypothetical. The fact is, we're stuck with conflicts in length measurement, so I've created four charts to assist you in converting from one system to another. From these you will see that when you need to drill a hole for a 5mm LED, a 3/16-inch bit is about right. (In fact, it results in a tighter fit than if you drill an actual 5mm hole.)

Figure 3-71 will help you to convert between 1/64ths and 1/100ths of an inch. The gray column is divided into 1/64ths, the blue column is in 1/32nds, the green column is in 1/16ths, and the orange column is in 1/8ths. Customarily, if a value can be precisely expressed in larger units, we use that option; thus instead of referring to 8/64ths of an inch, we would express this as 1/8th of an inch. This causes some confusion when you're trying to figure out whether one measurement is larger than another—for instance, are 11/32nds of an inch larger than 5/8ths of an inch? Check the diagram to make sure.

Because datasheets often express dimensions using decimal fractions of an inch, a second chart in Figure 3-72 converts between decimals and 64ths. I think you're quite likely to find a measurement such as 0.375", and if you know that this is the same as 3/8", the knowledge can be useful.

Many datasheets provide measurements in both millimeters and inches, but some now use millimeters only. If you are still thinking in inches, or if you want to know if a component will fit the 1/10" hole spacing in a breadboard or perforated board, it's helpful to remember that 1/10" is equivalent to 2.54mm. Provided a component is small, pin spacing in multiples of 2.5mm is acceptable. However, when pins are 25mm or more apart, they may not fit into holes that are 25.4mm apart (that is, one inch or more).

Figure 3-73 enables conversion between millimeters, 1/100ths of an inch, and 1/64ths of an inch.

Figure 3-74 is a magnified version of the previous chart, showing tenths of millimeters and 1/1,000ths of an inch.

Some progress has been made during the past four decades toward adopting the metric system in the United States, but more decades will pass before this transition is complete. In the meantime, anyone using parts or tools manufactured or sold in the United States should be familiar with both systems. There's no way around it.

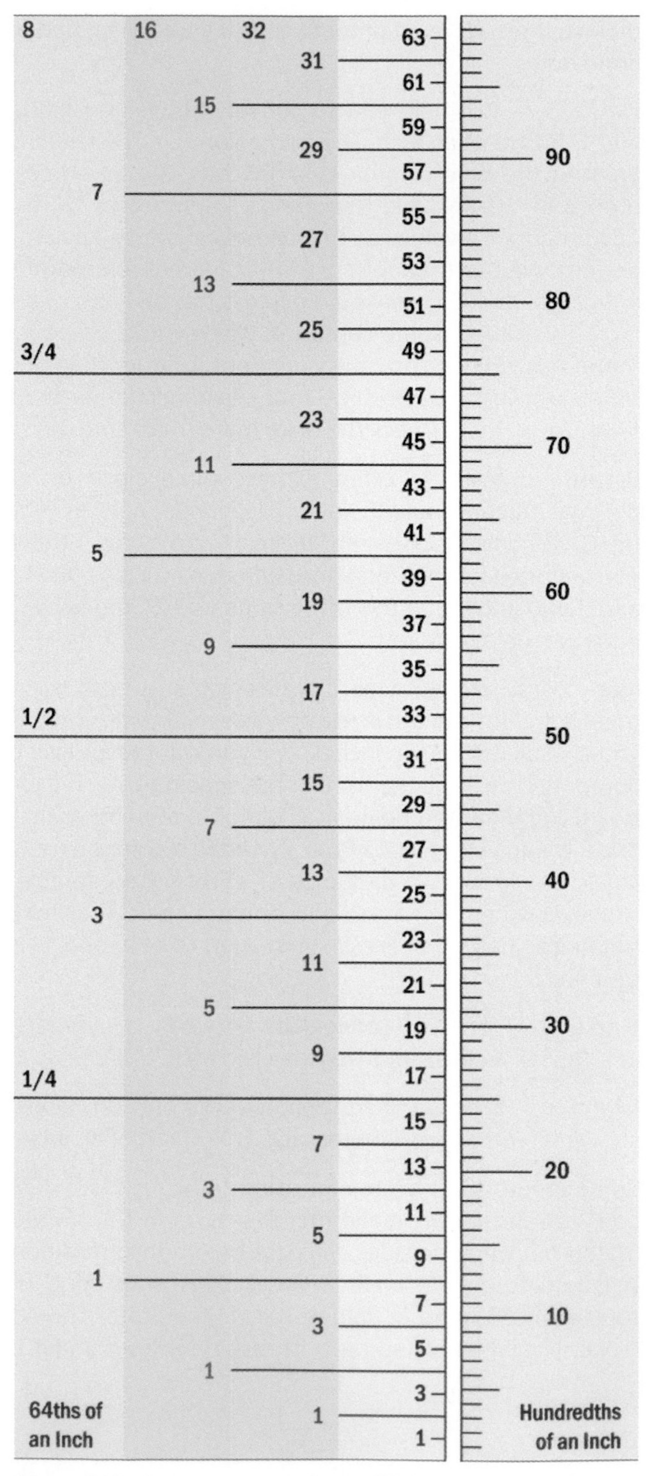

Figure 3-71 *For conversion between 64ths of inch and 100ths.*

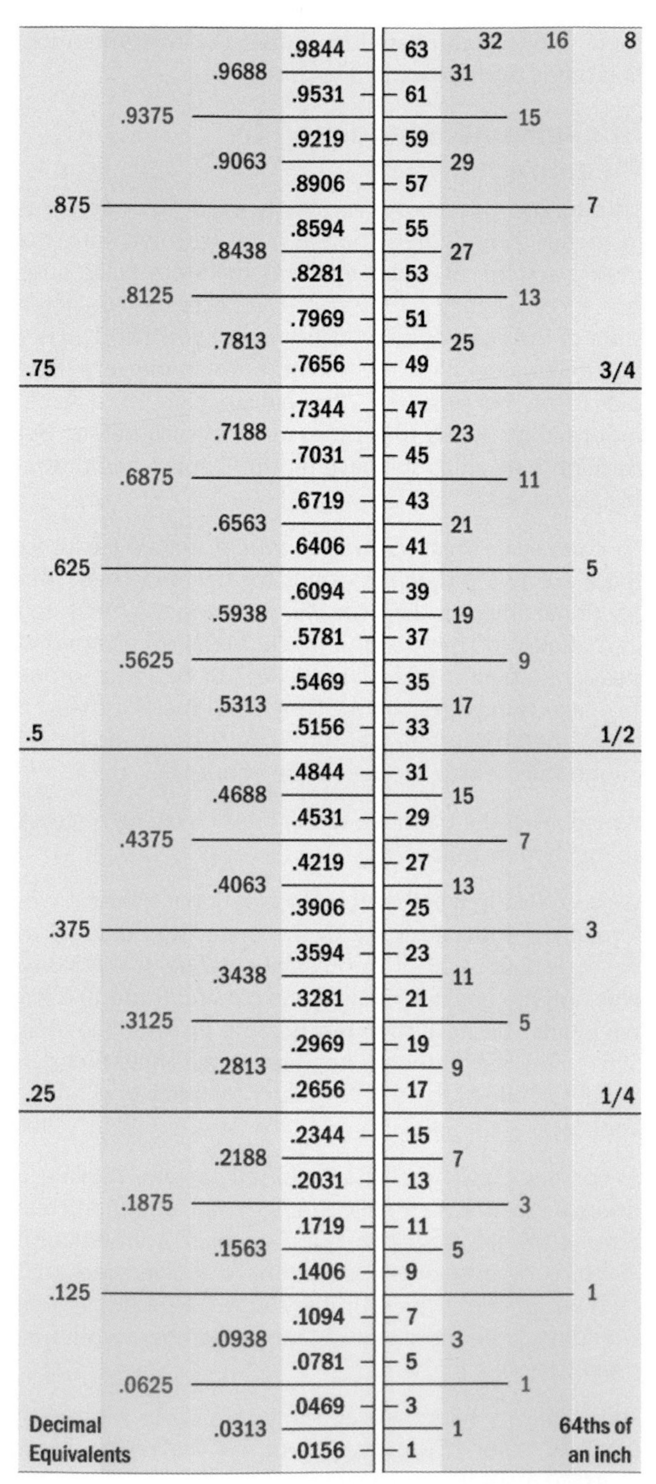

Figure 3-72 *For conversion between decimal inch values and 64ths.*

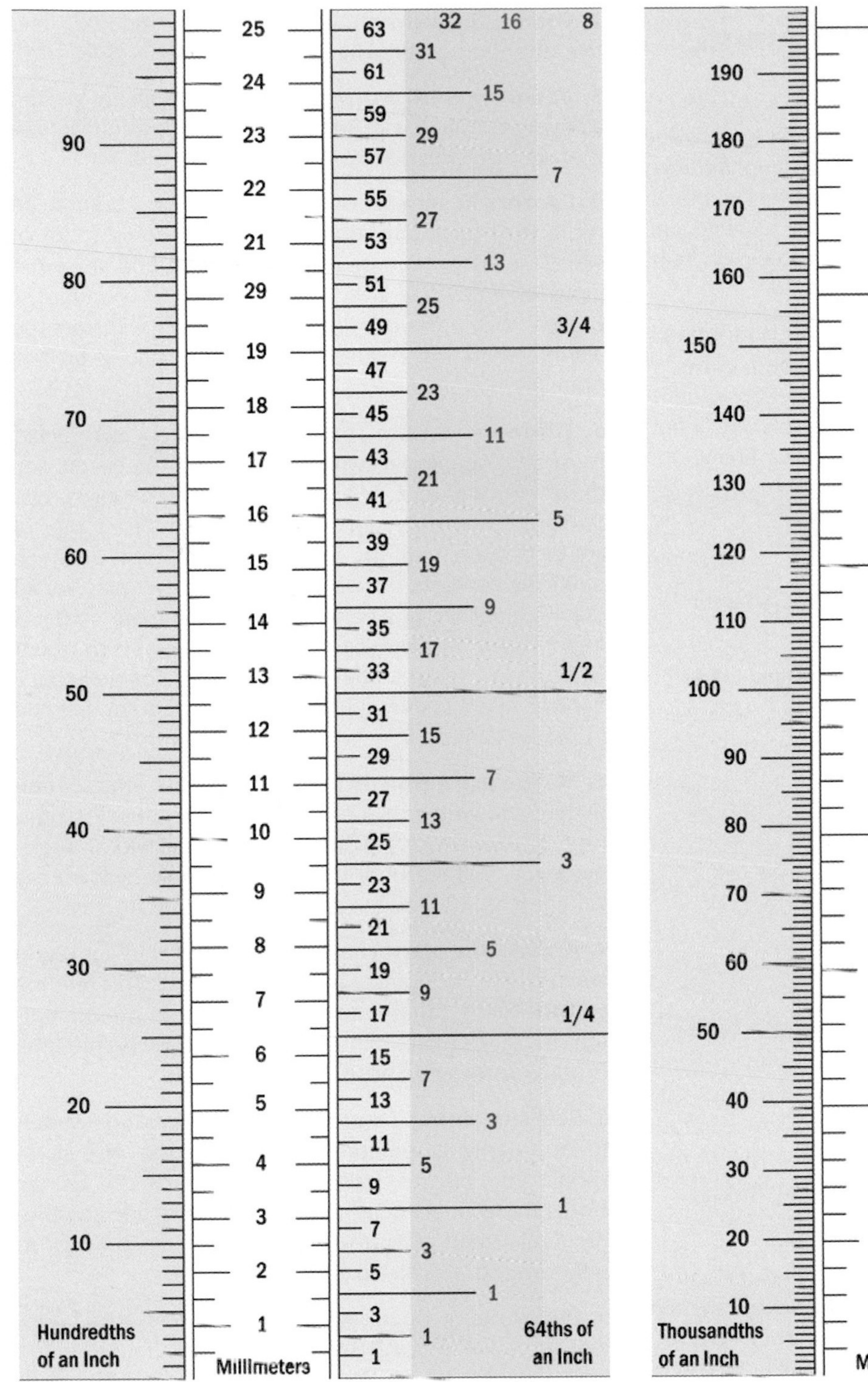

Figure 3-73 *For conversion between US measurements and metric (millimeters).*

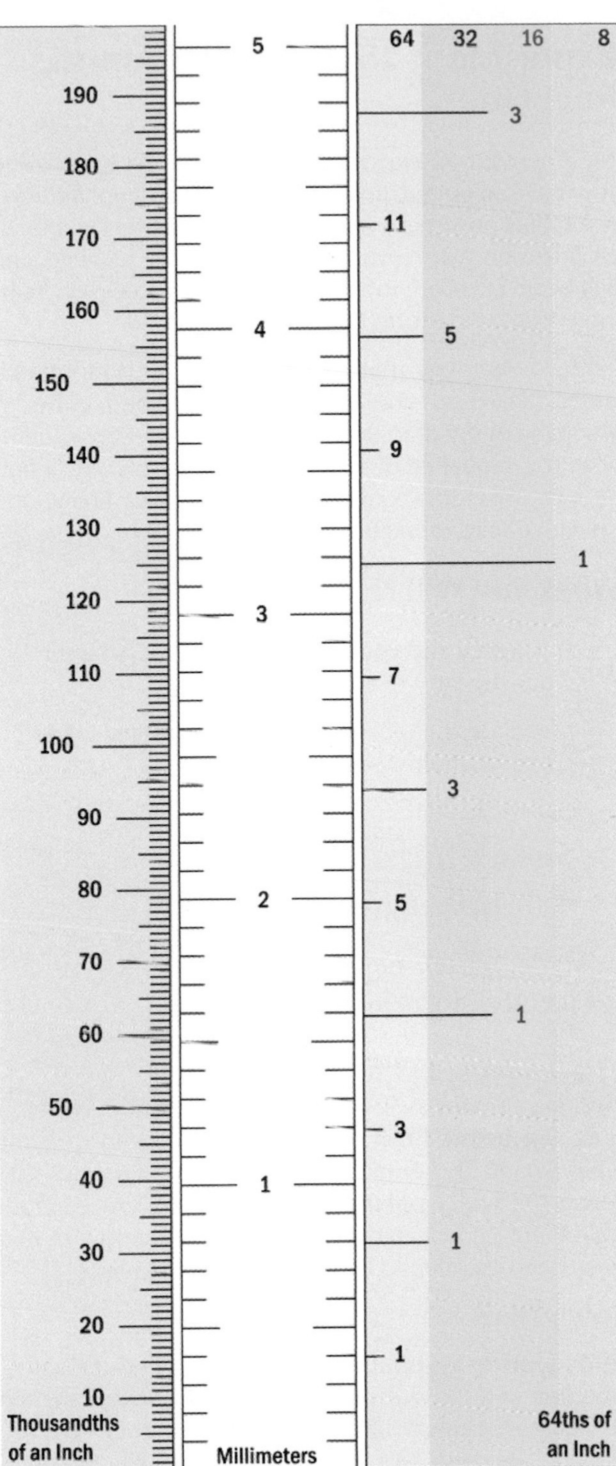

Figure 3-74 *For conversion between smaller values of US measurements and metric (tenths of millimeters).*

Experiment 15: Intrusion Alarm, Part One

Time now for an experiment that takes the knowledge you have acquired and applies it to a simplified but workable consumer product. You may not feel you need an intrusion alarm, but figuring out how to build one will be an excellent introduction to the process of creating circuits to perform tasks in the real world.

I should warn you that designing a circuit from scratch usually leads to unexpected problems and errors. It would be misleading to pretend otherwise. Consequently, in the sequence of steps described below, you'll find at least one setback and reversal—until we finally end up with a solid, workable system.

What You Will Need

- 9V battery and connector, or 9V AC-DC adapter (your choice)
- Breadboard, hookup wire, wire cutters, wire strippers, multimeter
- Generic LED (1)
- Transistor, 2N2222 (1)
- DPDT 9VDC relay (1)
- 1N4001 diode (1)
- Resistors: 470 ohms (1), 1K (1), 10K (1)

Wish List

This experiment is sufficiently complex; it requires a plan. But before I can develop a plan, I need to know what I want. This entails writing what I call a "wish list." Along the way, I will also try to visualize how each requirement can be satisfied with the components that have been mentioned in previous experiments.

So, what does an intrusion alarm require?

1. Triggering system. The device has to detect if someone has entered the property. A sophisticated system using laser beams or ultrasound would be cool—but too difficult. Because this is a first attempt, I'll stick with widely available magnetic sensor switches for windows and doors.

2. Sound. The alarm should make some kind of distinctive, attention-getting, fluctuating sound.

3. Tamper-proof. No one should be able to kill the alarm by cutting a wire. In fact, tampering should actually make the alarm go off.

4. Sensors in series. To tamper-proof the system I can run a very small but constant current through a lot of sensor switches that are normally closed, and are wired in series. If any one switch opens, or if the wire itself is broken, this interrupts the current, which will start the alarm. I think most wired alarms are designed on this principle.

5. Off-to-on. If I use sensors in series, an "off" event, caused by opening a switch or breaking the circuit, has to turn the alarm on. Maybe a double-throw relay could do this. Current through the relay coil holds a pair of contacts open, until the flow of current stops, at which point the contacts close by default. But the relay would draw significant power while it's holding its contacts open. I want my alarm system to draw very little current while it's in "ready" mode, so that it can be powered by a battery. Alarm systems should never depend entirely on AC house current.

6. Maybe use a transistor? If I don't use a relay, a transistor could switch on the alarm when the circuit is interrupted. The base of the transistor can be held at a relatively low voltage until the circuit is broken. Then the voltage goes up, and the transistor switches on.

7. Arming the alarm. I need a little light that comes on when all the doors and windows are closed. This tells me that the alarm can be used. Then I want to press a button that starts a one-minute countdown, giving me time to leave. After a minute, the alarm is armed.

8. Self-sustaining. Once the alarm starts, I don't want it to stop easily. If someone opens a window, the alarm should continue to make a noise even if the person closes the window again. Maybe the transistor can trigger a relay, and when the relay switches on, it keeps itself powered? Or can a transistor do this?

9. Initial delay. I don't want the alarm to start whooping immediately, each time I walk into the protected area. I want it to wait for a minute, to give me time to reach it and switch it off. If I fail to deactivate it within that time window, then it can start making noise.

10. Deactivation with a code. Some kind of secret-code keypad would be good, to switch off the alarm.

Implementing the Wish List

This wish list sounds a bit ambitious, bearing in mind that the only thing you have built so far is a small oscillator using three transistors. But in fact most of the functions can be implemented fairly easily. I will leave some of the harder ones until later in the book, when I have established a broader knowledge base. In the end, I will be able to deal with everything on the list, and the components will all fit on a single breadboard (with the exception of a noisemaking circuit, which will be optional).

Magnetic Sensor Switches

Let's start with the component that triggers the alarm. A typical sensor switch consists of two modules: the magnetic module and the switch module. These are shown side by side in Figure 3-75.

Figure 3-75 *A typical alarm sensor consists of a magnet in a plastic pod (bottom left), and a magnetically activated reed switch in a similar pod (top right).*

The magnetic module contains a permanent magnet, and nothing else. The switch module contains a *reed switch*, which makes or breaks a connection (like a contact inside a relay) under the influence of the magnet.

You attach the magnetic module to the moving part of a door or window, and attach the switch module to the window frame or door frame. When the window or door is closed, the magnetic module is almost touching the switch module. The magnet keeps the switch closed—until the door or window is opened, at which point the switch opens. See Figure 3-76 for a cutaway diagram of the magnet-switch combination.

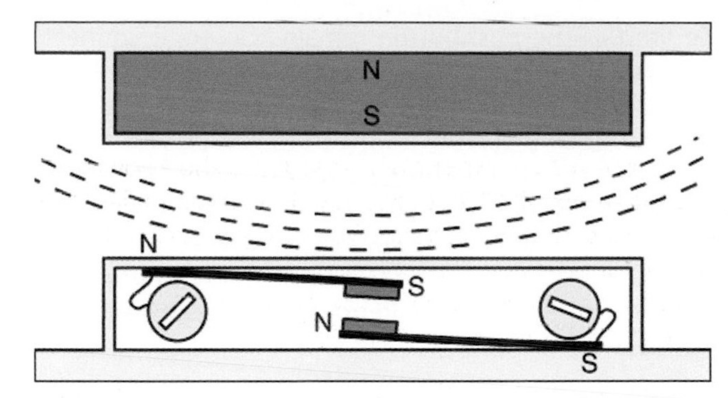

Figure 3-76 *This cutaway diagram shows the two components in a typical sensor for an alarm system: a reed switch (bottom) and the magnet that activates it (top).*

The switch consists of two flexible magnetized strips terminating in electrical contacts. Each strip connects with an external screw to which a wire can be attached.

When the magnet approaches the switch, it magnetizes the flexible strips, causing them to attract each other until the contacts close.

From my description, you can see that the reed switch is normally open (abbreviated NO), but is kept closed by the magnetic field. If you decide to buy alarm sensors, you should know that some of them contain reed switches that work the other way around. They are normally closed (abbreviated NC), but are opened by the magnetic field. Those are not the ones you want for this project.

A Break-to-Make Transistor Circuit

Now, how can we switch on the noisemaking part of the alarm? Remember, we will have a series of switches that are all normally closed, and when any one of them opens, the alarm must start.

Recall how an NPN transistor works. When the base is not so positive, the transistor blocks current between its collector and emitter. When the base is more positive, the transistor passes current.

Take a look at the schematic in Figure 3-77, which is built around our old friend the 2N2222 NPN transistor. For testing purposes, I have shown a normally closed pushbutton to represent an alarm sensor. I realize that you don't have a normally closed pushbutton in your supply of parts to build this circuit, but just use your imagination until we're ready to breadboard it.

So long as the pushbutton remains closed, it connects the base of the transistor to the negative side of the power supply through a 1K resistor. At the same time, the base is connected with the positive side of the power supply through a 10K resistor. Because of the difference in resistances, the base is closer to zero volts than nine volts, holding the transistor below its turn-on threshold. As a result, the transistor will not pass much current, and the LED should not have enough voltage to light up.

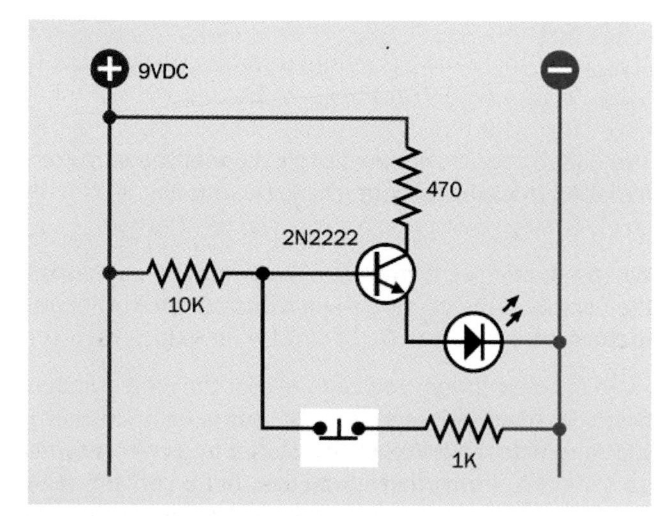

Figure 3-77 *A basic circuit in which an LED lights up when a normally closed pushbutton is opened.*

Now what happens when the pushbutton is opened? The base of the transistor loses its negative power supply and has only a positive power supply. It becomes much more positive, and tells the transistor to lower its resistance and pass more current. The LED now glows brightly. Thus, when the pushbutton breaks the connection, the LED is turned on.

This seems to be a workable system. Multiple sensors will be needed, for various doors and windows, but that's OK—we can wire as many as we like in series, as shown in Figure 3-78, where an alarm sensor can be substituted for each pushbutton. The wiring can be laid all around the house, as its total resistance should be low relative to the 10K resistor.

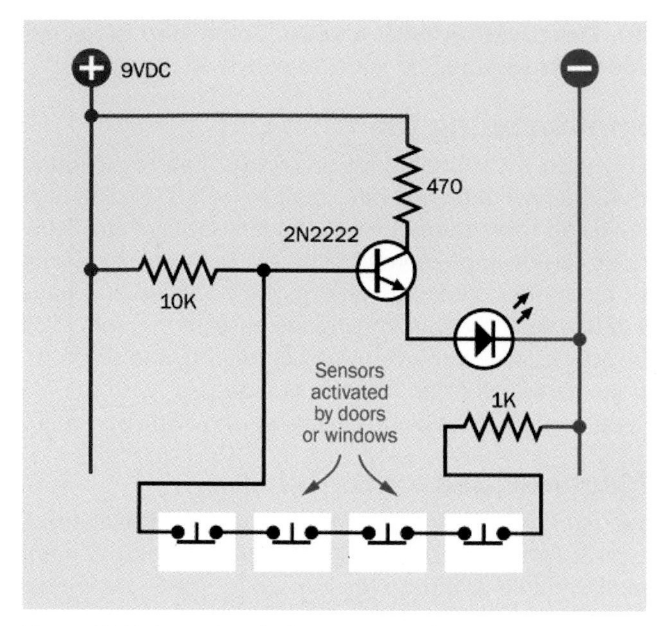

Figure 3-78 *In a network of sensors, wired in series, any one sensor will break continuity and trigger the transistor.*

While all the sensors remain closed, the transistor is drawing very little current—probably about 1mA. For development and demonstration purposes, you can run it with a 9-volt battery. For actual use, you would really want a 12-volt alarm battery that is maintained by an automatic charging system. That's outside the scope of this book, but bear in mind that alarm batteries and chargers are widely available if you ever want them.

Now suppose we swap out the LED and put a relay in there instead, as shown in Figure 3-79. (I've shown a double-pole relay, even though we have no use for the second pole right now.) So long as all the pushbuttons remain closed, the base of the transistor is held at a relatively low voltage, so the transistor does not apply power to the coil of the relay, and its contacts remain in the state shown.

When any sensor is opened, the higher voltage on the base of the transistor causes it to conduct current to the relay coil, which starts the alarm, as in Figure 3-80. (It's OK to use a relay in this mode, because the relay will not be "always on." It will normally be off, and will draw power only when the alarm is triggered.) Note that I eliminated the 470-ohm resistor from the circuit, because the relay doesn't need any protection from the power supply.

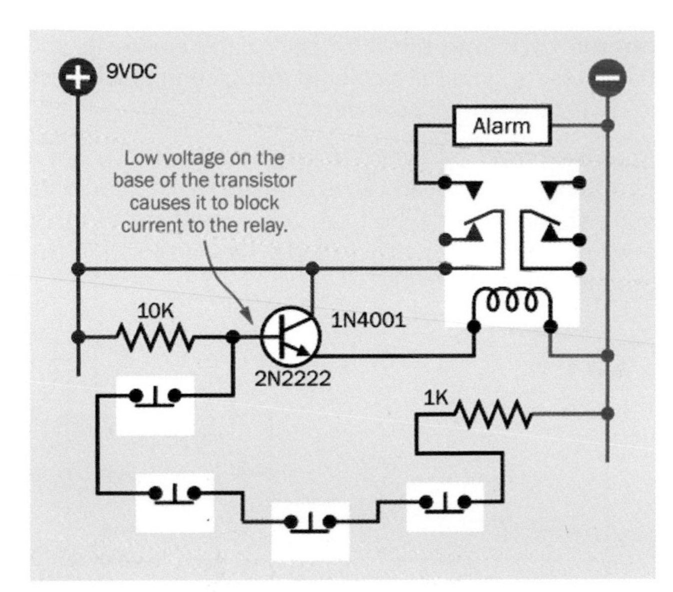

Figure 3-79 *In this circuit, the relay will be activated when any switch in the sensor network is opened.*

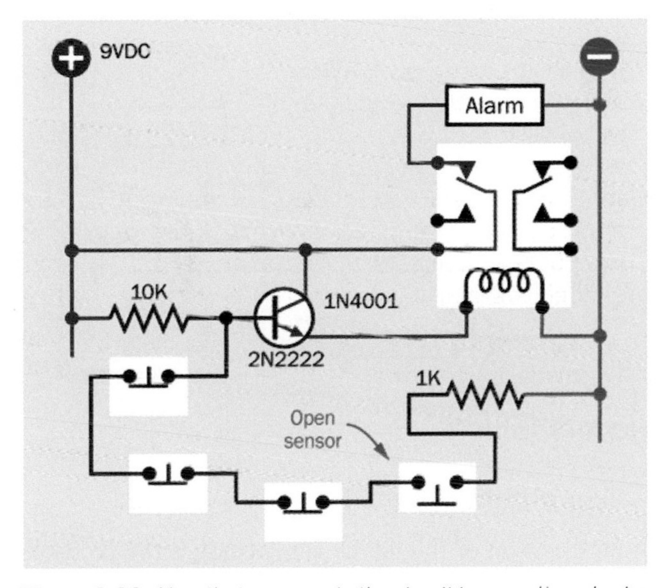

Figure 3-80 *Now that a sensor in the circuit is open, the relay is activated by the transistor.*

You can build this circuit yourself, using the same relay that you used in Experiment 7 (see "Experiment 7: Investigating a Relay" on page 60). But maybe you should wait until I develop it a little further.

A couple of things that you may want to consider:

- Will the relay overload the transistor? You can find the answer by looking at the datasheets for these two components.
- Remember that a transistor imposes a small voltage drop, even when it is "on." Will the voltage still be sufficient to activate a 9-volt relay? The datasheet for the relay will tell you the minimum operating voltage for its coil. You can verify this by testing it.

Self-Locking Relay

The circuit that I have developed so far will activate the alarm when any sensor is opened. That's good, but what happens if the sensor goes back into its closed state? The low voltage is reapplied to the base of the transistor, so it switches off the alarm. This is not good.

According to item #8 on my wish list, the alarm should be self-sustaining. It must continue making noise even after someone who has opened a door or window closes it again quickly. Therefore, the relay must lock itself on, somehow.

One way to do this would be by using a *latching relay*, which remains in either of two states, and only requires power to flip from one to the other. But a latching relay has two coils, and would require extra circuitry to unlatch it when you want to turn the alarm off. Really, it's easier to use a nonlatching relay, and I can think of a way to keep the relay switched on indefinitely, after it has received just one jolt of power.

The secret is revealed in Figure 3-81. In this view, the far-right pushbutton has closed again after being open, so the transistor has switched off—but the relay is still on, because a wire now connects its contacts back to its own coil. When the relay activates the alarm, it also activates itself.

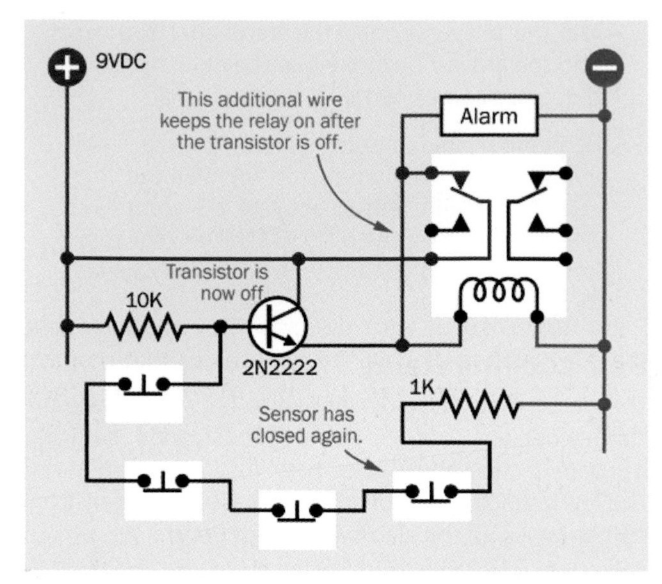

Figure 3-81 *The sensor has closed again. The transistor is no longer active, but the alarm is locked on.*

Figure 3-82 clarifies this concept by showing the path that current can take. So long as the contacts of the relay are closed, the coil of the relay is energized via its own contacts. In this way, it keeps itself switched on.

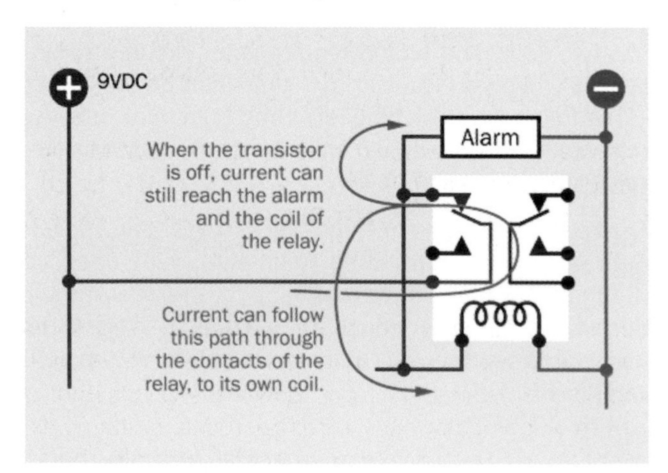

Figure 3-82 *Close-up from the previous schematic, showing how the relay keeps itself switched on.*

Blocking Bad Voltage

This looks promising, but there's a problem. The picture in Figure 3-81 was not totally accurate. Take a look at Figure 3-83. The top part of this figure is another close-up of the relevant part of the circuit. When the alarm has locked itself on, but the transistor has turned off, current can run back from the relay coil to the emitter of the transistor. I should have colored this section of wire red, as it will be relatively positive.

Applying power backward through a transistor is not a nice thing to do. It can cause damage. What should be done about this? Maybe I can use something to block this reverse flow: a rectifier diode. This is shown in the bottom half of Figure 3-83.

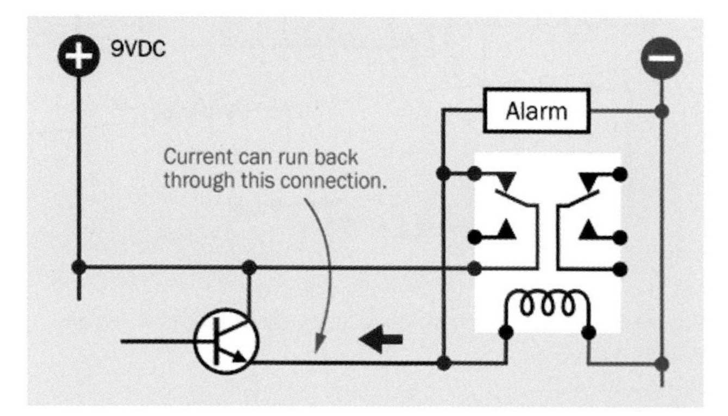

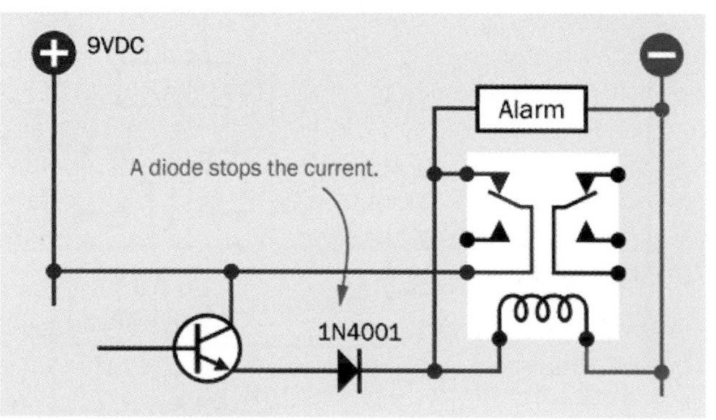

Figure 3-83 *A diode can be added to prevent current from forcing its way back into the transistor when the alarm has locked itself on and the transistor switches off.*

A new version of the full circuit, including the diode, is shown in Figure 3-84.

But what exactly is a diode? Is it the same as a light-emitting diode (LED)? Well, yes and no.

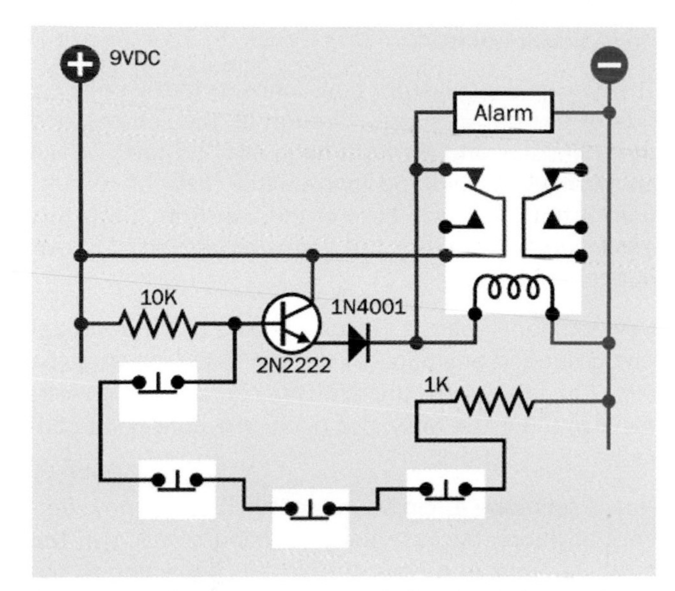

Figure 3-84 *The full circuit, now including the diode.*

Fundamentals: All About Diodes

A diode is a very early type of semiconductor. It allows electricity to flow in one direction, but blocks it in the opposite direction. Like its more recent cousin, the LED, a diode can be damaged by reversing the voltage and applying excessive power, but most diodes generally have a much greater tolerance for this than LEDs. In fact, they are designed to block reverse voltage, up to a limit specified by the manufacturer.

The end of the diode that stops positive voltage is always marked, usually with a circular band, as shown in Figure 3-25. The marked end is called the *cathode*. The other end is the *anode*, and remains unmarked. Diodes are sometimes useful in logic circuits, and can also convert alternating current (AC) into direct current (DC). If a diode isn't strong enough to withstand the current that you want it to block, you simply use a bigger diode. They come in many sizes.

It's good practice to use diodes at less than their rated capacity. Like any semiconductor, they can overheat and burn out if they are subjected to mistreatment.

The symbol for a diode looks like the heart of an LED, with the circle and the arrows removed. Three variants are shown in Figure 3-85.

Figure 3-85 *Three schematic symbols that are used to represent a diode. They are functionally identical.*

One Problem Creates Another

Previously, I had to solve the problem of how to make the relay keep itself switched on. I solved that problem by adding an extra wire, but the wire created a new problem, in which current could flow back to the transistor. I solved that problem by adding a diode, but this creates yet another problem.

We have to pay a fee for the service that the diode provides, just as we have to pay a fee for the service that the transistor provides. In fact, because both of these components are semiconductors, the fee is very similar. It consists of a reduction in voltage.

When the relay is off, current must switch it on by passing through the transistor and then through the diode. After the relay is on, it keeps itself on, and that's not a problem. But a transistor imposes a penalty of around 0.7V, and the diode imposes an additional penalty of about 0.7V, making 1.4V total. This voltage penalty is fixed, regardless of the supply voltage.

I think 9-volt relays should work reliably at 7.6V. My Omron datasheet tells me that the G5V-2 series, which I have recommended, needs 75% of its supply voltage, which would be only 6.75V. That seems like a reasonable margin of error.

But what if someone substitutes a different relay? Some have narrower specifications than others. Or what if someone is using a battery to power the circuit, and its voltage drops below 9V? A designer should always expect the unexpected, and as a general principle, should use components as close to their rated values as possible.

A couple of readers wrote to me about the issue of voltage reduction when the circuit appeared in the first edition of the book. (Yes, I do pay attention to reader feedback.) At that time, I had specified a 12VDC power source, and felt that the 1.4V penalty imposed by the transistor and the diode would be acceptable. But in this edition, I decided that all the experiments should work with 9VDC power supplies, so that you don't have to buy an AC adapter and can just use 9V batteries if that's what you prefer. Unfortunately a deduction of 1.4V from 9V is not so acceptable.

You see how a decision leads to consequences. Now that I am using a 9VDC power supply, I think I need a better way to make the relay lock itself on.

Solving the Problem

The first step to solving a problem is to be very clear about what's happening.

The task of controlling the alarm is shared by two components: the transistor and the relay. The transistor starts the alarm. After that, the transistor doesn't do anything. It switches off, and the relay has the task of keeping itself locked on. The weakness in this system is that when a task is shared by two components, they can interfere with each other. A better plan would be to have one component in charge of everything. I should maintain the transistor in a controlling role. It should keep itself turned on, and so long as it is on, it will keep the relay on.

Ah—now I see how to fix it. All I need is to use the second pole of the relay (which is the same relay that you already used in Experiment 7). I can use the contacts in the second pole, which are normally closed, to ground the chain of sensors, as shown in Figure 3-86.

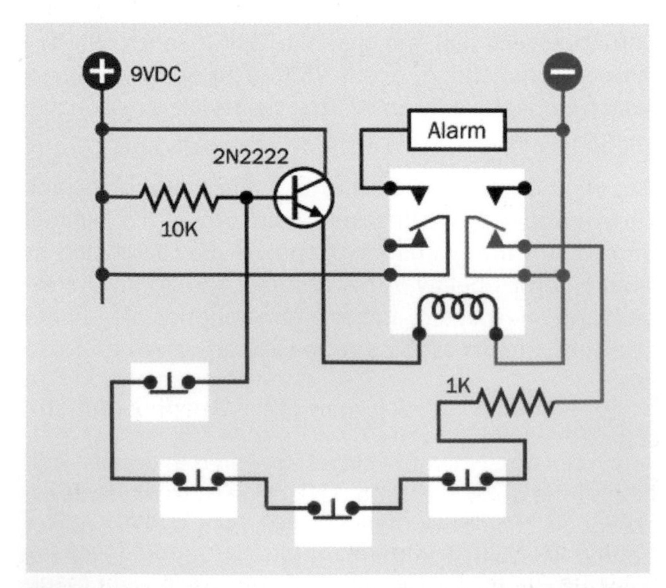

Figure 3-86 *The chain of sensors is now grounded through the right-hand contacts in the relay, which are normally closed.*

Here's how it will work:

The base of the transistor now connects to the negative side of the power supply through all the sensors, and the 1K resistor, and the right-hand relay contacts (which are normally closed). So long as this chain of connections is unbroken, the base of the transistor is at a low enough voltage to stop the transistor from passing current.

Now someone opens a sensor. The base of the transistor isn't grounded anymore, so the transistor activates the relay. The relay closes the left-hand contacts, which start the alarm. But the relay also opens the right-hand contacts.

Now if someone re-closes the sensor, it makes no difference anymore, because the right-hand contacts of the relay are open, and have cut off the connection to the negative side of the power supply. The transistor continues to pass current, and the relay remains active. This is shown in Figure 3-87.

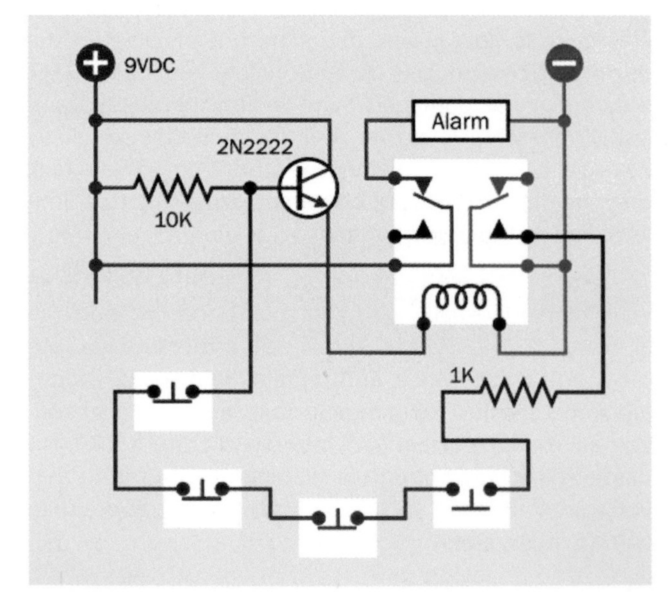

Figure 3-87 *Now that a sensor has been opened, the transistor remains energized even if the sensor is subsequently closed.*

This solves the problem.

Protection Diode

As you saw above, I eliminated the diode from the circuit. But if you take a look at Figure 3-88 (which I promise is the last version, at least for now) you will notice that the diode has found its way back in again, although it's doing something very different. It is now in parallel with the coil of the relay. What's it doing there, of all places?

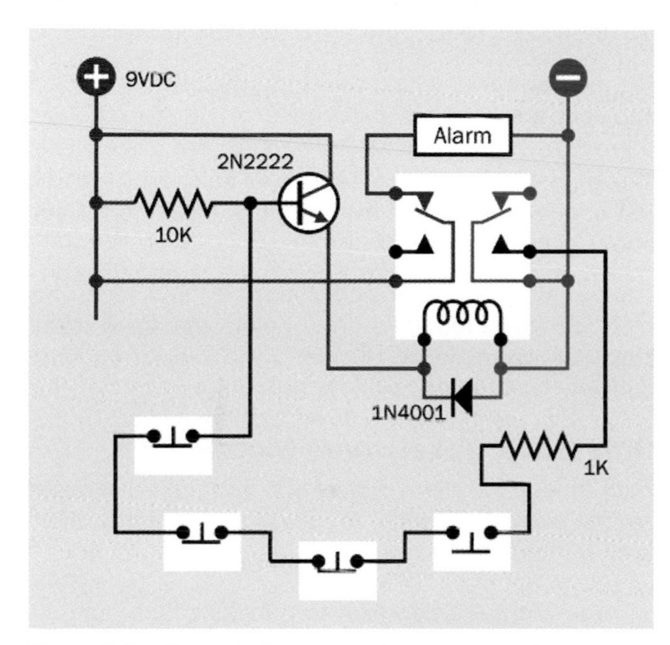

Figure 3-88 *The diode has returned, now serving a function as a protection diode.*

Much later in the book, I'll be discussing coils. One thing I can tell you right now is that a coil of wire stores energy when you apply power, and releases the energy when you disconnect the power. The release of energy creates a surge of current that can harm some types of components, especially semiconductors.

Therefore, it is standard procedure to add a *protection diode* across a relay coil. The diode is oriented so that it blocks the normal flow of current, forcing it to go through the coil, which is where we want it. But when the flow stops, and the coil tries to release its energy, the diode is there, saying to the relay, "I have a very low resistance in that direction. Why don't you shunt the current through me, instead of hassling the other components with it?"

And that is precisely what happens.

If you're only using a small relay with a small coil that doesn't take much current, you can get by without using a protection diode. Still, it's good practice, and you should always be in the habit of using one.

Time to Breadboard

I've given you a lot of explanations in this experiment, which is not what I normally like to do. But I had to show how a circuit is developed from the ground up. Now, finally, I'd like you to build it—otherwise, how will you know if it really works?

Figure 3-89 shows a breadboard layout. Instead of a noisemaker for the alarm, I've used an LED for demonstration purposes. I'll discuss noisemaking options in a moment.

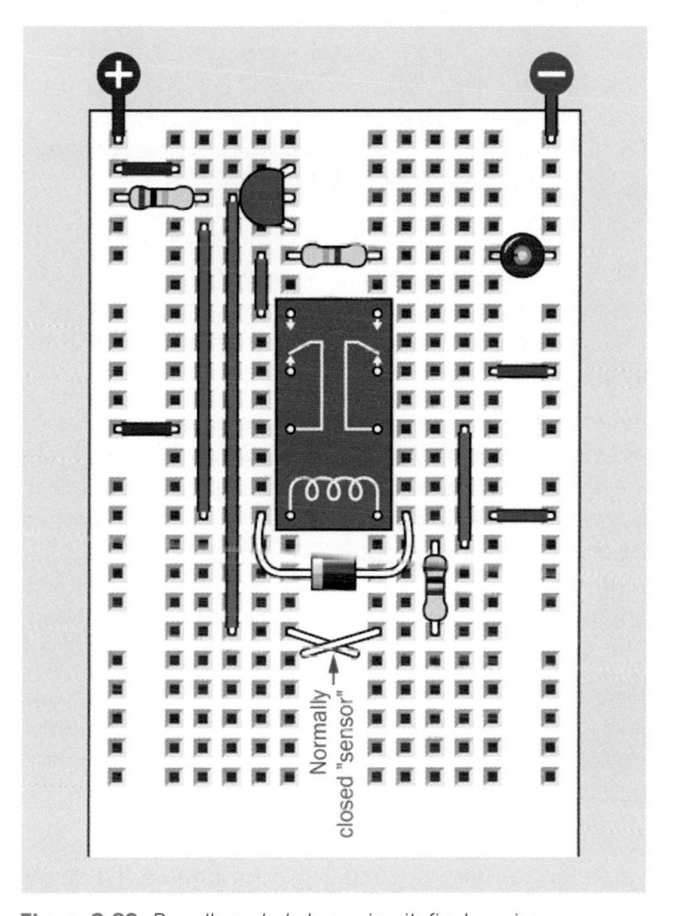

Figure 3-89 *Breadboarded alarm circuit, final version.*

Figure 3-90 shows an x-ray view of the breadboarded circuit

To simulate the alarm sensors on the breadboard, I should have used normally closed pushbuttons. But I wanted to minimize the component costs, and if you actually decide to use this alarm circuit, you'll want magnetic sensors, not pushbuttons. Consequently, as a substitute, I used two normally closed pieces of wire. This is adequate for testing. I will call them the "sensor wires." You can see them crossing over each other below the relay.

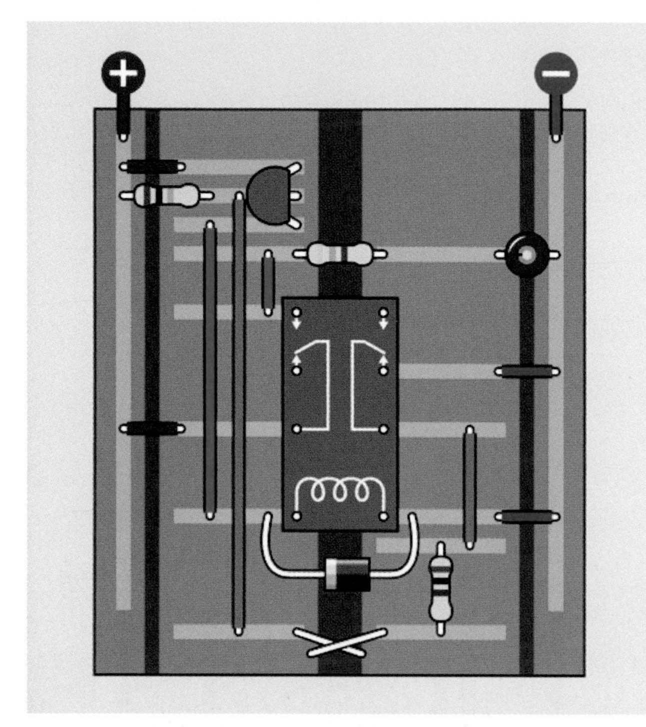

Figure 3-90 *X-ray view of the breadboarded alarm circuit.*

Make sure the wires are touching each other when you apply power to the circuit. Initially, nothing should happen.

Now disconnect the sensor wires. The LED comes on, and if you build the next version of this circuit, a noise-maker will sound, showing that the alarm has been triggered.

Now reconnect the sensor wires, imitating a situation where an intruder opened a window, heard the alarm, and quickly closed the window again. If you wired your circuit correctly, the LED will stay on.

So far so good. We have a functional circuit here. The alarm locks itself on.

But in that case—how do you ever get it to stop?

No problem. Just disconnect the power. The relay relaxes back to its default position, and the next time you apply the power, it will be in standby mode again. In a finished version of this project, you should have to enter some kind of secret code on a keypad to turn the alarm off. In Experiment 21, I'll be suggesting a way to create a passcode-protected system. You'll need to use logic chips, which we haven't dealt with yet.

Adding the Sound

For the alarm sound, you could use the oscillator circuit and loudspeaker from Experiment 11. Really, though, there is a better way. A little integrated circuit chip known as a 555 timer is a better tool for the job—and it just happens to be the next thing that I want to tell you about, in Experiment 16.

The 555 timer can also satisfy items #7 and #9 on my wish list, which require a delay before the alarm starts. Therefore, I'm going to put the alarm project on hold, and we'll finish it completely in Experiment 18.

Reference: Take-Home Messages

Even though the alarm project isn't finished yet, it raised several important points. I'm going to summarize them here, for future reference.

- You can use a transistor to provide a high output in response to a low input, and vice versa.
- You can wire a relay to lock itself into "on" mode, simply by feeding current back to its own coil.
- A diode can stop current from flowing into places where you don't want it to go.
- When forward current passes through a diode, the voltage is reduced by about 0.7V.
- A transistor also reduces voltage by about 0.7V.
- Voltage reduction imposed by semiconductors remains the same regardless of the supply voltage. Consequently, the reduction is more significant when the supply voltage is lower.
- A relay coil can create back-EMF (a pulse of reverse current) when it is switched off.
- A protection diode in parallel with a relay coil can suppress back-EMF. The diode should be oriented so that it blocks the normal flow of current but passes the reverse pulse created by the coil.

4 Chips, Ahoy!

Before I get into the fascinating topic of integrated circuit chips (often referred to as *ICs* or simply *chips*), I have to make a confession. Some of the things I asked you to do in previous experiments could have been done a bit more simply, if we had used chips.

Does this mean you have been wasting your time? Absolutely not! I firmly believe that by building circuits with individual components such as transistors and diodes, you acquire the best possible understanding of the principles of electronics. Still, you are going to find that chips containing dozens, hundreds, or thousands of transistor junctions will enable some shortcuts.

You may also find chips curiously addictive to play with —although you may not become quite as excited as the character in Figure 4-1.

The tools, equipment, components, and supplies described below will be useful in Experiments 16 through 24, in addition to items that have been recommended previously.

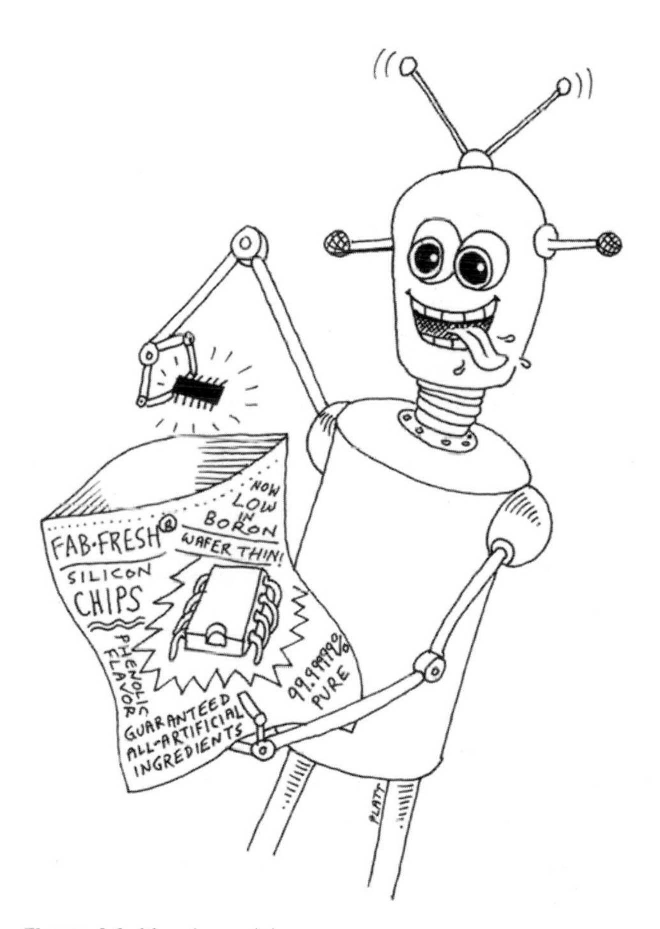

Figure 4-1 *My role model.*

Necessary Items for Chapter Four

The only new tool that you might consider using in conjunction with chips is a logic probe. This tells you whether a single pin on a chip has a high or low voltage, which can be helpful in figuring out what your circuit is doing. The probe has a memory function so that it will light its LED, and keep it lit, in response to a pulse that may have been too quick for the eye to see.

Some of my readers disagree with me, but I regard the logic probe as optional, not essential. Search online and buy the cheapest one you can find. I don't have any specific brand recommendations.

Components

As before, if you want kits containing components, see "Kits" on page 311. If you prefer to buy your own components from online sources, see "Components" on page 317. For supplies, see "Supplies" on page 316.

Fundamentals: Choosing Chips

Figure 4-2 shows two integrated circuit chips. The one at the top is of the old-school, *through-hole* design, with pins spaced at 1/10" so that they will fit through the holes in your breadboard or perforated board. These are the chips I will be using exclusively, because they're easy to handle. The smaller chip is a *surface-mount* design, which I will not be using, because they don't fit breadboards or perforated boards and are difficult to handle.

Many through-hole and surface-mount chips are functionally identical. The only difference is the size (although some surface-mount versions use a lower voltage).

The body of a chip is usually made of plastic or resin and is often referred to as the *package*. The traditional chip is usually sold in a *dual-inline package*, meaning that it has two (i.e., dual) rows of pins. The acronym for this package is *DIP* or (when it is made of plastic) *PDIP*.

Figure 4-2 *A through-hole chip (top) and surface-mount chip (bottom).*

Surface-mount packages often are identified with acronyms beginning with letter S, as in SOIC, meaning small-outline integrated circuit. Numerous surface-mount variants exist, with different pin spacing and other specifications. They are all outside the scope of this book, and if you buy your own components, you should be careful not to select them by mistake.

Inside the package, the circuit is etched on a tiny wafer of silicon, which is where the term "chip" comes from, although the whole component is now usually referred to as a chip, and I will follow that convention here. Tiny wires inside the package link the circuit with the rows of pins that protrude on either side.

The PDIP chip in Figure 4-2 has seven pins in each row, making a total of 14. Other chips may have 4, 6, 8, 16, or more pins.

Just about every chip has a part number printed on it. Notice in the photograph that even though the chips look quite different from each other, they both have "74" in their part numbers. This is because both of them are members of the family of logic chips that were assigned part numbers from 7400 and upward when they were introduced several decades ago. This is often referred to as the 74xx family, and I'll be using these chips a lot.

Take a look at Figure 4-3. The initial letters identify the manufacturer, which you can ignore, as it really makes no difference for our purposes. (If you are wondering why "SN" identifies Texas Instruments, it's because the company used to call their chips "semiconductor networks" in the early days.)

Skip the letters until you get to the "74." After that, you find two more letters, which are important. The 7400 family has evolved through many generations, and the letter(s) inserted after the "74" tell you which generation you're dealing with. Generations have included: 74L, 74LS, 74C, 74HC, and 74AHC. There are many more.

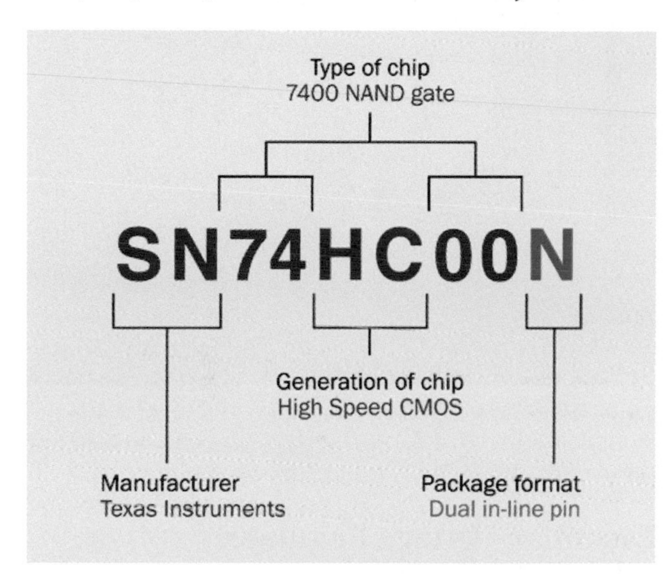

Figure 4-3 *How to decode the part number of a chip in the 74xx family.*

Generally speaking, later generations tend to be faster or more versatile than earlier generations. In this book, I am using the HC generation of the 7400 family exclusively, because almost all 7400 chips are available in it, the cost is moderate, and the chips don't use a lot of power. For our purposes, the extra speed offered by later generations is not relevant—although you can certainly use the HCT generation if you prefer.

Following the letters identifying the generation, you'll find a sequence of two, three, four, or (sometimes) five digits. These identify the specific function of the chip. Following the digits is another letter, or two letters, or more. For our purposes, those terminating letters are not important.

Looking back at Figure 4-2, the DIP chip part number, M74HC00B1, tells you that it is a chip made by STMicroelectronics, in the 74xx family, HC generation, with its function identified by numerals 00.

The purpose of this long explanation is to enable you to interpret catalog listings if you go chip shopping. You can search for "74HC00" and the search engines at online vendors are usually smart enough to show you appropriate chips from multiple manufacturers, even though there are letters preceding and following the term that you're searched for.

Just be sure that they will fit your breadboard. Limit your search results to DIP, PDIP, or through-hole packages. If the part number begins with SS, SO, or TSS, it's absolutely definitely surface-mount, and you don't want it. For a lot more information on searching and shopping, see "Searching and Shopping Online" on page 311.

All the chips needed for the experiments in this chapter of the book are listed in Figure 6-7. You will need a few other types of components, which I will list here.

Optional: IC Sockets

If you plan to immortalize any of your circuits in solder, I suggest you avoid soldering chips directly, because if you make a wiring error or damage the chip, you have to desolder multiple pins in order to remove it. This is very difficult. To avoid the problem, buy some DIP sockets, solder the sockets onto the board, and then plug the chips into the sockets. You can use the cheapest sockets you can find (you don't need gold-plated contacts for our purposes). You will need 8-pin, 14-pin, and 16-pin sockets. Quantity of each: 5 minimum. Two sockets are shown in Figure 4-4.

Figure 4-4 *To avoid the risk of damaging a chip by soldering it directly, it can be mounted in an IC socket after the socket has been soldered into a circuit board.*

Essential: Subminiature Slide Switch

A *slide switch* has a tiny lever that you slide to and fro with the tip of your finger, making and breaking an electrical contact inside the switch, as shown in Figure 4-5. It has three pins spaced 0.1" apart (2.54mm in metric). If you buy your own components, see "Other Components" on page 319 and go down to the subhead "Components for Chapter Four" for more information about switches.

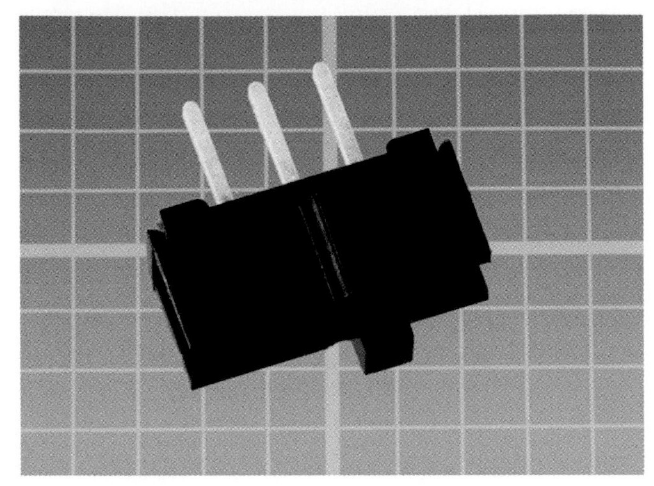

Figure 4-5 *The subminiature slide switch recommended for projects in this book.*

Caution: Switching Overload

A very small slide switch is not designed to switch significant currents or voltages. It is designed for low-powered circuits. A limit may be as low as 100mA at 12VDC. This is sufficient for our purposes. Check the manufacturer's datasheet if you want a slide switch to do more than this.

Essential: Low-Current LEDs

HC series logic chips are not designed to deliver current much beyond 5mA. You can take as much as 20mA from them to drive an LED, but this will pull down the output voltage, making it unsuitable as an input to other logic chips. I am suggesting low-current LEDs for all your experiments with logic chips.

Remember that low-current LEDs require higher-value series resistors, because they don't tolerate as much current as generic LEDs. I will mention this wherever it is important.

Essential: Numeric Displays

One of the chip projects will display its output using seven-segment numeric displays—the simple kind of digits that you still find on digital clocks and microwave ovens. See Figure 4-6. For purchasing information, see "Other Components" on page 319 and go down to the subhead "Components for Chapter Four."

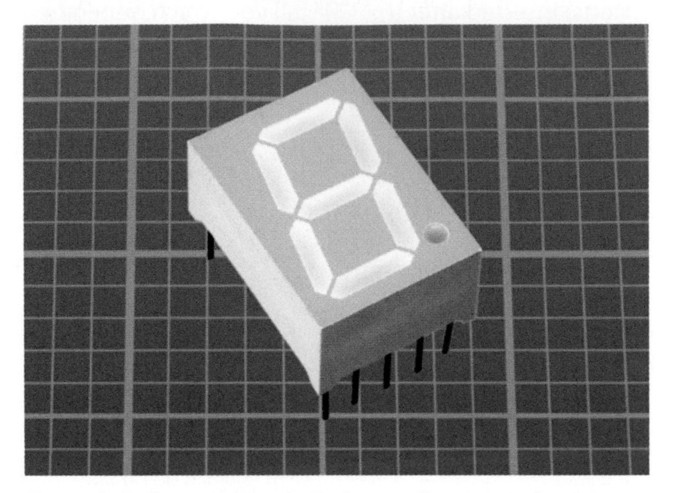

Figure 4-6 *Seven-segment displays are the cheapest way to show a numeric output and can be driven directly by some CMOS chips.*

Essential: Voltage Regulator

Because many logic chips require precisely 5 volts DC, you need a voltage regulator to guarantee this. The LM7805 does the job. The chip number will be preceded or followed with an abbreviation identifying the manufacturer and package style, as in the LM7805CT from Fairchild. Any manufacturer will do, but the regulator should look like the one in Figure 4-7. (This is known as the TO220 package style.) You need them in any logic circuit, so five would be a good number to have available.

Optional Extras

To complete the alarm system in Experiment 18, you will require magnetic sensors that you can apply to doors or windows, such as the Directed model 8601, available from dozens of sources online.

If you expect to move a project from a breadboard to a permanent enclosure, the tactile switches that you have been using will be insufficiently sturdy or accessible. For Experiment 18, you will need a full-size DPDT pushbutton switch, ON-(ON) type, with solder terminals. If you search eBay for "DPDT pushbutton" there will be no shortage of options.

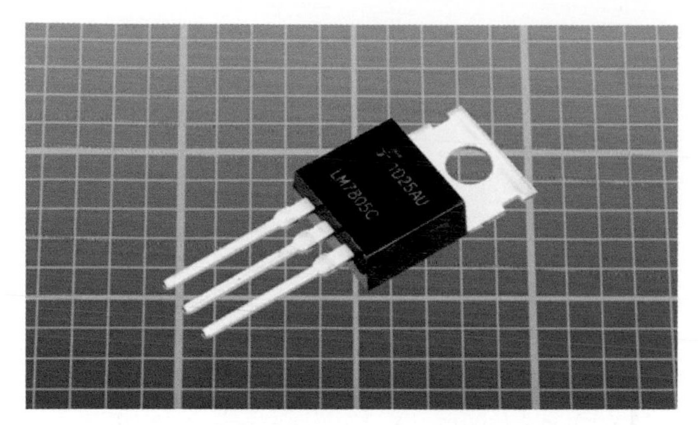

Figure 4-7 *Many integrated circuit chips require a controlled power supply of 5 volts, which can be delivered by this regulator from a supply of 7.5 to 12 volts.*

Background: How Chips Came to Be

The concept of integrating solid-state components into one little package originated with British radar scientist Geoffrey W. A. Dummer, who talked about it for years before he attempted, unsuccessfully, to build one in 1956. The first true integrated circuit wasn't fabricated until 1958 by Jack Kilby, working at Texas Instruments. Kilby's version used germanium, as this element was already in use as a semiconductor. (You'll encounter a germanium diode when I deal with crystal radios in Experiment 31.) But Robert Noyce, pictured in Figure 4-8, had a better idea.

Figure 4-8 *Robert Noyce, who patented the integrated circuit chip and cofounded Intel.*

Born in 1927 in Iowa, Noyce moved to California in the 1950s, where he found a job working for William Shockley. This was shortly after Shockley had set up a business based around the transistor, which he had coinvented at Bell Labs.

Noyce was one of eight employees who became frustrated with Shockley's management and left to establish Fairchild Semiconductor. While he was the general manager of Fairchild, Noyce invented a silicon-based integrated circuit that avoided the manufacturing problems associated with germanium. He is generally credited as the man who made integrated circuits possible.

Early applications were for military use, as Minuteman missiles required small, light components in their guidance systems. These applications consumed almost all chips produced from 1960 through 1963, during which time the unit price fell from around $1,000 to $25 each, in 1963 dollars.

In the late 1960s, MSI (medium-scale integration) chips emerged, each containing hundreds of transistors. LSI (large-scale integration) enabled tens of thousands of transistors on one chip by the mid-1970s, and today's computer chips can contain several billion transistors.

Robert Noyce eventually cofounded Intel with Gordon Moore, but died unexpectedly of a heart attack in 1990. You can learn more about the fascinating early history of chip design and fabrication at the Silicon Valley Historical Association (*http://www.siliconvalleyhistorical.org*).

Experiment 16: Emitting a Pulse

I'm going to begin our experiments with chips by introducting you to the most successful one ever made: the 555 timer. You can find numerous guides to it online, so why do I need to discuss it here? I have three reasons for doing so:

It's unavoidable. You simply have to know about this chip. Some sources estimate that more than 1 billion are still being manufactured annually. It will be used in one way or another in most of the remaining circuits in this book.

It's useful. The 555 is probably the most versatile chip that exists, with endless applications. Its relatively powerful output (rated at up to 200mA) is extremely useful, and the chip itself is hard to damage.

It's misunderstood. After reading literally dozens of guides, beginning with an early Signetics datasheet and making my way through various hobby texts, I concluded that the inner workings of the chip are seldom explained on an introductory level. I want to give you a graphic understanding of what's happening inside it, because if you don't have this, you won't be in a good position to use the chip creatively.

What You Will Need

- Breadboard, hookup wire, wire cutters, wire strippers, multimeter
- 9VDC power supply (battery or AC adapter)
- Resistors: 470 ohms (1), 10K (3)
- Capacitors: 0.01µF (1), 15µF (1)
- Trimmer potentiometers: 20K or 25K (1), 500K (1)
- 555 timer chip (1)
- Tactile switches (2)
- Generic LED (1)

Know Your Chips

The pins of a 555 timer are numbered counterclockwise (seen from above), as shown in Figure 4-9. The package has a notch, or a dimple, or both, at the end which is considered the top. The pin spacing is 1/10".

All other through-hole chips have the same specification, although they may have more pins. Usually (not always) the horizontal spacing between the two rows of pins is 3/10", which means that the chip neatly straddles the channel down the middle of a breadboard, and the conductors inside the breadboard allow you to have access to each pin of the chip. Yes, *that's* why the breadboard is designed that way.

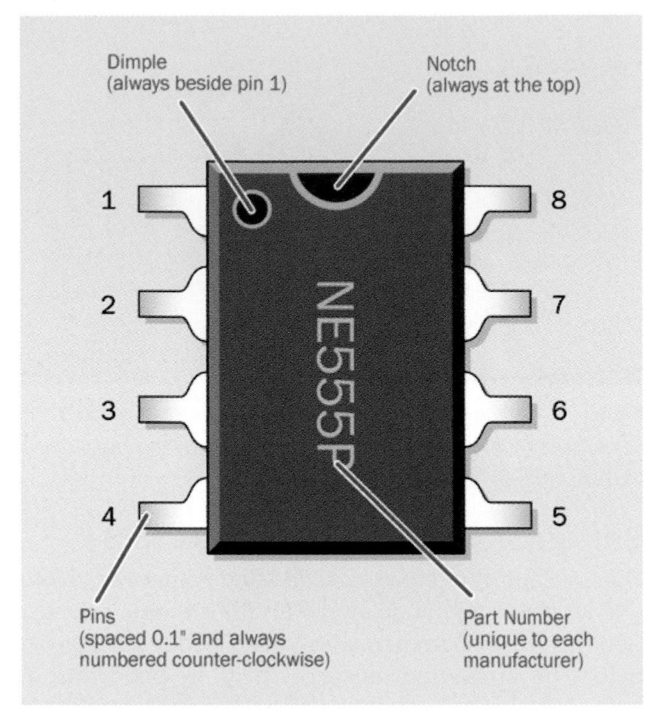

Figure 4-9 *The package design of an eight-pin chip. While virtually all chips have a semicircular notch at the top, some do not have a dimple beside pin 1.*

Monostable Test

The pins on a 555 timer also have names, as shown in Figure 4-10. A diagram like this tells you the *pinouts* of the chip. I'll be explaining the function of each pin—but as usual, I prefer you to make a preliminary investigation of your own.

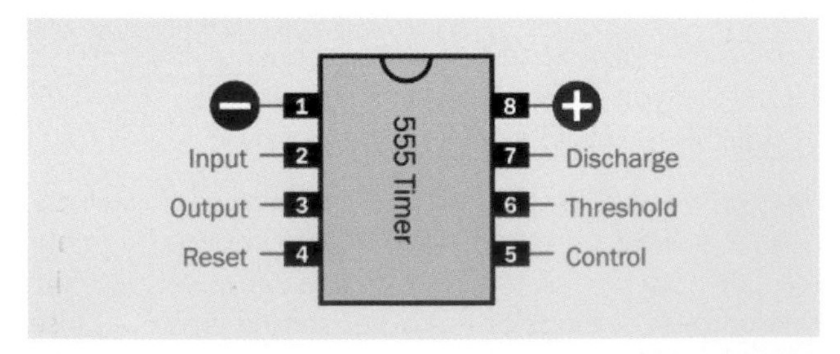

Figure 4-10 *The pinouts of the 555 timer chip.*

The schematic of a test circuit for the timer is shown in Figure 4-11.

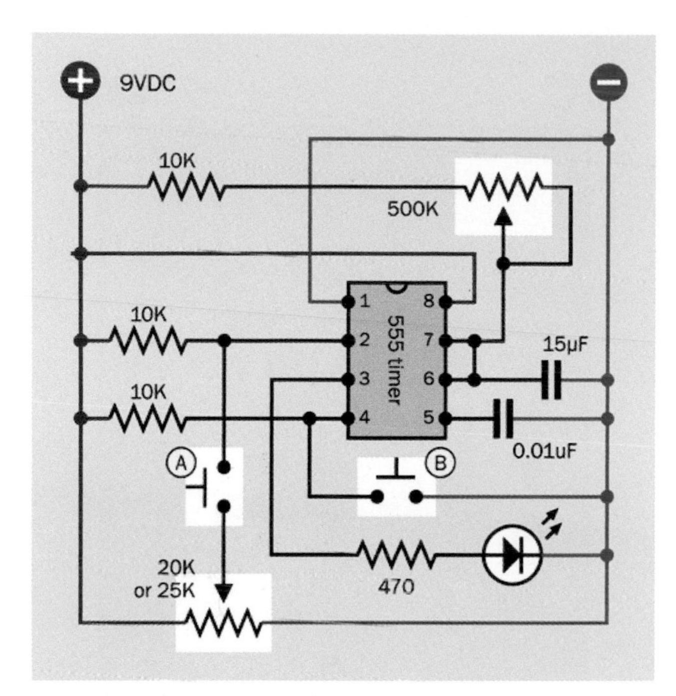

Figure 4-11 *A circuit to assist in your investigation of the 555 timer chip.*

You can set up this circuit on a breadboard as shown in Figure 4-12. Note that near the bottom-left corner there is a short red jumper connecting the top section of the positive bus to the section immediately below it. The jumper is there in case your breadboard is the type that has a break in its bus.

The component values are shown in Figure 4-13. To assist you in visualizing the connections, an x-ray version is shown in Figure 4-14.

Apply some power, and nothing happens. The timer is waiting for you to trigger it. Set it up by turning the 500K trimmer to the middle of its range.

Now rotate the 20K trimmer all the way counterclockwise, and press button A. If still nothing happens, rotate the 20K trimmer all the way clockwise, and try again. One of these settings or the other should create a pulse from the LED, depending on which way around you plugged in your trimmer. If you don't get anything, there is an error in your circuit.

Check the schematic, and you see that pin 2 of the timer —the trigger pin—is hardwired through a 10K resistor to the positive side of the power supply. But a purple wire also connects with the trigger pin, and runs down, through a tactile switch, to the trimmer. If the trimmer is rotated so that the wiper connects directly to the negative ground side of the power supply, this will allow the tactile switch to overwhelm the 10K resistor and apply low voltage to pin 2. This triggers the timer.

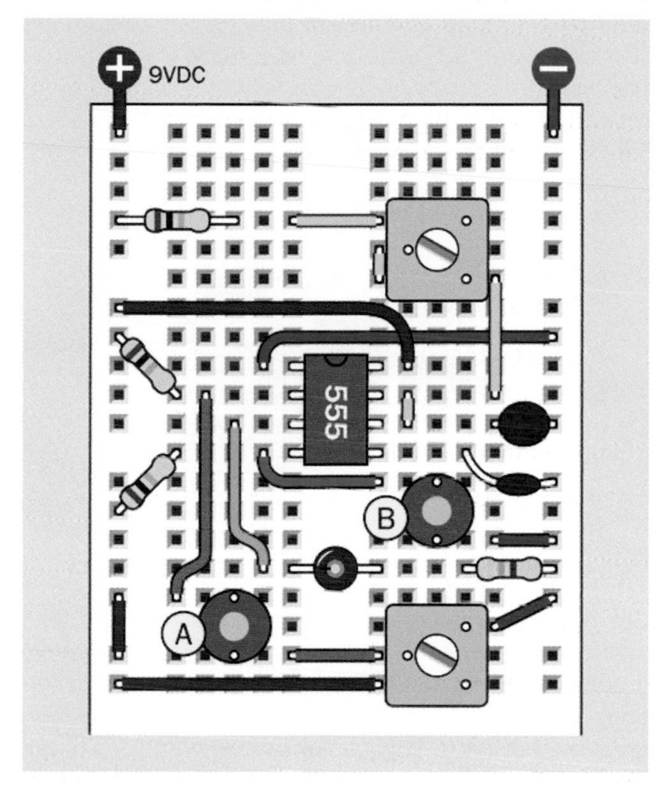

Figure 4-12 *Breadboarded version of the timer test circuit.*

If the 20K trimmer is turned all the way in the opposite direction, button A will apply positive voltage directly to pin 2, and because pin 2 already has positive voltage through the 10K resistor, an additional positive voltage through button A makes no difference.

- Positive voltage on the trigger pin is ignored by the chip.
- A drop in voltage on the trigger pin will trigger the chip.

But, how positive is positive, and how much of a drop will be low enough to act as a trigger? Let's find out.

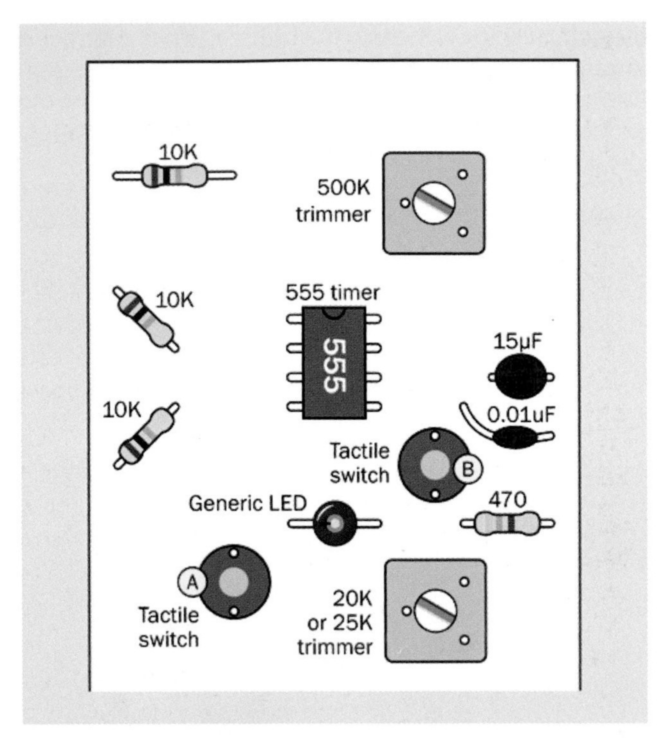

Figure 4-13 *Component values for the timer test circuit.*

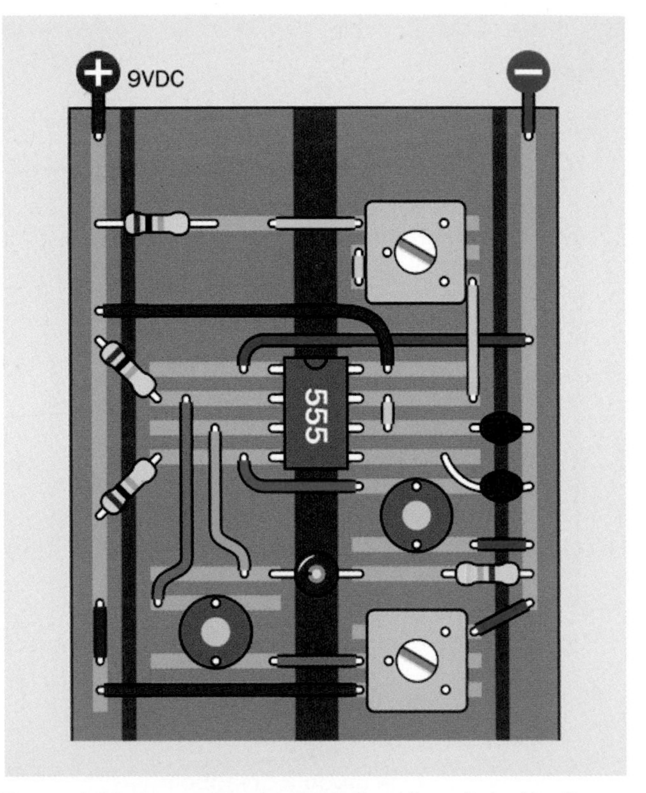

Figure 4-14 *Connections inside the breadboard wired for the timer test.*

Get out your meter, set it to measure DC volts, and measure the voltage between pin 2 and negative ground while you adjust the 20K trimmer to various settings and press button A. I'm betting that when you press the button to apply a voltage below 3 volts to pin 2, the timer will flash the LED. Above 3 volts, I doubt that anything will happen.

- The timer is triggered by a voltage on its trigger pin that is one-third of the supply voltage (or less).
- The LED will continue to glow after you release the button.
- You can press the button for any length of time that is less than the timer's cycle time, and the LED always emits the same length of pulse.

Figure 4-15 shows how the timer is behaving, in graphical format. The 555 converts the imperfect world around it into a precise and dependable output. It doesn't switch on and off absolutely instantly, but is fast enough to *appear* instant.

Now try triggering the timer while you turn the 500K trimmer to different positions. You will find that this adjusts the length of the pulse.

- The resistance between pin 7 and the positive side of the power supply determines how long the pulse from the timer will last (in conjunction with the capacitor on pin 6).

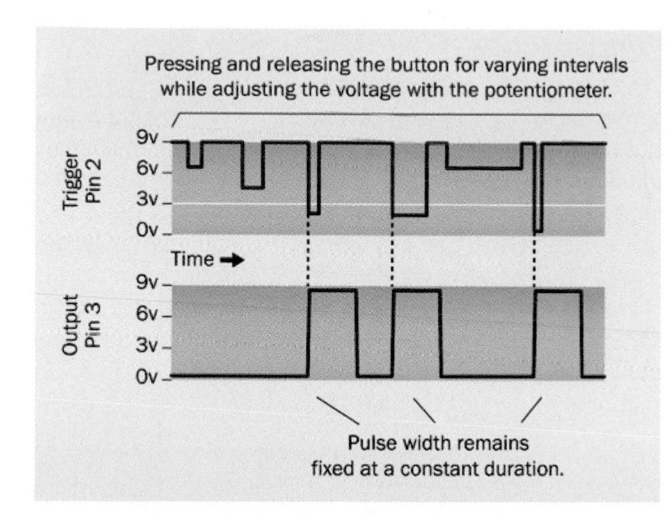

Figure 4-15 *How a 555 timer responds to different durations and voltages on its trigger pin.*

Here's another thing to try. Set the 500K trimmer to give a long pulse. Press button A, and then quickly press button B, which will stop the pulse before it completes. Hold down button B while you try to trigger the timer again with button A, and nothing happens.

- Pin 4 is the reset pin. When you ground it, you force the timer to interrupt whatever it is doing, and it is immobilized until you release pin 4 from its connection with negative ground.

Lastly, let go of button B, hold down button A, and continue to hold it down. This prolongs the pulse from the timer, until you let go of button A.

- Maintaining a low voltage on the trigger pin of the timer will *retrigger* it indefinitely.

Regarding the 10K resistors attached to pins 2 and 4—these are known as *pullup* resistors, because they hold the pins at a positive level. A more direct connection to negative ground will overwhelm the pullup resistor.

The concept of a pullup resistor is important when you are dealing with chips, because you must never allow an input pin to remain unconnected. An unconnected pin is said to be *floating*, and can cause trouble, as it may pick up stray electromagnetic fields, and we won't know what voltage is on it from one moment to the next.

Is there such a thing as a *pulldown* resistor? Absolutely. But the 555 timer needs pullup resistors, because pin 2 or pin 4 is kept in a normal state by positive voltage, and is activated by a low voltage.

- The 555 timer is triggered or reset by negative voltages on pins 2 and 4 respectively.

Timing the Pulse

If you study the schematic in Figure 4-11, you can see that positive current reaches pin 7 (the discharge pin) by passing through a 10K resistor and a 500K trimmer. (The 10K resistor is there because pin 7 should not be connected directly to the positive side of the power supply.)

You can also see that after passing through the 500K trimmer, the current can reach a 15µF capacitor. Hmmmm, a resistor followed by a capacitor—does this look like an RC network? Is the timer using the combination of the resistance and the 15µF capacitor to determine the length of the output pulse?

Yes, that's exactly what is going on. Inside the timer chip, some clever electronics are sensing the voltage on the 15µF capacitor, and the timer uses this to terminate its output pulse.

You can measure this yourself. Set the 500K trimmer to create a long pulse, and use your meter to measure the voltage on the left side of the 15µF capacitor. You should see it climbing up—until it reaches about 6 volts. The timer uses this as a signal to stop its output pulse, and the voltage quickly goes down again, because the timer is grounding it internally. This is why pin 7 is known as the discharge pin: the timer discharges the capacitor through it.

- When the voltage on the timing capacitor reaches two-thirds of the supply voltage, the timer ends its output pulse.

But why are the discharge pin and the threshold pin tied together? You'll learn about that in the next experiment, when the timer is rewired to deliver a series of pulses, instead of just one. At that time, the timer will be running in *astable* mode. Currently, you are using it in *monostable* mode.

- In monostable mode, the timer delivers only one pulse in response to a triggering event.
- In astable mode, the timer delivers a continuing series of pulses.

Finally, you may be wondering about the purpose of the 0.01µF capacitor attached to pin 5. This pin is the "control" pin, which means that if you apply a voltage to it, you can control the sensitivity of the timer. Because we are not using this function yet, it's good practice to put a capacitor on pin 5 to protect it from voltage fluctuations and prevent it from interfering with normal functioning.

Caution: Beware of Pin-Shuffling!

In all of the schematics in this book, chips are shown exactly as you would see them on a breadboard, with the pins in numerical sequence.

Other schematics that you may find, on websites or in books, do things differently. For convenience in drawing circuits, people often resequence the pin numbers. Also, there is no attempt to replicate a breadboard layout with a positive bus and a negative bus at each side. To give you an example, in Figure 4-16 the circuit is identical to that in Figure 4-11, but the pins have been shuffled to simplify connections and minimize wiring crossovers.

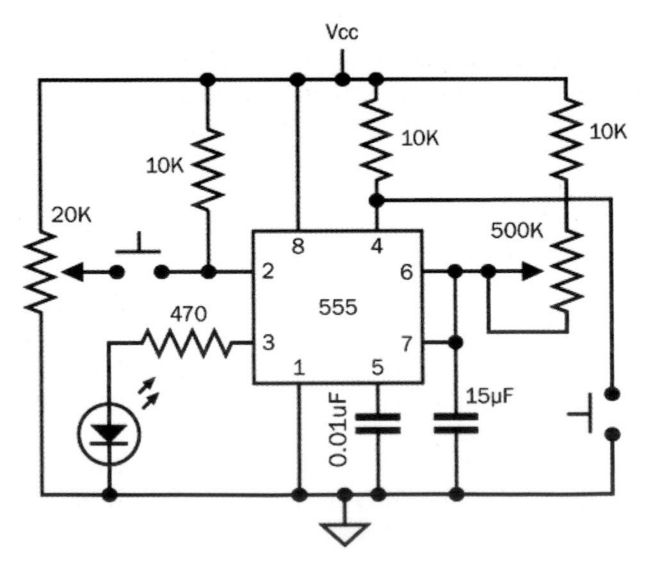

Figure 4-16 *This circuit is identical in function to the test circuit shown previously, but pins on the chip have been resequenced to simplify the schematic.*

Pin shuffling can create a circuit that is easier to understand in some ways (especially if positive power is at the top and ground at the bottom), but you have to convert the layout, often using pen and paper, before you can build it on a breadboard.

Fundamentals: Timer Duration

When you made your own RC network in Experiment 9, some annoying calculations were necessary to figure out how long a capacitor would take to reach any particular voltage. Using a 555 timer, everything is much easier. You just look up the duration of its output pulse in a table such as the one in Figure 4-17. The resistance between pin 7 and the positive side of the power supply is shown along the top of the table, the value of the timing capacitor is shown down the left side, and the numbers in the table tell you the approximate pulse duration in seconds.

	10K	22K	47K	100K	220K	470K	1M
1000µF	11	24	52	110	240	520	1100
470µF	5.2	11	24	52	110	240	520
220µF	2.4	5.2	11	24	52	110	240
100µF	1.1	2.4	5.2	11	24	52	110
47µF	0.52	1.1	2.4	5.2	11	24	52
22µF	0.24	0.53	1.1	2.4	5.3	11	24
10µF	0.11	0.24	0.52	1.1	2.4	5.2	11
4.7µF	0.052	0.11	0.24	0.52	1.1	2.4	5.2
2.2µF	0.024	0.052	0.11	0.24	0.53	1.1	2.4
1.0µF	0.011	0.024	0.052	0.11	0.24	0.52	1.1
0.47µF		0.011	0.024	0.052	0.11	0.24	0.52
0.22µF			0.011	0.024	0.052	0.11	0.24
0.1µF				0.011	0.024	0.052	0.11
0.047µF					0.011	0.024	0.052
0.022µF						0.011	0.024
0.01µF							0.011

Figure 4-17 *Pulse duration, in seconds, of a 555 timer running in monostable mode, for values of the timing resistor and timing capacitor. Times are rounded to two digits.*

- Resistor values below 1K should not be used.
- Resistor values below 10K are undesirable, as they increase power consumption.
- Capacitor values above 100µF may produce inaccurate results because leakage in the capacitor becomes comparable with its charging rate.

What if you want a time longer than 1,100 seconds or shorter then 0.01 seconds? Or what if you want a pulse duration that falls somewhere between the values in the table?

You can use this simple formula, where T is the pulse time in seconds, R is the resistance in *kilohms*, and C is the capacitance is *microfarads*.

```
T = R × C × 0.0011
```

Bear in mind that the result may not be exact, because resistor and capacitor values can be inaccurate, and because of other factors such as ambient temperature.

Theory: Inside the 555 in Monostable Mode

The plastic body of the 555 timer contains a wafer of silicon on which are etched dozens of transistor junctions in a pattern that is too complex to be explained here. However, I can summarize their function by dividing them into groups, as shown in Figure 4-18.

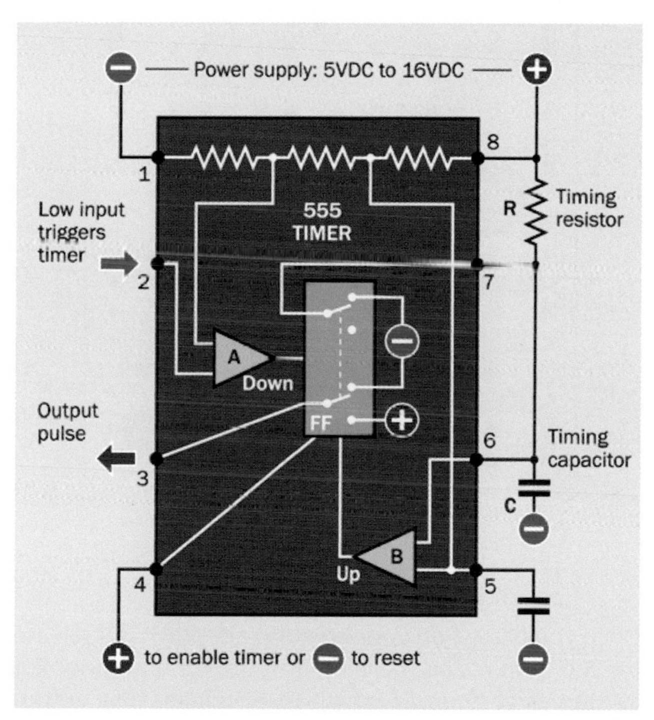

Figure 4-18 *Simplified representation of internal functions of a 555 timer, wired in monostable mode.*

The negative and positive symbols inside the chip are power sources that actually come from pins 1 and 8, respectively. I omitted the internal connections to those pins for the sake of clarity.

The two yellow triangles are *comparators*. Each comparator compares two inputs (at the base of the triangle) and delivers an output (from the apex of the triangle) depending on whether the inputs are similar or different. FF is a *flip-flop*, a logic component that can rest in one state or the other. I have depicted it as a double-throw switch, although in reality it is solid-state.

Initially when you power up the chip, the flip-flop is in its "up" position, which delivers low voltage through the output, pin 3. If the flip-flop receives a signal from comparator A, it flips to its "down" state, and flops there. When it receives a signal from comparator B, it flips back to its "up" state, and flops there. The "Up" and "Down" labels on the comparators will remind you how each one changes the switch when it is activated. Some people feel that the term "flip-flop" is derived from it having two states named "flip" and "flop." But I prefer to think of it flipping and flopping.

Notice the external wire that connects pin 7 with capacitor C. As long as the flip-flop is "up," it sinks the positive voltage coming through resistor R to pin 7, and prevents the capacitor from charging positively.

If the voltage on pin 2 drops to 1/3 of the supply, comparator A notices this, and flips the flip-flop to its "down" position. This sends a positive pulse from pin 3, the output pin, and also disconnects the negative power from pin 7. Now the capacitor can start charging through the resistor. While this is happening, the positive output from the timer continues.

As the voltage increases on the capacitor, comparator B monitors it through pin 6. When the capacitor accumulates 2/3 of the supply voltage, comparator B sends a pulse to the flip-flop, flipping it back into its original "up" state. This discharges the capacitor through pin 7. Also, the flip-flop ends the positive output through pin 3 and replaces it with a negative voltage. This way, the 555 returns to its original state.

I'll sum up this sequence of events:

- Initially, the flip-flop grounds the capacitor and grounds the output (pin 3).
- A drop in voltage on pin 2 to 1/3 of the supply voltage or less makes the output (pin 3) positive and allows capacitor C to start charging through resistor R.

- When the capacitor reaches 2/3 of the supply voltage, the chip discharges the capacitor, and the output at pin 3 goes low again.

Fundamentals: Pulse Suppression

When power is first applied to a timer that is configured in monostable mode, the timer tends to emit one pulse spontaneously before going dormant and waiting to be triggered again. This can be annoying in many circuits.

One way to prevent it is by putting a 1μF capacitor between the Reset pin and negative ground. The capacitor sinks current from the Reset pin when the power is first turned on, and holds the pin low for a fraction of a second—just long enough to stop the timer from emitting its waking-up pulse. After the capacitor is charged, it doesn't do anything more, and a 10K resistor holds the reset pin positive, so it won't interfere with the running of the timer.

I'll be using the pulse-suppression concept in subsequent experiments.

Fundamentals: Why the 555 Is Useful

In its monostable mode, the 555 will emit a single pulse of fixed (but programmable) length. Can you imagine some applications? Think in terms of the pulse from the 555 controlling some other component. A motion sensor on an outdoor light, perhaps. When an infrared detector "sees" something moving, the light comes on, but only for a specific period—which can be controlled by a 555.

Another application could be a toaster. When someone lowers a slice of bread, a switch will close, triggering the toasting cycle. To change the length of the cycle, you could use a potentiometer and attach it to the external lever that determines how dark you want your toast. At the end of the toasting cycle, the output from the 555 would pass through a power transistor, to activate a solenoid (which is like a relay, except that it has no switch contacts) to release the toast.

Intermittent windshield wipers could be controlled by a 555 timer—and on earlier models of cars, they were. The repeat rate of keys on a rather basic computer keyboard could be controlled by a 555 timer—and on the Apple II, it was.

What about the intrusion alarm in Experiment 15? One of the features on my wish list was that it should wait long enough for you to shut it off, before it starts to make a noise. The output from a 555 timer can take care of that.

The experiment that you just performed seemed trivial, but it implies a huge range of possibilities.

Fundamentals: Bistable Mode

There's another way to use the timer, known as bistable mode. This entails disabling its fundamental features. Why would you want to do this? I will explain.

Figure 4-19 shows a circuit that you can build in just a few minutes. Give it a try. The two resistors on the left are pullup resistors, 10K each. The resistor at the bottom is 470 ohms, to protect the LED. Add two tactile switches, and the timer chip itself, and you're done.

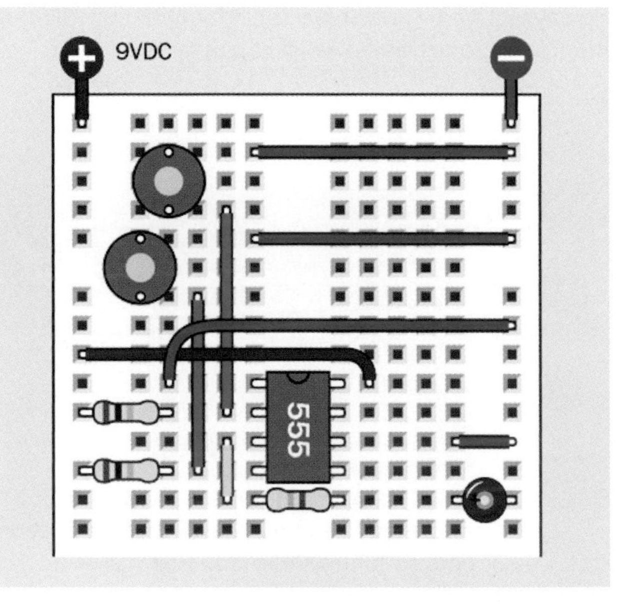

Figure 4-19 *Breadboard circuit to make the 555 timer function as a flip-flop.*

Once you have it on your breadboard, press and release the top button, and the LED lights up. For how long? For as long as you supply power to the circuit. The output from the timer continues indefinitely.

Now press and release the bottom button, and the LED goes off. For how long? For as long as you like. It won't come back on until you press the top button again.

I mentioned that there is a flip-flop inside the timer. This circuit makes the timer into one big flip-flop. It flips into its "on" state when you ground pin 2—and it flops there. It flips into its "off" state when you ground pin 4—and it flops there. Flip-flops are very important in digital circuits, as I will explain a bit later, but right now, how does this work, and why would you need it?

Take a look at the schematic in Figure 4-20. You may notice that there is no resistor or capacitor on the right-hand side. The RC network is missing. So—this timer circuit has no timing components! Normally, when you trigger the timer, its output pulse ends when the timing capacitor on pin 6 accumulates 2/3 of supply voltage. But pin 6 is grounded, so it can never reach the 2/3 value. Consequently, when you trigger the timer, the output pulse will never end.

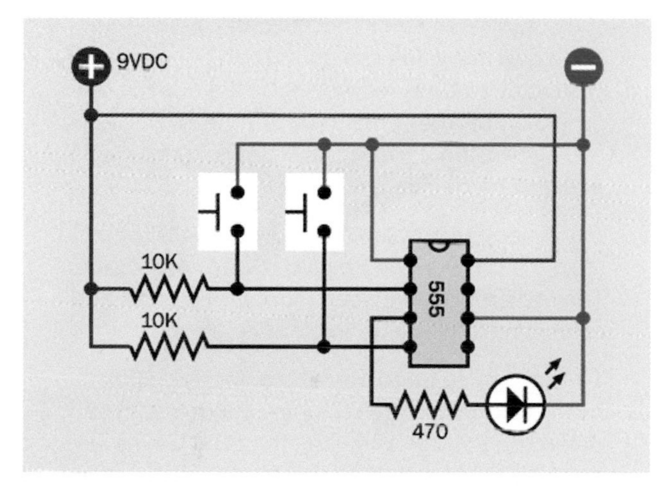

Figure 4-20 *A test circuit for the 555 timer in bistable mode.*

Of course, you can stop the output by applying a low voltage to the reset pin. But once the output stops, it will remain stopped, so long as you don't trigger the timer again.

This configuration is called bistable because it is stable when the output is high, and it is stable when the output is low. A simple flip-flop like this can also be called a *latch*.

- A negative pulse to pin 2 turns the output positive, and latches it.
- A negative pulse to pin 4 turns the output negative, and latches it.

You do have to keep pins 2 and 4 high when you are not triggering them. That's what the pullup resistors in the schematic are for.

It's OK to leave pin 5 of the timer unconnected, because we're pushing it into extreme states where any random signals from those pins will be ignored.

As for why you would need to use a timer in this way—you'll be surprised how useful it can be. I'm going to use it in three experiments in the remainder of this book. The 555 was not really designed to function in bistable mode, but it can be convenient.

Background: How the Timer Was Born

Back in 1970, when barely a half-dozen corporate seedlings had taken root in the fertile ground of Silicon Valley, a company named Signetics bought an idea from an engineer named Hans Camenzind (pictured in Figure 4-21). It wasn't a huge breakthrough concept—just 23 transistors and a bunch of resistors that would function as a programmable timer. The circuit would be versatile, stable, and simple, but these virtues paled in comparison with its primary selling point. Using the emerging technology of integrated circuits, Signetics could reproduce the whole thing on a silicon chip.

Figure 4-21 *Hans Camenzind, inventor, designer, and developer of the 555 timer chip for Signetics.*

This entailed some trial and error. Camenzind worked alone, building the whole thing initially on a large scale, using off-the-shelf transistors, resistors, and diodes on a breadboard. It worked, so then he started substituting slightly different values for the various components to see whether the circuit would tolerate variations during production, and other factors such as changes in temperature when the chip was in use. He made at least 10 different versions of the circuit. This took months.

Next came the crafts work. Camenzind sat at a drafting table and used a specially mounted X-Acto knife to scribe his circuit into a large sheet of plastic. Signetics then reduced this image photographically by a ratio of about 300:1. They etched it into tiny wafers, and embedded each of them in a half-inch rectangle of black plastic with the product number printed on top. Thus, the 555 timer was born.

It turned out to be the most successful chip in history, both in the number of units sold (tens of billions and counting) and the longevity of its design (not significantly changed in almost 40 years). The 555 has been used in everything from toys to spacecraft. It can make lights flash, activate alarm systems, put spaces between beeps, and create the beeps themselves.

Today, chips are designed by large teams and tested by simulating their behavior using computer software. Thus, the chips inside a computer enable the design of new chips. The heyday of solo designers such as Hans Camenzind is long gone, but his genius lives inside every 555 timer that emerges from a fabrication facility. If you'd like to know more about chip history, visit the Transistor Museum (*http://semiconductormuseum.com/Museum_Index.htm*).

A personal note: in 2010, when I was writing *Make: Electronics*, I looked up Hans Camenzind online and found that he maintained his own website, which included a phone number. On impulse, I called him. This was a strange moment, to be talking to the man whose chip design I had used for more than 30 years. He was friendly (although he didn't waste words), and readily agreed to review the text of my book. Even more kindly, after he read it, he gave it a strong endorsement.

Subsequently I bought his own short history of electronics, *Much Ado About Almost Nothing*, which is still available online, and which I highly recommend. I felt honored to have had the opportunity to talk to one of the pioneers in integrated circuit design. I was sad when I heard of his death in 2012.

Fundamentals: 555 Timer Specifications

- The 555 can run from a reasonably stable voltage source ranging from 5VDC to 16VDC. The absolute maximum is 18VDC. Many datasheet specifications are measured at 15VDC. The voltage does not have to be controlled by a voltage regulator.
- Most manufacturers recommend a range from 1K to 1M for the resistor attached to pin 7, but values below 10K draw a more significant amount of current. It's a better idea to reduce the value of the capacitor than to reduce the value of the resistor.
- The capacitor value can go as high as you like, if you want to time really long intervals, but the accuracy of the timer will diminish because leakage in the capacitor becomes comparable with its charge rate.
- The timer imposes a voltage drop that is greater than the drop created by a transistor or a diode. The difference between supply voltage and output voltage will be 1V or more.
- The output is rated to source or sink 200mA, but an output current above 100mA will pull down the voltage and can affect the timing accuracy.

Caution: Not All Timers Are Equal

Everything I have said so far applies to the old original "TTL" version of the 555 timer. TTL is an acronym for *transistor–transistor logic*, which preceded modern CMOS chips that use much less power. The TTL version of the timer is also referred to as the *bipolar version*, as it contains bipolar transistors.

The advantage of the original 555 is that it is cheap and robust. You can't damage it easily, and its output is powerful enough to connect directly with a relay coil or a small loudspeaker. However, the 555 is not efficient, and tends to generate voltage spikes that sometimes interfere with the operation of other chips.

To address these disadvantages, a newer version of the 555 timer was developed using CMOS transistors, which draw less power. This chip also doesn't create voltage spikes. But its output is more limited. How much more? That depends on the particular manufacturer.

Unfortunately, there is a lack of standardization among CMOS versions of the 555 timer. Some claim to deliver 100mA while others are limited to 10mA.

Confusingly, the CMOS versions have a variety of part numbers. The 7555 is clearly identified as a CMOS chip, but others merely precede the 555 number with a different group of letters, and it's up to you to notice, and to understand what they mean.

In this book, to avoid confusion and keep things simple, I am only using the TTL version of the 555 timer, also known as the bipolar version. If you are buying your own, see "Other Components" on page 319 and go down to the subhead "Components for Chapter Four" where you will find timer buying advice.

Experiment 17: Set Your Tone

Now that you're familiar with the 555 timer in monostable mode and bistable mode, I want you to get acquainted with it in *astable* mode—so called because the output fluctuates constantly between high and low, and does not remain stable in either of those states.

This resembles the output of the transistor oscillator that you built in Experiment 11, except that it is much more versatile and easier to control, and instead of requiring two transistors, four resistors, and two capacitors to create the oscillation, you only need one chip, two resistors, and one capacitor.

What You Will Need

- Breadboard, hookup wire, wire cutters, wire strippers, multimeter
- 9-volt power supply (battery or AC adapter)
- 555 timer chips (4)
- Miniature loudspeaker (1)
- Resistors: 47 ohms (1), 470 ohms (4), 1K (2), 10K (12), 100K (1)
- Capacitors: 0.01µF (8), 0.022µF (1), 0.1µF (1), 1µF (3), 3.3µF (1), 10µF (4), 100µF (2)
- 1N4148 diode (1)
- Trimmer potentiometer, 100K (1)
- Tactile switch (1)
- Generic LEDs (4)

Astable Test

A generic astable circuit is shown in Figure 4-22. I've put a loudspeaker on the output, because the timer will be running at an audio frequency. The loudspeaker is driven through a resistor, to limit the current, and a coupling capacitor, which passes audio frequencies while blocking DC. You'll see the values for these components in the next schematic. Right now I just want you to see the general layout.

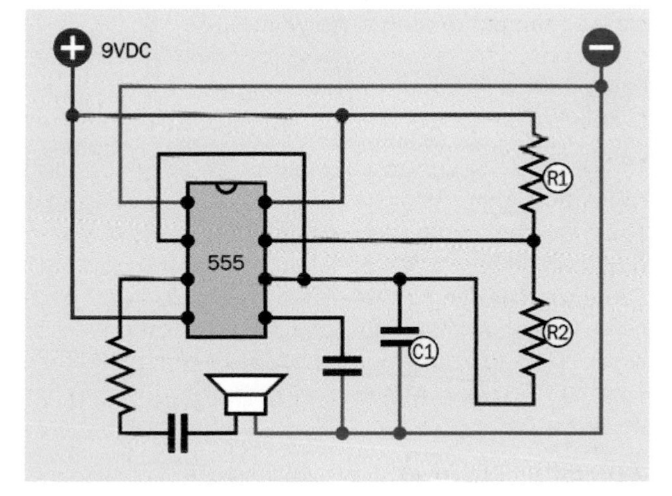

Figure 4-22 *A basic, generic circuit that runs a 555 timer in astable mode.*

The components labeled R1, R2, and C1 control the speed of the timer. These labels are always used in manufacturers' datasheets and other sources, so I'm following the same convention.

C1 does the same thing as the timing capacitor in the monostable circuit in Figure 4-11. The need for two resistors, instead of one, will be explained below.

See if you can get a sense of how this circuit may work, using the knowledge that you gained in Experiment 16. The first thing you may notice is that there's no input. Pin 2 (the trigger pin) is connected back to pin 6 (the threshold pin). Can you see how that's going to work? C1 will accumulate a charge, as it did when the timer was in monostable mode, until it reaches 2/3 of the power supply, at which point it will discharge through R2 into pin

7, and its voltage will drop. Its connection with pin 2 will mean that the trigger pin senses the drop in voltage on C1. And what does the trigger pin do when the voltage on it drops suddenly? It triggers the timer. So, in this configuration, the timer will retrigger itself.

How fast will that happen? I think you should build a test version of the circuit to find out. In Figure 4-23 I have suggested values for the components and have redrawn the schematic to include a trimmer potentiometer, so that you'll see (or, rather, hear) the effect of varying this resistance. The trimmer, plus the 10K resistor that precedes it, add up to R2. The timing capacitor, C1, is 0.022µF, and R1 is 10K.

Figure 4-24 shows the breadboard layout, while Figure 4-25 shows the component values.

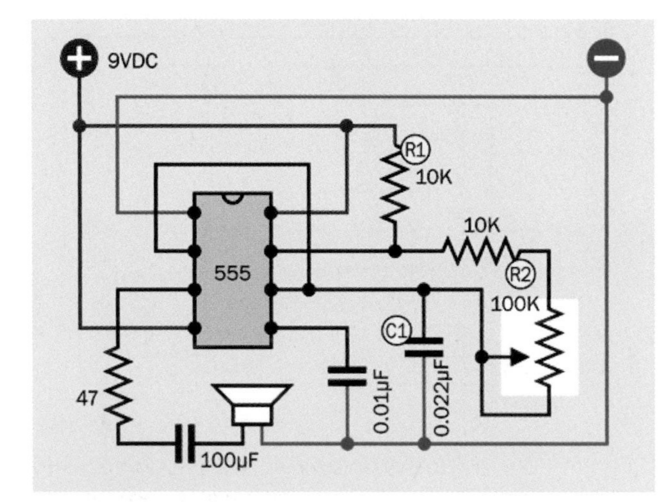

Figure 4-23 *This test circuit allows you to adjust the performance of the timer in astable mode.*

Now what happens when you apply power? Immediately, you should hear noise through the loudspeaker. If you don't hear anything, you almost certainly made a wiring error.

Notice that you don't have to activate the chip with a pushbutton anymore. The 555 timer is triggering itself, as predicted.

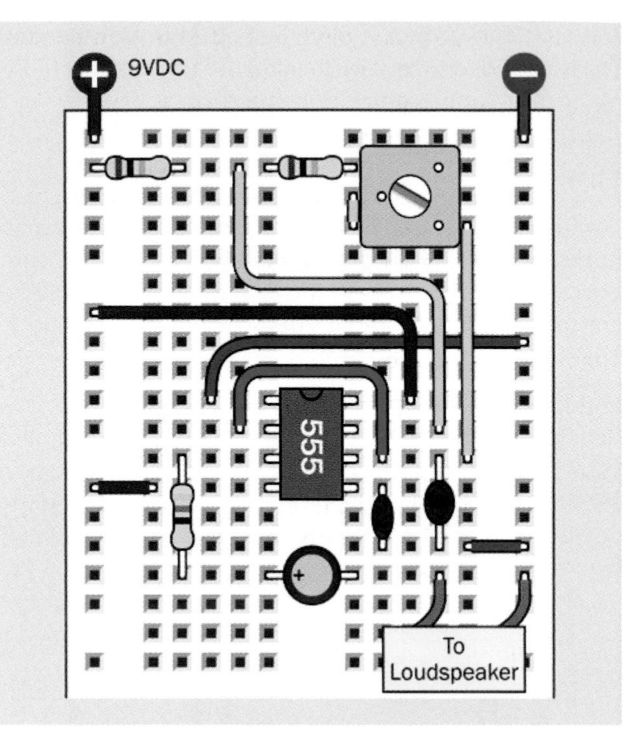

Figure 4-24 *Breadboard layout for the astable timer test.*

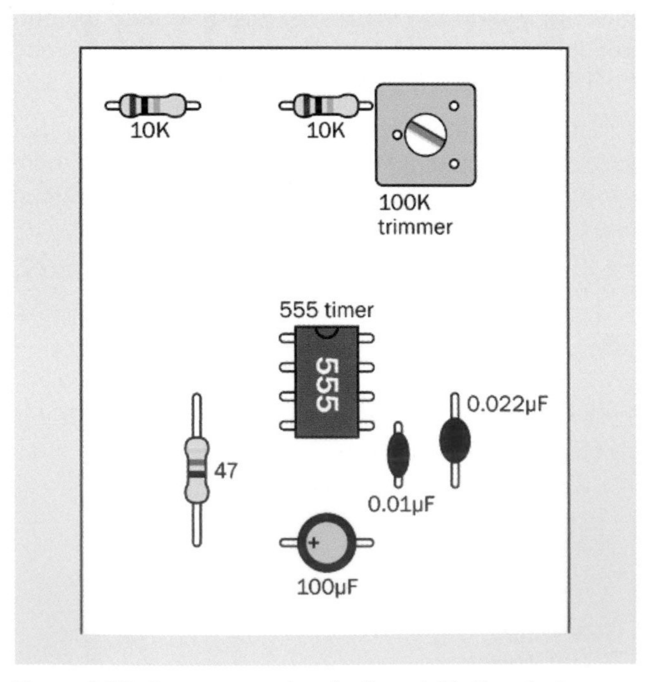

Figure 4-25 *Component values for the astable timer test.*

Rotate the screw on the trimmer potentiometer, and the pitch of the sound varies. The trimmer adjusts how quickly C1 charges and discharges, and this determines the length of one "on" cycle relative to the next "off" cycle in the audio signal. With these component values, the stream varies between approximately 300 and 1,200 pulses each second. These pulses are sent by the timer to the loudspeaker. They move its cone up and down, creating pressure waves in the air, and your ear responds to those waves, hearing them as sound.

Theory: Output Frequency

The *frequency* of a sound is its number of full cycles per second, including the high-pressure pulse and the low-pressure pulse that follows.

The term *hertz* is a unit of frequency, meaning the same thing as "cycles per second." It was introduced in Europe, named after yet another electrical pioneer, Heinrich Hertz. The abbreviation for hertz is Hz, so the output of your 555 timer in its test circuit will range approximately between 300Hz and 1,200Hz.

As in most standard units, a k can be inserted to mean "kilo," so 1,200Hz is normally written as 1.2kHz.

How do the values for the timing capacitor and the resistors determine the frequency of the timer? If R1 and R2 are measured in *kilohms*, and C1 is measured in *microfarads*, the frequency f, in hertz, is given by:

```
f = 1,440 / ( ( ( 2 × R2 ) + R1 ) × C1 )
```

Doing this calculation is a hassle, so I have provided you with a lookup table in Figure 4-26. In this table, *I am assuming that the value for the resistor labeled R1 in the schematic is fixed at 10K.* The values across the top of the table are for R2. The values on the left side of the table are for the timing capacitor, C1.

You may remember than the abbreviation pF means "picofarad," which is one-millionth of a microfarad. Nanofarads are halfway between microfarads and picofards, but the term is not used so frequently in the US, so I didn't use it in the table.

	10K	22K	47K	100K	220K	470K	1M
47µF	1	0.57	0.3	0.15	0.068	0.032	0.015
22µF	2.2	1.2	0.63	0.31	0.15	0.069	0.033
10µF	4.8	2.7	1.4	0.69	0.32	0.15	0.072
4.7µF	10	5.7	3.0	1.5	0.68	0.32	0.15
2.2µF	22	12	6.3	3.1	1.5	0.69	0.33
1.0µF	48	27	14	6.9	3.2	1.5	0.72
0.47µF	100	57	30	15	6.8	3.2	1.5
0.22µF	220	120	63	31	15	6.9	3.3
0.1µF	480	270	140	69	32	15	7.2
0.047µF	1K	570	300	150	68	32	15
0.022µF	2.2K	1.2K	630	310	150	69	33
0.01µF	4.8K	2.7K	1.4K	690	320	150	72
4,700pF	10K	5.7K	3K	1.5K	680	320	150
2,200pF	22K	12K	6.3K	3.1K	1.5K	690	330
1,000pF	48K	27K	14K	6.9K	3.2K	1.5K	720
470pF	100K	57K	30K	15K	6.8K	3.2K	1.5K
220pF	220K	120K	63K	31K	15K	6.9K	3.3K
100pF	480K	270K	140K	69K	32K	15K	7.2K

Figure 4-26 *For a 555 timer running in astable mode, values across the top refer to R2 in a standard circuit, so long as R1 has a fixed value of 10K. Numbers in the table show the frequency of the timer in Hz (cycles per second).*

Theory: Inside the 555 in Astable Mode

To gain a better understanding of what happens when the timer runs in astable mode, take a look at Figure 4-27. The internal configuration is exactly the same as in monostable mode, but the external connections are different.

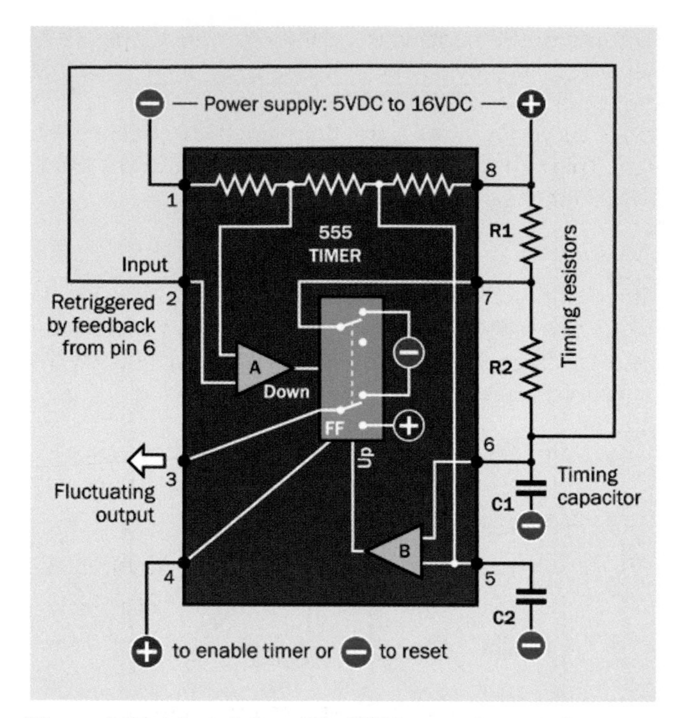

Figure 4-27 *Internal view of the 555 timer with external connections running it in astable mode.*

Initially, the flip-flop grounds C1, the timing capacitor, as before. But now the low voltage on the capacitor is connected from pin 6 to pin 2 through an external wire. The low voltage tells the chip to trigger itself. The flip-flop obediently flips to its "on" position and sends a positive pulse to the loudspeaker, while removing the negative voltage from pin 6.

Now C1 starts charging, as it did when the timer was in monostable mode, except that it is being charged through R1 + R2 in series. Because C1 has a low value, it charges quickly. When it reaches two-thirds of full voltage, comparator B takes action as before, discharging the capacitor and ending the output pulse from pin 3.

The capacitor discharges through R2 to pin 7, the discharge pin. While the capacitor is discharging, its voltage diminishes. The voltage is still linked to pin 2. When it drops to one-third of full power or less, comparator A kicks in and sends another pulse to the flip-flop, starting the process all over again.

Fundamentals: Unequal On-Off Cycles

When the timer is running in astable mode, C1 charges through R1 and R2 in series. But when C1 discharges, it dumps its voltage into the chip through R2 only. Because the capacitor charges through two resistors but discharges through only one of them, it charges more slowly than it discharges. While it is charging, the output on pin 3 is high; while it is discharging, the output on pin 3 is low. Consequently the "on" cycle is always longer than the "off" cycle. Figure 4-28 shows this as a simple graph.

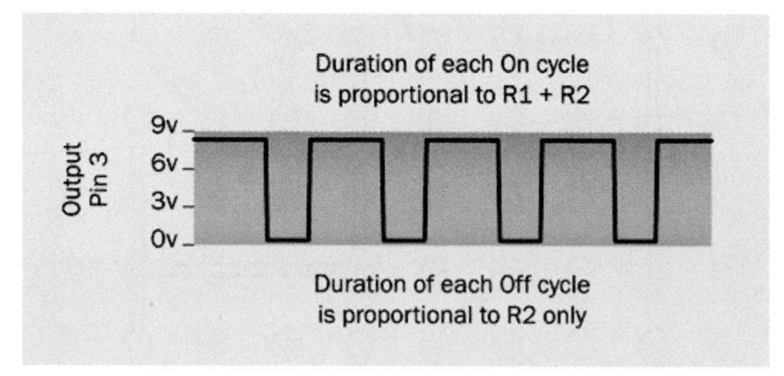

Figure 4-28 *High pulses are always longer than gaps between them in the output from a 555 timer chip, when the chip is wired in the standard way for astable operation.*

If you want the on and off cycles to be equal, or if you want to adjust the on and off cycles independently (for example, because you want to send a very brief pulse to another chip, followed by a longer gap until the next pulse), all you need to do is add a diode, as shown in Figure 4-29. (Because a diode deducts some voltage, this circuit works best with a supply voltage greater than 5VDC.)

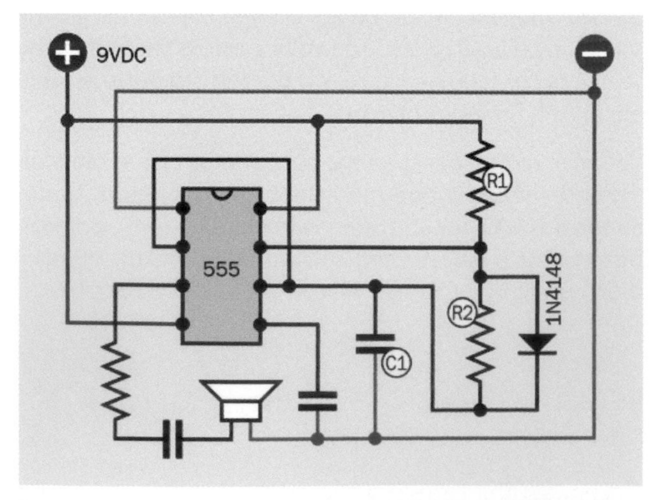

Figure 4-29 *Adding a diode to bypass R2 allows the high and low output cycles of a timer to be adjusted independently.*

Now when C1 charges, the electricity flows through R1 as before but takes a shortcut around R2, through the diode. When C1 discharges, the diode blocks the flow of electricity in that direction, and so the discharge goes back through R2.

R1 now controls the charge time on its own, while R2 controls the discharge time. The formula for calculating the frequency is now approximately:

```
Frequency = 1,440 / ( ( R1 + R2 ) × C1 )
```

where R1 and R2 are in *kilohms* and C1 is in *microfarads*. (I use the word "approximately" because the diode adds a small amount of effective resistance to the circuit, which is not factored into the formula.)

If you set R1 = R2, you should get almost equal on/off cycles.

Astable Modifications

Instead of using a potentiometer to vary the value of R2, the frequency of the timer can be changed to a limited extent by using pin 5, the control pin. This is shown in Figure 4-30.

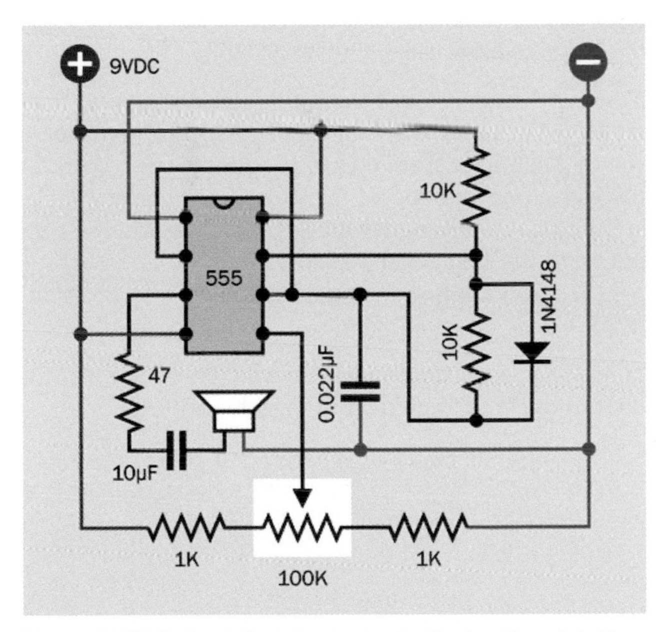

Figure 4-30 *A circuit that demonstrates the function of pin 5, the control pin, on the 555 timer.*

Disconnect the capacitor that was attached to that pin and substitute the series of resistors shown. They ensure that pin 5 always has at least 1K between it and the positive side or the negative side of the power supply. Connecting it directly to the power supply won't damage the timer, but will prevent it from generating audible tones. As you turn the potentiometer, the frequency will vary. This happens because you are changing the reference voltage on comparator B inside the chip.

Chaining Chips

Timer chips can be chained together in four possible ways. Note that these configurations are workable regardless of whether each timer is running in monostable or astable mode, except where noted.

- If you are using 9V to power a 555 timer, the output from that timer can be sufficient to power another 555 timer.
- The output from one timer can trigger the input of another timer. This only works if the second timer is running in monostable mode. In astable mode, it would be self-triggering.
- The output from one timer can unlock the reset pin of another timer.
- The output from one timer can be connected through a suitable resistor to the control pin of another timer.

These options are illustrated in Figure 4-31, Figure 4-32, Figure 4-33, and Figure 4-34.

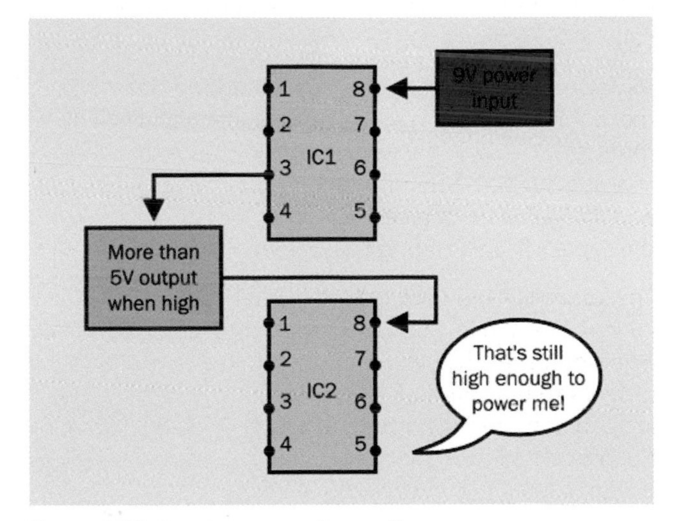

Figure 4-31 *One timer powering another.*

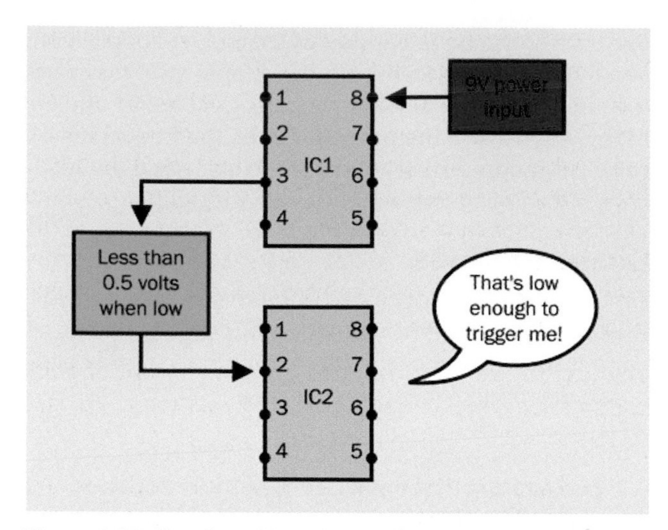

Figure 4-32 *One timer triggering another.*

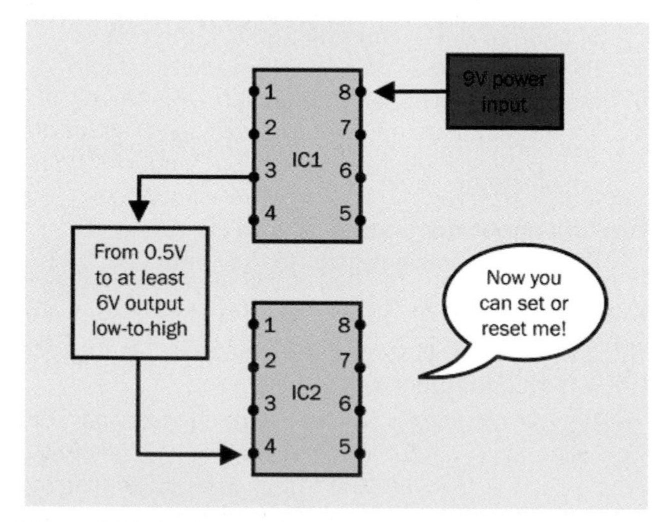

Figure 4-33 *One timer setting or resetting another.*

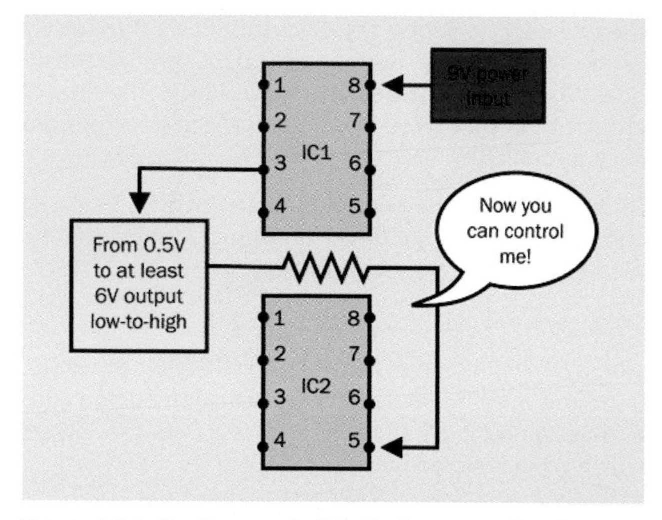

Figure 4-34 *One timer controlling the frequency of another.*

Why would you want to chain timers together? Well, you might want to have a couple of them running in monostable mode, so that the end of a high pulse from the first one triggers the start of a high pulse in the second one, and vice versa. In fact, you could chain together as many timers as you like, with the last one feeding back and triggering the first one, and they could flash a series of LEDs in sequence, like Christmas lights.

Figure 4-35 shows four timers chained in this way. They are linked through coupling capacitors, because we only want a brief pulse from one to trigger the next. Without the capacitors, the end of the pulse from the first timer in the chain would trigger the second timer, but the output from the first timer would remain in its low state, which would continue triggering the second timer indefinitely.

In addition, each timer must have a 10K pullup resistor on its trigger pin to hold it normally high.

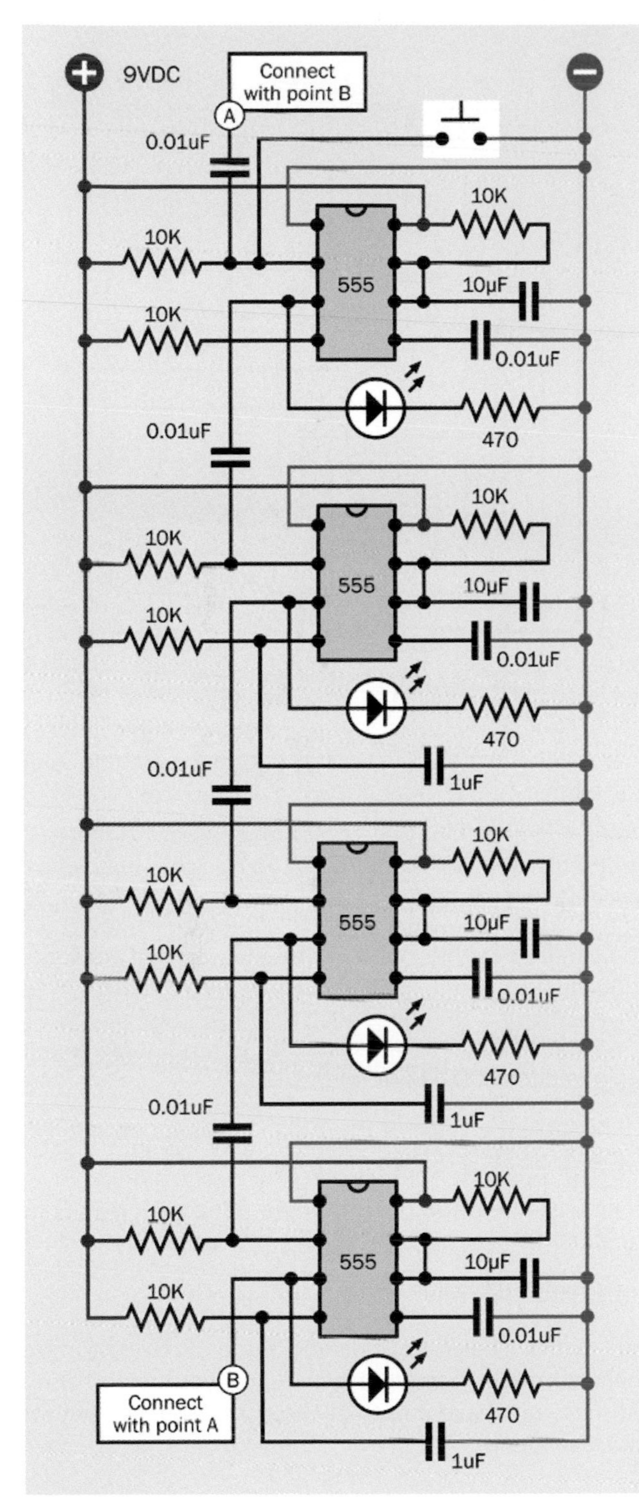

Figure 4-35 *Four timers wired to trigger each other in a recirculating sequence.*

An interesting question comes to mind when monostable timers are chained together. That is: how do they start? I mentioned in Experiment 16 that a 555 timer in monostable mode will usually emit one spontaneous pulse when it is first powered up. When multiple timers are chained together, they will all try to do this at about the same time, and because of small manufacturing differences, the outcome will be unpredictable. Sometimes they will settle down to a nice orderly sequence, but other times they will end up flashing in pairs.

The way to deal with this is to use the concept of pulse suppression that I mentioned in Experiment 16 (see "Fundamentals: Pulse Suppression" on page 152).

A 1μF capacitor between the reset pin and negative ground will hold the reset pin in a low state just long enough to suppress the timer's initial pulse. A 10K pull-up resistor sharing the reset pin will then keep it stable while the timer is running.

In my experience, this works well, although timers from a different manufacturer could conceivably behave differently, as the behavior of the reset pin is not well documented. If you have trouble with pulse suppression, try substituting a larger or smaller capacitor.

In a chain of timers, the only remaining problem is that pulse suppression works too well. You apply power, and —nothing happens, because all the timer outputs have been suppressed.

The way around this is to omit pulse suppression from just one timer. It will almost certainly emit an initial pulse when it receives power, and that will trigger the rest in sequence. This setup is illustrated in Figure 4-35.

But—wait a minute. What is this "almost certainly" phrase? Electronic circuits should always work, all the time, not "almost" all the time.

I agree. But I can't control the tendency of 555 timers to do unpredictable things when they are powered up. Therefore I also added a button, at the top of the circuit, that can be used to start the cascade if it doesn't start itself.

There is an alternative, which is for the first timer in the chain to run in astable mode. It sends a series of pulses that ripple down through the others configured in monostable mode, and the last one does not feed back to the first. In electronics terms, we say that the first one is the master and the others are the slaves.

I like this arrangement because it is totally predictable. The trouble is, you have to adjust the speed of the master timer so that it emits its next pulse exactly when the last slave in the chain has just finished emitting its pulse. Otherwise, the first timer will emit another pulse before the last pulse has ended, or there will be a gap between the last pulse and the next first pulse.

Whether this is important will depend on the application. Flashing lights won't be a problem, but if you increase the speed to drive a stepper motor, getting the timing right will be difficult.

Sounding Like a Siren

The fourth option that I listed for chaining chips, in Figure 4-34, is of special interest, because it can create a siren sound very like the noise made by a typical burglar alarm. In fact, this could be used for the audio output in the alarm project that I left unfinished in Experiment 15.

The circuit is shown in Figure 4-36. Timer 1 is wired in a basic astable circuit, which you can recognize as being similar to the circuit in Figure 4-22. The component values are larger, so the timer oscillates more slowly, around 1Hz. You can compare this circuit with the one that I suggested in Figure 2-120. The principle is similar.

Timer 2 is also wired in a basic astable circuit, running around 1kHz. The idea is that the slow fluctuations in voltage from Timer 1 are applied to the control pin of Timer 2, forcing it to modulate its sound up and down in that annoying way which we associate with alarm systems.

I encourage you to build this circuit, because you may want to use it in the final version of the intrusion alarm coming right up in Experiment 18. The breadboard layout for the siren circuit is in Figure 4-37, and the component values for the layout are shown in Figure 4-38.

Once you get it running, you may find it interesting to remove and then replace the 100µF capacitor connected between pin 6 and ground. The capacitor makes the frequency slide higher and lower instead of switching sharply up and down. I'm using it in the same way that I used a capacitor to make an LED fade smoothly in and out in Experiment 11.

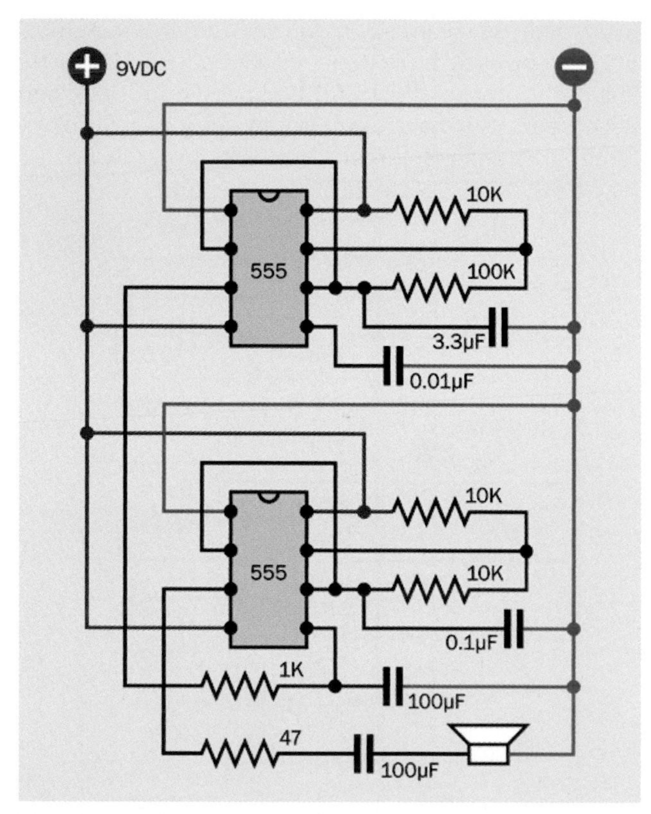

Figure 4-36 *When one timer runs relatively slowly, modulating another through its control pin (pin 5), the result is a wavering sound like an alarm siren.*

You can modify the sound in other ways. Here are some suggestions:

- Vary the 0.1µF timing capacitor to raise or lower the pitch of the basic sound.
- Double the value of the 100µF capacitor on pin 6, or divide it by two.
- Substitute a 10K potentiometer for the 1K resistor.
- Change the value of the 3.3µF capacitor.

Part of the pleasure of building things is to customize them, to make them your own. Once your siren sound satisfies you, make a note of the component values for future reference.

Incidentally, you can reduce the chip count (the number of chips) by using a 556 timer instead of two 555 timers. The 556 contains a pair of 555 timers in one package. Because you still have to make the same number of external connections (other than the power supply), I haven't bothered to use this variant.

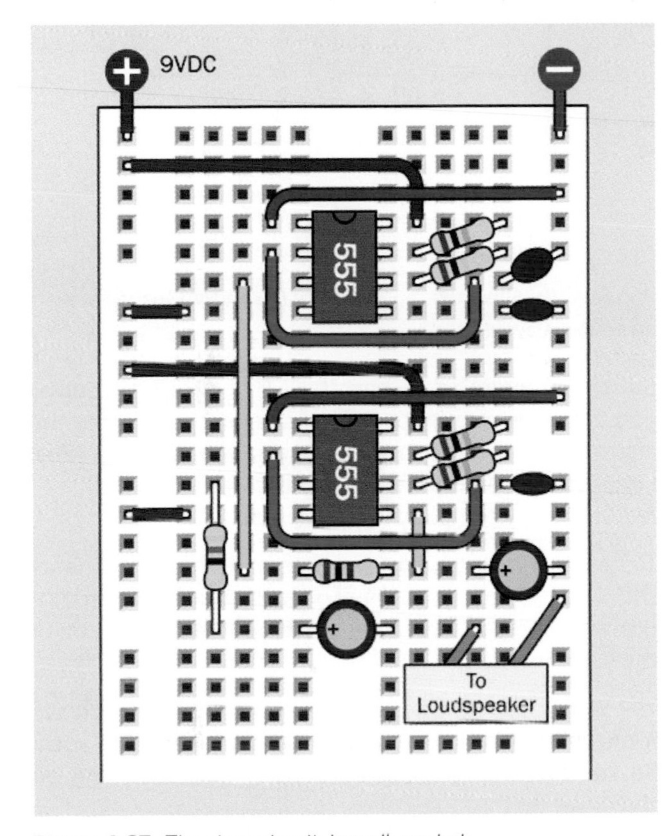

Figure 4-37 *The siren circuit, breadboarded.*

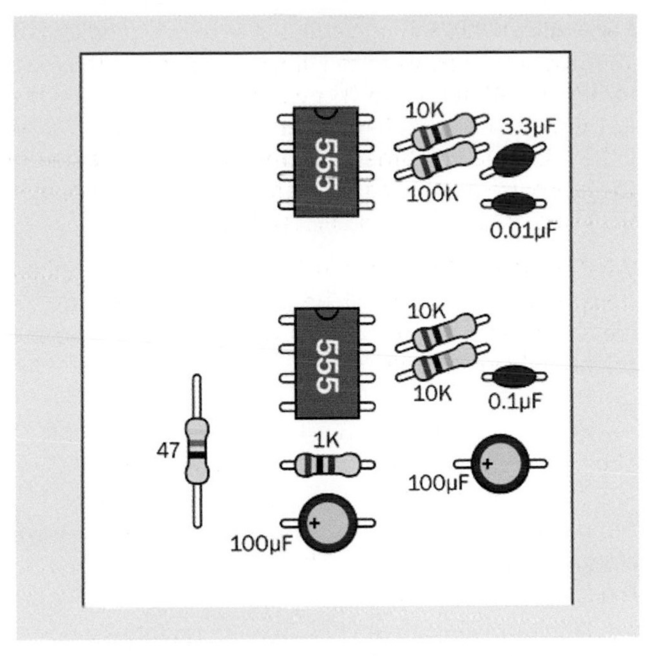

Figure 4-38 *Component values for the siren circuit.*

Experiment 18: Intrusion Alarm (Almost) Completed

Now that you've seen what the 555 timer can do, you can fulfill the remaining requirements on the intrusion-alarm wish list.

What You Will Need

- Breadboard, hookup wire, wire cutters, wire strippers, multimeter
- 9-volt power supply (battery or AC adapter).
- 555 timers (2)
- DPDT 9VDC relay (1)
- Transistors, 2N2222 (2)
- LEDs: Red, green, yellow (1 of each)
- Slide switch, SPDT, for breadboard (2)
- Tactile switch (1)
- Capacitors: 0.01µF (1), 10µF (2), 68µF (2)
- Resistors: 470 ohms (4), 10K (4), 100K (1), 1M (2)

- Diode, 1N4001 (1)

Optional (for audio output):

- Components shown in Figure 4-36

Optional (for permanent fabrication of this project):

- 15-watt soldering iron
- Thin solder
- Perforated board plated with copper in a breadboard layout
- SPDT or DPDT toggle switch (1)
- Pushbutton, SPST (1)
- Project box, at least 6″ × 3″ × 2″ (1)
- Power jack and matching power socket (1 each)
- Magnetic sensor switches, in pairs, quantity sufficient for your home
- Alarm network wiring, sufficient for your home

Three Steps to a Functional Device

This is a larger and more complicated circuit than anything you have tackled so far, but it's relatively easy to build, because you can assemble it in three parts that you can test individually. Eventually you'll end up with a breadboard looking like Figure 4-45, with component values in Figure 4-46. The equivalent schematic is shown in Figure 4-47. But we'll begin with just a small timer circuit.

Step 1

Take a careful look at Figure 4-39. Notice there are no timing components on the righthand side of the 555 timer. You may conclude that this is a version of the bistable circuit that I described in Experiment 16 (see Figure 4-20). When the timer is triggered, its output will go on indefinitely—which seems appropriate for an alarm system.

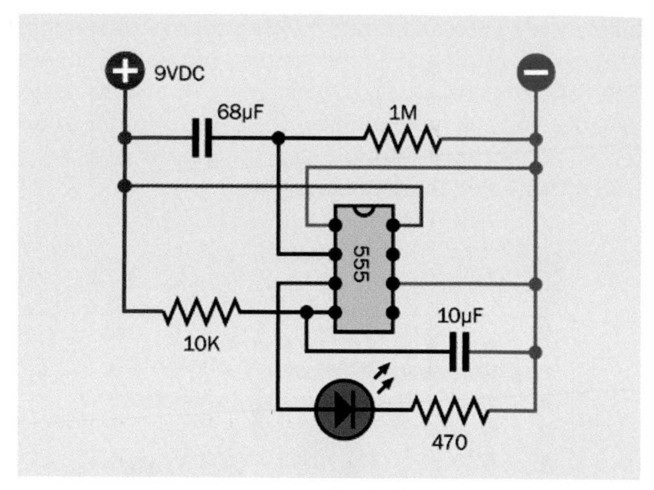

Figure 4-39 *Schematic equivalent of the bottom section of the breadboard layout.*

But there's more to it than that. This circuit also gives you a one-minute grace period in which to disable the alarm before it starts to sound, when you enter the area. (You may remember that this was #9 on the wish list of features that I compiled in Experiment 15.)

To see how it works, you can assemble the components shown in Figure 4-40. Their values are shown in Figure 4-46, and you'll see their placement at the bottom of the board in Figure 4-45.

The placement is important, because you have to leave room for the additional sections of the circuit that you'll be adding. One of those sections will be switching on the power to this one.

To make sure that you have everything in the right place, the 1M resistor on the right is on the 29th row in the board, counting down from the top. Also note that power is being applied near the components, not at the top of the board, and the positive bus is not being used, yet.

Don't power up the circuit just yet. Set your meter to measure at least 10VDC and attach it at the points shown in Figure 4-40, with the negative probe on the negative bus and the positive probe at the left end of the 1M resistor.

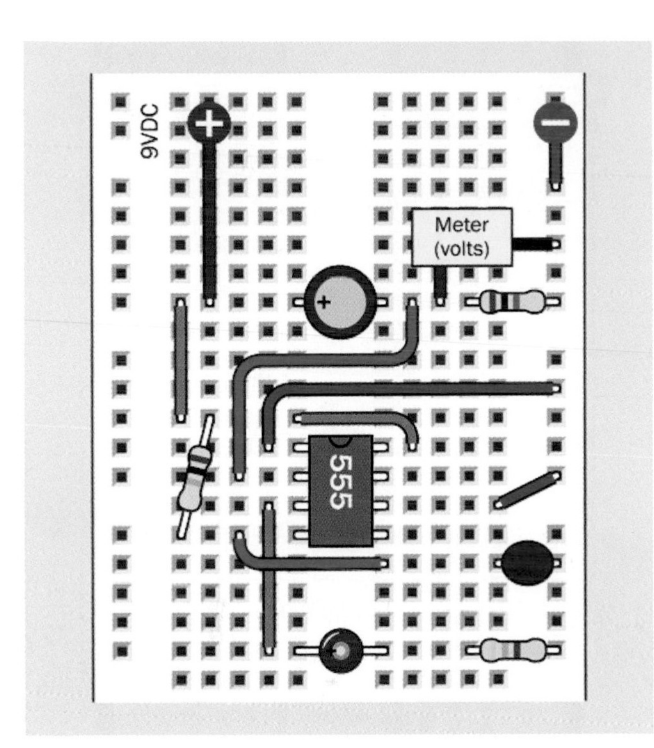

Figure 4-40 *Placement of components at the bottom of the breadboard for testing.*

Now apply power to the circuit, and you should see your meter slowly counting down from 9V. When it gets to one-third of the supply voltage, this triggers the 555 timer, and the red LED comes on. The LED is included for testing purposes; in the final version, you would substitute a noisemaking circuit.

The large 68μF capacitor delays the response of the timer. When you first apply power to the circuit, the capacitor passes the initial pulse to the point between it and the 1M resistor. The green wire runs from this point to the trigger pin of the timer. So, the pin starts off high, and you'll remember (I hope) that the timer does nothing until the trigger pin goes low.

Voltage on the righthand side of the capacitor leaks out very slowly through the 1M resistor. Eventually, it goes low enough to trigger the timer.

As for the rest of the circuit, in Experiments 16 and 17 I explained how to do "pulse suppression" to stop the timer from emitting a pulse when it is first powered up. This is why I have a 10μF capacitor and a 10K resistor applied to pin 4, the reset pin. I'm using a 10μF capacitor this time, instead of a 1μF capacitor, because this circuit reacts a bit more slowly than the one in Experiment 17.

- You can use a 555 timer with these components any time you want to delay the timer output in response to a trigger pulse.
- Use a larger or smaller value for the 68μF capacitor if you want to make the delay longer or shorter.

So far, so good. This part of the circuit will introduce a delay when it is powered up, and after the delay, it will activate the alarm for an indefinite period.

Step 2

Figure 4-41 and Figure 4-42 show the next step in building the circuit. The components that you placed previously are still there, but they have been grayed out to focus your attention on the new additions.

Don't forget to install S2, the slide switch at the bottom, and the 470-ohm resistor beside it, and the two long yellow wires. This slide switch is included for testing purposes. It represents the alarm sensors that you would use in an actual application.

The relay has the same function as in Experiment 15. In fact if you trace the connections in the circuit, you'll find that it works the same way as the one in Figure 3-88, with a couple of minor revisions. The only differences are that a 470-ohm resistor has been substituted for a 1K resistor, and I have added switch S1 at the top, with a green LED. Why? I'll get to that in a moment.

Place all the components carefully. Don't overlook the three red wires on the left and the three blue wires on the right. Make sure the pins of the relay are aligned with the wires that serve them.

Double-check that S1 is in the down position, and S2 is in the up position. For testing purposes, remove the 68μF capacitor so that the red LED will respond right away instead of waiting for a minute.

Attach power—and if you made all the connections correctly, nothing should happen. Switch S2 represents the alarm sensors, and in its up position, it simulates them being closed. Slide it down to simulate a sensor being opened, and the test LED at the bottom of the circuit should light up immediately. Slide the switch up, and the LED stays on. The circuit has locked the alarm on, regardless of a sensor being reset.

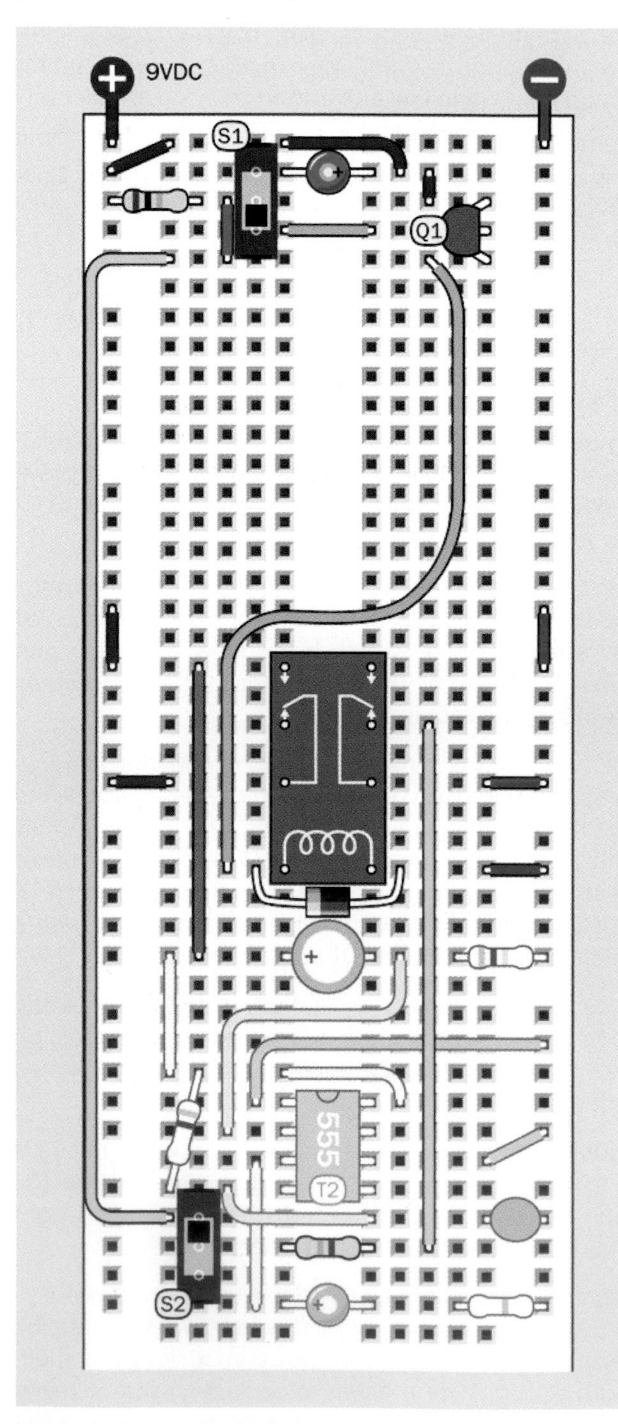

Figure 4-41 *The second step in building this circuit incorporates the relay configuration that was used in Experiment 15.*

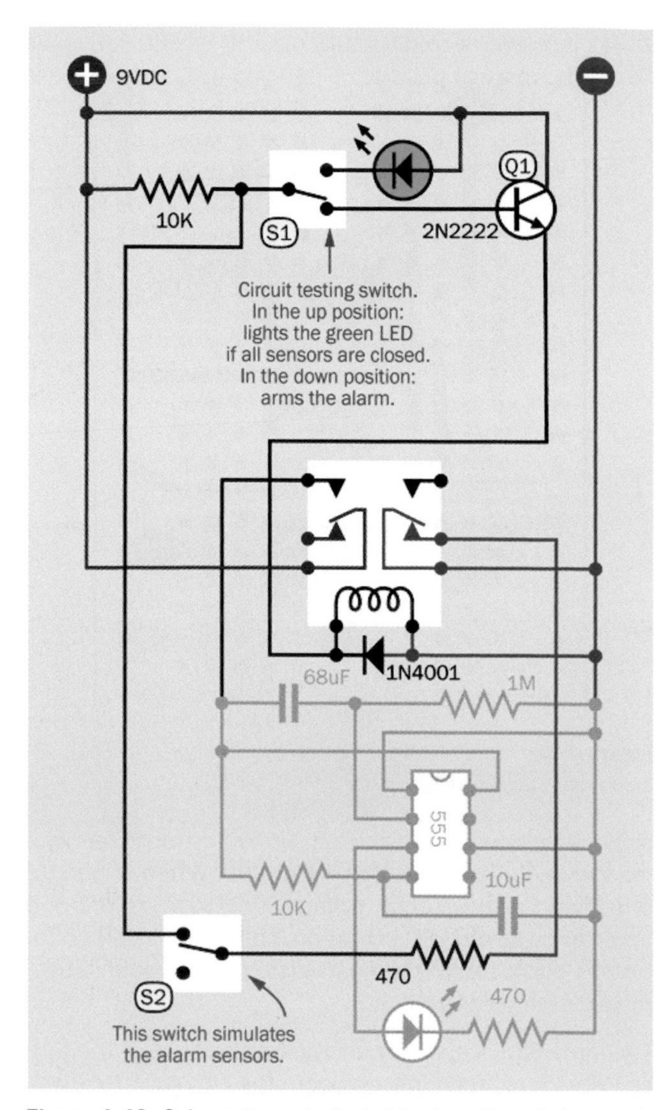

Figure 4-42 *Schematic equivalent of the breadboarded second stage in the circuit.*

Disconnect the power, keep S2 in its up position (simulating the sensors being closed), and reapply the power. Now move S1, at the top, to its up position, and the green LED comes on. This is a circuit-testing feature. It checks that all the sensors are closed. When you use the alarm, you will want to conduct this test before you leave the area. This satisfies the first part of item #7 in the wish list from Experiment 15.

Keep S1 in its up position while you move S2 to its down position, to simulate an open sensor. The green LED goes out. Move S2 to its up position, and the green LED comes back on. So, the testing procedure works.

Here's how you would actually use this circuit. You leave S1 in the up (test) position. When you are ready to leave the area, you apply power to the circuit. If the green LED doesn't come on, there is a door or window open somewhere. Find the source of the problem, and correct it. When the green LED is on, you know that all the sensors are closed. Now you can arm the alarm. Move S1 down. The green LED goes out, and the alarm is now armed. When you come home, the 555 timer gives you one minute to disable the alarm, to prevent it from going off (so long as you have replaced the 68μF capacitor in the circuit). You can disable the alarm by moving S1 to its up (test) position.

Now, how and why does the circuit work?

The 10K resistor at top-left connects with the base of transistor Q1 through switch S1, when the switch is in its down position. Meanwhile, the righthand pole contact inside the relay is connected to negative ground. This connection runs through the yellow wire on the right, through a 470-ohm resistor, through switch S2 (which simulates the sensors), and back up through the other long yellow wire. It holds the base of the resistor at a low voltage (through the orange wire). So long as the base is low, the transistor doesn't conduct.

If a sensor opens, the base of the transistor isn't held low anymore, and the 10K resistor pulls it up so that the transistor starts to conduct. It triggers the relay, through the long curving orange wire. The relay will supply power to the bistable timer, which will eventually activate the alarm. At the same time, the relay breaks the negative-ground connection on the right, so now the transistor is going to continue conducting even if the sensor is closed again.

This is exactly the same circuit concept that I ended up with in Figure 3-88. The significant difference is the green LED. When you move S1 to its "test" position, it cuts off positive power to the transistor (so that the transistor cannot start the alarm). If all the sensors are closed, the LED connects through them, and the 470-ohm resistor, to negative ground, and it lights up to tell you that the system is ready.

Step 3

What else could we need in this project? Well, imagine you are using the alarm system. You want to set it before leaving the area. At this point, you suddenly realize that if you set it, and then open a door to leave, you will trigger the alarm.

The bistable timer with its 68μF capacitor added a feature to suppress the alarm for a minute, to give you time to turn it off when you arrive. Now we need another timer to suppress the alarm for a minute when you leave.

This is a little more difficult to arrange. The key is to have the extra timer pull down the voltage on transistor Q1, so that it can't trigger the relay.

The problem is, a timer output goes high, not low, during its "on" cycle. I'll have to add another transistor to convert the high output so that it pulls down the voltage on the base of transistor Q1.

Figure 4-43 and Figure 4-44 show the components that will make it happen. Once again I have grayed out the components that you placed previously.

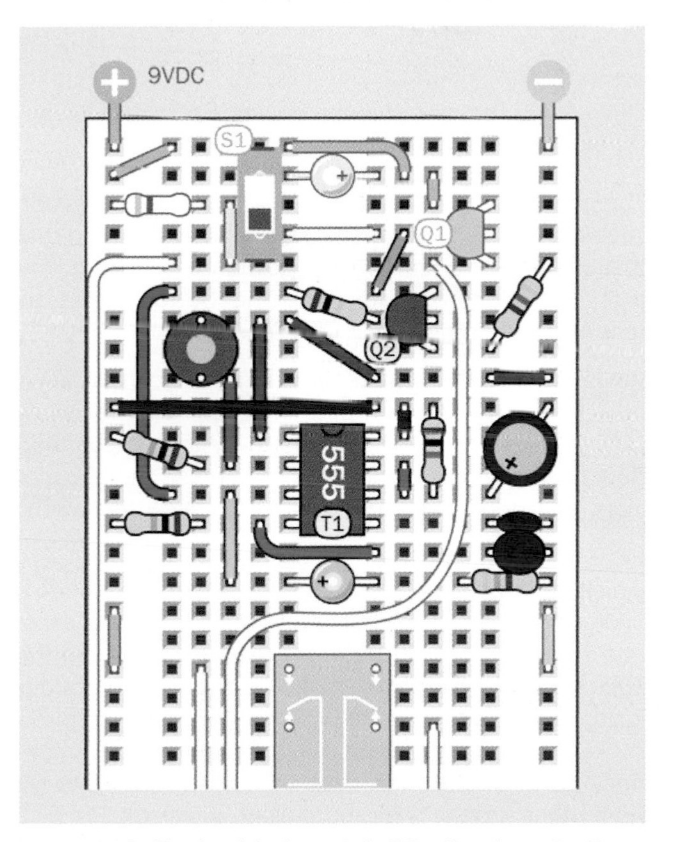

Figure 4-43 *Third and final step in building the alarm circuit.*

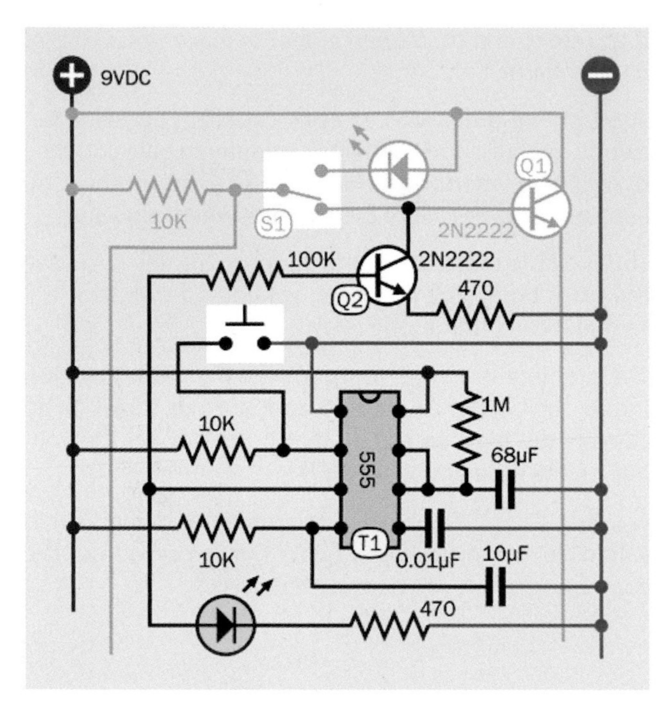

Figure 4-44 *The schematic version of the third step in building the circuit.*

The new 555 timer, labeled T1, has a pulse-suppression circuit on pin 4, its reset pin, like the other timer, so that it won't emit a pulse when you power up the circuit. You press the button to start T1. The button works by grounding the trigger pin of the timer.

While the output of the timer is high, the current flows out of pin 3 (the output pin) and lights up the yellow LED. This tells you that the alarm system is counting down to being armed. So long as you see the LED, the alarm will ignore any activity that opens a sensor switch.

Pin 3 also connects through a green wire, on the left, which is like an elongated letter C. This curls around to a 100K resistor, which is attached to the base of Q2, a second transistor. The output from the timer, through the 100K resistor, is enough to make Q2 conduct. Its emitter is grounded through a 470-ohm resistor, while its collector is attached to the base of Q1. So long as Q2 is conducting, it grounds the base of Q1, and prevents Q1 from triggering the relay and starting the alarm.

In this way, timer T1 stops the alarm from going off. When the one-minute grace period ends, T1 stops conducting, doesn't pull down the voltage on the first transistor anymore, and the alarm can go off—provided you remembered to move the switch at the top out of "test" mode, of course.

You would now use the circuit like this.

1. First put switch S1 into its "test" position and close all the doors and windows until the green LED comes on.
2. Move S1 into its lower position, so that the alarm is ready.
3. Press the button, and leave, closing the door behind you, while the yellow LED is on.

Does your version do what it is supposed to do? It should, so long as you wired it carefully. Timer T1 should light the yellow LED under any circumstances, making it easy to test. You can also touch your meter probe on the base of Q1, to verify whether the voltage is relatively high or low. So long as the voltage is relatively low, the alarm will not be triggered. When the voltage goes high, the alarm is triggered.

Don't forget to replace the 68μF capacitor in the circuit, just below the relay, to reactivate the delay timer, when your alarm is ready for prime time.

The complete breadboarded circuit is shown in Figure 4-45, the component values are in Figure 4-46, and the schematic is in Figure 4-47.

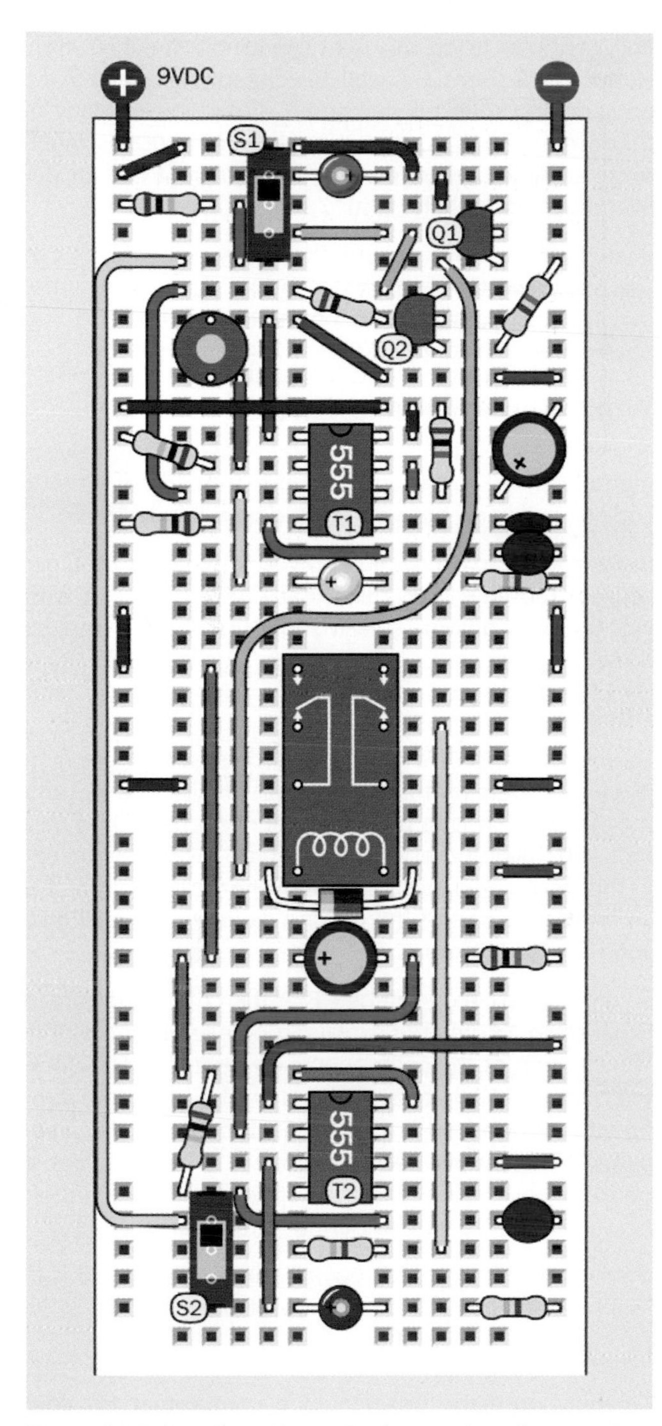

Figure 4-45 *Breadboard layout for the complete alarm circuit.*

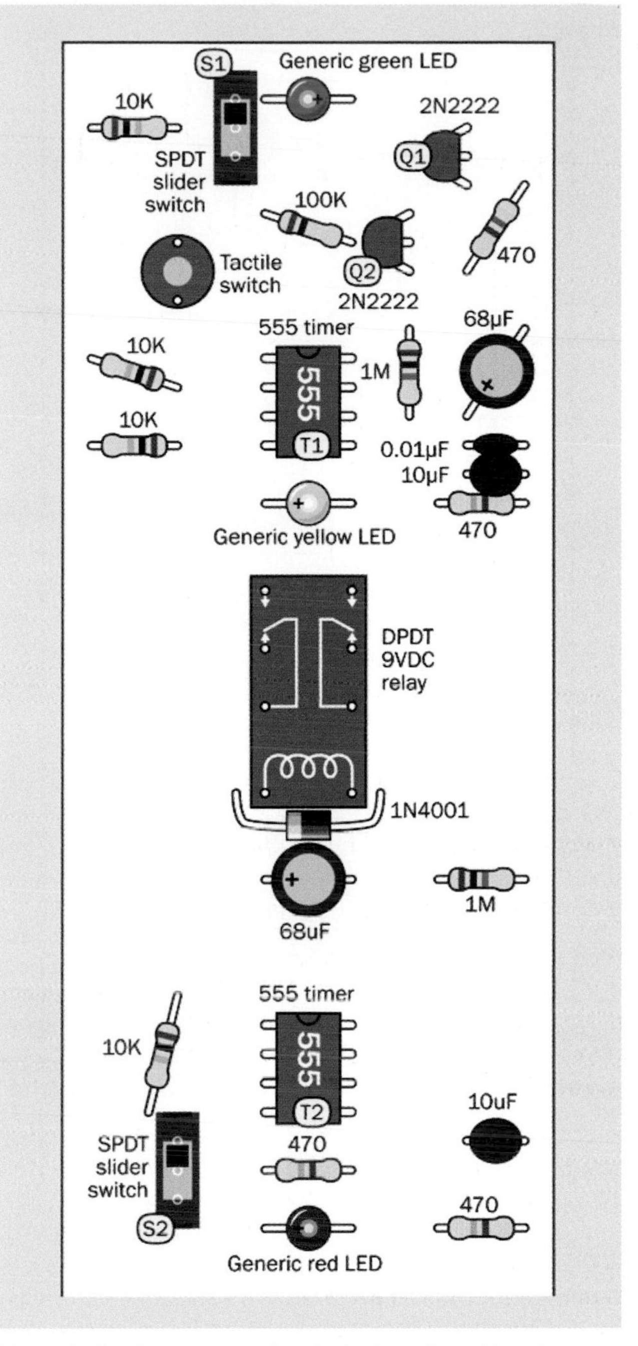

Figure 4-46 *Component values in the breadboard layout.*

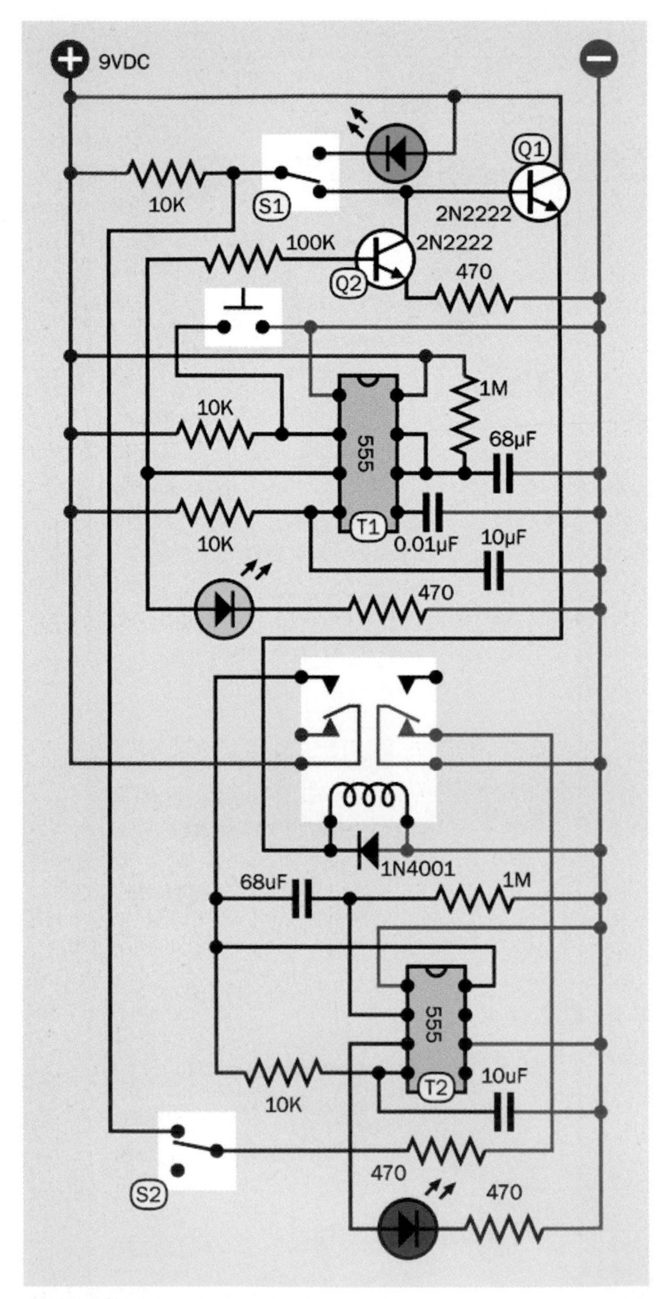

Figure 4-47 *Schematic equivalent of the breadboarded alarm circuit.*

What About the Noise?

When you want the alarm to make noise, you'll need to substitute an audio circuit or device for the red LED that you have been using for testing purposes.

The easy way to do this is to use an off-the-shelf item. Hundreds of sirens are available cheaply, ready to make an annoying sound if you simply apply power. Many of them require 12VDC, but they deliver almost as much noise at 9VDC. Just remember that timer T2 cannot deliver much more than 150mA.

If you prefer to have a sound that's all your own, you can use the circuit that I showed in Figure 4-36. Just use the output from your relay to power this circuit, and you have your own sound.

What About On and Off?

You've been testing the circuit by applying and removing power. You can add an on-off switch, but a numeric code to switch off the alarm would be more desirable.

Right now, I can't show you how to implement that, because it requires logic chips that I have not dealt with yet. But Experiment 21 will show you how it can be done.

Finalizing

Meanwhile, since the alarm circuit does actually work in its current form, I want to talk about finalizing it. By this I mean soldering it to a board, mounting the board in a box, and making everything look nice. My main concern in this book is electronics, but still, finalizing a project is an important part of the making experience, so I'll give you some suggestions.

Soldering the circuit can be easier than the procedure in Experiment 14, where I explained point-to-point wiring. You can mount the components on the type of perforated board that has copper traces on the back, in a configuration identical to the connectors inside a breadboard. Simply move each component to its comparable position, and solder it to the copper conductor underneath. No wire-to-wire soldering is necessary.

Guidance for finding and buying this kind of board is in "Supplies" on page 316 near the end of the book.

Now, how to proceed:

Carefully note the position of a component on your breadboard, then move it to the same relative position on the perforated board, poking its wires through the little holes.

Turn the perforated board upside down, make sure that it's stable, and examine the hole where the wire is poking through, as shown in Figure 4-48, which shows the *underside* of the board (the component is on the other side). A copper trace surrounds this hole and links it with others. Your task is to melt solder so that it sticks to the copper and also to the wire, forming a solid, reliable connection between the two of them.

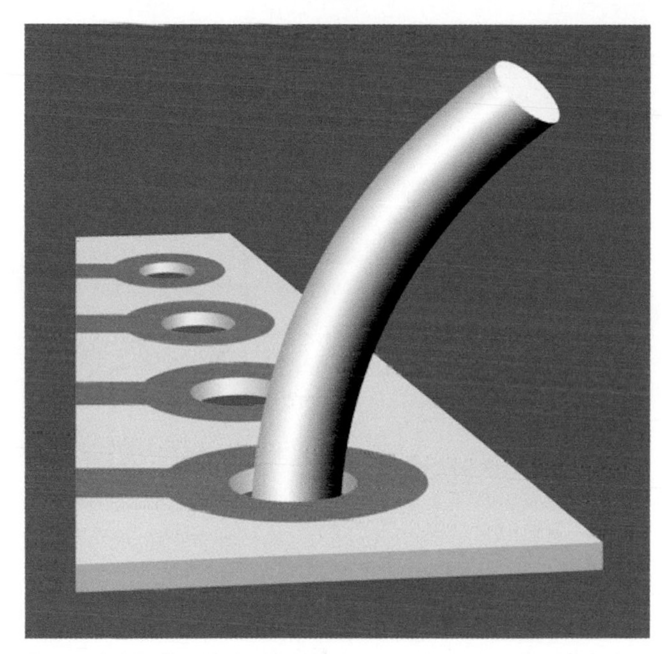

Figure 4-48 *The underside of perforated board, with a wire sticking through.*

Clamp the perforated board or rest it on a surface where it won't skid around easily. Take your low-wattage soldering iron in one hand and some solder in your other hand. Hold the tip of the iron against the wire and the copper, and feed some thin solder to their intersection. After two to four seconds, the solder should start flowing.

Allow enough solder to form a rounded bump sealing the wire and the copper, as shown in Figure 4-49. Wait for the solder to harden thoroughly, and then grab the wire with pointed-nosed pliers and wiggle it to make sure you have a strong connection. If all is well, snip the protruding wire with your cutters. See Figure 4-50.

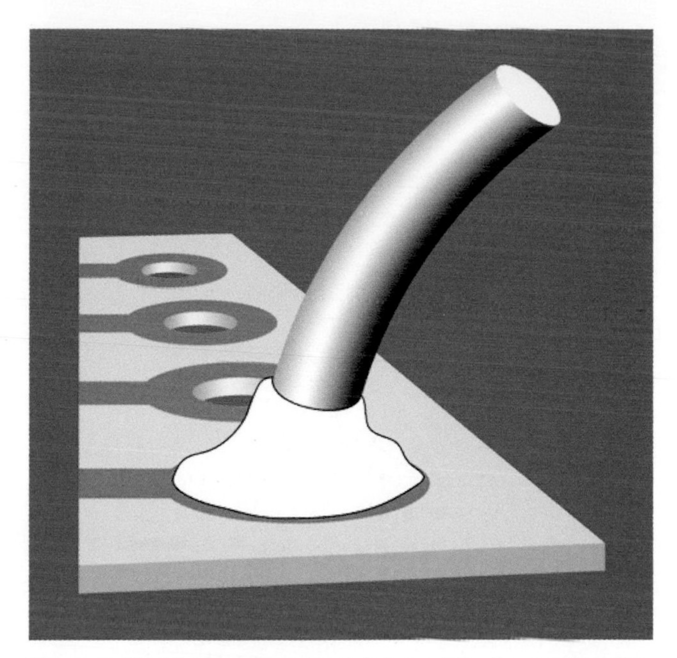

Figure 4-49 *Ideally, your solder joint should look something like this.*

Figure 4-50 *After the solder has cooled and hardened, you snip off the projecting wire.*

Because solder joints are difficult to photograph, I'm using drawings to show the wire before and after making a reasonably good joint. The solder is shown in pure white, outlined with a black line.

The process of actually soldering components into perforated board is illustrated in Figure 4-51 and Figure 4-52.

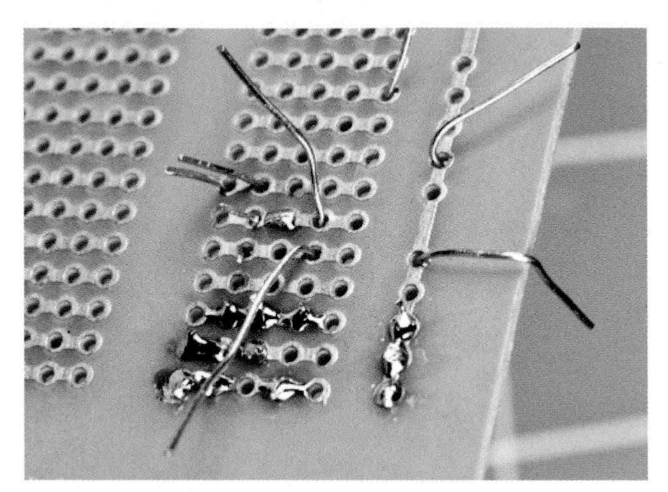

Figure 4-51 *This photograph was taken during the process of transferring components from breadboard to perforated board. Two or three components at a time are inserted from the other side of the board, and their leads are bent over to prevent them from falling out.*

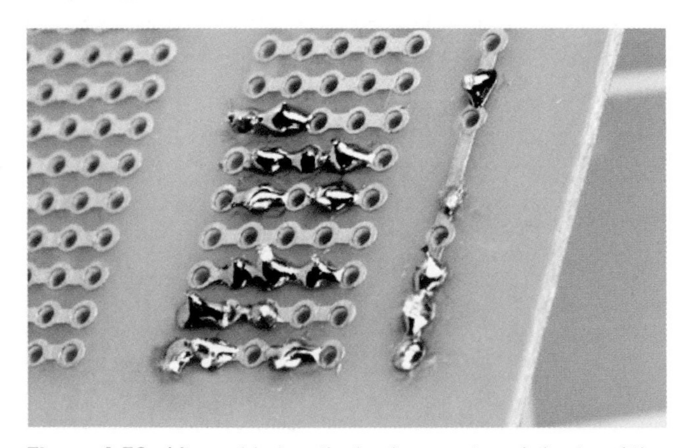

Figure 4-52 *After soldering, the leads are snipped short and the joints are inspected under a magnifying glass. Another two or three components can now be inserted, and the process can be repeated.*

Most Common Perfboarding Errors

1. Too much solder. Before you know it, solder creeps across the board, touches the next copper trace, and sticks to it, as depicted in Figure 4-53. When this happens, you can either try to suck it up with a desoldering kit, or carve it away with a knife. Personally I prefer to use a knife, because if you suck it up with a rubber bulb or solder wick, some of it will tend to remain.

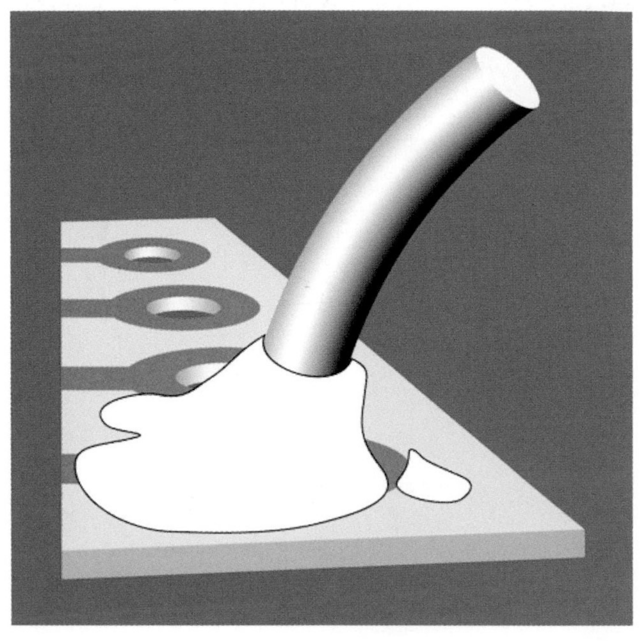

Figure 4-53 *If you use too much solder, it will tend to end up in places where you don't want it.*

Even a microscopic trace of solder is enough to create a short circuit. Check the wiring with a magnifying glass while turning the perforated board so that the light strikes it from different angles.

2. Not enough solder. If the joint is thin, the wire can break free from the solder as it cools. Even a microscopic crack is sufficient to stop the circuit from working. In extreme cases, the solder sticks to the wire, and sticks to the copper trace around the wire, yet doesn't make a solid bridge connecting the two, leaving the wire encircled by solder yet untouched by it, as shown in Figure 4-54. You may find this undetectable unless you observe it with magnification.

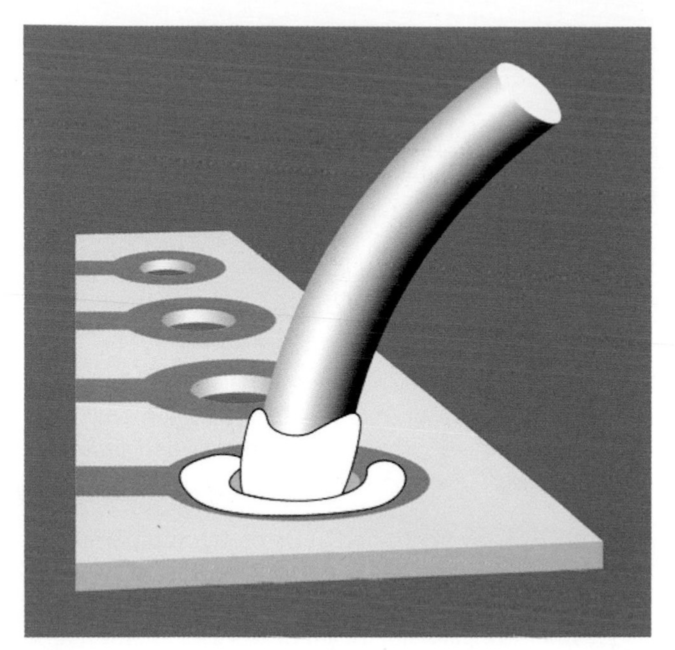

Figure 4-54 *Too little solder (or insufficient heat) can allow a soldered wire to remain separate from the soldered copper on the perforated board. Even a hair-thin gap is sufficient to prevent an electrical connection.*

You can add more solder to any joint that may have insufficient solder, but be sure to reheat the joint thoroughly.

3. Components incorrectly placed. It's very easy to put a component one hole away from the position where it should be. It's also easy to forget to make a connection.

I suggest that you print a copy of the schematic, and each time you make a connection on the perforated board, you eliminate that wire on your hardcopy, using a highlighter.

4. Debris. When you're trimming wires, the little fragments that you cut don't disappear. They start to clutter your work area, and one of them can easily get trapped under your perforated board, creating an electrical connection where you don't want it.

Clean the underside of your board with an old toothbrush before you apply power to it. Dip the toothbrush in rubbing alcohol to remove flux residues. Keep your work area as neat as possible. The more meticulous you are, the fewer problems you'll have later.

Once again, be sure to check every joint with a magnifying glass.

Fundamentals: Perforated Board Fault Tracing

If the circuit that worked on your breadboard doesn't work after you solder it to perforated board, your fault tracing procedure is a little different from that which I outlined previously.

First look at component placement, because this is the easiest thing to verify.

If all the components are placed correctly, flex the board gently while applying power. If you now get an intermittent response from the circuit, you can be virtually certain that solder didn't stick where it was supposed to stick, or a joint has a tiny crack in it.

Anchor the black lead of your meter to the negative side of the power supply, then switch on the power and go through the circuit point by point, from top to bottom, checking the voltage at each point with the red lead of the meter while you continue to flex the board. In most circuits, almost every part should show at least some voltage. If there is a dead zone, or if your meter responds intermittently, you can zero in on a joint that has something wrong with it, even though it looks good superficially.

A bright desk lamp and a magnifier are indispensable for this procedure. A gap of 1/1,000" or less is quite sufficient to stop your circuit from working. But you'll have difficulty detecting it without magnification, and even then, sometimes the light has to be exactly right.

Dirt, water, or grease can prevent solder from sticking properly to wires or copper traces. This is just another reason to be as meticulous as possible in your work habits.

The Project Box

The easiest way to enclose your perforated board is to put it in a project box. (I mentioned this item in the list of components and supplies at the beginning of Chapter 3). Hundreds of variants are available. Aluminum boxes look cool and professional, but you have to protect the circuit board from short-circuiting itself on the inside of the box. Plastic boxes are easier, and cheaper, too.

To make everything look professional, you should not just start drilling holes arbitrarily for your switches and LEDs. You should draw a layout on paper (or use a drawing program, and then print the image on paper). Just make sure there's room for the components to fit together, and try to place them similarly to the schematic, to minimize the risk of confusion.

Tape your sketch to the inside of the top panel, as shown in Figure 4-55, and then use a sharp, pointed tool (such as an awl or a needle) to press through and mark the plastic at the center of each hole. The indentations will help to center your bit when you drill the holes.

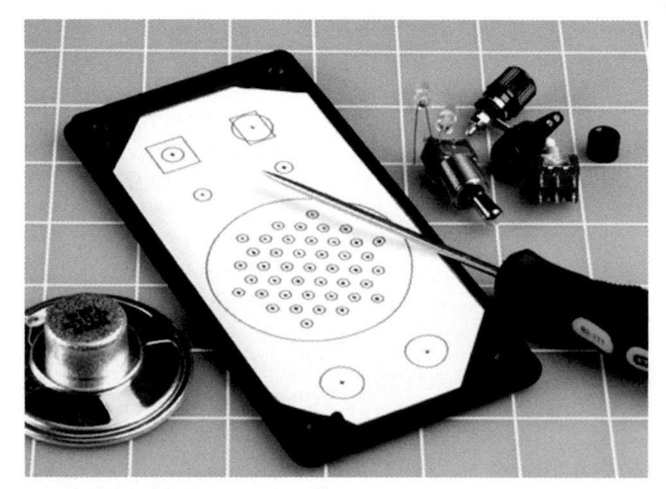

Figure 4-55 *A printed layout for the switches, LEDs, and other components has been taped to the underside of the lid of the project box. An awl is pressed through the paper to mark the center of each hole to be drilled in the lid.*

If you're using an audio circuit driving a loudspeaker (instead of a ready-made siren) you'll need to make multiple holes to vent the sound from the loudspeaker, which will be beneath the top panel of the box. The panel that I made is shown in Figure 4-56.

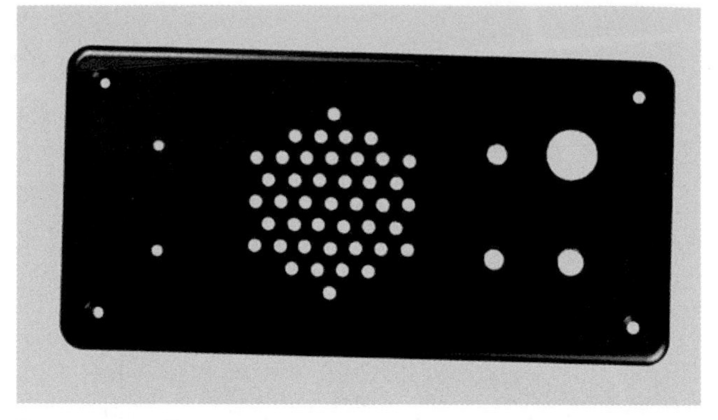

Figure 4-56 *The exterior of the panel after drilling. A small handheld cordless drill can create a neat result if the holes were marked carefully.*

I placed all the switches and LEDs on the top panel. The power input jack is positioned at one end of the box. Naturally, each hole has to be sized to fit its component, and if you have calipers, they'll be very useful for taking measurements and selecting the right drill bit. Otherwise, make your best guess, too small being better than too large. A *deburring tool* is ideal for slightly enlarging a hole so that a component fits snugly. This may be necessary if you drill 3/16-inch holes for your 5-inch mm LEDs. Fractionally enlarge each hole, and the LEDs should push in very snugly.

If your loudspeaker lacks mounting holes, you'll have to glue it in place. I used five-minute epoxy to do this. Be careful not to use too much. You don't want any of the glue to touch the speaker cone.

Drilling large holes in the thin, soft plastic of a project box can be a problem. The drill bit tends to dig in and create a mess. You can approach this problem in one of three ways:

- Use a Forstner drill bit if you have one. It creates a very clean hole. A hole saw can also be used.
- Drill a series of holes of increasing size.
- Drill a smaller hole than you need, and enlarge it with a deburring tool.

Regardless of which approach you use, you'll need to clamp or hold the top panel of the project box with its outside surface face-down on a piece of scrap wood. Then drill from the inside, so that your bit will pass through the plastic and into the wood.

Finally, mount the components in the panel, as shown in Figure 4-57, and turn your attention to the underneath part of the box.

Soldering the Switches

Your first step is to decide which way up the switch should be. Use your meter to find out which terminals are connected when the switch is flipped. You'll probably want the switch to be on when the toggle is flipped upward. The underside of my control panel is shown in Figure 4-57. I used a DPDT switch because I just happened to have one. You only need a SPST switch for the project.

Figure 4-57 *Components have been added to the control panel of the project box (seen from the underside). The loudspeaker has been glued in place. Spare glue was dabbed onto the LEDs, just in case.*

Remember, the center terminal of any double-throw switch is almost always the pole of the switch.

Stranded wire is appropriate to connect the circuit board with the components in the top panel, because the strands flex easily and impose less stress on solder joints. Twisting each pair of wires together helps to minimize the mess.

When you connect wires or components with the lugs on the switches, your pencil-style soldering iron may not deliver enough heat to make good joints. You can use your higher-powered soldering iron in these locations, but you absolutely must apply a good heat sink to protect the LEDs when you attach them, and don't allow the iron to remain in contact with anything for more than 10 seconds. It will quickly melt insulation, and may even damage the internal parts of the switches.

In projects that are more complex than this one, it would be good practice to link the top panel with the circuit board more neatly. Multicolored ribbon cable is ideal for this purpose, with plug-and-socket connectors that attach to the board. For this introductory project, I didn't bother. The wires just straggle around, as shown in Figure 4-58.

Figure 4-58 *Twisted wire pairs have been connected on a point-to-point basis, without much concern over neatness, as this is a relatively small project.*

Fixing the Board

The circuit board will sit on the bottom of the box, held in place with four #4-size machine screws (bolts) with washers and nylon-insert locknuts. I prefer using nuts and bolts, rather than glue, in case I ever need to remove the circuit to make a repair. You need to use locknuts to eliminate the risk of a nut working loose and falling among components, where it can cause a short circuit.

You'll have to cut the perforated board to fit, taking care not to damage any of the components on it. I use a band saw for cutting, but a hacksaw will work. Remember that perforated board often contains glass fibers that can blunt a wood saw.

Check the underside the board for loose fragments of copper traces after you finish cutting.

Drill bolt holes in the board, taking care again not to damage any components. Then mark through the holes to the plastic bottom of the box, and drill the box. Countersink the holes (i.e., bevel the edges of a hole so that a flat-headed screw will fit into it flush with the surrounding surface), push the little bolts up from underneath, and install the circuit board. Because you're using locknuts, which will not loosen, there's no need to make them especially tight. Indeed, excessive tightness must be avoided.

After mounting the board, test the circuit again, just in case.

Caution: Avoid Board Stress

Be extremely careful not to attach the circuit board too tightly to the project box. This can impose bending stresses, which may break a joint or a copper trace on the board.

Final Test

When you've completed the circuit, if you don't have your network of magnetic sensor switches set up yet, you can just use a piece of wire instead. I used a pair of *binding posts* on my box, just for convenience. You can equally well run a pair of wires from the circuit board out through a small hole in the lid of the box.

If everything works the way it should, it's time to screw the top of the box in place, pushing the wires inside. Because you're using a large box, you should have no risk of metal parts touching each other accidentally, but still, proceed carefully. My finished product is shown in Figure 4-59.

Figure 4-59 *The completed alarm box.*

Alarm Installation

If you are going to complete this project with magnetic sensor switches, you should test each one by moving the magnetic module near the switch module and then away from it, while you use your meter to test continuity between the switch terminals. The switch should close when it's next to the magnet, and open when the magnet is removed.

Now draw a sketch of how you'll wire your switches together. Always remember that they have to be in series, not in parallel! Figure 4-60 shows the concept in theory. The two terminals are the binding posts on top of your control box (which is shown in green), and the dark red rectangles are the magnetic sensor switches on windows and doors. Because the wire for this kind of installation usually has two conductors, you can lay it as I've indicated but cut and solder it to create branches. The solder joints are shown as orange dots. Note how current flows through all the switches in series before it gets back to the control box.

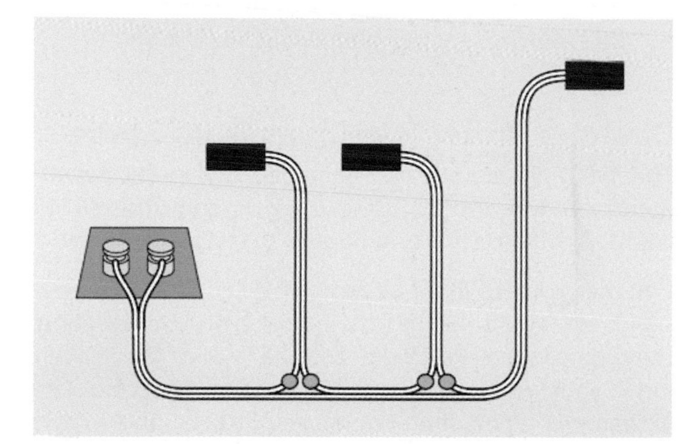

Figure 4-60 *Dual-conductor, white insulated wire can be used to connect the terminals on the alarm control box with magnetic sensors (shown in dark red). Because the sensors must be in series, the wire is cut and joined at positions marked with orange dots.*

Figure 4-61 shows the same network as you might actually install it in a situation where you have two windows and a door. The blue rectangles are the magnetic modules that activate the switch modules.

You'll need a large quantity of wire, obviously. The type of white, stranded wire that is sold for doorbells or furnace thermostats is good. Typically, it is 20-gauge or larger.

After you install all the switches, clip your meter leads to the wires that would normally attach to the alarm box. Set your meter to test continuity, and open each window or door, one at a time, to check whether you're breaking the continuity. If everything is OK, attach the alarm wires to the binding posts on your project box.

Now deal with the power supply. Use your AC adapter, with its output set to 9 volts, hooked up to your DC power plug, or attach the power plug to a 9-volt battery. The circuit will run off a 12-volt alarm battery too, but you will have to substitute a 12-volt relay for the one that I specified.

Figure 4-61 *In an installation involving two windows and a door, the magnetic components of the sensors (blue rectangles) could be placed as shown, while the switches (dark red) are located alongside them.*

The only remaining task is to label the switch, button, power socket, and binding posts on the alarm box. You know that the switch puts the alarm in and out of continuity testing mode, and the button gives you a minute to leave before the system alarms itself, but no one else knows, and you might want to allow a guest to use your alarm while you're away. For that matter, months or years from now, you may forget some details yourself.

Conclusion

The alarm project has taken you through the basic steps that you will usually follow any time you develop something:

- Make a wish list.
- Decide which types of components are appropriate.
- Draw a schematic and make sure that you understand it.

- Modify it to fit the pattern of conductors on a breadboard.
- Install components on the breadboard and test the basic functions.
- Modify or enhance the circuit, and retest.
- Transfer to perforated board, test, and trace faults if necessary.
- Add switches, buttons, a power jack, and plugs or sockets to connect the circuit with the outside world.
- Mount everything in a box (and add labeling).

Experiment 19: Reflex Tester

Because the 555 timer can run at thousands of cycles per second, you can use it to measure human reflexes. You can compete with friends to see who has the fastest response—and note how your response changes depending on your mood, the time of day, or how much sleep you got last night.

This circuit is not conceptually difficult, but requires quite a lot of wiring, and will only just fit on a breadboard that has 60 rows of holes (or more). Still, it can be tested in sections, like the circuit in Experiment 18. If you avoid making errors, the whole project can be assembled in a couple of hours.

What You Will Need

- Breadboard, hookup wire, wire cutters, wire strippers, multimeter
- 9-volt power supply (battery or AC adapter)
- 4026B chips (3)
- 555 timers (3)
- Resistors: 470 ohms (2), 680 ohms (3), 10K (6), 47K (1), 100K (1), 330K (1)
- Capacitors: 0.01µF (2), 0.047µF (1), 0.1µF (1), 3.3µF (1), 22µF (1), 100µF (1)
- Tactile switches (3)
- Generic LEDs: one red, one yellow
- Trimmer potentiometer, 20K or 25K (1)
- Single-digit numeric LED displays, height 0.56", low-current red preferred, able to function at 2V forward voltage and 5mA forward current (3) (Avago HDSP-513A preferred, or Lite-On LTS-546AWC, or Kingbright SC56–11EWA, or similar)

Caution: Protecting Chips from Static

The 555 timer is not easily damaged, but in this experiment you will also be using a CMOS chip (the 4026B counter), which is more vulnerable to static electricity.

Whether you are likely to zap a chip by handling it depends on factors such as the humidity in your location, the type of shoes you wear, and the type of floor covering in your work area. Some people seem to accumulate a charge of static more easily than others, and I don't have an explanation for this. Personally, I have never damaged a chip with static, but I know people who have.

If static is a risk for you, you'll probably know about it, because you'll suffer from sudden little jolts when you reach for a metal door handle or a steel faucet. If you really feel you need to protect chips from this kind of discharge, the most thorough precaution is to ground yourself. The best way to do this is by using an anti-static wrist strap. The conductive strap is secured to your wrist with velcro and connects through a high-value resistor (typically 1M) to an alligator clip that can be attached to any large metal object.

When you receive chips via mail order, they are usually shipped in channels of conductive plastic, or with their legs embedded in conductive foam. The plastic or the foam protects the chips by insuring that all the pins have an approximately equal electrical potential. If you want to repackage your chips, but you don't have any conductive foam, you can poke their legs through aluminum foil.

Caution: Be Careful When Grounded

The resistor built into an anti-static wrist strap protects you from electrocuting yourself if you happen to touch a source of relatively high voltage with your other hand. This is an important feature, as an electric shock that travels from one hand to the other will pass through your chest and can stop the heart.

If you use just a plain piece of wire to ground yourself, you lose this protection. The small cost of a proper wrist strap is a sensible investment.

Now, back to the experiment.

A Quick Demo

In the previous edition of this book I suggested using a single three-digit display for this project. In this edition, I have switched to three individual digits. The cost is fractionally higher, but the wiring is much simpler, and the project can be built more easily. Also, I believe there is a better chance of single LED digits remaining available for many years to come.

I am specifying digits that are 0.56" high because this is an industry standard, with pinouts that are also standardized. If you use a smaller size, the pinouts will be different. If you use a larger size, they won't fit with the other components on the breadboard.

Let's begin by getting acquainted with one of the numerals, and the 4026B chip that will be driving it.

The first module of this circuit is shown in Figure 4-62. (If you feel that a schematic is easier to understand, see Figure 4-63, which shows the same components.)

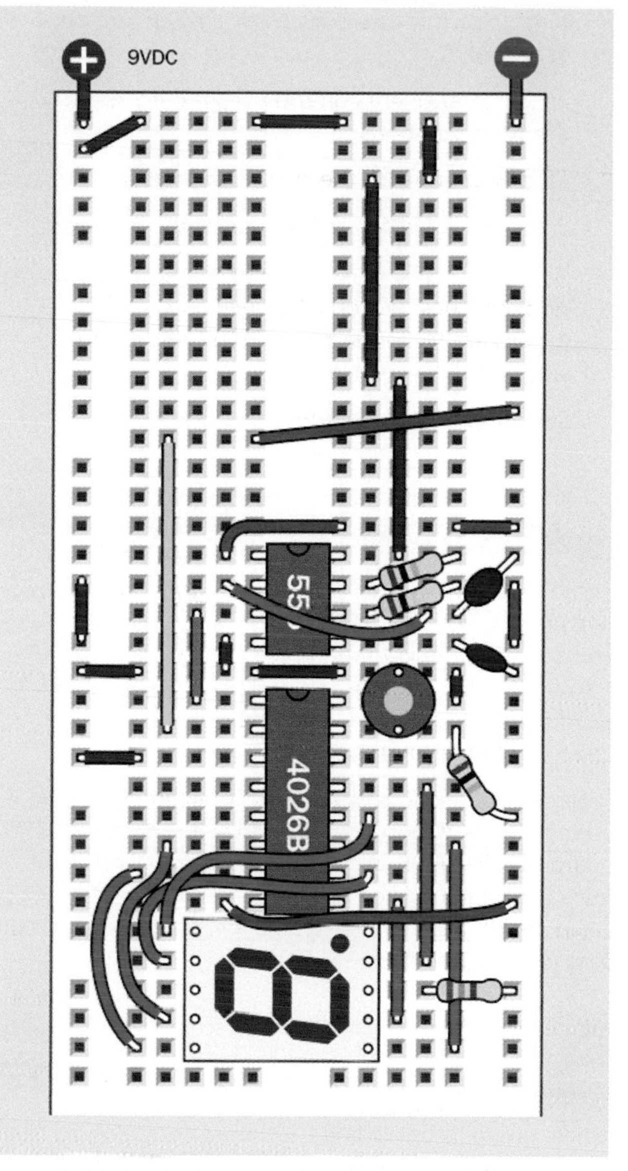

Figure 4-62 *The first module of the reflex tester demonstrates how a timer can run a counter chip that powers a single-digit LED display.*

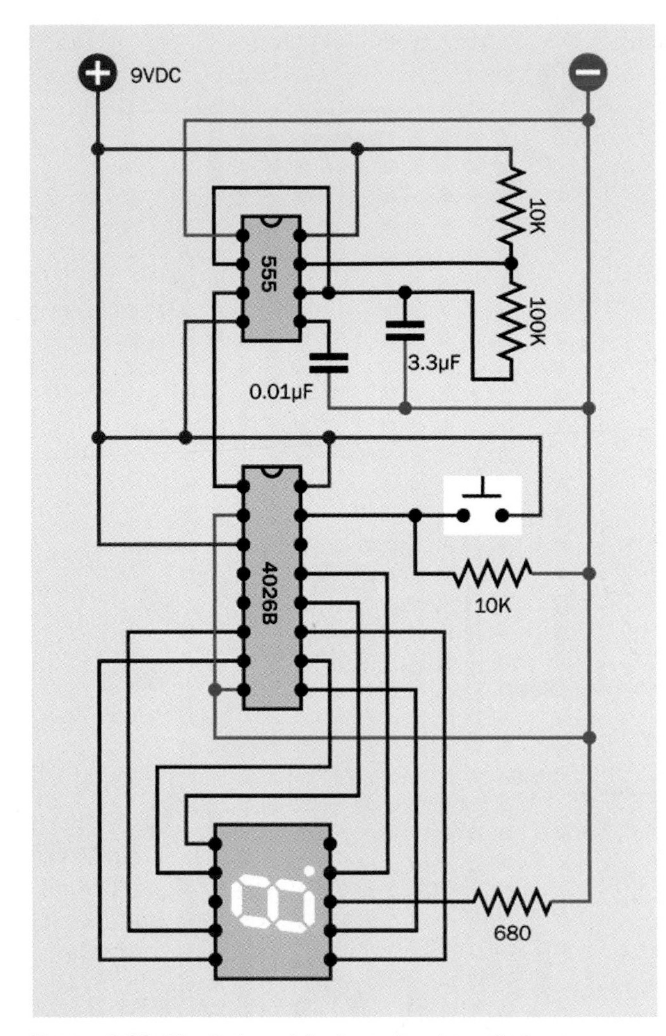

Figure 4-63 *The first module shown in schematic form.*

In addition, component values are shown in Figure 4-64.

You're going to be adding a lot more components to the board (in fact, it will be packed full by the time you're finished), so you need to position everything exactly as I have shown it in the figure. Count the rows of holes carefully! You'll see some wires that don't make sense right now (all those red pieces—what are they for?) but they will enable you to add and activate a couple more 555 timers as you proceed.

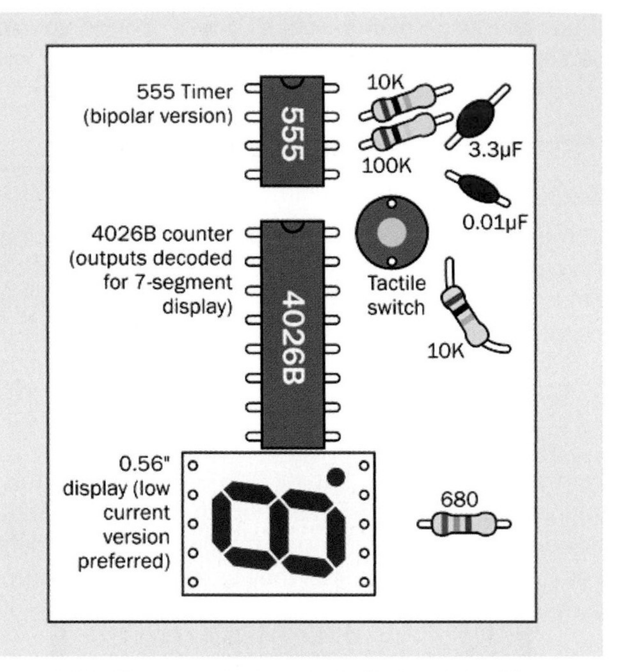

Figure 4-64 *Component values for the first module of the circuit.*

Apply power from a 9-volt battery or AC adapter, and you should see the numeric display counting repeatedly from 0 through 9.

If you don't see any numerals at all, set your meter to measure DC volts, clip the black probe to the negative side of the power supply, and use the red probe to test for voltages at key locations in the circuit, such as the power input pins of the chips. If the voltages look okay, make sure the resistor at bottom-right is 680 ohms (not 68K or 680K, which are colored similarly).

If the display shows fragments of numerals, or numerals that are not counting sequentially, you made an error in the green wires connecting with the 4026B chip.

If the display shows a 0 that does not change, you wired the 555 timer wrongly, or failed to connect the timer with the 4026B chip correctly.

Once you have the numbers counting upward, hold down the tactile switch, and notice that it forces the counter back to 0. As soon as you let go of the tactile switch, the numbers start counting again.

We have the basis for a reflex tester right here. We just need to add a couple more digits, increase the counting speed, and make some more refinements. First, however, I'll explain what's going on.

Fundamentals: LED Displays

The term "LED" is slightly confusing. The type of component that you have used in previous experiments is properly known as a *standard LED*, a *through-hole LED*, or an *LED indicator*: a little rounded component with two long leads sticking out of the base. These became so common, people started calling them simply "LEDs." But LEDs are used in other components, too, such as the glowing numeral currently plugged into your breadboard. It is properly known as an *LED display*. More precisely, it is a *seven-segment single-digit LED display*.

In Figure 4-65 the dimensions of the display are shown, along with the pin positions that are hidden underneath it. The important things to notice are that the numeral does consist of seven segments, plus a decimal point; and the pin spacings are all measured in multiples of 0.1", which is helpful for your breadboard.

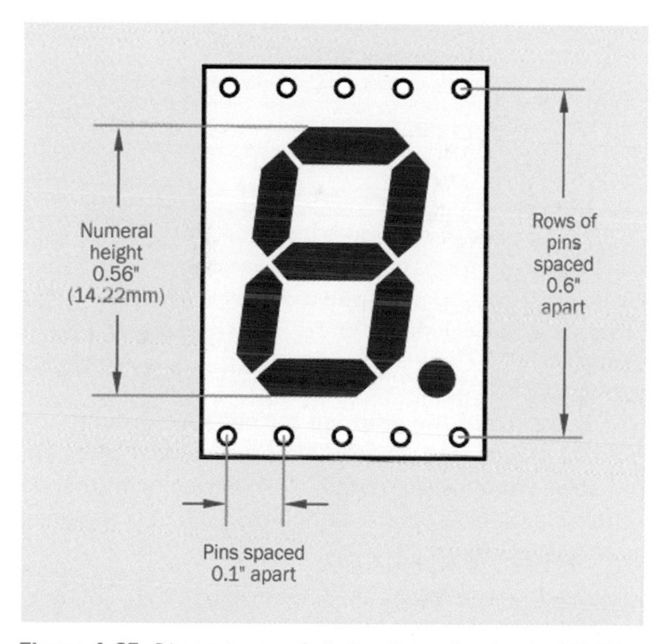

Figure 4-65 *Dimensions and pin locations of a standard 0.56" seven-segment LED display.*

Now take a look at Figure 4-66. This shows you the internal connections between the pins and the segments of the digit. Notice that pins 3 and 8 have blue centers, indicating that they should be connected with negative ground. All the other pins are designed to receive positive power to activate the LED segments. This is called a *common cathode* type of LED, because the negative sides of the internal diodes (the cathodes) are all tied together.

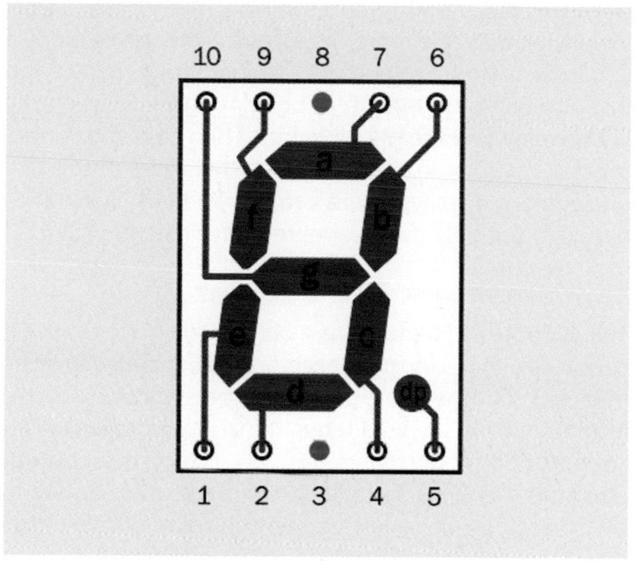

Figure 4-66 *The scheme that is used for pin numbering, and the internal connections that are hidden inside the component.*

In a *common anode* display, the situation is reversed, and you activate the segments by applying negative power to each one, while they all share a positive internal connection. You can choose whichever type of display is convenient in a circuit, but common cathode displays are more widely used.

Note that the segments are identified with lowercase letters a through g, plus dp for the decimal point. This system is common on almost all datasheets (although a few use the letter h for the decimal point).

So far so good, but I have left out a crucial piece of information: like all LEDs, the segments in the numeral must be protected with series resistors. This is a hassle, and you may wonder why the manufacturer doesn't build resistors in. The answer is that the display must be usable on a wide variety of voltages, and the values of the resistors will depend on the voltage.

Well—why can't we use just one resistor that all the segments share, perhaps between pin 3 and negative ground? Actually we can do that, but the resistor will be dropping the voltage and restricting the current for different numbers of segments, depending on which numeral is being displayed. A numeral 1 lights up only two segments, while a numeral 8 uses all seven of them. Consequently, some numerals will be brighter than others.

Does this really matter? I'm thinking that for this demo, simplicity may be more important than perfection. If you look at Figure 4-62 you'll see that I have, in fact, put just one 680-ohm resistor at bottom-right, between the LED display and the negative bus. This is not the correct procedure, but you are going to install three of the seven-segment displays in this project, and I think you'll be happy to use only three series resistors instead of 21.

Fundamentals: The Counter

The 4026B chip is known as a *decade counter,* because it counts in tens. Most counters have a *coded output*, meaning that they output the numbers in binary-coded format (which I will discuss in a later project). This counter doesn't do things this way. It has seven output pins, and it powers them in patterns that just happen to be correct for a seven-segment display. While other counters require a *driver* to convert a binary output to the seven-segment patterns, the 4026B gives you everything in one package.

This is very convenient, except that the 4026B is an old-fashioned CMOS chip that has limited power. The datasheet advises you to draw less than 5mA from any one pin when powering the chip with 9 volts.

Ideally you should pass the outputs from the counter through an array of transistors, to amplify them. You can buy a chip containing seven transistor pairs for exactly this purpose. It's called a *Darlington array*. (What if you want to display the decimal point? No problem. You can buy a different Darlington array containing eight transistor pairs.)

I could have used three Darlington-array chips to drive the three LED displays in this project, but that would have added to the complexity and expense, and I would have needed two breadboards. Therefore I decided it was acceptable to use low-current LED displays that can be driven directly by a counter. They're not as bright, but they do the job. I chose the 680-ohm resistor because it should limit the current to less than 5mA from any single pin on the counter chip, and it imposes a voltage drop of around 2V across the LEDs (which varies depending how many segments are lit up).

Now I'll give you some details about the internal workings of the 4026B. Counter chips always have several useful features built in. Take a look at Figure 4-67, which shows the pinouts of the chip. The pins with labels such as "To segment a" are easy to understand. You simply run a wire from each pin to the appropriate pin on your LED display. If you look at Figure 4-62 you'll see each of the green wires connecting the output pins from the counter with the input pins on the numeral.

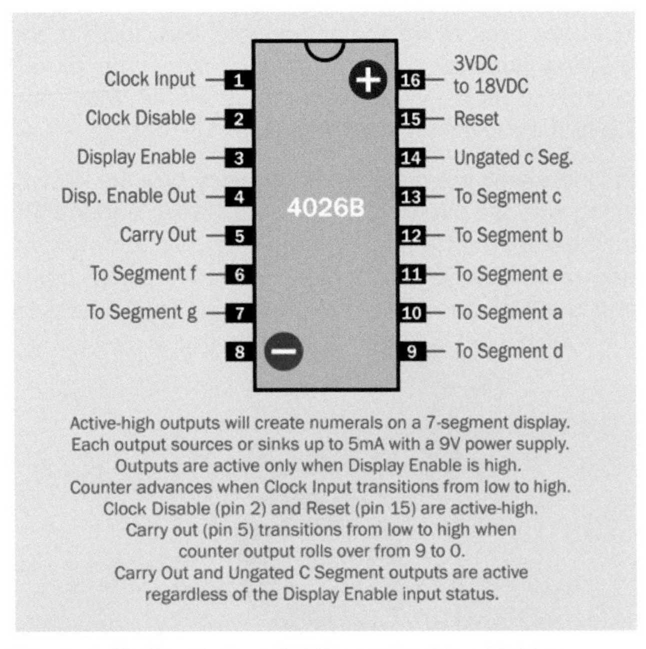
Clock Input
1
Clock Disable
2
Display Enable
3
Disp. Enable Out
4
Carry Out
5
To Segment f
6
To Segment g
7
8
4026B
16
3VDC
to 18VDC
15
Reset
14
Ungated c Seg.
13
To Segment c
12
To Segment b
11
To Segment e
10
To Segment a
9
To Segment d
Active-high outputs will create numerals on a 7-segment display.
Each output sources or sinks up to 5mA with a 9V power supply.
Outputs are active only when Display Enable is high.
Counter advances when Clock Input transitions from low to high.
Clock Disable (pin 2) and Reset (pin 15) are active-high.
Carry out (pin 5) transitions from low to high when
counter output rolls over from 9 to 0.
Carry Out and Ungated C Segment outputs are active
regardless of the Display Enable input status.

Figure 4-67 *Pinouts for a 4026B counter chip, which has decoded outputs to drive a seven-segment single-digit LED display.*

Pins 8 and 16 of the chip are for negative-ground and positive power, respectively. Almost all digital chips apply power to opposite corners in this way (with the exception of the 555 timer—although, really, it is classified as an analog chip).

Because Figure 4-67 contains information that you may not need and is just included for future reference, I have also created a simplified view of the counter and the display in Figure 4-68, ignoring the pins that we won't be using, and showing the relationship of output pins with the pins on the display.

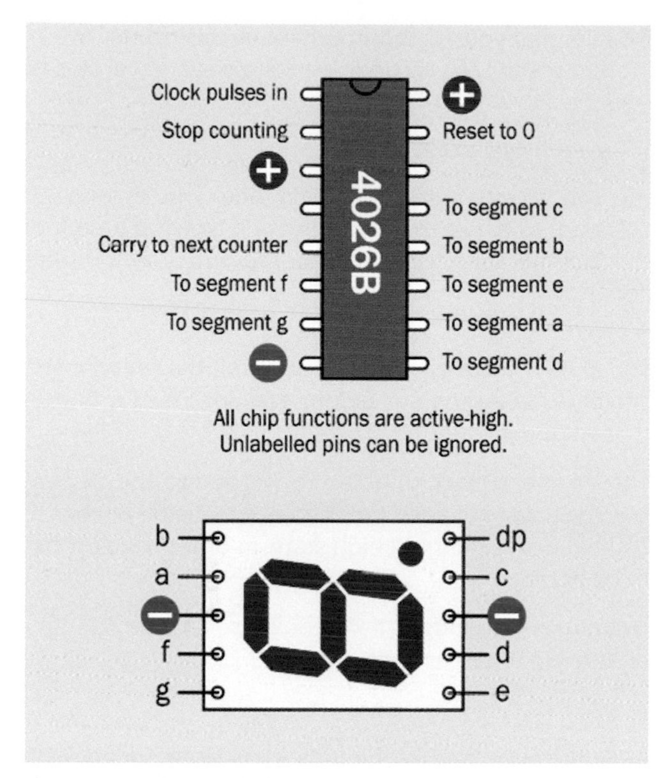

Figure 4-68 *A simplified view of the chip and the display, in the style that they will be shown on the breadboard.*

Look at pin 15, the reset pin. Now look at Figure 4-62. The pushbutton, more properly known as a tactile switch, is placed so that it will apply positive voltage to pin 15 when you press the button. (The voltage makes its way across the board to the pushbutton via the red segments of wire that I mentioned a moment ago.)

When you're not pressing the button, no positive voltage is applied to the reset pin of the counter. However, a 10K resistor permanently connects pin 15 to the negative bus on the breadboard. This is a *pulldown resistor*. It pulls the voltage on the pin to near zero—until you press the button, at which point the positive input overwhelms the negative power supplied through the resistor. Remember, if you don't apply a defined voltage to each input pin of a digital chip, you will get random, inexplicable, and confusing results. I've mentioned this before, but I have to emphasize the topic, because it's such a common cause of errors.

- To keep an input pin normally high, connect it to the positive bus through a 10K resistor (at least, for the circuits in this book). When you need to pull it low, use a switch or other device that overrides the resistor by making a more direct connection to the negative bus.
- To keep an input pin normally low, connect it to the negative bus through a 10K resistor. When you need to push it high, use a switch or other device that overrides the resistor by making a more direct connection to the positive bus.
- All the inputs on a counter chip must have some kind of connection. Never allow any floating input pins!
- Unused output pins should be left unconnected.

One more thing. Sometimes a chip has an input that we won't need at all. The 4026B, for instance, tells us that pin 3 is a display enable input. I want the display to be enabled all the time, so I connected pin 3 directly to the positive bus on a set-it-and-forget-it basis.

- If you won't be using an input pin, it must still have a defined state. You can deal with it by wiring it directly to the positive or negative side of the power supply.

Now I'll run through the remaining features of the 4026B.

The *clock input* (pin 1) accepts a stream of high and low pulses. The chip doesn't care how long the pulses are. It just responds by adding 1 to its count, each time it senses the input voltage rising from low to high.

The *clock disable* (pin 2) tells the counter to block the clock input. Like all the other pins on the chip, this one is *active-high*, meaning it performs its function when it has a high state. On your breadboard, I ran in a temporary blue wire and yellow wire to hold pin 2 low. In other words, I disabled the clock disable pin. This is confusing, so I will summarize the situation:

- When the clock disable pin is in a high state, it stops the counter from counting.
- When the clock disable pin is pulled down to negative ground, it allows the counter to count.

The *display enable* (pin 3) I already mentioned.

The *display enable out* (pin 4) will not be used here. It takes the state of pin 3 and connects it with pin 4 so that you can pass it along to other 4026B timers.

The *carry out* (pin 5) is essential if you want to count higher than 9. The pin state shifts from low to high when the counter has reached 9 and goes back to 0. If you take this output and connect it with the clock input pin of a second 4026B timer, the second timer will count in tens. You can use its carry output pin to signal a third timer, which will count in hundreds. I will be using this feature at the end of this project.

Lastly, Pin 14 can be used to restart the counter after it counts through 0, 1, and 2. This is useful in a digital clock that only counts up to 12 hours, but not relevant to us here. It is an output pin that we will not use, so it can be left unconnected.

Perhaps all the features seem confusing, but if you ever find yourself confronted with a counter chip that you've never seen before, you can figure it out (if you are patient and methodical) by looking up the manufacturer's datasheet. Then you can test it with LEDs and tactile switches, to make sure there are no misunderstandings. In fact, this is how I figured out the 4026B, myself.

Pulse Generation

Because the 555 timer accepts a power supply ranging from 5 to 15 volts, just like the 4026B, the output from the timer, on pin 3, can be connected directly to the input of the 4026B. That's the function of the purple wire in the breadboard layout. The 555 supplies pulses, and the 4026B counts them.

The rest of the wiring around the 555 timer should look familiar to you by now. You can see that it's running in astable mode. Your only question may be why it's running so slowly. We're not going to measure someone's reflexes at that speed.

True, but for this demo, I didn't want the digits to look like a featureless blur. We'll crank up the speed a bit later.

Time for a Plan

How should the reflex timer work? Here's my wish list:

1. I need a start button.
2. After the start button is pressed, there is a delay in which nothing happens. Suddenly there is a visual prompt, asking the player to respond.
3. Simultaneously, the count starts from 000, in 1/1,000ths of a second.
4. The user has to press a button to stop the counting process.
5. The count freezes, showing how much time elapsed between the prompt and the stop time. This measures the user's reflexes.
6. A reset button sets the count back to 000.

You already installed a reset button on the breadboard, which is necessary. But before that, we need a button that stops the counting process.

The counter's clock disable pin will freeze the display, but if you want to keep the display frozen, the pin has to be held in a continuing high state. In other words, it has to be latched.

Hmmm, that sounds as if we could use another 555 timer, wired in bistable mode.

A Control System

In Figure 4-69 the bistable timer has been added, with two new buttons. The diagonal blue wire in Figure 4-62 has been removed (to make room for the new timer). The other parts that you installed previously are still there, but are shown in gray.

Figure 4-70 shows the new part of the circuit in schematic form, and values of the additional parts are shown in Figure 4-71.

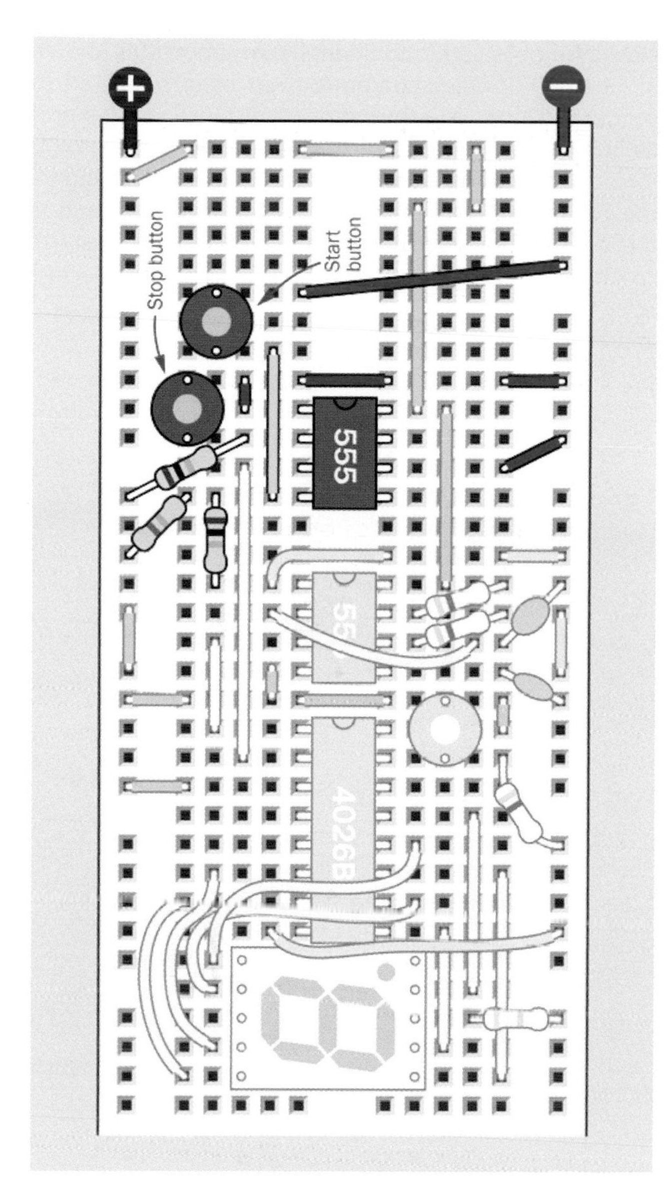

Figure 4-69 *A bistable 555 timer has been added. Previously wired components are grayed out.*

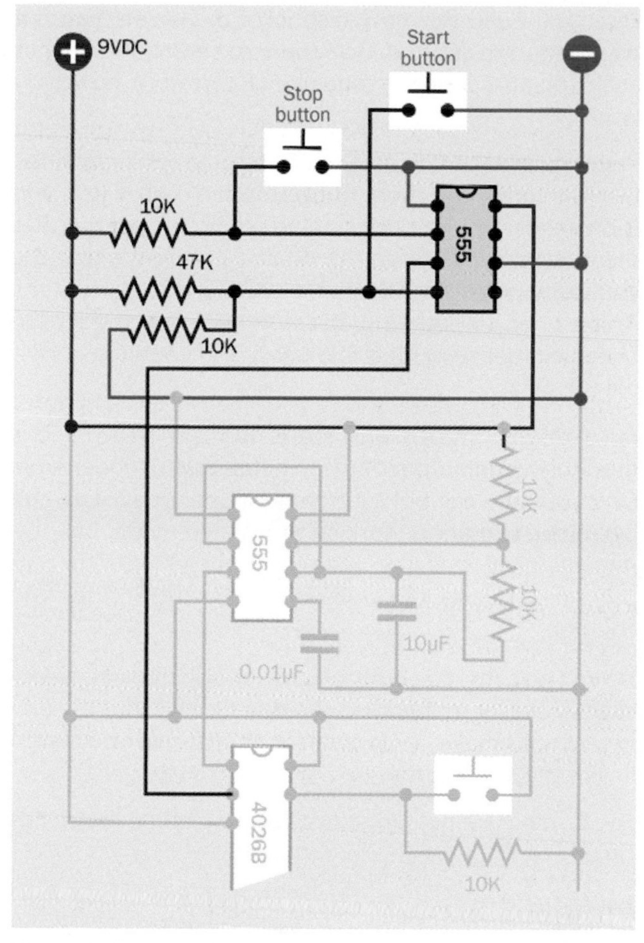

Figure 4-70 *This schematic shows the second timer and its associated components. Previous components are grayed out.*

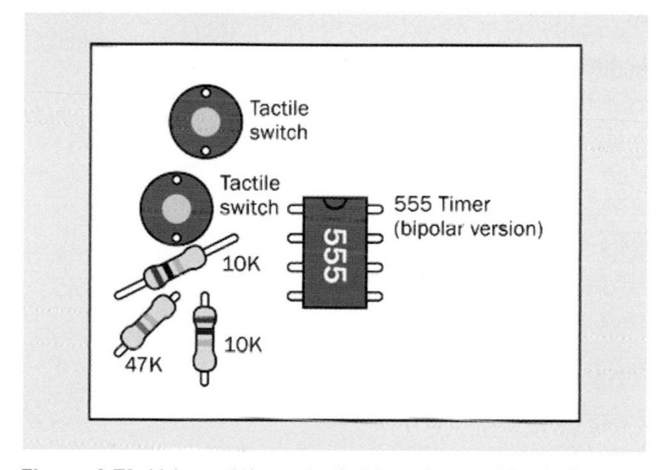

Figure 4-71 *Values of the parts that have been added in this module of the project.*

After you build this new section of the circuit, you can try it. You should find that the two new buttons start and stop the counting process. Do you see how they work?

Press the start button, and it grounds the reset pin of the bistable timer. The timer output on pin 3 goes low, and is connected to the clock disable pin on the counter. Remember, a *low* state on the disable pin means that the counter is *not* disabled. So, the counter starts counting. And it goes on counting, because once you trigger the bistable timer, its output is latched, and continues indefinitely.

But, you can stop it. Just press the stop button. This grounds the input pin of the bistable timer, triggering it. Consequently, the output from the timer goes high, and because the timer is working in bistable mode, the output is latched and stays high indefinitely. The high output goes to the clock disable pin, which stops the counter.

When you press the button at lower-right, which you installed initially, it still resets the timer to 000. But the timer remains latched in its disabled mode until you restart it with the start button.

The bistable 555 was exactly what we needed to run this circuit.

Progress Report

Let's see how far we've proceeded in satisfying the wishlist. It looks to me as if we're almost there. You press a button to start it, press a second button to stop it, and then once it's stopped, you press another button to set it to zero.

The only thing missing is the element of surprise. Really the person using it shouldn't know when the count will start. The idea is to measure the speed of his reflexes when he responds.

Why not add yet another timer, in monostable mode, to insert a delay before the action begins? This way, the start will be unexpected.

The Delay

First remove the start button, and the diagonal piece of blue wire connecting it with the negative bus. Leave the vertical yellow piece of wire where it is.

Now assemble some additional components as shown in Figure 4-72. The start button has been relocated to trigger the input of a third timer, which will insert a preliminary delay. The output from this timer will go high for 5 or 10 seconds, and then when it goes low it triggers the bistable timer, sending its output low, to suppress the clock disable feature of the 4026B, so that it starts counting.

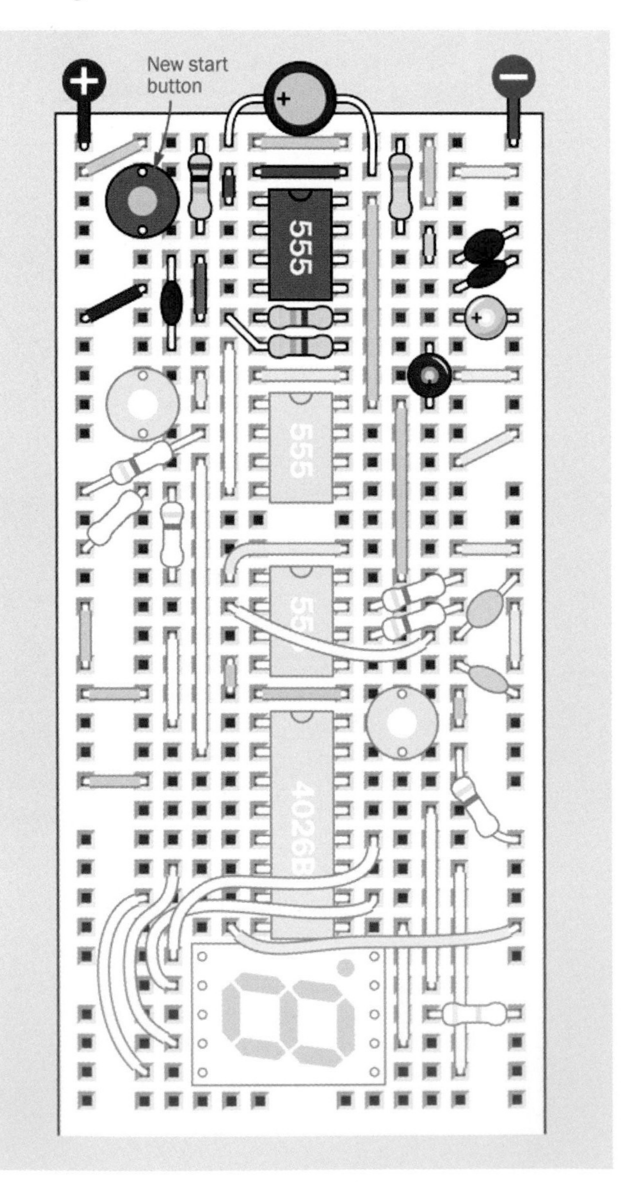

Figure 4-72 *The upper part of the reflex timer circuit is now complete.*

Be careful when installing the red and yellow LEDs. The red one is the opposite way around from what you might expect, because it is tied in to a positive source. So, its long, positive lead is at the bottom, not the top.

A schematic version of the new part of the circuit appears in Figure 4-73.

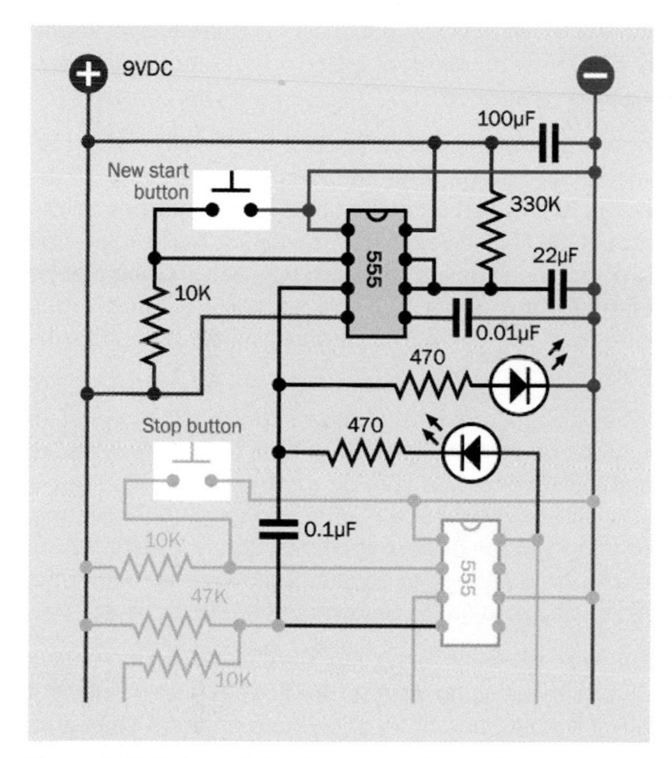

Figure 4-73 *Schematic for the new and final addition to the control circuit.*

The values of the parts that you added to the breadboard are shown in Figure 4-74.

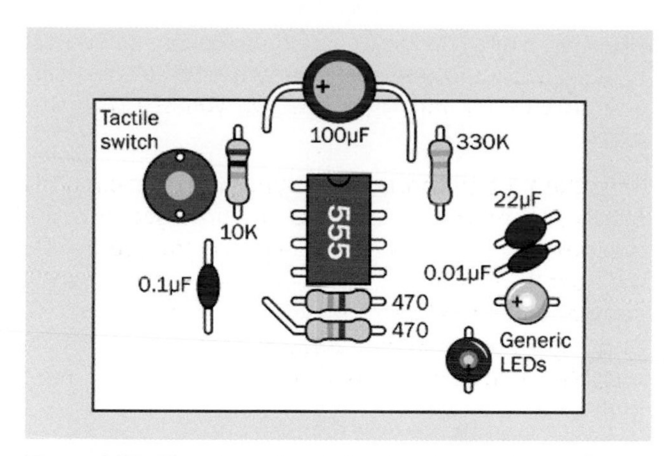

Figure 4-74 *The values of parts that have been added to the breadboard.*

Testing

When you apply power to this circuit, the counter immediately starts counting, without you asking it to. This is annoying, but easily dealt with. Press the stop button to stop the count. Press the bottom-right pushbutton to reset the timer to zero. Now you're ready for action.

Press the new start button, which creates an initial delay. During this delay, the yellow LED lights up. The delay lasts for about seven seconds, at which point the yellow LED goes out and the red LED comes on. Simultaneously, the counter starts counting—until you press the stop button.

The 100µF capacitor at the top of the breadboard looks like an afterthought, but is actually quite important. The 555 timer has a bad habit of creating voltage spikes when it switches its output, and in this circuit, the spikes can trigger the second timer without waiting for the delay. The 100µF capacitor suppresses this unfortunate tendency.

We now have all the features installed, except for the need to increase the counter speed and add a couple more counters and displays for fractions of a second.

How Does It Work?

Figure 4-75 shows you how the components are communicating with each other.

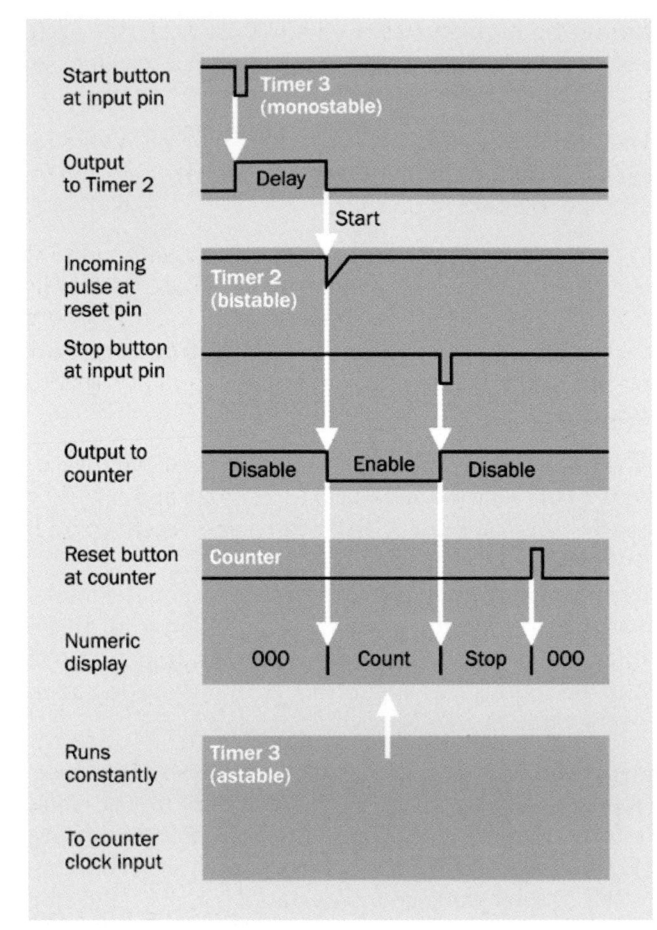

Figure 4-75 *Interaction between the components in the timer control circuit.*

I'll describe this figure from top to bottom. The start button (at the top, connected with Timer 3) pulls the input of the timer low to trigger it.

The output of Timer 3 goes high for about seven seconds. This creates your initial delay.

At the end of the delay, the output from Timer 3 drops back down. This transition passes through a 0.1µF coupling capacitor to Timer 2, which is bistable. The capacitor allows only a brief pulse to reach the reset pin of Timer 2. The pulse sends the output of Timer 2 low. This low output goes to the clock disable pin of the 4026B counter. A low state enables the counter, and the count begins.

Now we wait for the user to respond. The user presses the stop button for Timer 2, connected with its input on pin 2. The brief low input to Timer 2 sends its output high, which enables the clock disable pin on the counter, and stops the count.

Background: Development Issues

This project created a problem. When I built the original circuit several years ago, it ran fine. When some interns at Make: magazine built the circuit, it ran fine. Little did we know that the reset pin of a 555 timer behaves slightly differently on different brands. This is not documented on the datasheet.

After my book had been in print for literally years, I received notification from a reader that his version of the circuit functioned erratically, and sometimes not at all. I rebuilt the circuit, connected it with an oscilloscope, and saw that the coupling capacitor was transmitting a pulse faithfully from Timer 3 to the reset pin of Timer 2. But, sure enough, Timer 2 sometimes didn't recognize the pulse.

What was the problem? Either the pulse was too brief, or it didn't go low enough. Either way, the solution was to apply a lower pullup voltage to pin 4 of Timer 2. This is why you see two resistors attached to pin 4. They function as a voltage divider, applying slightly less than 2V to pin 4. This is enough to keep it functional, but allows the reset voltage to go down lower to enable the reset.

This now works fine—for me. We'll test the circuit again before this edition of the book goes into print. If it doesn't work for you, try applying a different voltage to pin 4 of Timer 2, using a substitute for the 47K resistor, either lower or higher. You can also try a larger coupling capacitor. And let me know. Obviously I want all the circuits in this book to work properly, every time. But I can't foresee all the manufacturing variations that may affect the outcome.

Extra Digits

Adding two more digits is pretty easy, as each will be controlled by its own 4026B counter, and all the counters and digits will be wired in basically the same way. This is shown in Figure 4-76.

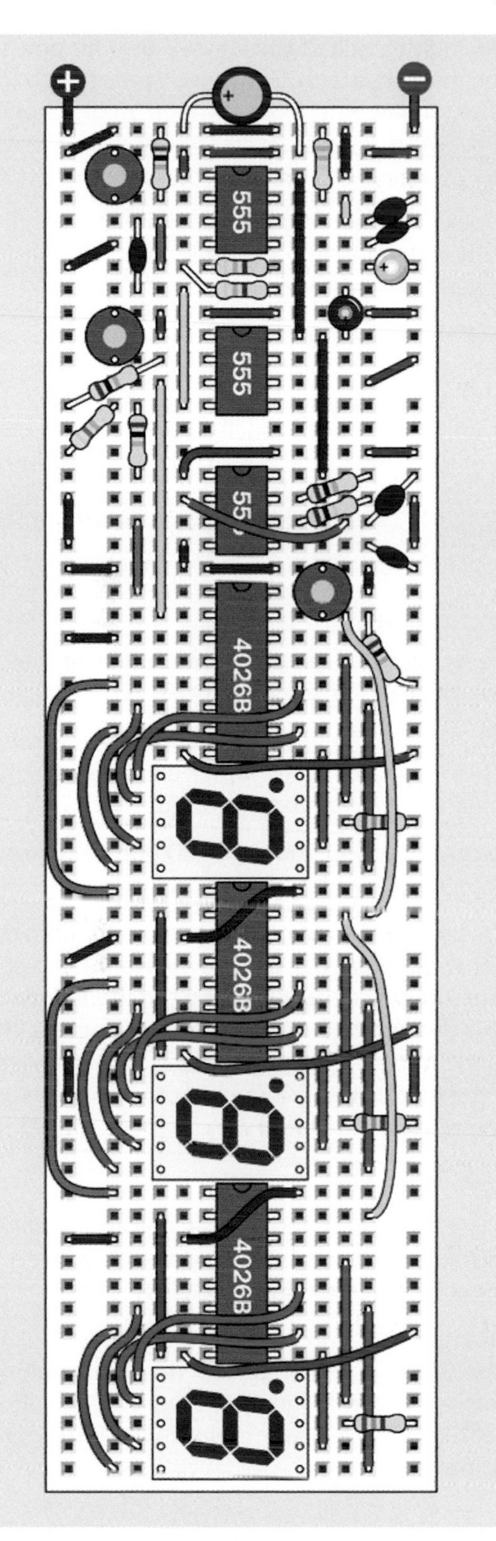

Figure 4-76 *The completed reflex timer barely fits on one 60-row breadboard.*

Notice the purple wires on the left. Each of them connects the carry output from one counter to the clock input of the next.

The yellow wires down the right side connect all the reset inputs of the counters together, so that when you reset one, you reset them all.

In the second and third counters, blue wires have been added to ground pin 2 of each. Remember, pin 2 is the clock disable pin. We never need to stop the second and third counters from counting, because they are completely under the control of the first counter. When it stops, they stop.

Don't forget to supply positive voltage to pin 16 (the power input pin) of the second counter and the third counter, by running a red wire across each chip, as shown.

Calibration

How do you make the circuit run at the right speed?

Begin by substituting a 10K resistor for the 100K resistor on the first timer that you installed, and replace the 3.3µF capacitor with a 47nF capacitor (0.047µF). In theory this should create a timer frequency of 1,023Hz, which is very close to the 1,000Hz that you want.

To fine-tune it, you'll have to substitute a trimmer potentiometer for one of the 10K resistors on the first timer that you installed. This circuit is so densely packed, there's barely enough room, but I figured out a way to squeeze in a trimmer. This is shown in Figure 4-77, which is a close-up of the immediate area around the lowest of the three timers.

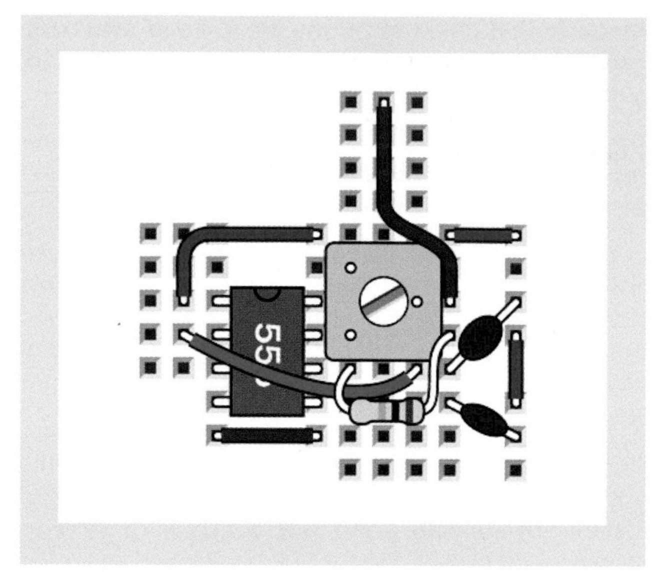

Figure 4-77 *How to squeeze in a trimmer to enable adjustment of your reflex tester.*

First move the blue wires up by one row. Snake the vertical red wire around the righthand side. Extend the lead from the remaining 10K resistor, taking care not to allow it to touch any other exposed conductors. You can now insert a trimmer, with its wiper pin making contact with the positive supply and another pin connecting with pin 7 of the timer. The third trimmer pin plugs into a vacant row on the breadboard, and can be ignored.

You should use a trimmer rated at 20K or 25K and start with it around the middle of its range. You now have three options to fine-tune the circuit so that it runs at 1kHz.

If you happen to have a multimeter with a setting on its dial to measure kHz, simply ground the black meter probe, touch the red probe to pin 3 of the first timer that you installed, and turn the trimmer until the meter shows 1kHz. Job done!

If you don't have a meter than measures frequency, perhaps you have a digital guitar tuner. They only cost a few dollars on eBay. Attach a loudspeaker to the output from the 555 timer (including a 10µF coupling capacitor and a 47-ohm series resistor), and the tuner should tell you the frequency of the note being generated by the timer.

If you don't have a suitable meter or guitar tuner, you can use any watch, clock, or phone that will display whole seconds. When the timer is running at 1kHz, the second counter will advance every 1/100th of a second, and the third counter will advance every 1/10th of a second. The third counter goes through 10 digits before repeating, which means it will display a zero once each second.

The problem is, each digit is illuminated so briefly, you'll have a hard time seeing precisely when the zero comes up. So here's what you do.

Cover all the segments of the slowest display except the segment at its bottom-right corner. This is illuminated all the time except when a 2 is displayed, at which point it blinks off. You'll have a much easier time counting the blinks of this single segment than if you try to recognize whole numbers. Adjust the trimmer potentiometer that you added, and gradually you should be able to synchronize the slowest display with a timekeeping device.

Enhancements

Whenever I finish a project, I see opportunities to improve it. Here are some ideas:

No counting at power-up. It would be nice if the circuit begins in its "ready" state, rather than already counting. I'm going to leave you to think about this.

Audible feedback when the red LED lights up. Not essential, but might be a nice feature.

A random delay interval before the count begins. Making electronic components behave randomly is very difficult, but one way to do it would be to require the user to hold his finger on a couple of metal contacts. The skin resistance of the finger would determine the delay. Because the finger pressure would not be exactly the same each time, the delay would vary.

Next?

A counter such as the 4026B is technically a logic chip. It contains *logic gates* that enable it to count. Every digital computer works on similar principles.

Because logic is so fundamental in electronics, I'm going to delve into it at much greater depth, beginning with the next experiment. The magic words AND, OR, NAND, NOR, XOR, and XNOR will open a whole new world of digital intrigue.

Experiment 20: Learning Logic

When you deal with logic gates individually, they're extremely easy to understand. When you chain them together, they become more challenging. So I'll start with them one at a time.

This experiment contains a significant amount of explanation. I don't expect you to remember it all. The purpose here is to provide a store of information that you can refer back to, later.

What You Will Need

- Breadboard, hookup wire, wire cutters, wire strippers, test leads, multimeter
- 9-volt power supply (battery or AC adapter)
- SPDT slide switch (1)
- 74HC00 quad 2-input NAND chip (1)
- 74HC08 quad 2-input AND chip (1)
- Low-current LEDs (2)
- Tactile switches (2)
- LM7805 voltage regulator (1)
- Resistors: 680 ohms (1), 2.2K (1), 10K (2)
- Capacitors: 0.1µF (1), 0.33µF (1)

The Regulator

Logic gates are much fussier than the 555 timer or the 4026B counter that you used previously. The versions that we will be using demand a precise 5VDC, with no fluctuations or "spikes" in the flow of current.

This is easy and inexpensive to achieve. Just set up your breadboard with an LM7805 voltage regulator. It delivers a well-controlled voltage of 5 volts if you supply it with a DC voltage of 7 volts or more.

For a diagram of the regulator showing its three pin functions, see Figure 4-78. For a schematic showing how to use the regulator, see Figure 4-79. For a suggestion of how to place the regulator and its two capacitors at the top of a breadboard, to occupy minimum space, see Figure 4-80. I have added a miniature on-off slide switch at top-left, and a low-current LED to show when power is connected. I think a visual indicator is useful to provide assurance that the power is on, especially when you're searching for a fault in a circuit. The high-value 2.2K resistor for the LED is chosen to consume as little current as possible, in case you are still using a 9-volt battery as your power supply.

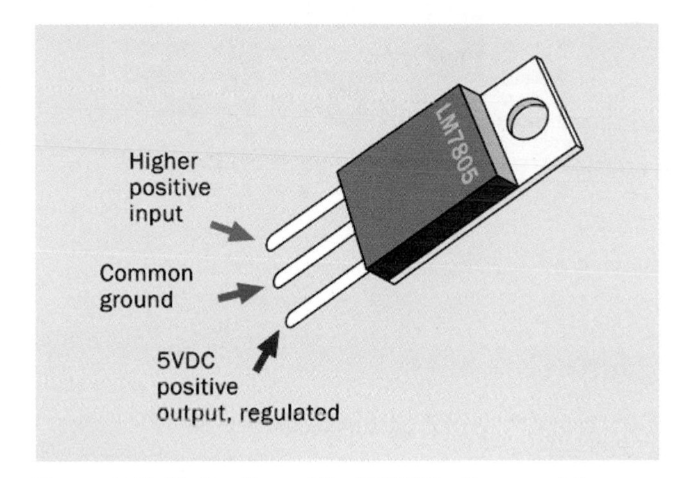

Figure 4-78 *Pin functions of the LM7805 voltage regulator, shown with the metal back facing away from you.*

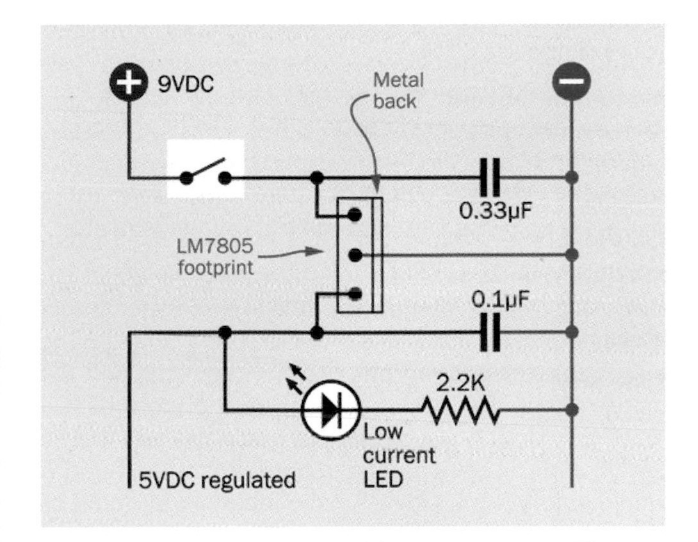

Figure 4-79 *How to use the LM7805 voltage regulator. The capacitors are mandatory.*

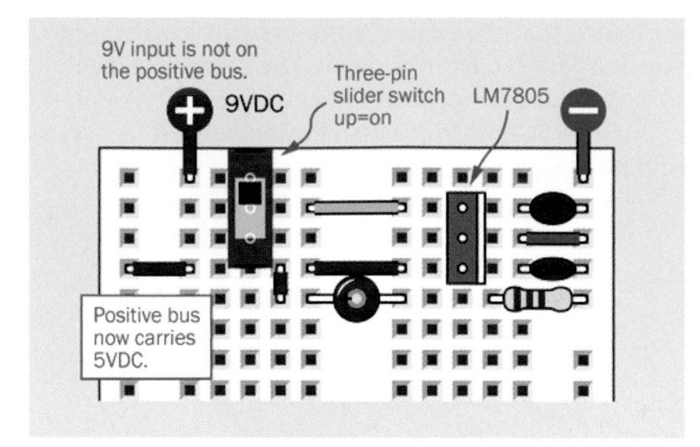

Figure 4-80 *Placing a voltage regulator at the top of a breadboard, to occupy minimal room while also allowing for an on-off slide switch and a low-current LED to show that power is switched on.*

Caution: Inappropriate Inputs

DC, not AC. Remember that the LM7805 is a DC-to-DC converter. Do not confuse it with an AC adapter, which uses alternating current from an outlet in your home, and converts it to DC. Do not apply AC to the input of your voltage regulator.

Maximum Current. The LM7805 does a wonderful job of maintaining its output at an almost constant voltage, regardless of how much current you draw through it—so long as you stay within its rated range. Don't try to pull more than one amp through the voltage regulator.

Maximum Voltage. Although the voltage regulator is a solid-state device, it behaves a little like a resistor in that it radiates heat in the process of reducing a voltage. The higher the voltage you put into the regulator, and the more current that passes through it, the more heat it must get rid of. Theoretically you could use an input of 24VDC and still get a regulated output of 5VDC, but this would not be a good idea. A good input range would be 7VDC to 12VDC.

Minimum Voltage. Like all semiconductor devices, the regulator delivers a voltage that is lower than its input voltage. This is why I suggest a minimum input of 7VDC.

Heat Sinking. The purpose of the metal back with a hole in the top is to radiate heat, which it will do more effectively if you bolt it to a piece of aluminum, since aluminum conducts heat very effectively. The aluminum functions as a heat sink, and you can buy fancy ones that have multiple cooling fins. If you're not planning to draw more than 200mA through the regulator, a heat sink is unnecessary. The circuits in this book will require less than that.

Usage

When building circuits using 5-volt logic chips, you want 5VDC to be available down the positive bus of your breadboard. Note very carefully that the 9-volt input to the circuit in Figure 4-80 is *not* on the positive bus, but is applied only to the upper pin on the voltage regulator. The 5VDC output from the lower pin of the voltage regulator is connected with the positive bus.

The negative bus on the breadboard is shared by the voltage regulator and your external voltage supply. This is known as a "common ground."

After you install your voltage regulator, set your meter to measure DC volts and measure the voltage between the two buses on the breadboard, just to make sure. Logic chips will be easily damaged by incorrect or reversed voltage.

Your First Logic Gate

Now that you have your 5VDC breadboard ready, take a couple of tactile switches, two 10K resistors, a low-current LED, and a 680-ohm resistor, and set them around a 74HC00 logic chip as shown in Figure 4-81. (Because you are using a low-current LED, a 680-ohm resistor is appropriate.)

You may notice that many of the pins of the chip are shorted together and connected to the negative side of the power supply. I'll explain that in a moment.

When you connect power, the LED should light up. Press one of the tactile switches, and the LED remains illuminated. Press the other tactile switch, and again the LED stays on. Now press both switches, and the light should go out.

Pins 1 and 2 are logic inputs for the 74HC00 chip. By default, the circuit holds them at a low voltage, being connected to the negative side of the power supply through 10K pulldown resistors. But each pushbutton overrides its resistor and forces the input pin to rise near to the value of the 5V positive bus.

- When an input or output associated with a 5V logic chip is near 0VDC, we say it is *logic-low*.

- When an input or output associated with a 5V logic chip is near 5VDC, we say it is *logic-high*.

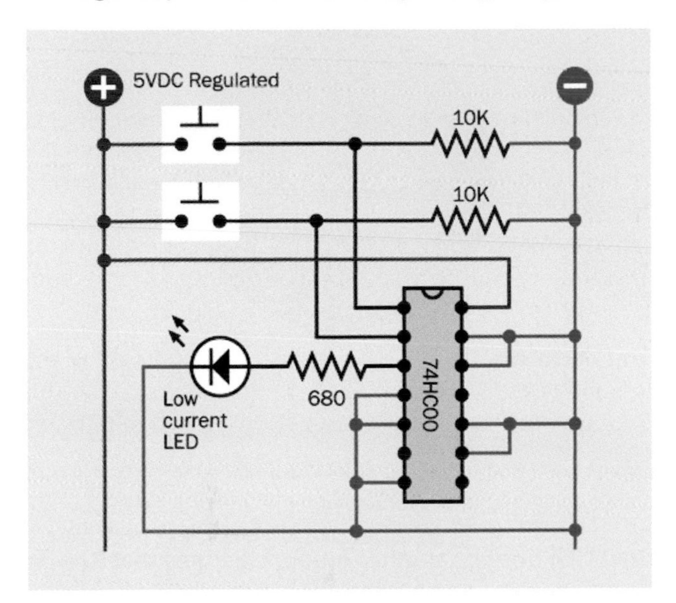

Figure 4-81 *Figuring out the logical function of the NAND gate.*

The logic output from the chip, as you saw, is normally high—but *not* if the first input *and* the second input are high. Because the chip does a "Not AND" operation, we say that it contains a *NAND logic gate*.

Logic gates can be represented with special symbols that are used in a kind of schematic known as a *logic diagram*. The logic diagram that corresponds with the circuit in Figure 4-81 is shown in Figure 4-82, where the U-shaped thing with a circle at the bottom is the logic symbol for a NAND gate. No power supply is shown for it in a logic diagram, but if you refer back to the schematic in Figure 4-81, the chip does in fact require power, at pins 7 (negative ground) and 14 (positive). This enables the chip to put out more current than its inputs take in.

- Any time you see a symbol for a logic chip, remember that it needs power to function.

The 74HC00 chip actually contains four separate NAND gates, each of which has two logical inputs and one output. They are arrayed as shown in Figure 4-83, in the diagram on the right. Because only one gate was needed for the simple test, the input pins of the unused gates were shorted to the negative side of the power supply, to stop them from floating.

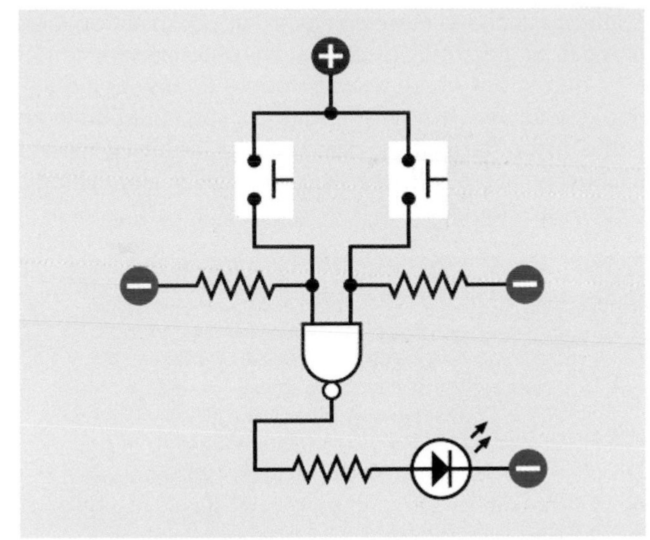

Figure 4-82 *A logic diagram can be easier to understand than a schematic showing a logic chip.*

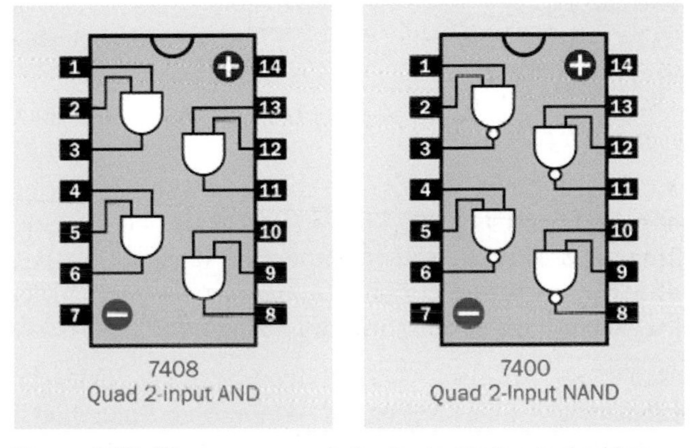

Figure 4-83 *The arrangement of gates inside two logic chips.*

Many logic chips are interchangable. In fact, let's try that right now. First, disconnect the power. Carefully pull out the 74HC00 and put it away with its legs embedded in conductive foam (or aluminum foil, if you don't have any foam). Substitute a 74HC08 chip, which is an AND chip. Make sure you have it the right way up, with its notch at the top. Reconnect the power and use the pushbuttons as you did before. This time, you should find that the LED comes on if the first input AND the second input are both positive, but it remains dark otherwise. Thus, the AND chip functions exactly opposite to the NAND chip. Its pinouts are shown in Figure 4-83, on the left.

You may be wondering why these things are useful. Soon you'll see that we can put logic gates together to do things such as create an electronic combination lock, or a pair of electronic dice, or a computerized version of a TV quiz show where users compete to answer a question. If you were insanely ambitious, you could build an entire computer out of logic gates. A hobbyist named Bill Buzbee actually built a web server from vintage logic chips—see Figure 4-84.

Figure 4-84 *This computer motherboard hand-made by Bill Buzbee is built from 74xx series logic chips, and functions as the hearts of a web server.*

Background: Logical Origins

George Boole was a British mathematician, born in 1815, who did something that few people are ever lucky enough or smart enough to do: he invented an entirely new branch of mathematics.

Interestingly, it was not based on numbers. Boole had a relentlessly logical mind, and he wanted to reduce the world to a series of true-or-false statements that could overlap in interesting ways.

Venn diagrams, conceived around 1880 by a man named John Venn, can be used to illustrate some logical relationships of this type. Figure 4-85 shows the simplest possible Venn diagram, where I have established one very large group (all the creatures in the world) and have defined a subgroup (consisting only of those creatures that live in the water). The Venn diagram illustrates that all creatures that live in water also live in the world, but only a subset of creatures in the world live in the water.

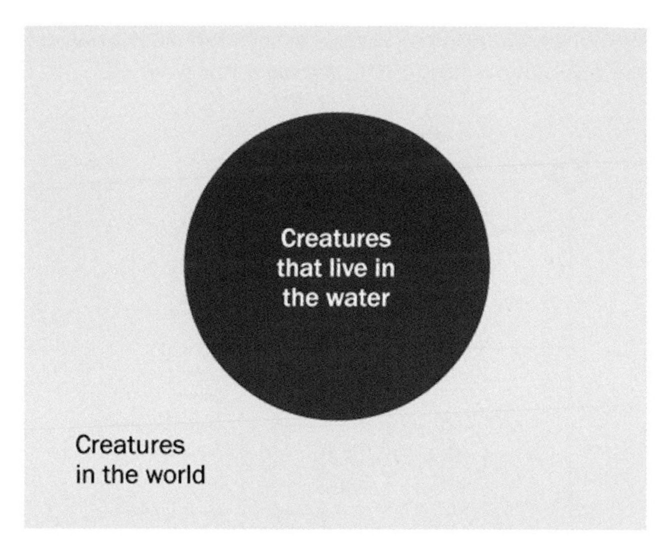

Figure 4-85 *The simplest possible relationship between a group and the larger world that contains it.*

Now I'll introduce another group: creatures that live on land. But wait—some creatures are able to live on both land and water. Frogs, for instance. These amphibians are members of both groups, and I can show this with another Venn diagram in Figure 4-86, where the groups overlap.

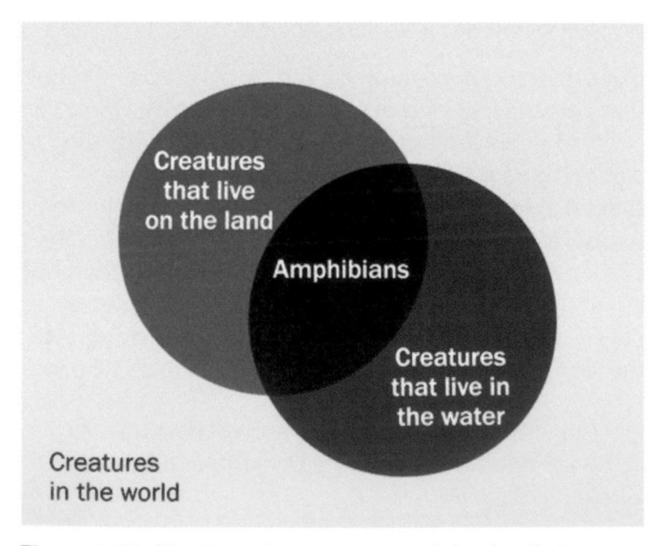

Figure 4-86 *This Venn diagram is a way of showing that some creatures in the world live on land, some live in water, and some inhabit both land and water.*

However, not all groups overlap. In Figure 4-87 I have created one group of creatures with hooves, and another group of creatures with claws. Are there any creatures that have hooves *and* claws? I don't think so. I could express this by creating a *truth table*, as in Figure 4-88.

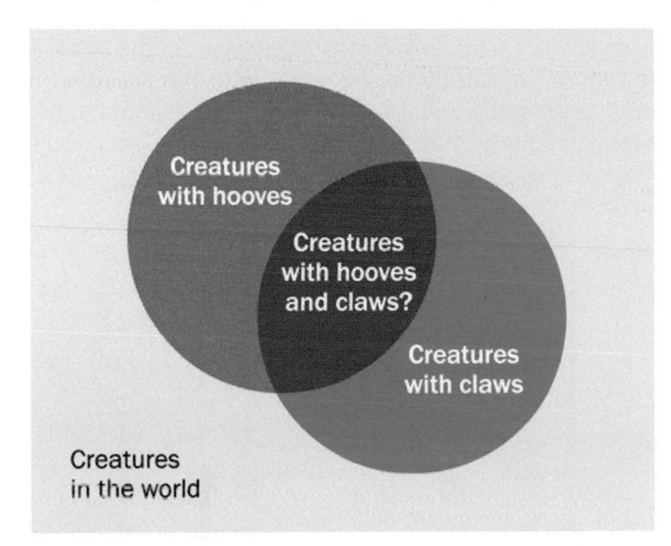

Figure 4-87 *Some subgroups do not overlap. I don't think any creatures have claws as well as hooves.*

This creature has hooves	This creature has claws	This combination can be
NO	NO	TRUE
NO	YES	TRUE
YES	NO	TRUE
YES	YES	FALSE

Figure 4-88 *The simplest form of a truth table tabulates the validity of pairs of inputs, each of which can have one of two states.*

A NAND gate could be used to represent this table, because its pattern of inputs and outputs is exactly the same, as shown in Figure 4-89.

If NAND input A is:	And if NAND input B is:	Then NAND output will be:
LOW	LOW	HIGH
LOW	HIGH	HIGH
HIGH	LOW	HIGH
HIGH	HIGH	LOW

Figure 4-89 *This truth table for a NAND gate has exactly the same pattern as the previous table.*

Beginning with these very simple concepts, Boole developed his logic language to a very high level. He published a treatise on it in 1854, long before it could be applied to electrical or electronic devices. In fact, during his lifetime, his work seemed to have no practical applications at all. But a man named Claude Shannon encountered Boolean logic while studying at MIT in the 1930s, and in 1938 he published a paper describing how Boolean analysis could be applied to circuits using relays. This had immediate practical applications, as telephone networks were growing rapidly, creating complicated switching problems.

A very simple telephone problem could be expressed like this. Long ago, it was common for two customers living in separate homes in a rural area to share one telephone line. If one of them wanted to use the line, or the other wanted to use it, or neither of them wanted to use it, there was no problem. But they could not both use it simultaneously. Once again, this is the same logical pattern as described in Figure 4-89, if we interpret "high" to mean that one person wants to use the line, and "low" meaning that the person doesn't want to use it.

But now there's an important difference. The NAND gate doesn't just illustrate the network. Because the telephone network uses electrical states, a NAND gate can *control* the network. (Actually, in the early days of networks, everything was done by relays; but an assembly of relays can still function as a logic gate.)

After Shannon's application of Boolean logic to the telephone system, the next step was to see that if you used an "on" condition to represent numeral 1 and an "off" condition to represent numeral 0, you could build a system of logic gates that could count. And if it could count, it could do arithmetic.

When vacuum tubes were substituted for relays, the first practical digital computers were built. Then transistors took the place of vacuum tubes, and integrated circuit chips replaced transistors, leading to the desktop computers that we now take for granted. But deep down, at the lowest levels of these incredibly complex devices, they still use the laws of logic discovered by George Boole.

Incidentally, when you use a search engine online, if you use the words AND and OR to refine your search, you're actually using *Boolean operators*.

Fundamentals: Logic Gate Basics

The NAND gate is the most fundamental building block of digital computers, because you can do addition sums using nothing but NANDs. If you want to learn more about this, search online for topics such as "binary arithmetic" and "half-adder." You can also find circuits that do addition sums using logical operators in my book *Make: More Electronics*.

Generally, there are seven types of logic gates:

```
AND, NAND, OR, NOR, XOR, XNOR, NOT
```

Their names are usually printed entirely in capital letters. Of the first six, XNOR is hardly ever used.

All of the gates have two inputs and one output, except for the NOT gate, which has only one input and one output, and is more often referred to as an *inverter*. If it has a high input, it gives a low output, and if it has a low input, it gives a high output.

Symbols that represent the seven types of gates are shown in Figure 4-90. Notice that the little circles at the bottom of some of the gates invert the output. (These circles are called *bubbles*.) Thus, the output of a NAND gate is the inverse of an AND gate.

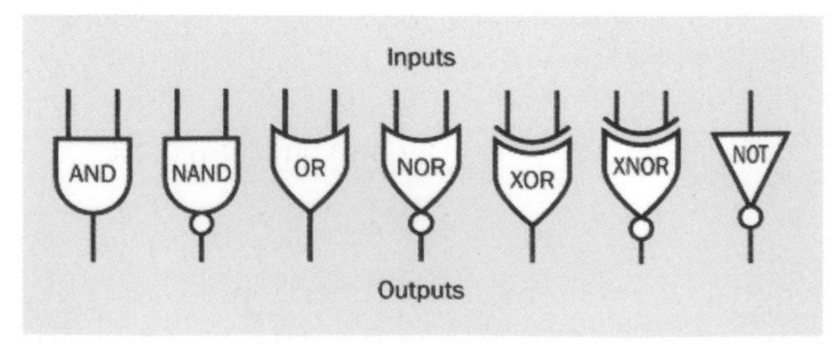

Figure 4-90 *Symbols for the six two-input logic gates, plus the NOT gate.*

What do I mean by "inverse"? This should become clear if you look at the truth tables for logic gates that I have drawn in Figure 4-91, Figure 4-92, and Figure 4-93. In each of the tables, two inputs are shown on the left followed by an output on the right, with red meaning a high-logic state and blue meaning a low-logic state. Compare the outputs of each pair of gates, and you'll see how the patterns are inverted.

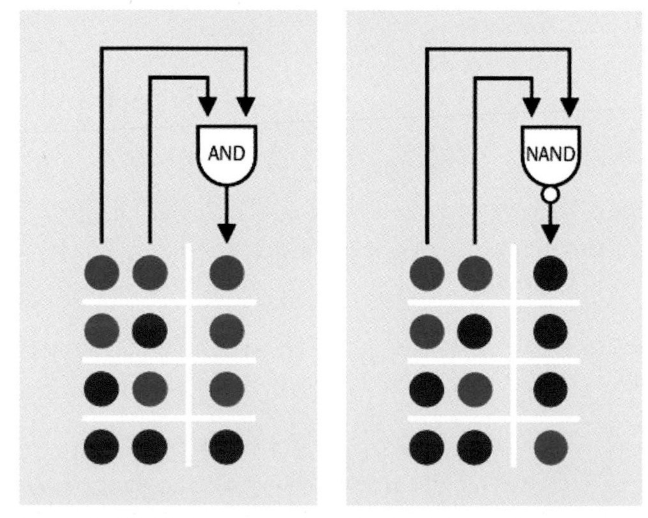

Figure 4-91 *The inputs shown on the left produce the outputs shown on the right.*

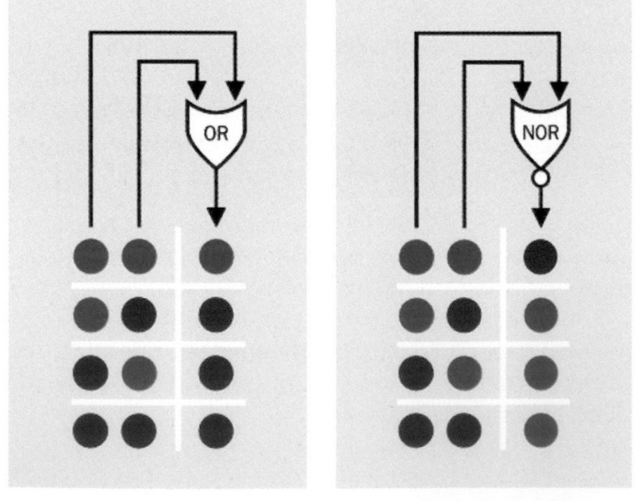

Figure 4-92 *The inputs shown on the left produce the outputs shown on the right.*

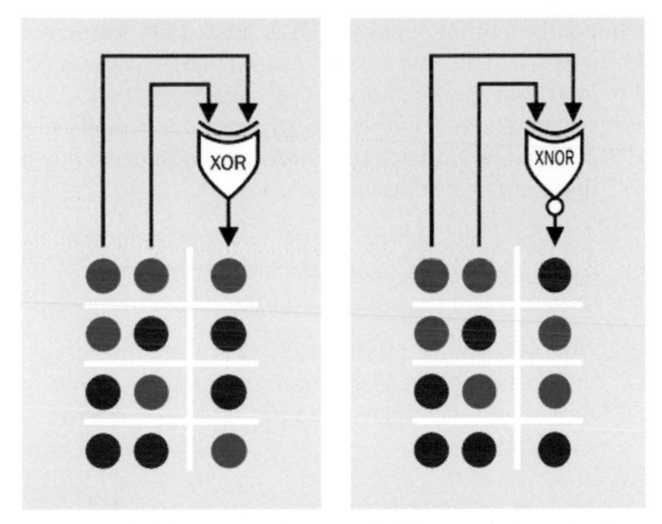

Figure 4-93 *The inputs shown on the left produce the outputs shown on the right.*

Background: The Confusing World of TTL and CMOS

Back in the 1960s, the first logic gates were built with transistor–transistor logic, abbreviated TTL, meaning that tiny bipolar transistors were etched into a single wafer of silicon. Soon, these were followed by complementary metal oxide semiconductors, abbreviated CMOS. The 4026B chip that you used in Experiment 19 is an old CMOS chip.

You may remember that bipolar transistors amplify current. Thus, TTL circuits require a significant flow of electricity to function. But CMOS chips are voltage-sensitive, enabling them to draw hardly any current while they are waiting for a signal, or while they are pausing after emitting a signal.

The table in Figure 4-94 summarizes the original advantages and disadvantages of the two types of chips. The CMOS series, with part numbers from 4000 upward, were slower and more easily damaged by static electricity, but were valuable because of their meager power consumption. The TTL series, with part numbers from 7400 upward, used much more power but were less sensitive and very fast. So, if you wanted to build a computer, you used the TTL family, but if you wanted to build a little gizmo that would run for weeks on a small battery, you used the CMOS family.

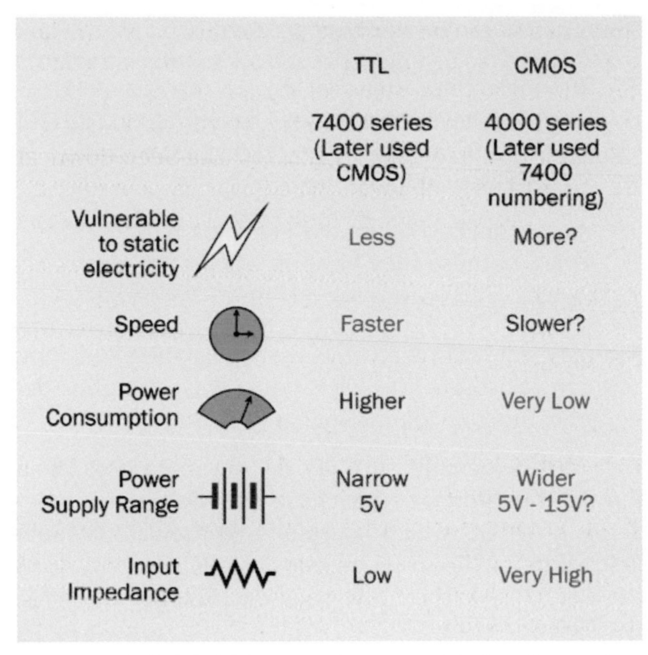

	TTL	CMOS
	7400 series (Later used CMOS)	4000 series (Later used 7400 numbering)
Vulnerable to static electricity	Less	More?
Speed	Faster	Slower?
Power Consumption	Higher	Very Low
Power Supply Range	Narrow 5v	Wider 5V - 15V?
Input Impedance	Low	Very High

Figure 4-94 *In this table comparing early CMOS and TTL chips, the CMOS attributes with a question mark were eventually reconciled with the TTL attributes.*

From this point on everything became confusing, because CMOS manufacturers wanted to grab market share by emulating the advantages of TTL chips. Newer generations of CMOS chips even changed their part numbers to begin with "74" to emphasize their compatibility, and the functions of pins on CMOS chips were swapped around to match the functions of pins on TTL chips. The voltage requirements of CMOS were also changed to match TTL.

Today, you can still find some of the old TTL chips around, especially in the LS series (with part numbers such as 74LS00 and 74LS08). However, they have become uncommon.

It's more common to find the 4000 series CMOS chips—such as the 4026B that you used in the previous experiment. They are still being made because their wide range of acceptable power-supply voltages is useful.

Over the years, CMOS chips became faster and less vulnerable to static electricity, which explains why I added a question mark in these categories in Figure 4-94. Modern CMOS chips also mostly reduced their maximum power supply voltages to 5VDC—which is why I added a question mark in this category, too.

The situation can be summed up like this:

- Any logic chips still available in the old 4000 series will have the features I have listed in Figure 4-94. You are quite likely to find a use for the 4000 series of chips.
- You are unlikely to use old TTL chips in the 7400 series, because they have no significant advantages.

You may still find schematics that specify 74LSxx chips. You can substitute the 74HCTxx chips, which are designed to function identically.

The 74HCxx generation is by far the most popular in through-hole format. They have the high input impedance of CMOS, which is useful, and they are cheaper than some of the more modern, exotic versions. All of the logic chips in this book are of the HC type.

Now for the part numbering. Where you see a letter "x" in the list that follows, it means that various letters and numbers may appear in that location. Thus "74xx" includes the 7400 NAND gate, the 7402 NOR gate, the 74150 16-bit data selector, and so on. A combination of letters preceding the "74" identifies the chip manufacturer, while letters following the part number may identify the style of package, may indicate whether it contains heavy metals that are environmentally toxic, and other details. This is explained visually in Figure 4-3.

Here's the history of the TTL family:

- 74xx: The old original generation, now obsolete.
- 74Sxx: Higher speed "Schottky" series, now obsolete.
- 74LSxx: Lower power Schottky series, still used occasionally.

The CMOS family:

- 40xx: The old original generation, now obsolete.
- 40xxB: The 4000B series was improved but still susceptible to damage from static electricity. These chips are still commonly used, especially in hobby-electronics applications.
- 74HCxx: Higher-speed CMOS, with part numbers matching the TTL family, and pinouts matching the TTL family. I've used this generation extensively in this book, because it's widely available, and the circuits here have no need for greater speed or power.
- 74HCTxx: Like the HC series but matching the old TTL standard for maximum and minimum logic-low and logic-high voltages, respectively.
- 74xx series with other letters in the middle of the part number: More modern, faster, usually surface-mount, often designed for lower operating voltages.

What You Don't Need

Speed differences are irrelevant from our point of view, as we're not going to be building circuits running at millions of cycles per second.

Price differences between chip families are usually minor when purchased in low quantities.

Lower-voltage chips are not appropriate for our purposes, as they are almost all surface-mount format, and we would have to create a lower-voltage power supply. Because surface-mount chips are so much more difficult to deal with, and their only major advantage is miniaturization, I won't be using them. Through-hole equivalents have the same logic functions.

Fundamentals: Part Numbers and Functions

The interior connections of currently available through-hole 14-pin logic chips in the HC series are shown in Figure 4-83, Figure 4-95, Figure 4-96, Figure 4-97, Figure 4-98, Figure 4-99, Figure 4-100, and Figure 4-101.

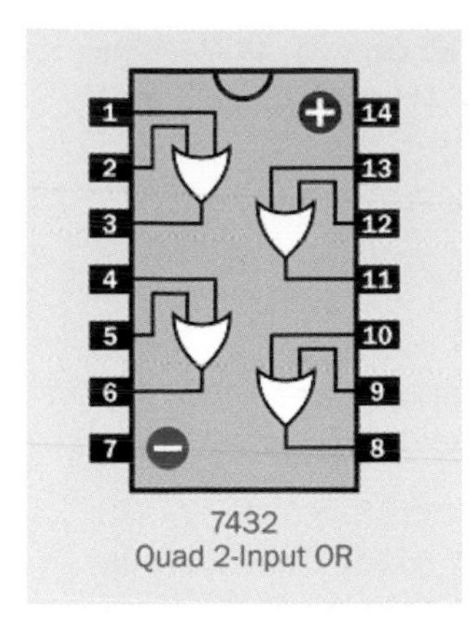

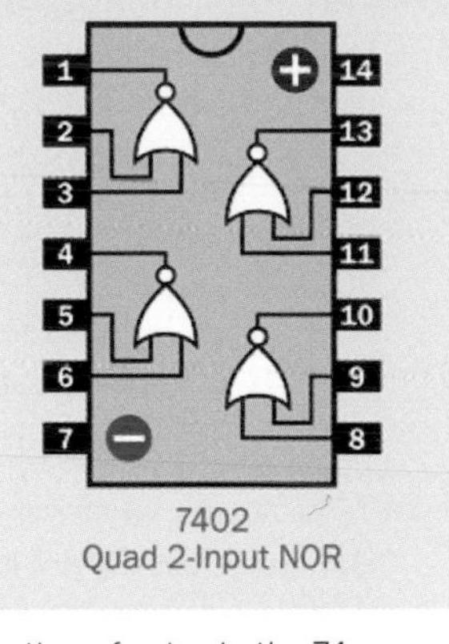
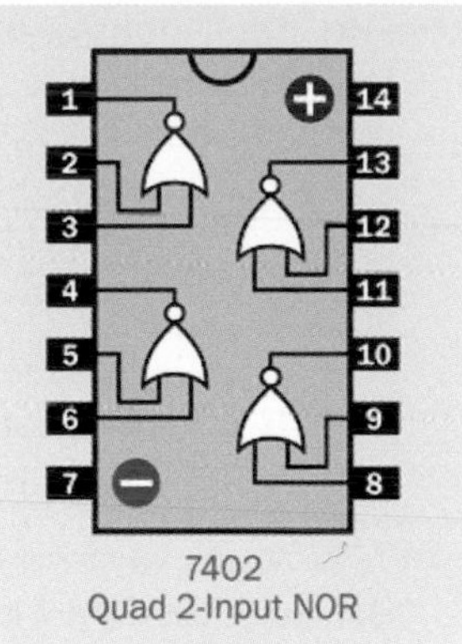

Figure 4-95 *Standardized configuration of gates in the 74xx family of logic chips.*

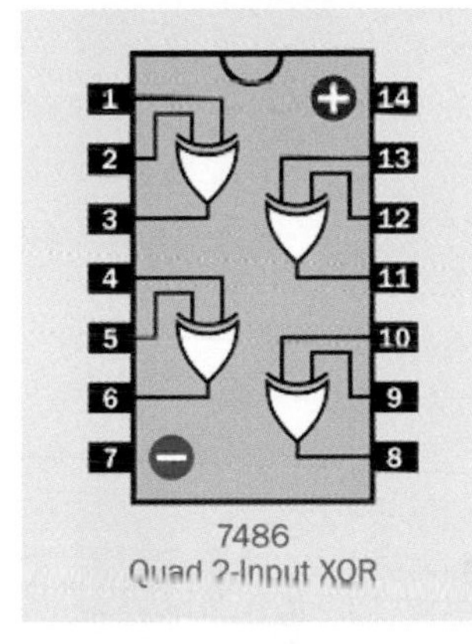

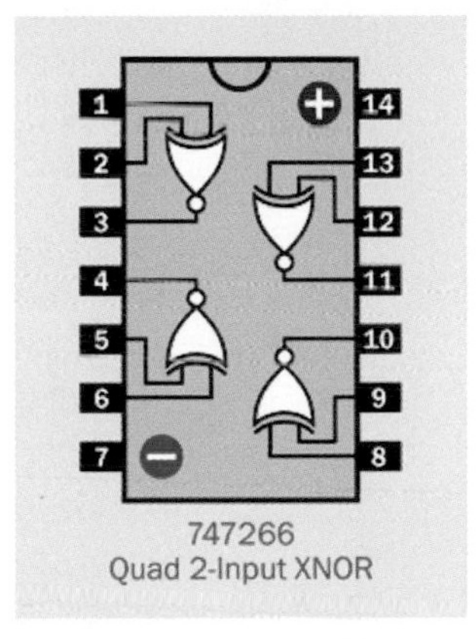

Figure 4-96 *Standardized configuration of gates in the 74xx family of logic chips.*

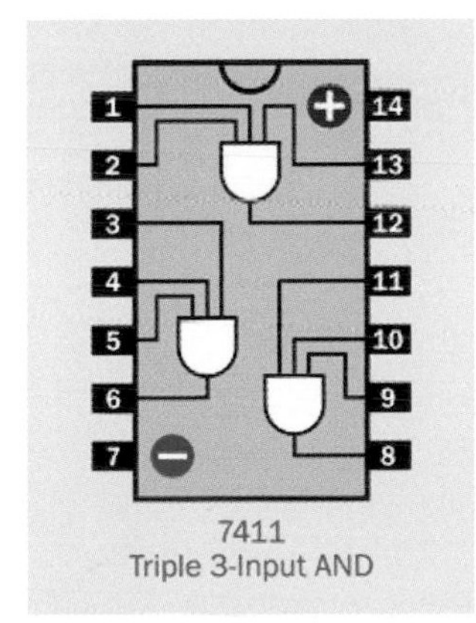

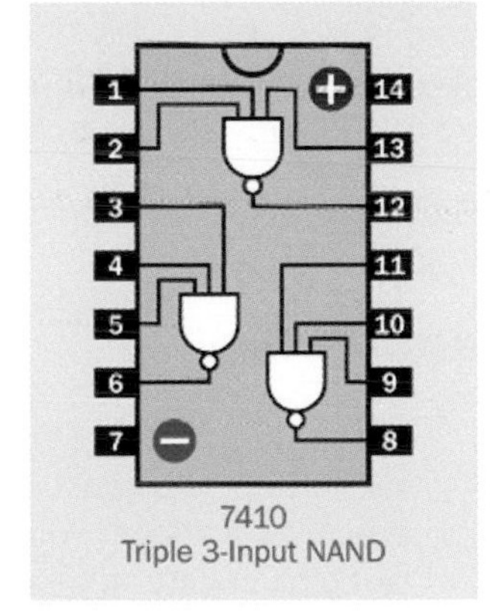

Figure 4-97 *Standardized configuration of gates in the 74xx family of logic chips.*

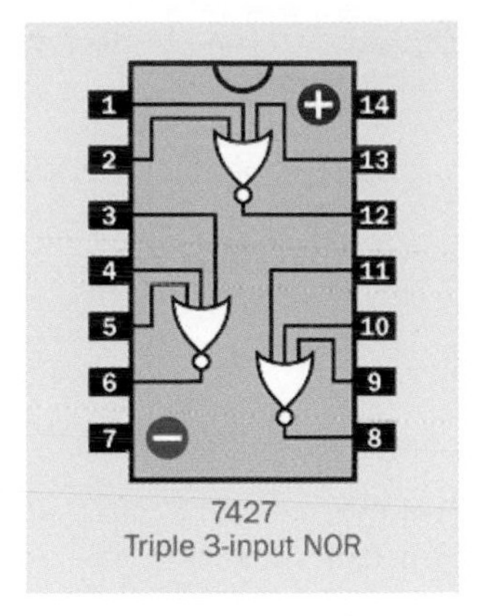

Figure 4-98 *Standardized configuration of gates in the 74xx family of logic chips.*

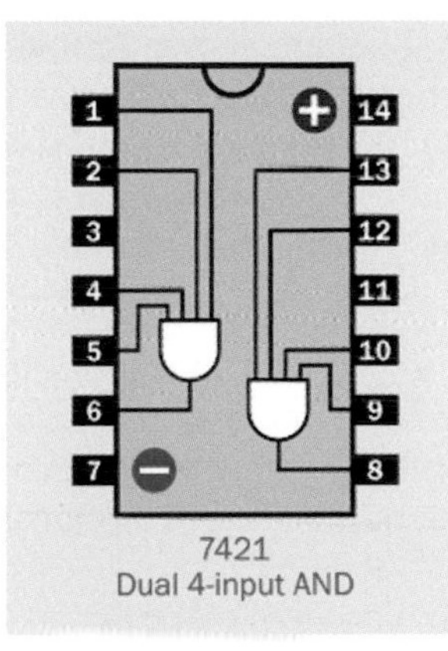

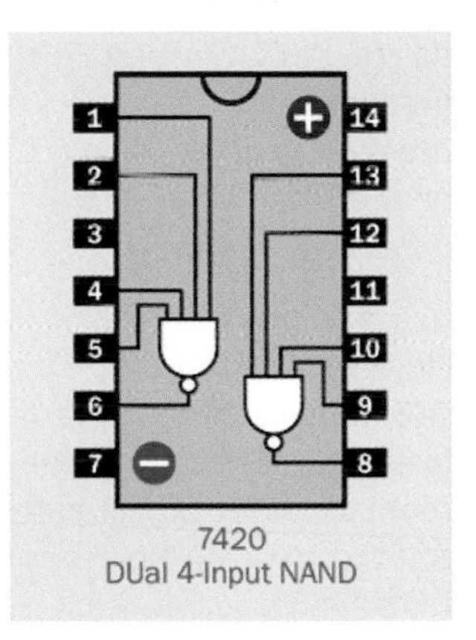

Figure 4-99 *Standardized configuration of gates in the 74xx family of logic chips.*

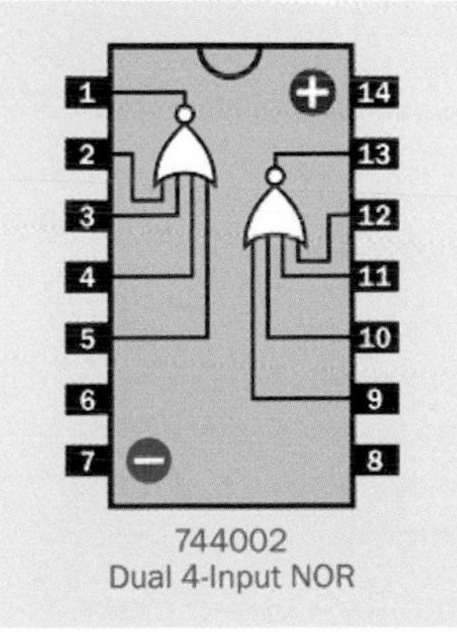

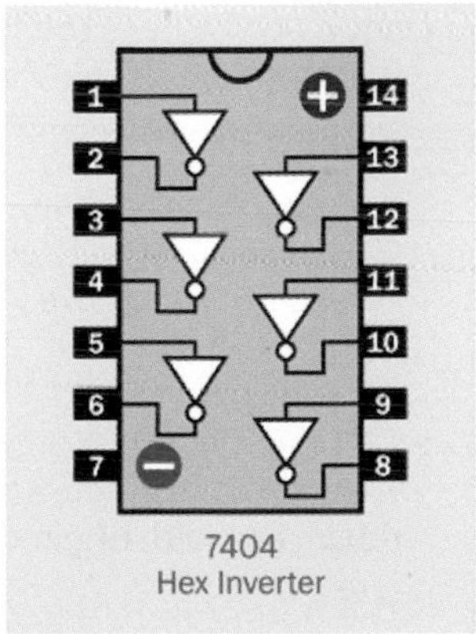

Figure 4-100 *Standardized configuration of gates in the 74xx family of logic chips.*

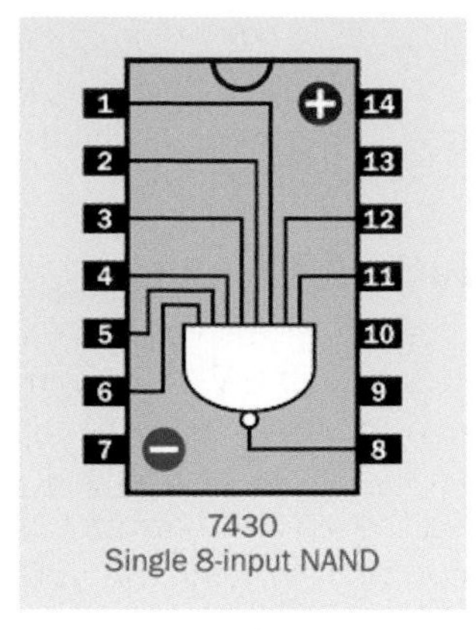

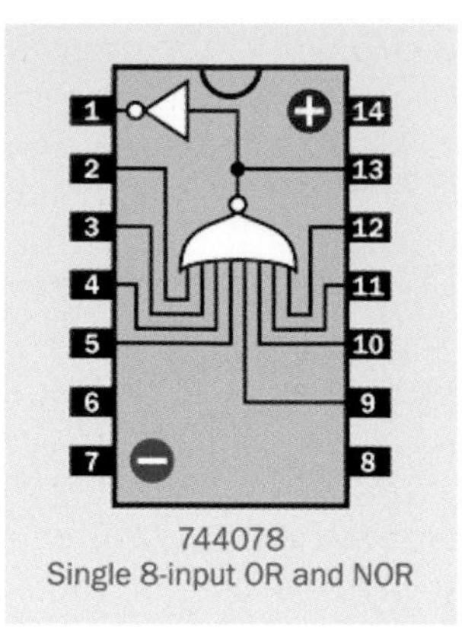

Figure 4-101 *Standardized configuration of gates in the 74xx family of logic chips.*

All the part numbers in these chips are shown in their minimal form. Thus the 7400 chip may have actual part numbers such as 74HC00, 74HCT00, and so on, and will be preceded and followed by other letter codes; but it is generically referred to as a 7400 chip, so that's how I present it here.

It's very important to check the pin functions of logic chips against the diagrams here or a manufacturer's datasheet before using them. Interior connections may seem to follow similar patterns, but there are exceptions.

Fundamentals: Rules for Connecting Logic Gates

Permitted:

- You can connect the input of a gate directly to your regulated power supply, either positive side or negative side.
- You can connect the output from one gate directly to the input of another gate.
- The output from one gate can power the inputs of many other gates (this is known as "fanout"). The exact ratio depends on the chip, but in the 74HCxx series, you can always power at least ten inputs with one logic output.
- The output from a logic chip can drive the trigger (pin 2) of a 555 timer, so long as the timer shares the same 5VDC power supply.
- Low input doesn't have to be zero. A 74HCxx logic gate will recognize any voltage up to 1 volt as "low."
- High input doesn't have to be 5 volts. A 74HCxx logic gate will recognize any voltage above 3.5 volts as "high."

The acceptable ranges for inputs, and the minimum guarantees for outputs, are shown in Figure 4-102.

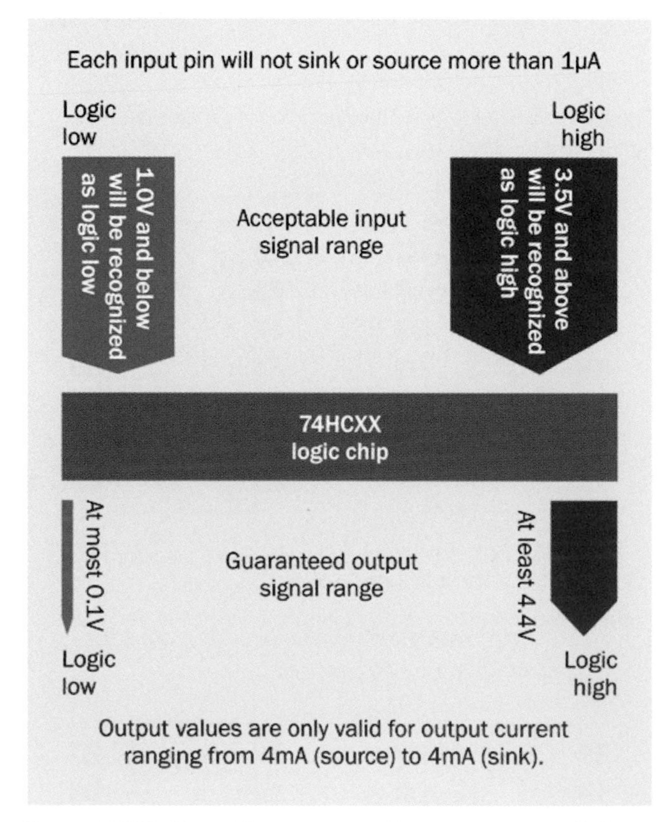

Figure 4-102 *To avoid errors, stay within the recommended input ranges for logic chips.*

Not permitted:

- No floating inputs! On CMOS chips, such as the HC family, you must always connect all input pins with a known voltage. This includes the input pins of gates that are unused on a chip.
- Any single-throw switch or pushbutton should be used with a pullup or pulldown resistor, so that when the contacts are open, the input to the chip is not floating.

- Don't use an unregulated power supply, or more than 5 volts, or less than 5 volts, to power 74HCxx logic gates.
- Be careful when using the output from a 74HCxx logic chip to power an LED. You can draw as much as 20mA from the chip, but this will pull down the output voltage. If you are also connecting that voltage to the input of a second chip, the voltage may be pulled down so much by the LED that the second chip doesn't recognize it as "high" anymore. Generally, try not to use a logic output to power an LED *at the same time as* another logic chip. Always check currents and voltages when modifying a circuit or designing a new one.
- Throughout this book, I am using low-current LEDs in conjunction with logic-chip outputs, because I think this is a good habit to acquire, in case an output powering an LED is also used as a logic input for another chip at some point in the future.
- Never apply a significant voltage or current to the output pin of a logic gate. In other words, *don't force an output into another output*.
- For this reason, don't tie the outputs together from two or more logic gates.

So much for the dos and don'ts. Now it's time for your first real logic-chip project.

Experiment 21: A Powerful Combination

Suppose you want to prevent other people from using your computer. I can think of two ways to do this: using software, or using hardware. The software would be some kind of startup program that intercepts the normal boot sequence and requests a password. It would be a little more secure than the password protection that is a standard feature of Windows and Mac operating systems.

You could certainly do it that way, but I think it would be more fun (and more relevant to this book) to do it with hardware. What I'm imagining is a numeric keypad requiring the user to enter a secret combination before the computer can be switched on. I'll call it a "combination lock," even though it won't actually lock anything. It will disable the power-on button that you normally use to switch on your computer.

Caution: The Warranty Issue

If you follow this project all the way to its conclusion, you'll open your desktop computer, cut a wire, and insert your own little circuit. You won't be going near any of the boards inside the computer, and you will only access the wire to the computer's "power on" button, but still, if you bought your computer new, you will void your warranty. I don't take this too seriously, myself, but if that makes you uneasy, here are three options:

- Breadboard the circuit for fun, and leave it at that.
- Use the circuit on some other device.
- Use it on an old computer.

What You Will Need

- Breadboard, hookup wire, wire cutters, wire strippers, multimeter
- 9-volt power supply (battery or AC adapter)
- Low-current LED (1)
- Generic LED (1)
- LM7805 voltage regulator (1)
- 74HC08 logic chip (1)
- 555 timer chip (1)
- 2N2222 transistor (1)
- DPDT 9VDC relay (1)
- Diodes: 1N4001 (1), 1N4148 (3)
- Resistors: 330 ohms (1), 470 ohms (1), 1K (1), 2.2K (1), 10K (6), 1M (1)
- Capacitors: 0.01µF (1), 0.1µF (1), 0.33µF (1), 10µF (2)
- Tactile switches (8)
- Optional: Tools to open your computer, drill four holes, and make saw cuts between the holes, to create a rectangular opening for the

keypad (if you want to take this project to its conclusion). Also, four small bolts to attach the keypad to the computer cabinet after you create the opening for it

A Three-Part Circuit

The entire breadboarded circuit is shown in Figure 4-107, but before you start to build it, let's take a look at the schematics.

The circuit is divided into three sections:

1. Power supply and three dummy pushbuttons.
2. Active pushbuttons and logic.
3. Output.

The schematic in Figure 4-103 shows part 1. This is simple enough. When you press button A, this supplies 9VDC to the voltage regulator, which delivers 5VDC down the lefthand bus. The button also supplies 9VDC down a magenta-colored wire on the right, for reasons which I'll get to in a moment.

In addition you will notice buttons B, C, and D, any one of which makes a connection with negative ground.

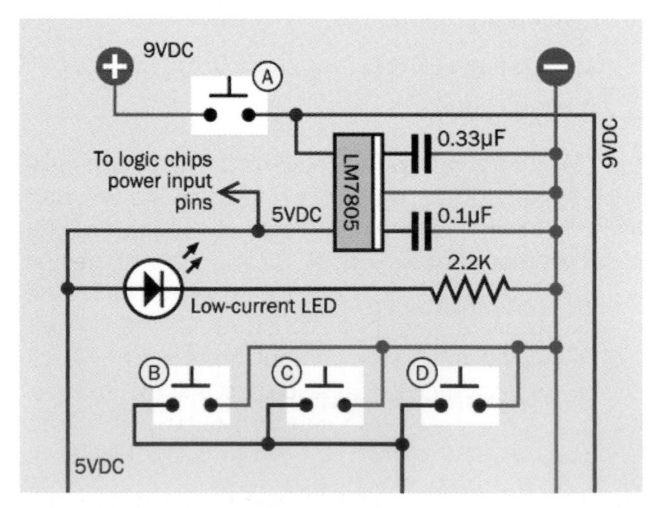

Figure 4-103 *The top section of the circuit.*

Now take a look at Figure 4-104, which shows the center section of the circuit using logic symbols. Imagine this attached to the top part of the circuit in Figure 4-103. Each of the buttons E through H can supply positive voltage to an AND gate, whose lefthand input is normally held low through a 10K pulldown resistor. The output from each gate feeds an input of the next gate.

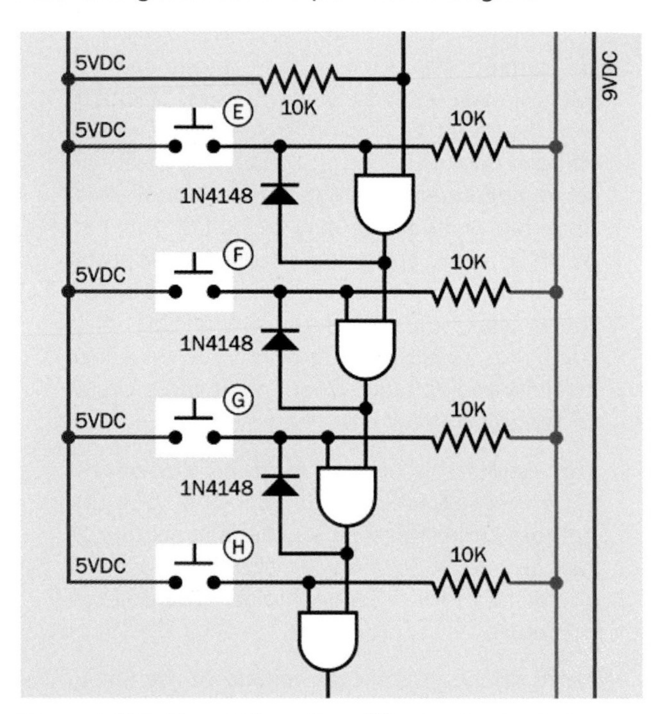

Figure 4-104 *The middle section of the circuit, using logic symbols.*

Lastly, Figure 4-105 shows the bottom section of the circuit, where the output from the last of the AND gates activates a transistor, which triggers a 555 timer. The timer controls a relay, and the relay will be used to lock and unlock your computer (or any other device that uses a simple on-off button).

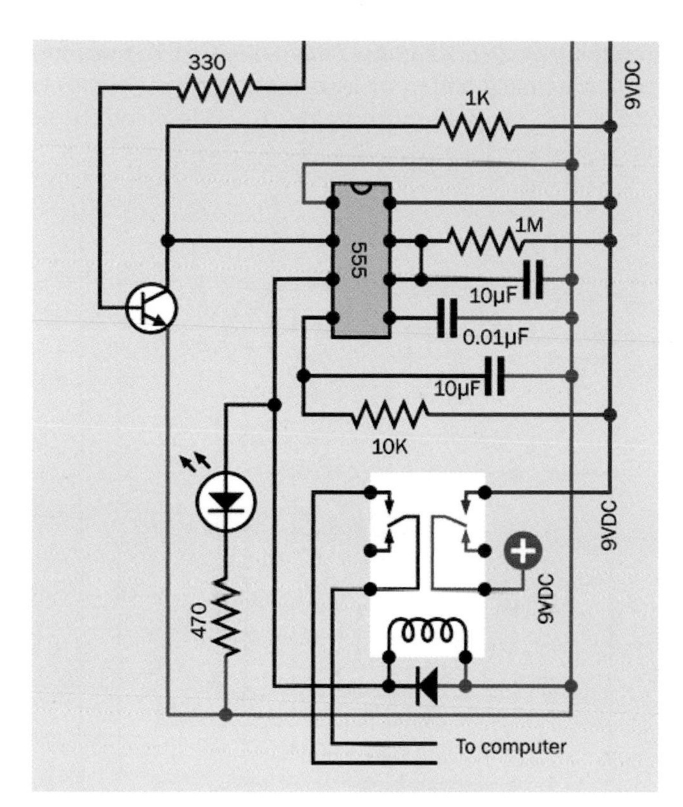

Figure 4-105 *The bottom section of the circuit.*

How It Works

I have set this up so that you have to hold down button A to activate the circuit, and you have to continue pressing it while you enter a secret sequence on the pushbuttons. This satisfies two purposes: the circuit draws no power while you're not using it, and you can't leave it switched on by mistake.

The secret sequence consists of pressing buttons E, F, G, and H in that order, while you are holding down button A. Of course if you actually install the circuit, you can scramble the positions of the buttons. I laid them out like this on the breadboard for the sake of simplicity.

Suppose you are holding down button A, and you press button E as the first in the sequence to unlock the circuit. In Figure 4-104 you can see that button E conducts 5 volts directly to the left input of the first AND gate. This overcomes the pulldown resistor, so the left input is now logic-high.

The righthand input of the AND gate is being held high through a 10K resistor. So, now both inputs of the AND gate are high, and consequently its output changes from low to high.

Current from the output circulates around through a diode, to the lefthand input. Consequently, you can let go of button E, and the output from the AND gate will maintain the lefthand input in a high state. The gate has latched itself, like the relay in Experiment 15. It is able to do this because it has its own power supply (not shown in the logic schematic) which sustains the output voltage regardless of small reductions in the input voltage.

The high output from the first AND gate also connects with the righthand input of the second AND gate. Now that the second AND gate has a high righthand input, if you press the button to pull up the lefthand input, the output from the second AND gate goes high. Note that the button wouldn't have worked before, because you needed a high output from the first AND gate to feed the second AND gate.

- After you press each button, the AND gate beside it latches itself, and you can release the button.
- You have to press the buttons in sequence. If you press four buttons out of sequence, nothing happens.
- You have to hold down button A throughout the whole procedure.

Now take a look at buttons C, D, and E. What happens if you press any of these buttons while you are entering the code to unlock the circuit? Any of the buttons will pull down the voltage on the righthand input of the first AND gate. Consequently, the output from the AND gate will go low. If the gate had latched itself, it comes unlatched. Moreover, the low output from the first AND gate will unlatch the second AND gate, and the low output from the second AND gate will unlatch the third AND gate.

Any of buttons C, D, and E will reset the whole circuit. I included them just to make entering the correct combination more difficult. Naturally I am assuming that if you install this system, all the buttons will look the same.

More than One Button?

What if someone presses more than one button simultaneously (while holding down button A)? The results will be unpredictable. If all the buttons E, F, G, and H are pressed together, the relay will be activated—except that if any of the buttons B, C, or D is also pressed, this will stop anything from happening. The possibility of simultaneous button presses can be seen as a flaw in this circuit, but the chance of someone pressing A, E, F, G, and H while not pressing B, C, or D is small. To reduce this risk further, you could add more "reset" buttons in parallel with B, C, and D.

Triggering the Relay

Suppose you enter the correct combination. The last AND gate applies about 5 volts to the base of transistor in Figure 4-105. The transistor turns on and starts conducting. This reduces the resistance between pin 2 of the 555 timer and negative ground, so the voltage on pin 2 is pulled down, and the timer is triggered.

The timer is being powered by 9 volts, which reaches it down the magenta-colored wire on the right. The timer output should be sufficient to activate the relay. Now see what the relay does: its righthand set of contacts provides an alternate source of power to the 9-volt bus.

So long as the pulse from the 555 timer continues, it holds the relay contacts closed. So long as the relay is closed, it powers the circuit—including the timer. Yes, the timer is energizing the relay, and the relay is powering the timer.

You can let go of button A now, and the relay will remain latched, so long as the pulse from the timer lasts. The pulse will end after about 30 seconds, cutting off power to the relay, so its contacts open. This turns off the timer, and also the rest of the circuit. The circuit now consumes no power at all.

The set of contacts on the left side of the relay are intended to provide power for the "on" button on your computer. So, during the brief time while the timer energizes the relay, you'll be able to switch on your computer. The rest of the time, the "on" button won't work.

The Logic Chip

Now take a look at Figure 4-106. This is the center section of the circuit, redrawn with an actual 74HC08 logic chip containing four two-input AND gates. It performs exactly the same function as the logic schematic in Figure 4-104. You can compare the two schematics to see how their functions are identical. The big difference is that the circuit showing the chip tells you how the component actually has to be installed—but, you are likely to find this diagram much harder to understand. Logic diagrams have their uses.

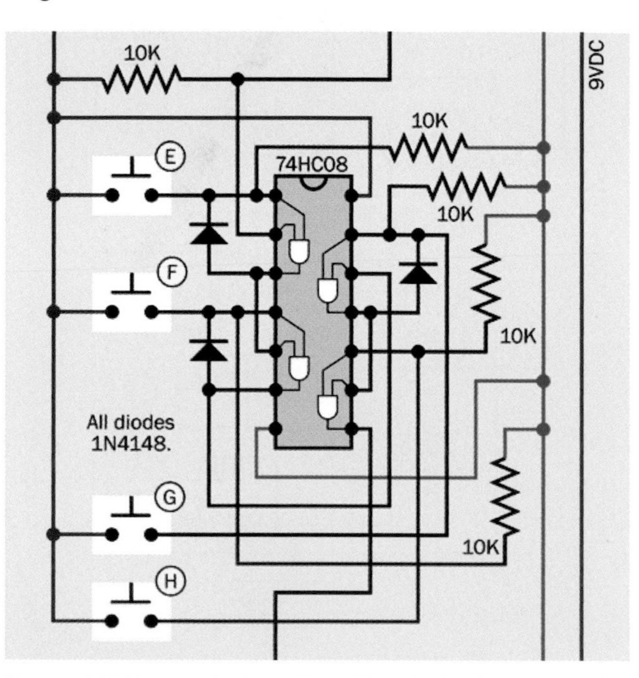

Figure 4-106 *The middle section of the circuit, showing actual components.*

Time to Build!

The fully breadboarded circuit is shown in Figure 4-107. For this project, you cannot really test it in stages; you have to build the whole thing. Component values are shown in Figure 4-108.

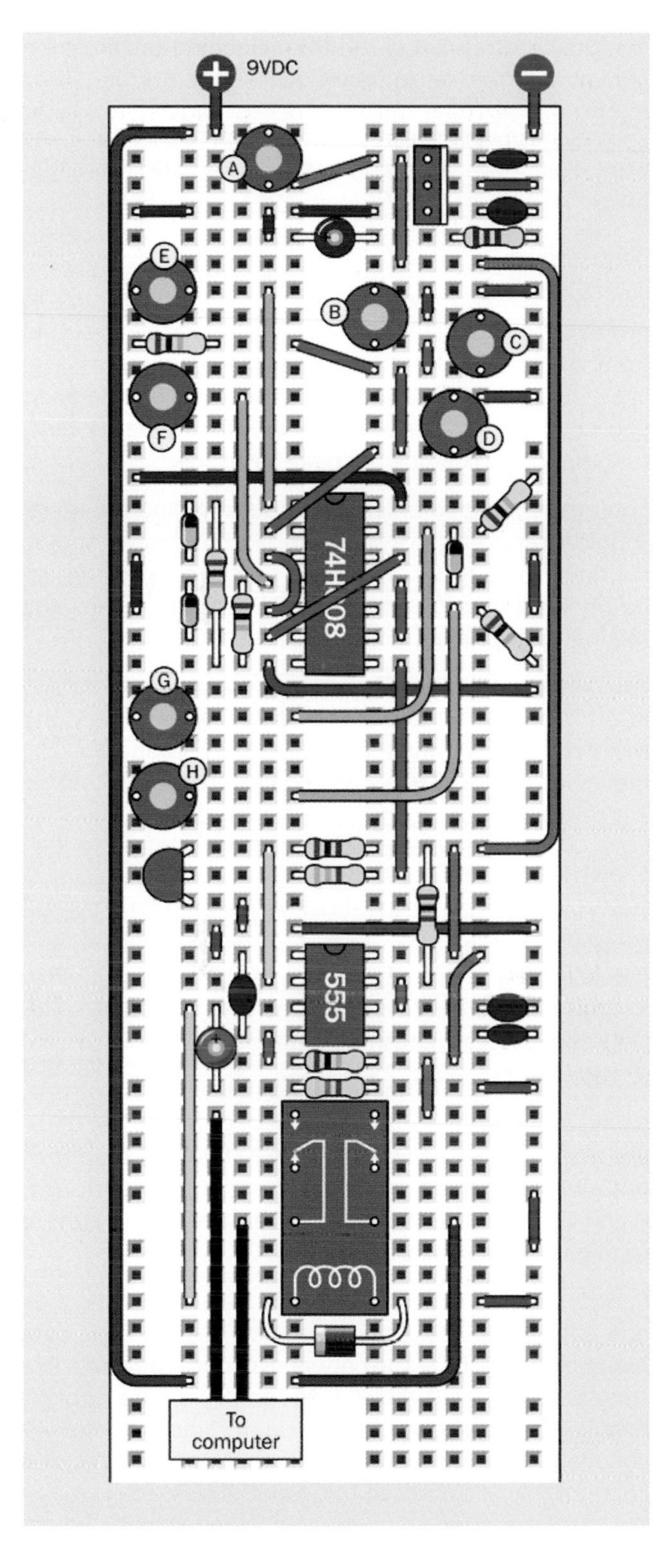

Figure 4-107 *The complete breadboarded circuit for the electronic combination lock.*

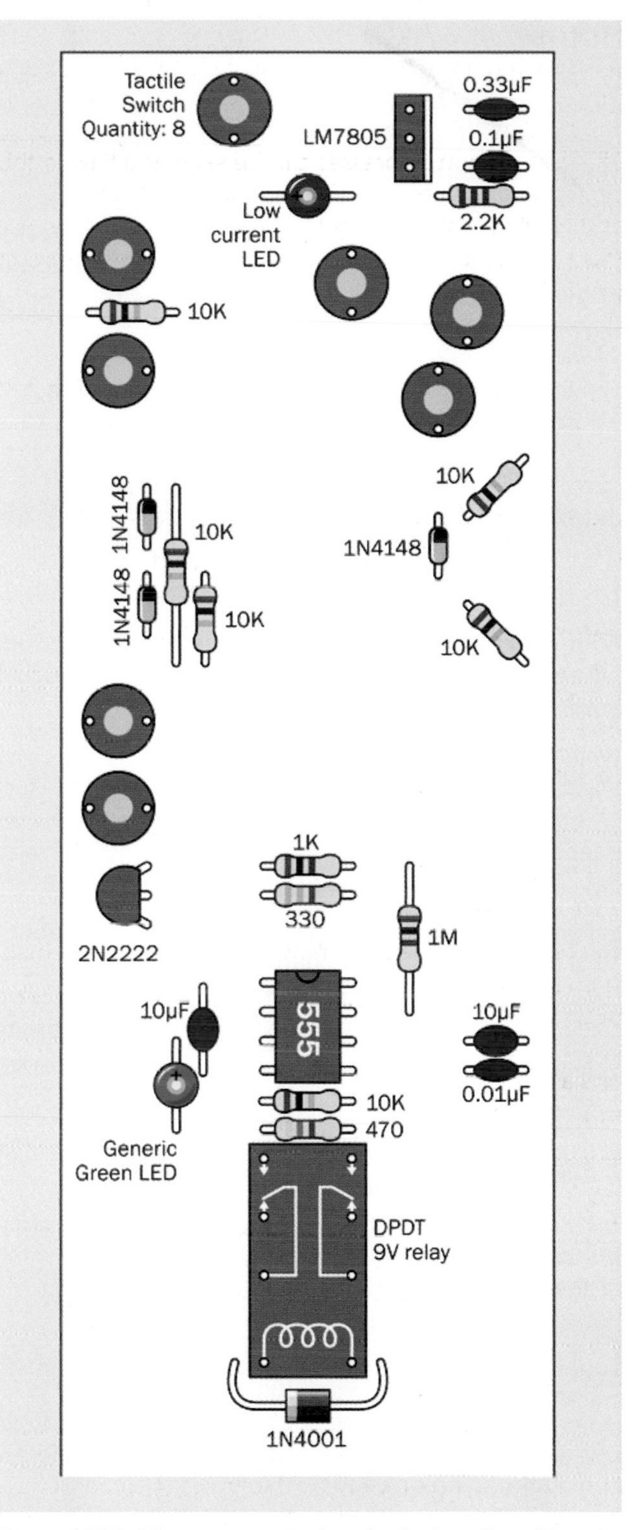

Figure 4-108 *The component values for the breadboard layout.*

Setup

Be careful to keep the two voltages in this circuit separate from each other. The 5VDC supply is insufficient to operate the relay, but 9VDC will burn out your logic chip. The lefthand bus on the breadboard is for the 5VDC supply. Unswitched 9VDC is taken down to the relay by the brown wire on the left of the breadboard in Figure 4-107. The purple or magenta wires on the right carry 9VDC, which is switched on either by button A or by the righthand relay contacts.

- Brown indicates the unswitched 9VDC supply from a battery or AC adapter.
- Magenta or purple is the 9VDC supply, which is switched, either by the relay or the button labelled A.
- Red is 5VDC, delivered by the voltage regulator.

After you build the circuit, attach your 9VDC power supply and hold down button A. The red LED comes on, but nothing else happens.

While you continue to hold down button A, press and release each of the buttons labeled E, F, G, and H in succession, from top to bottom. When you complete the sequence, the green LED comes on, showing that the relay has closed and you successfully unlocked the circuit.

Let go of button A, and the LED should stay on for about 30 seconds before the circuit automatically switches itself off. During this 30-second period, if this circuit is installed in your computer, you will have the opportunity to start the computer.

After the circuit switches itself off, it consumes no power at all. You can run it with a 9-volt battery, and the battery should literally last for years.

Try holding the power button again, and press the same buttons in a different sequence. Also, try including some of the buttons labeled B, C, and D. The green LED will not light up, and the relay will not be activated.

Suppose you install a finished version of this circuit. To crack the code, someone will have to know:

- You have to hold down button A while entering the correct sequence.
- If you press an incorrect button, the code must be re-entered from the start.
- Only buttons E, F, G, and H are active, and they must be pressed in that order.

That seems a very secure arrangement, to me. But if you want even more security, you can always add more buttons!

Testing

Set your meter to measure continuity, and attach its probes (using alligator test leads) to the lefthand output from the relay that is labeled "To Computer" in Figure 4-105. These two wires do not carry any voltage, so you need your meter in continuity-test mode to verify that the contacts inside the relay have closed.

Enter the correct combination of buttons, and the meter should beep. Let go of button A, and the meter should continue to beep while the 555 timer energizes the relay. At the end of the timer cycle, the relay opens and the meter stops beeping.

You can reconfigure your meter to measure current and insert it between the positive side of your battery and the 9VDC supply entry point on the breadboard. The meter should show no power consumption until button A is pressed.

Dealing with Diodes

This circuit contains two types of latching. The system for latching the relay is unusual but satisfies the requirement that the circuit should consume zero current when it is not being used. The system by which the AND gates latch themselves is another matter.

The fourth AND gate doesn't need to latch, because only a brief pulse (from button H) is needed to start the timer. But the first three AND gates have to latch, to keep their outputs positive after you release each of the buttons E, F, and G. The diodes take care of this by feeding current from gate outputs back to their inputs.

Can you see any problem with this? Remember that a diode takes about 0.7 volts. Remember that a logic gate has to distinguish clearly between a high state and a low state on its inputs. If you start to sprinkle diodes through a logic circuit without keeping careful track of voltages, you may end up with a logic gate that doesn't recognize an input that is supposed to be high. This is the same issue that I was concerned about in Experiment 15, where a voltage reduced by a transistor followed by a diode might have failed to trigger a relay.

When in doubt, verify voltages with your meter and check back to the input specifications illustrated in Figure 4-102.

In the combination-lock circuit that I've shown here, the output from each of the first three AND gates only circulates back through one diode to the input of the gate, so there's no reason why this should not work reliably. Just bear in mind that you have to exercise care and discretion when mixing diodes with logic chips.

Perhaps this makes you wonder—if a diode isn't the formally correct method to make a logic gate latch itself, what *is* the ideal way to do things?

One option might seem to be, replace each diode with a piece of wire, to feed the signal back to the gate input. What are the diodes really for, anyway?

They serve an important purpose. If a diode was replaced with a piece of wire, the positive voltage being applied through a pushbutton could also flow through that wire, down and around to the output of the logic gate.

- It is absolutely not a good idea to push voltage into any output of a gate.

The correct option for latching a logic state in a circuit is with a flip-flop. Previously I have used a 555 timer wired in bistable mode as a flip-flop, because we were already working with timers, and I wanted to demonstrate this application. But in this circuit, it wouldn't make sense to add four 555 timers just to serve that function. You can buy chips containing several flip-flops, and you can also make a flip-flop by combining two NAND gates or two NOR gates, as I will show you in Experiment 22.

For this little circuit creating a combination lock, I wanted to minimize the chip count and the complexity. Diodes were the simplest, easiest way to achieve this.

Questions

The output from the fourth AND gate was a single positive pulse. Why didn't I use that to activate the relay directly, instead of introducing a timer?

One reason is that the relay draws an initial surge of current exceeding the maximum 20mA that an AND gate can supply. Also of course I wanted a fixed-length pulse from the timer.

All right, but why did I add a transistor to the circuit? Because the AND gate gives a positive pulse, and the timer needs a negative transition on its trigger pin. The transistor provided a way to convert a positive to a negative. Adding a NOT gate (that is, an inverter) could have achieved the same goal, but would have increased the chip count.

In that case, why didn't I use NAND gates instead of AND gates? A NAND has a normally positive output that goes low when both of its inputs are high. That seems to be exactly what the 555 timer wants. With a NAND gate, I could have omitted the transistor.

This is true, but the preceding AND gates require positive outputs to feed back and sustain their positive inputs. So, I must retain those AND gates for the first three pushbuttons. I could only substitute a NAND gate for the last one, to create the right output for the timer. This means you would have still needed the 74HC08 chip, and you would have had to add a 74HC00 chip, just to use one of its gates. A transistor is easier and takes less room.

Here's another question. Why did I include two LEDs in the circuit? Because when you're punching buttons to unlock your computer, you need to know what's going on. The Power On LED reassures you that your battery isn't dead. The Relay Active LED tells you that the system is now unlocked, in case you are unable to hear the relay click.

Lastly, the big question: how can you actually install this circuit in a computer, assuming you are willing to give that a try? It's a lot easier than it sounds, as I will explain.

The Computer Interface

First, make sure you have wired the combination lock circuit correctly. A single wiring error can cause your circuit to deliver 9VDC through the lefthand relay contacts instead of merely closing a switch. This is important!

To make absolutely sure, change your meter to measure volts DC, and enter the correct combination on the tactile switches. If the green LED lights up but your meter measures no voltage, this is good. Anything else, and you have a wiring error.

Now let's consider how your computer normally functions when you want to switch it on.

Old computers used to have a big switch at the back, attached to the heavy metal box inside the computer that transformed house current to regulated DC voltages that a computer needs. Most modern computers are not designed this way; you leave the computer plugged in, and you touch a little button on the front of the computer (if it's not a Mac) or the keyboard (if it is a Mac). The button is connected by an internal wire to the motherboard.

This is ideal from our point of view, because we don't have to mess with high voltages. Inside your computer, don't even think of opening that metal box with the fan mounted in it, containing the computer power supply. Just look for the wire that runs from the "power up" button to the motherboard. Most often this wire will contain just two conductors, but on some computers it will be part of a ribbon cable. The key is to look at the contacts for the pushbutton, which will be attached to the conductors that you want.

First *make sure that your computer is unplugged*, ground yourself (because computers contain CMOS chips that are sensitive to static electricity), and very carefully snip just one of the two conductors from the pushbutton. Now plug in your computer and try to use the "power up" button. If nothing happens, you've probably cut the right wire. Even if you cut the wrong wire, it still prevented your computer from booting, which is what you want, so you can use it anyway.

Remember, we are not going to introduce any voltage to this wire. We're just going to use the relay as a switch to reconnect the conductor that you cut. You should have no problem if you maintain a cool and calm demeanor, and look for that single wire that starts everything. Check online for the maintenance manual for your computer if you're really concerned about making an error.

After you find the wire and cut just one of its conductors, keep your computer unplugged during the next steps.

Find where the wire attaches to the motherboard. Usually there's a small unpluggable connector. First, mark it so that you know how to plug it back in the right way around. Better still, photograph it. Then disconnect it while you follow the next couple of steps.

Strip insulation from the two ends of the wire that you cut, and solder an additional piece of two-conductor wire, as shown in Figure 4-109, with heat-shrink tubing to protect the solder joints. (This is very important!)

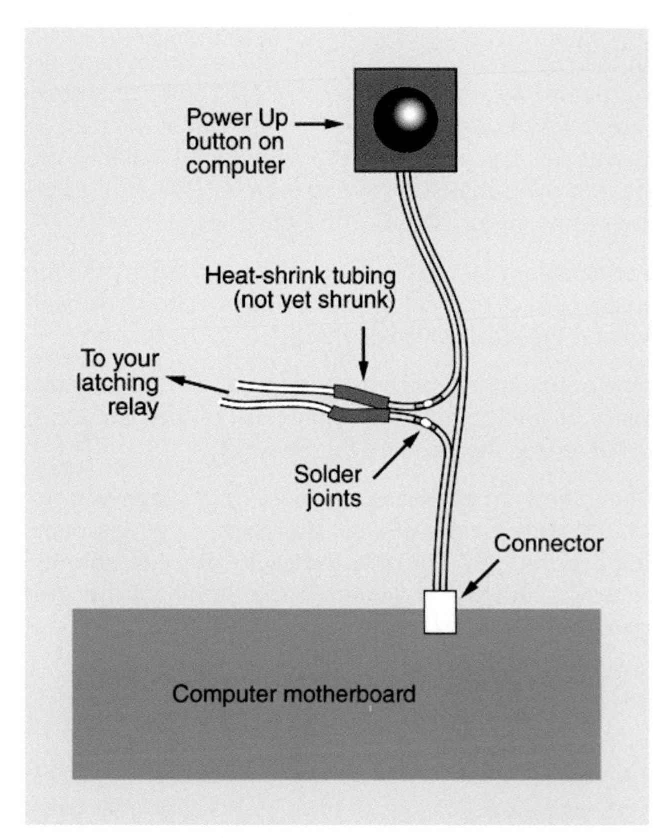

Figure 4-109 *The combination lock project can be interfaced with a typical desktop computer by cutting one conductor in the wire from the "power up" pushbutton, soldering an extension, and covering the joints with heat-shrink tubing.*

Run your new piece of wire to the relay, making sure you attach it to the pair of contacts that close, inside the relay, when it is energized by the unlocking operation. You don't want to make the mistake of unlocking your computer when you think you're locking it, and vice versa.

Reconnect the connector that you disconnected from your motherboard, plug in your computer, and press the computer's "start" button. If nothing happens, this is good! Now enter the secret combination on your keypad (while holding down the power button to provide battery power) and watch for the green LED to light up. Now try the "start" button again, and everything should work—so long as you press the button within the 30 seconds allowed by your circuit.

Having tested your circuit, the only remaining task is installation. Just remember to remove the case completely from the rest of the computer, if you are contemplating something along the lines shown in Figure 4-110.

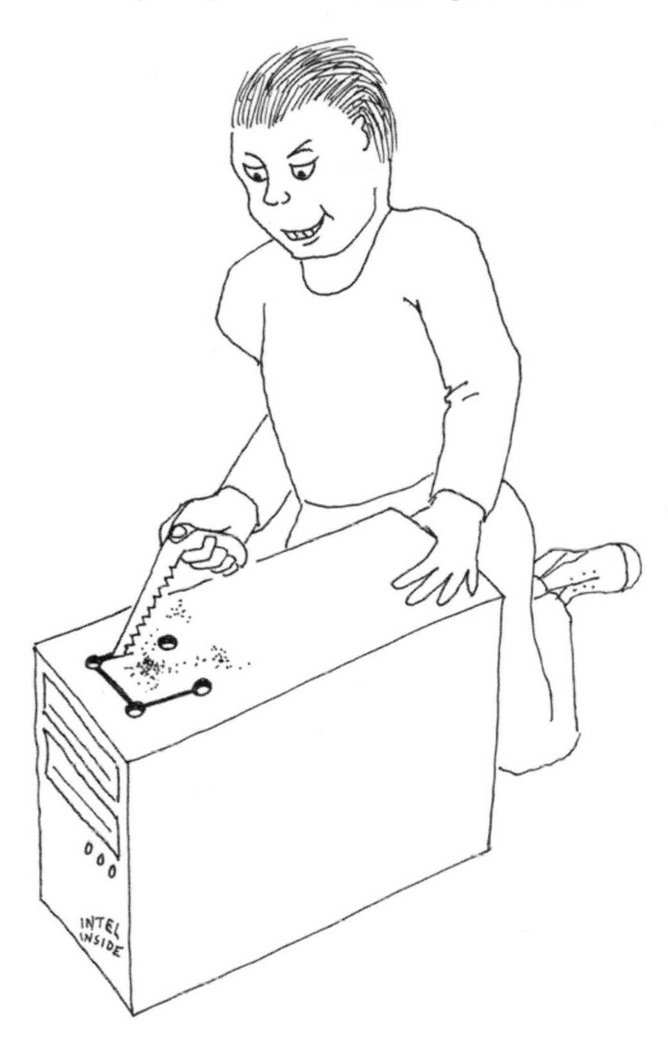

Figure 4-110 *An option for installing your keypad (not necessarily recommended).*

Enhancements

At the end of any project, there's always more you can do.

Using a Keypad. In the previous edition of this book, I suggested using a numeric keypad in this project. Some people felt that the keypad cost too much, and others had difficulty finding the right kind of keypad. After some thought, I decided just to use tactile switches this time around. They're easy to install on your breadboard, and if you decide to make a more permanent version of the circuit, you can simply mount eight pushbuttons on a square of metal or plastic. But a keypad is still an option, so long as it is *not matrix encoded*. That kind of keypad is really designed for use with a microcontroller. The kind you want will have as many pins as there are buttons, plus one.

Powering the Relay. You may be wondering if the voltage from the output of a 555 timer is sufficient to work the relay reliably. This is the same question that I discussed in Experiment 15, where I decided not to power a relay through a transistor-diode combination. The problem is, the voltage from a 555 timer varies depending on how heavily you load the output. This is why I recommended a high-sensitivity relay for this experiment. Typically it will draw less than one-third of the current of a standard type, which I felt was satisfactory for demonstration purposes. Bear in mind, I wanted to specify just one type of relay for all the experiments in this book. However, if you plan to install the circuit, and it absolutely positively has to work every time, even when your 9V battery is running low—you could consider substituting a 6VDC relay. Really? Won't the timer output overload it? Not necessarily. Some relays are designed to tolerate overvoltage. For instance, the datasheet for the Omron G5V-2-H1-DC6 6-volt relay allows a maximum voltage of 180% of the rated voltage. As always, the best recommendation is to test a circuit thoroughly, consider options, and read datasheets.

Safeguarding the Computer. To make this project more secure, you could remove the usual screws that secure the case of the computer, and replace them with tamper-proof screws. Naturally, you will also need the special tool that fits the screws, so that you can install them (or remove them, if your security system malfunctions for any reason).

Code Update. Another enhancement would be a way to facilitate changing your secret code if you feel the need. This will be difficult if you make a soldered version of the circuit, but you can install miniature plugs and sockets known as "headers" to allow you to swap the wires around.

Destructive Security. For those who are absolutely, positively, totally paranoid, you could fix things so that entering a wrong code flips a second high-amperage relay which supplies a massive power overload, melting your CPU and sending a big pulse through the hard drive. For a solid-state drive, you could consider installing a "suicide relay" that will apply a higher voltage to the 5VDC input. But I wouldn't recommend this myself.

There's no doubt about it. Messing up the hardware has major advantages compared with trying to erase data using software. It's faster, difficult to stop, and tends to be permanent. So, when the Recording Industry Association of America comes to your home and asks to switch on your computer so that they can search for illegal file sharing, just accidentally give them an incorrect unlocking code, sit back, and wait for the pungent smell of melting insulation—or a burst of gamma rays, if you go for the nuclear option (see Figure 4-111).

Figure 4-111 *For those who are absolutely, positively, totally paranoid: a meltdown/self-destruct system controlled by a secret key combination provides enhanced protection against data theft or intrusions by RIAA investigators asking annoying questions about file sharing.*

On a more realistic level, no system is totally secure. The value of a hardware locking device is that if someone does defeat it (for instance, by figuring out how to unscrew your tamper-proof screws, or simply ripping your keypad out of the computer case with metal shears), at least you'll know that something happened—especially if you put little dabs of paint over the screws to reveal whether they've been messed with. By comparison, if you use password-protection software and someone defeats it, you may never know that your system has been compromised.

Experiment 22: Race to Place

The next project using digital logic is going to get us into the concept of feedback, where the output is piped back to affect the input—in this case, blocking it. It's a small project, but quite subtle, and the concepts will be useful to you in the future.

What You Will Need

- Breadboard, hookup wire, wire cutters, wire strippers, multimeter
- 9-volt power supply (battery or AC adapter)
- 74HC32 logic chip (1)
- 555 timers (2)
- SPDT slide switches (2)
- Tactile switches (2)
- Resistors: 220 ohms (1), 2.2K (1), 10K (3)
- Capacitors: 0.01µF (2), 0.1µF (1), 0.33µF (1)
- LM7805 voltage regulator (1)
- Generic LEDs (2)
- Low-current LED (1)

The Goal

On quiz shows such as *Jeopardy!*, contestants race to answer each question. The first person who hits an answer button automatically locks out the other contestants, so that their buttons become inactive. How can we make a circuit that will do the same thing?

If you search online, you'll find several hobby sites where other people have suggested circuits to work this way, but they lack features that I think are necessary. The approach I'm going to use here is both simpler and more elaborate. It's simpler because it has a very low chip count, but it's more elaborate in that it incorporates "quizmaster control" to make a more realistic game.

I'll suggest some initial ideas for a two-player version. After I develop that idea, I'll show how it could be expanded to four or even more players.

A Conceptual Experiment

I want to show how this kind of project grows from an initial idea. By going through the steps of developing a circuit, I'm hoping I may inspire you to develop ideas of your own in the future, which is much more valuable than just replicating someone else's work.

First consider the basic concept: there are two people, each with a button to press, and whoever goes first locks out the other person.

Sometimes it helps me to visualize this kind of thing if I draw a sketch, so that's where I'll begin. In Figure 4-112, the signal from each button passes through an imaginary component that I'll call a "button blocker," activated by the other person's button. I'm not exactly sure what the button blocker will be or how it will work, yet, but it will be activated by one player going first, and it will block the other player.

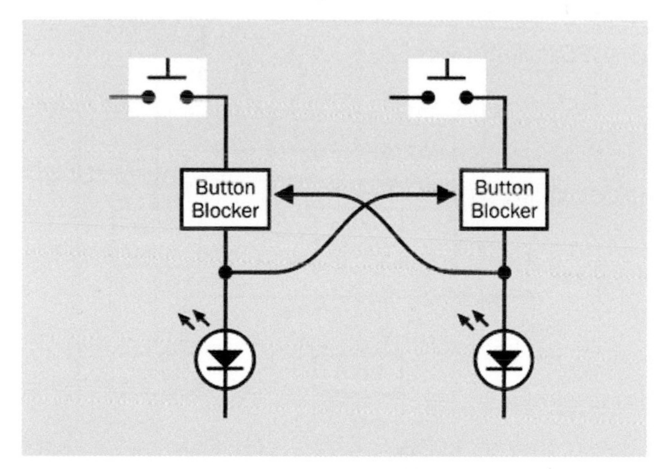

Figure 4-112 *The basic concept: whoever goes first blocks the other player.*

Now that I'm looking at it, I see a problem here. If I want to expand this to three players, it will get complicated, because each player must activate the "button blockers" of two opponents, and if there are four players, each much activate the "button blockers" of three opponents. The number of connections will become unmanageable. Figure 4-113 shows this.

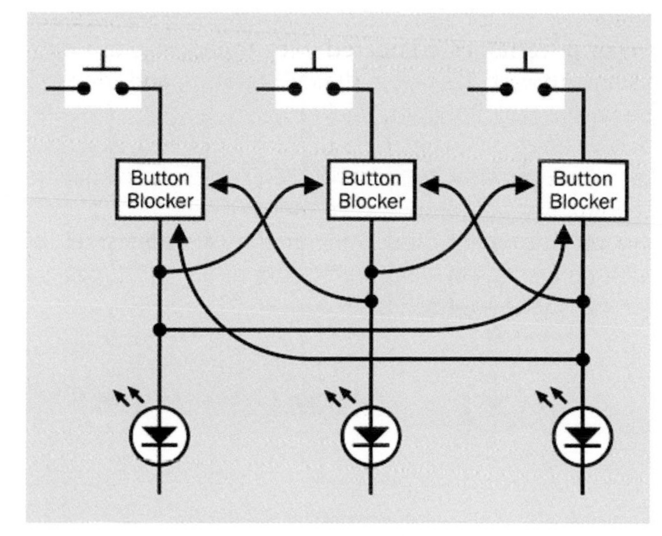

Figure 4-113 *Increasing the number of participants from two to three more than doubles the number of connections.*

Any time I see this kind of complexity, I think there has to be a better way.

Also, there's another problem. After a player lets his finger off the button, the other players' buttons will be unblocked again. This suggests to me that as in Experiments 15, 19, and 21, I need . . . a flip-flop, also known as a latch. Its purpose will be to hold the signal from the first player's button and continue to block the other players, even after the first player has released his button.

This now sounds even more complicated. But wait a minute. If the winning player's button triggers a latch, the latch now keeps the winning circuit energized, and the winner's button becomes irrelevant. So, the latch can block *all* the buttons. This makes things much simpler. I can summarize it as a sequence of events:

- First player presses his button.
- His signal is latched.
- The latched signal feeds back and blocks all the buttons.

The new sketch in Figure 4-114 shows this. Now the configuration is modular, and can be expanded to almost any number of players, just by adding more modules, without increasing in complexity.

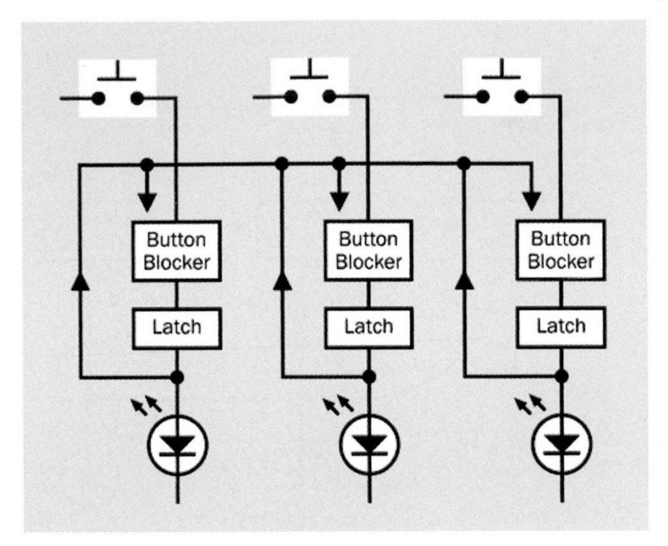

Figure 4-114 *Any latch now blocks all the buttons.*

There's something important missing, though: a reset switch, to put the system back to its starting mode after the players have had time to press their buttons and see who won. Also, I need a way to prevent players from pressing their buttons too soon, before the quizmaster has finished asking the question. Maybe I can combine this function in just one switch, which will be under the quizmaster's control.

See Figure 4-115. In its Reset position, the quizmaster switch can reset the system and remove power to the buttons. In its Set position, the switch stops holding the system in reset mode, and provides power to the buttons. I've gone back to showing just two players, to make everything as simple as possible, but the concept is still easily expandable.

Now I have to deal with a logic problem in the diagram. The way I've drawn it, everything is joined together. I've used arrows to show the direction of signals, but I don't know how I could actually stop signals from going the wrong way. If I don't deal with this, the signal from either player will light up both LEDs. How can I stop this from happening?

I could put diodes in the "up" wires to block current from running down them. But I have a more elegant idea: I'll add an OR gate, because the inputs to an OR gate are separated from each other electrically. Figure 4-116 shows this.

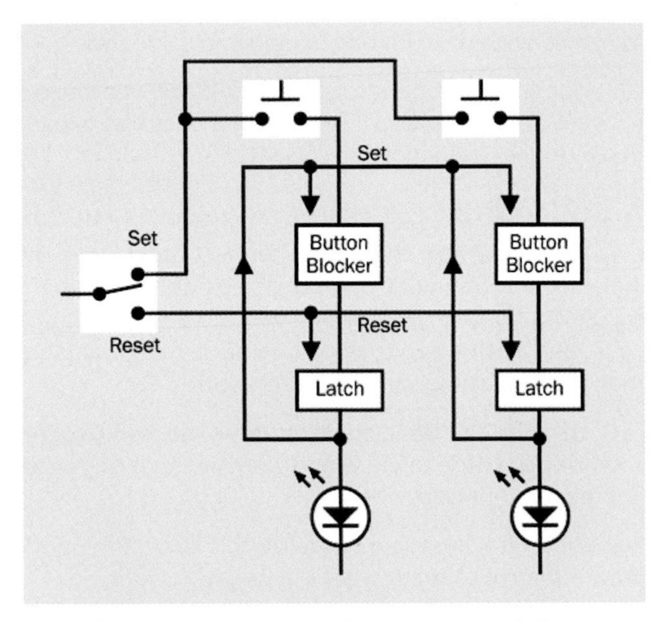

Figure 4-115 *Quizmaster control has now been added.*

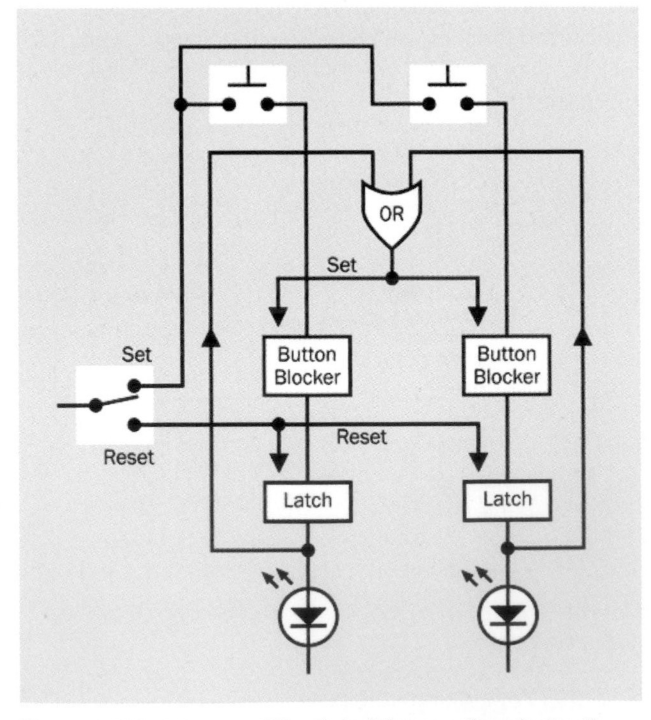

Figure 4-116 *Adding an OR gate isolates one player's circuit from the other.*

The basic OR gate has only two logical inputs. Will this prevent me from adding more players? No, because you can buy an OR that has three, four, or even eight inputs. If any one of them is high, the output is high. For fewer players than a gate can handle, you tie the unused inputs to ground and ignore them.

Now I'm getting a clearer idea of what the thing I've called a "button blocker" should actually be. I think it should be another logic gate. It should say, "If there's only one input, from a button, I'll let it through. But if there are any additional inputs, I won't let them through."

Before I start choosing gates, though, I have to decide what the latch will be. I can buy an off-the-shelf flip-flop, which flips "on" if it gets one signal and "off" if it gets another, but chips containing flip-flops tend to have more features than I need for a simple circuit like this. Therefore I'm going to use 555 timers again, in bistable mode. They require very few connections, work very simply, and can deliver a good amount of current to drive bright LEDs. The only problem with them is that 555s in bistable mode require:

- Negative trigger input to create a high output
- Negative reset input to create a low output

All right, so, each player's button will have to generate a negative pulse instead of a positive pulse. That will suit the requirements of the timers.

Finally, here's a simplified schematic, in Figure 4-117. I like to show the pins of the 555 timers in their correct positions, so I had to move the components around a little to minimize wire crossovers, but you can see that logically, it's the same basic idea.

I didn't have enough room to add positive and negative symbols to show which timer pins are being held in each state, so a red circle means that the pin is being held in a high state, and a blue circle shows that it is being held in a low state. Black circles mean that the pin states may change. White circles mean that the states of these pins are not important, and the pins can be left unconnected.

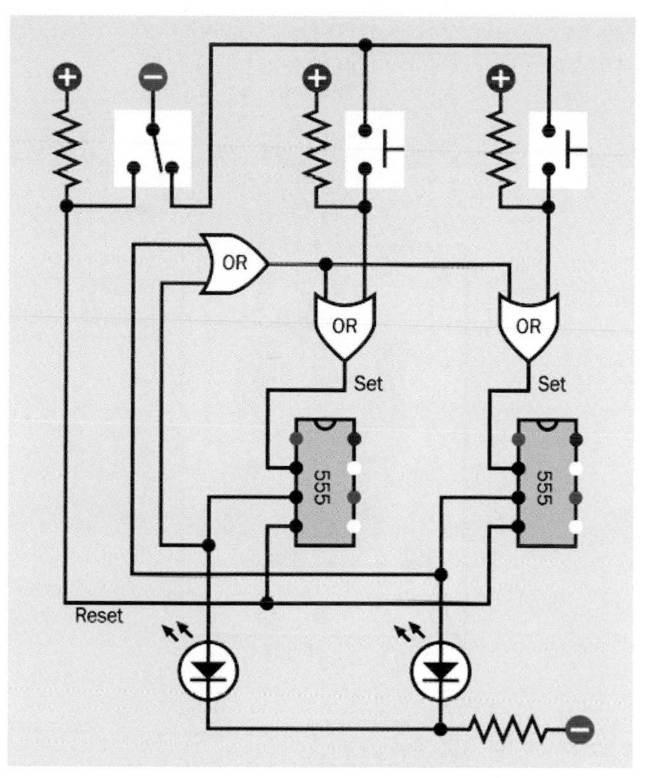

Figure 4-117 *A preliminary logic diagram. Blue pins on the timers are held in a low state, red pins are held in a high state, and white pins are not relevant.*

Before you try to build it, just run through the theory of it, because that's the final step, to make sure there are no mistakes. The important thing to bear in mind is the 555 needs a *negative* input on its trigger pin to create a *positive* output. This means that when any of the players presses a button, the button has to create a negative "flow" through the circuit.

This is a bit counterintuitive, so I'm including a four-step visualization in Figure 4-118, Figure 4-119, Figure 4-120, and Figure 4-121, showing how it will work.

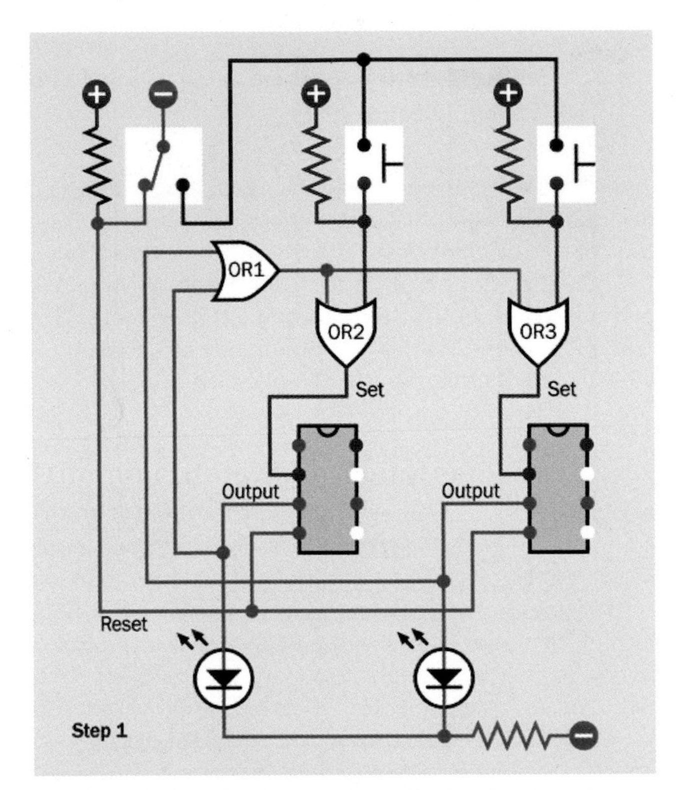

Figure 4-118 *Step 1 in the circuit visualization. Reset mode.*

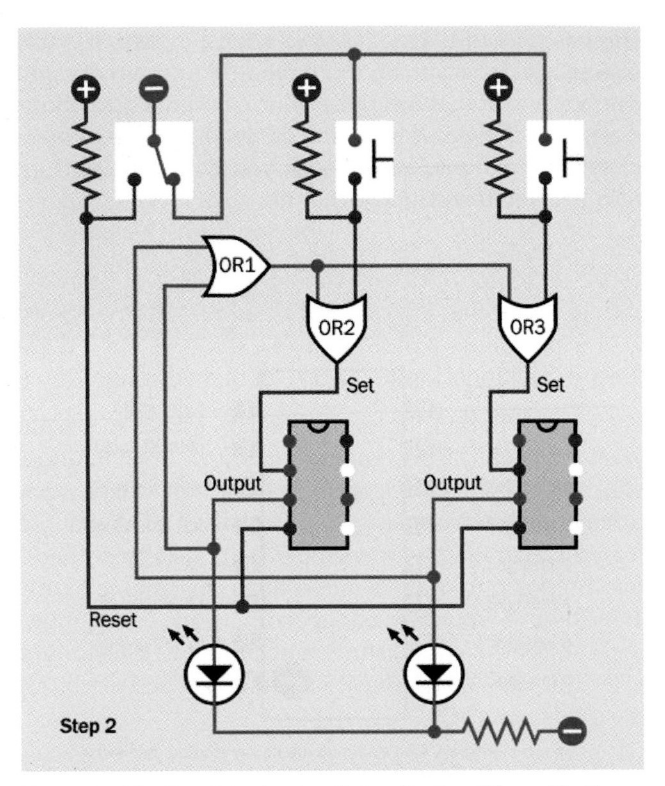

Figure 4-119 *Step 2 in the circuit visualization. Players' buttons are active, but no one has pressed a button yet.*

In Step 1, the quizmaster's switch is in reset mode. The low voltage on the reset pins of the timers has forced them both to create negative outputs. These outputs keep the LEDs dark and also go to the OR1 gate. Because it has negative inputs, it creates a negative output, although OR2 and OR3 ignore that because one input on each of these gates is high, provided by pullup resistors beside the buttons. Remember, if either of the inputs to an OR gate is high, the output is high. And so long as the trigger pin on a timer in bistable mode is high, the timer will not be triggered. So, the circuit is stable.

In Step 2, the quizmaster has asked a question and flipped his switch to the right, to supply (negative) power to the players' buttons. However, neither of the players has responded yet, so the pullup resistors maintain the circuit in a stable state with negative outputs from the timers.

In Step 3, Player 1 has pressed the lefthand button. This sends a low pulse to OR2. Now that OR2 has two low inputs, its output goes low. The low pulse goes to the trigger pin of the lefthand timer. But components do not respond instantaneously, and the timer has not processed the signal yet.

In Step 4, a few microseconds later, the timer has processed the negative input signal and created a positive output pulse that lights the LED and also circulates back to OR1. Now that OR1 has a positive input, it also has a positive output. This goes to the inputs of OR2 and OR3, and so their outputs become positive. As a result, both of the timers now have positive inputs on their trigger pins. Any button press by either player will now be ignored, because OR1 continues to supply positive current in the circuit.

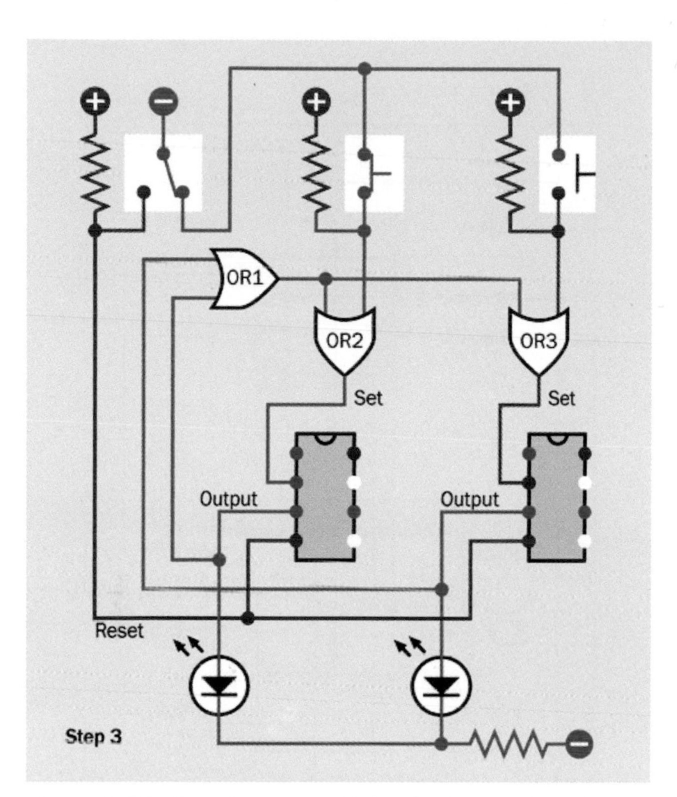

Figure 4-120 *Step 3 in the circuit visualization. The left player has pressed a button, but the 555 timer has not responded yet.*

- Remember, when a 555 is running in flip-flop mode, a low input on its trigger pin will flip it into having a high output, and the output will persist even if the trigger pin goes high again.
- The only thing that will end the 555's high output is a low state on its reset pin. That will only happen when the game master turns his switch back to reset mode.

There's only one situation that can upset this happy scenario. What if both players press their buttons absolutely simultaneously? In the world of digital electronics, this is highly unlikely. But if it somehow happens, both of the timers should react, and both of the LEDs will light up, showing that there has been a tie.

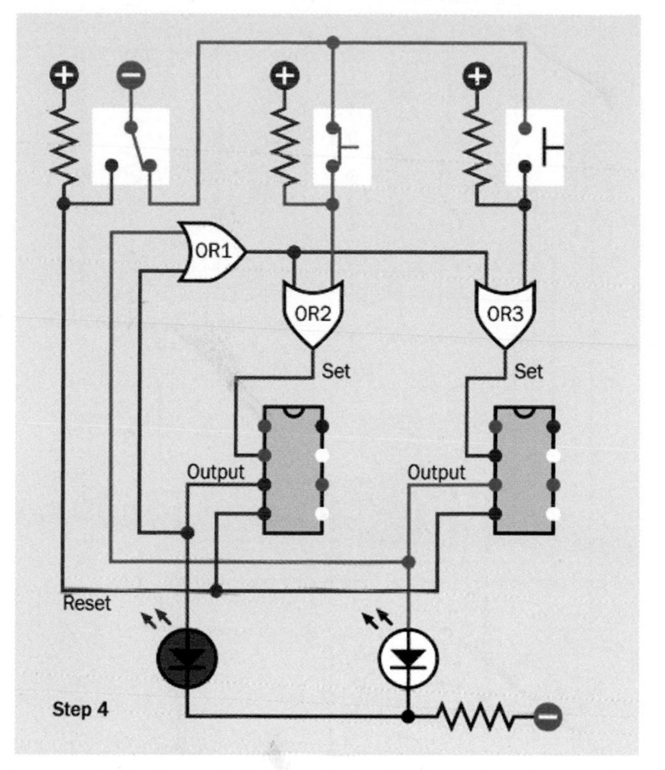

Figure 4-121 *Step 4 in the circuit visualization. The left player's action has flowed through the circuit and now blocks the right player.*

On the *Jeopardy!* TV show, you never see a tie. Absolutely never! I'm wondering, if the electronic system on the show registers a simultaneous response from two players, maybe it has a randomizing feature to select one of them. Just speculating, of course.

To demonstrate how a two-player circuit can be upgraded to handle extra players, I've included a simplified three-player schematic in Figure 4-122. The circuit could be extended outward indefinitely, the only limit being the number of inputs available on OR1.

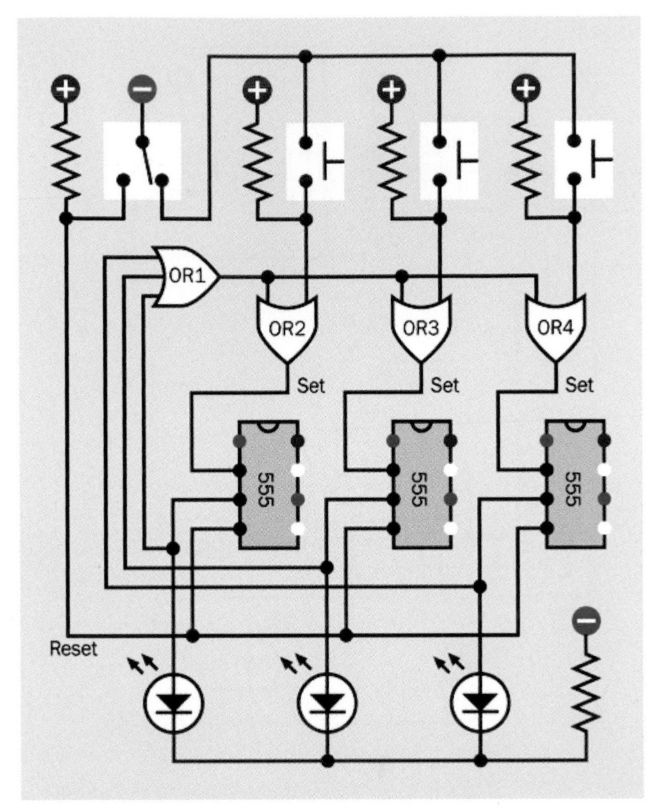

Figure 4-122 *The circuit can be expanded easily for more players.*

Breadboarding It

In Figure 4-123 I have revised the schematic using an actual OR chip, in a layout as close to a breadboard configuration as possible, so that you can build this thing easily. A breadboarded version is shown in Figure 4-124, and the component values are shown in Figure 4-125.

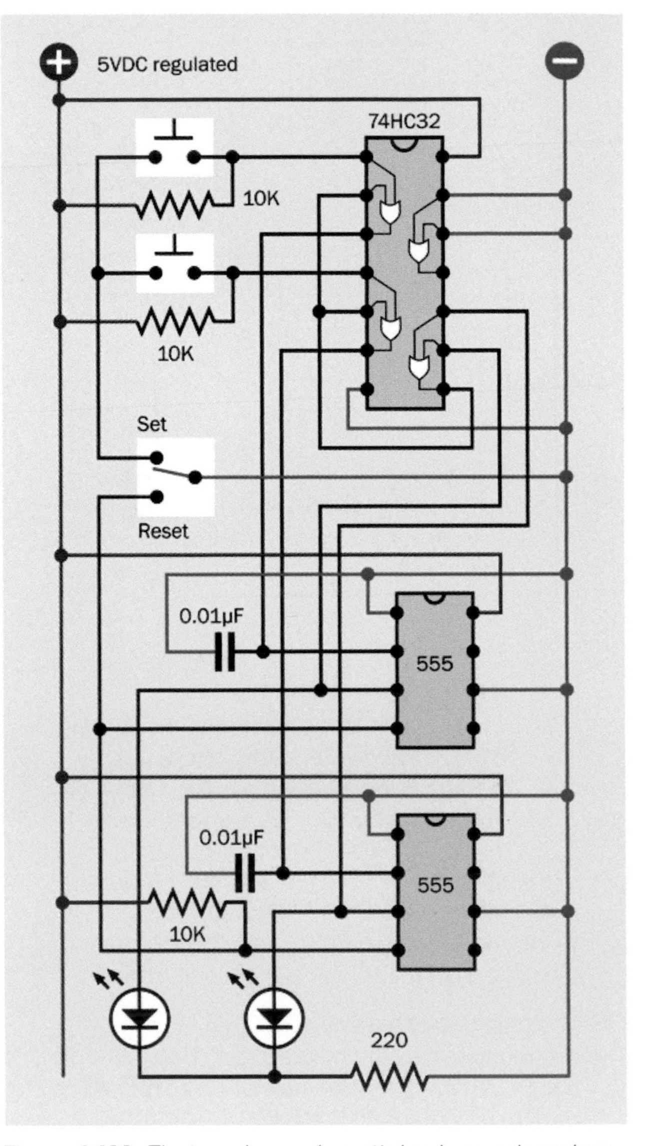

Figure 4-123 *The two-player schematic has been redrawn here, using a quad two-input OR chip.*

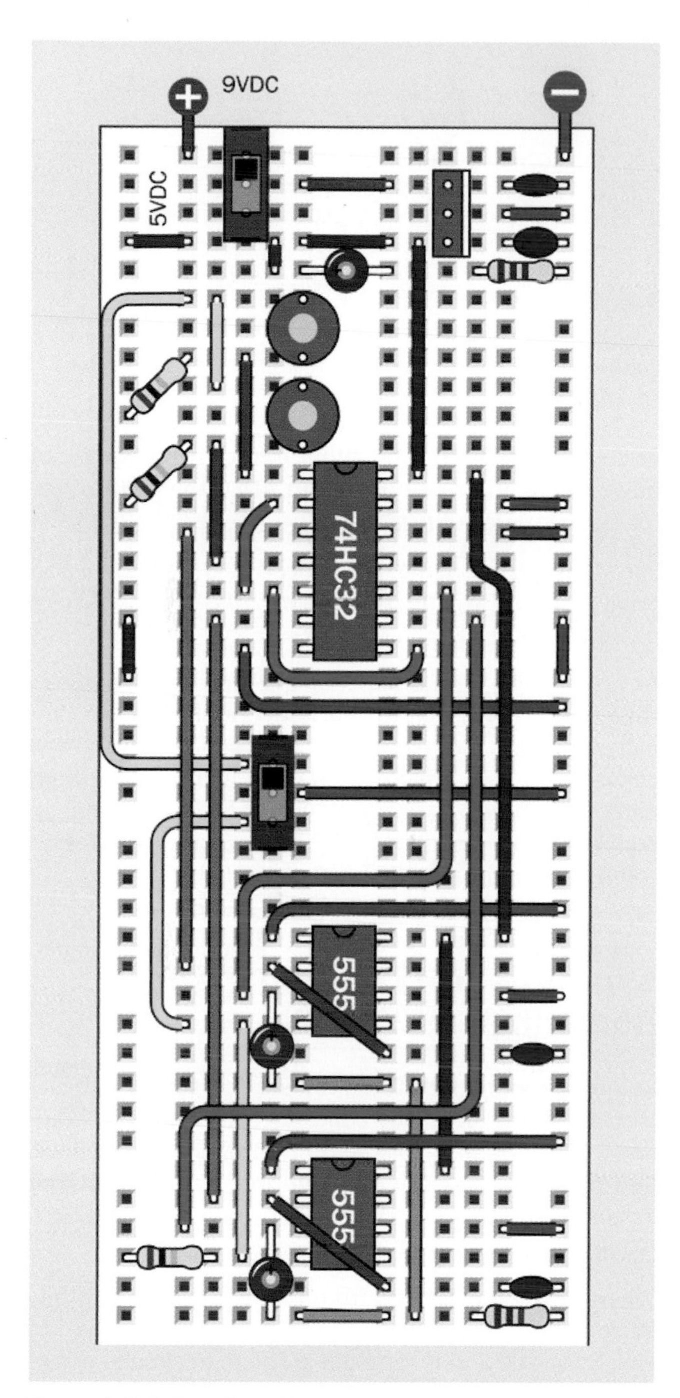

Figure 4-124 *Breadboard layout, equivalent to the schematic.*

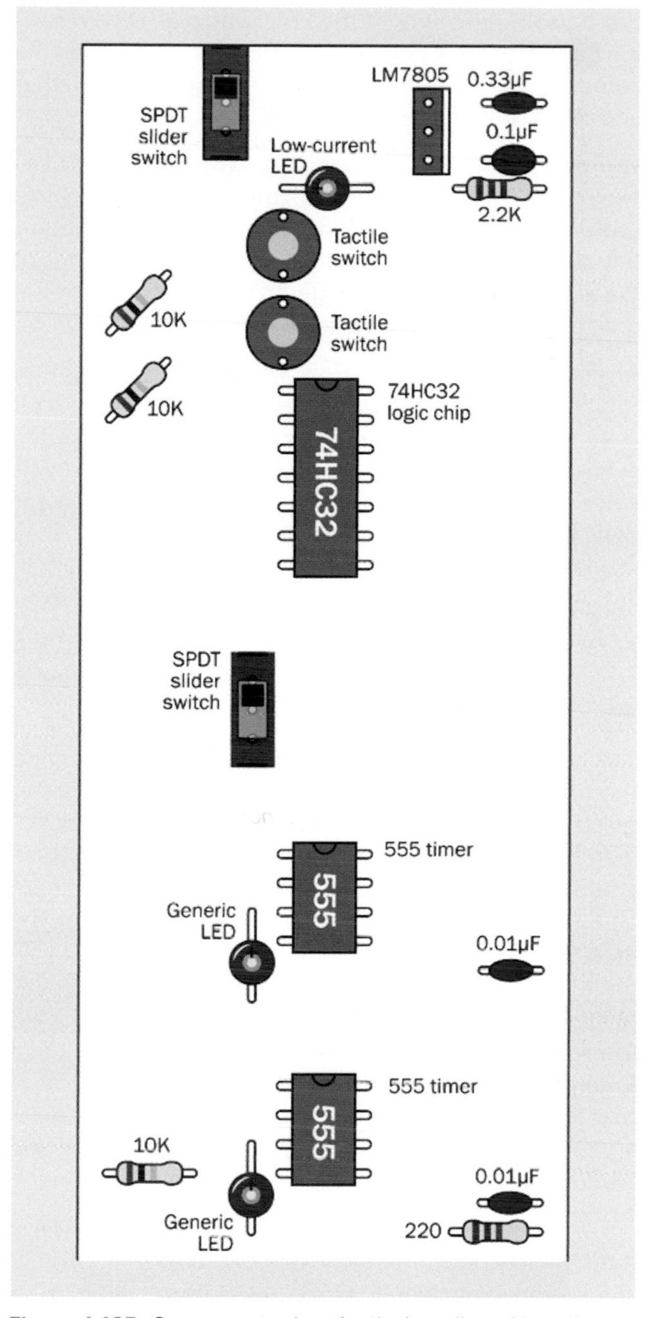

Figure 4-125 *Component values for the breadboard layout.*

Because the only logic gates that I've used are OR gates, and only three are needed, I just need one logic chip: the 74HC32, which contains four two-input OR gates. (I've grounded the inputs to the fourth). The two OR gates on the left side of the chip have the same functions as OR2 and OR3 in my simplified schematic, and the OR gate at the bottom-right side of the chip works as OR1, receiving input from pin 3 of each 555 timer. If you have all the components, you should be able to put this together and test it quite quickly.

You may notice that I've added a 0.01µF capacitor between pin 2 of each 555 timer (the input) and negative ground. Why? Because when I tested the circuit without the capacitors, sometimes I found that one or both of the 555 timers would be triggered simply by flipping the quizmaster switch, without anyone pressing a button.

At first this puzzled me. How were the timers getting triggered, without anyone doing anything? Maybe they were responding to "bounce" in the quizmaster switch —meaning tiny and very rapid vibrations in the contacts when the switch is moved. Sure enough, this was happening, and small capacitors solved the problem. They may also slow the response of the 555 timers fractionally, but not enough to interfere with slow human reflexes.

As for the buttons, it doesn't matter if they "bounce," because each timer locks itself on at the very first impulse and ignores any hesitations that follow.

You can experiment building the circuit, disconnecting the 0.01µF capacitors, and flipping the quizmaster switch to and fro a dozen times. Because I am recommending a small, cheap slide switch, I think you will see a number of "false positives." I'm going to explain more about switch bounce, and how to get rid of it, in the next experiment.

Enhancements

After you breadboard the circuit, if you proceed to build a permanent version, I suggest that you expand it so that at least four players can participate. This will require an OR gate capable of receiving four inputs. The 74HC4078 is the obvious choice, as it allows up to eight. Just connect any unused inputs to negative ground.

Alternatively, if you already have a couple of 74HC32 chips and you don't want to bother ordering a 74HC4078, you can gang together three of the gates inside a single 74HC32 so that they function like a four-input OR. Look at the simple logic diagram in Figure 4-126 showing three ORs, and remember that the output from each OR will go high if at least one input is high.

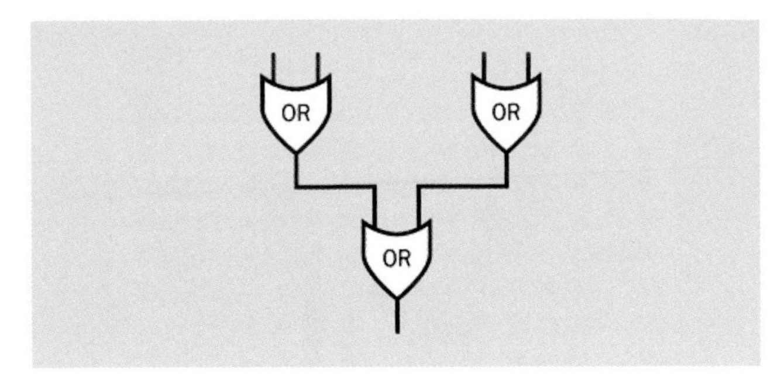

Figure 4-126 *Three two-input ORs can emulate a single four-input OR.*

And while you're thinking about this, can you figure out how to combine three two-input ANDs that will substitute for a four-input AND gate?

For a four-player game, you'll need two additional 555 timers, of course, and two more LEDs, and two more pushbuttons.

As for creating a schematic for the four-player game—I'm going to leave that to you. Begin by sketching a simplified version, just showing the logic symbols. Then convert that to a breadboard layout (which is the hard part). And here's a suggestion: pencil, paper, and an eraser can still be quicker, initially, than circuit-design software or graphic-design software, in my opinion.

Experiment 23: Flipping and Bouncing

In three experiments, now, I've used 555 timers in bistable mode. The time has come to deal with "real" flip-flops, including an explanation of how they work. I will also show how they can deal with the phenomenon that I mentioned briefly in the previous experiment: *switch bounce*.

When a switch is flipped from one position to another, its contacts vibrate very briefly. This is the "bounce" of which I speak, and it can be a problem in circuits where digital components respond so quickly, they interpret every tiny vibration as a separate input. If you connect a pushbutton to the input of a counter chip, for instance, the counter may register 10 or more input pulses from a single press of the button. A sample of actual switch bounce is shown in Figure 4-127.

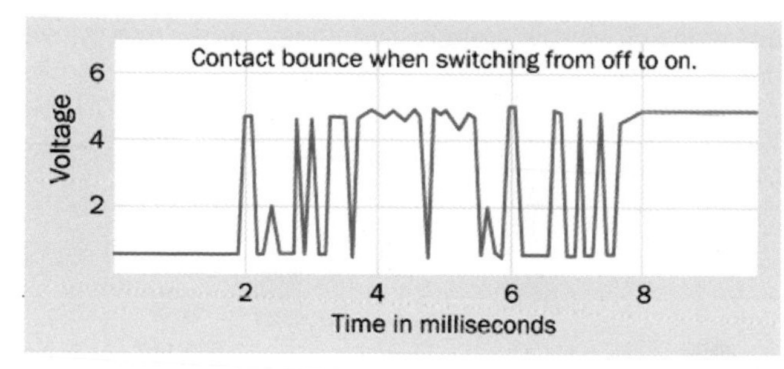

Figure 4-127 *Fluctuations created by vibrating contacts when a switch is closed. (Derived from a datasheet at Maxim Integrated corporation.)*

There are many techniques for debouncing a switch, but using a flip-flop is probably the most fundamental.

What You Will Need

- Breadboard, hookup wire, wire cutters, wire strippers, multimeter
- 9-volt power supply (battery or AC adapter)
- 74HC02 logic chip (1), 74HC00 logic chip (1)
- SPDT slide switches (2)
- Low-current LEDs (3)
- Resistors: 680 ohms (2), 10K (2), 2.2K (1)
- Capacitors: 0.1µF (1), 0.33µF (1)
- LM7805 voltage regulator (1)

Assemble the components on your breadboard, as shown in Figure 4-128. The same circuit is shown as a schematic in Figure 4-129, and the component values are shown in Figure 4-130. When you apply power, one of the LEDs at the bottom should be lit.

Now I want you to do something odd. Please disconnect the wire labeled A in Figure 4-128. Just pull it out of the board. If you refer to the schematic in Figure 4-129, you'll see that you have disconnected power to the pole of the switch, leaving the two NOR gates connected only with their pulldown resistors.

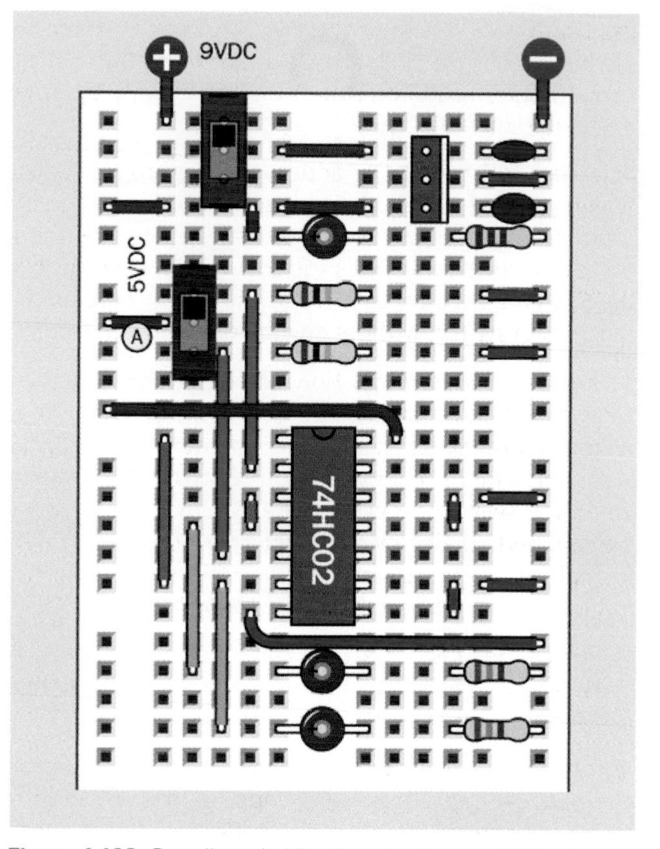

Figure 4-128 *Breadboarded flip-flop circuit using NOR gates.*

You may be surprised to find that the LED remains lit.

Push the wire back into the board, slide the switch to its opposite position, and the first LED should go out, while the other LED should come on. Once again, pull out the wire, and once again, the LED should remain lit.

Here's the take-home message:

- A flip-flop requires only an initial input pulse—for example, from a switch.
- After that, it ignores that input.

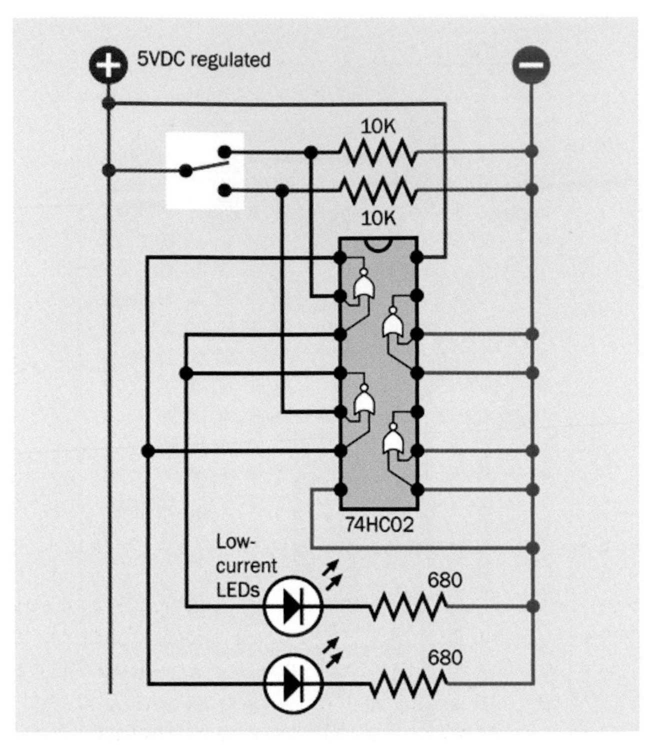

Figure 4-129 *Flip-flop schematic using NOR gates.*

How It Works

Two NOR gates or two NAND gates can function as a flip-flop.

- Use NOR gates when a double-throw switch delivers a positive input.
- Use NAND gates when a double-throw switch delivers a negative input.

Either way, you have to use a double-throw switch.

I've mentioned the double-throw switch three times (actually, four times if you count this sentence) because for some strange reason, most introductory books fail to emphasize this point. When I first started learning electronics, I went crazy trying to understand how two NORs or two NANDs could debounce a simple SPST pushbutton—until finally I realized that they can't. The reason is that when you power up the circuit, the NOR gates (or NAND gates) need to be told in which state they should begin. Their initial orientation comes from the switch being in one state or the other. A SPST pushbutton cannot do that, when it is not being pushed. So you have to use a double-throw switch. (Now I've mentioned it five times.)

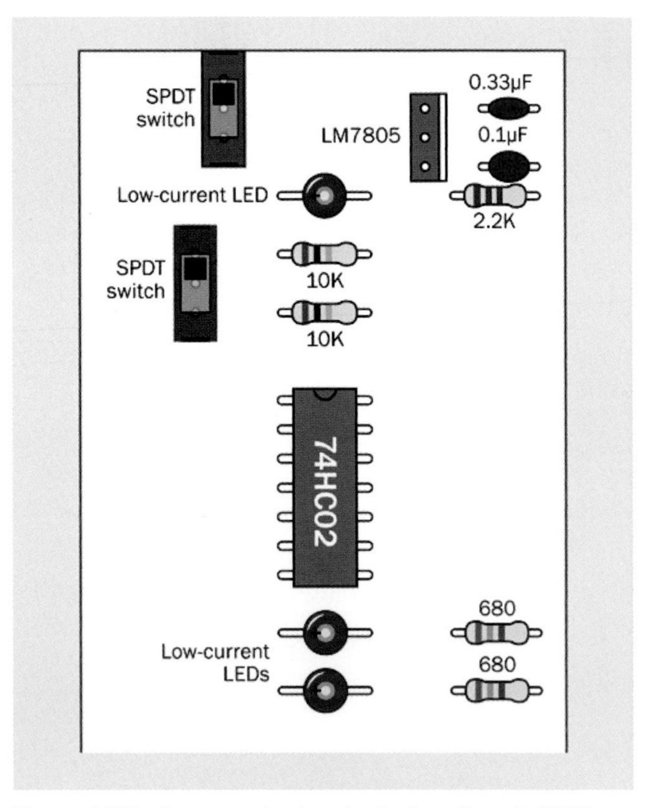

Figure 4-130 *Component values for the breadboarded NOR-based flip-flop.*

Debouncing with NORs

I've created a multiple-step schematic in Figure 4-131 and Figure 4-132 to show the changes that occur as a switch flips to and fro with two NOR gates. To refresh your memory, I've also included a truth table in Figure 4-133 showing the logical outputs from NOR gates for each combination of inputs.

Referring initially to Figure 4-131, in Step 1, the switch is supplying positive current to the lefthand side of the circuit, overwhelming the negative supply from the pull-down resistor, so we can be sure that the NOR gate on the left has one positive logical input. Because any positive logical input will make the NOR give a negative output (as shown in the truth table in Figure 4-133), the negative output crosses over to the righthand NOR, so that it now has two negative inputs, which make it give a positive output. This crosses back to the lefthand NOR gate. So, in this configuration everything is stable.

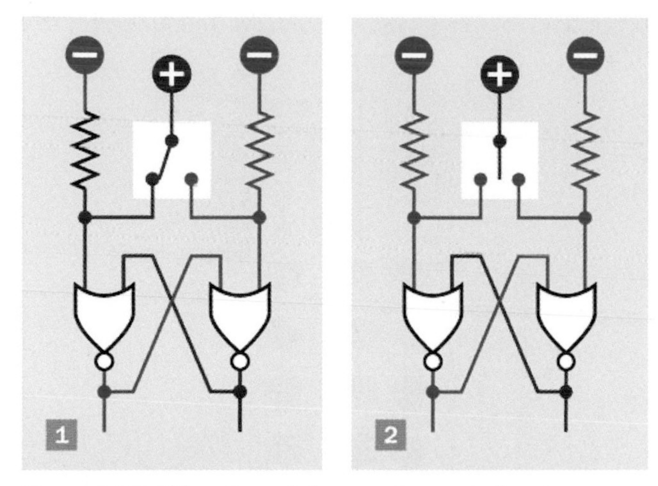

Figure 4-131 *When the switch moves to a neutral center position, the status of the NOR gates remains unchanged.*

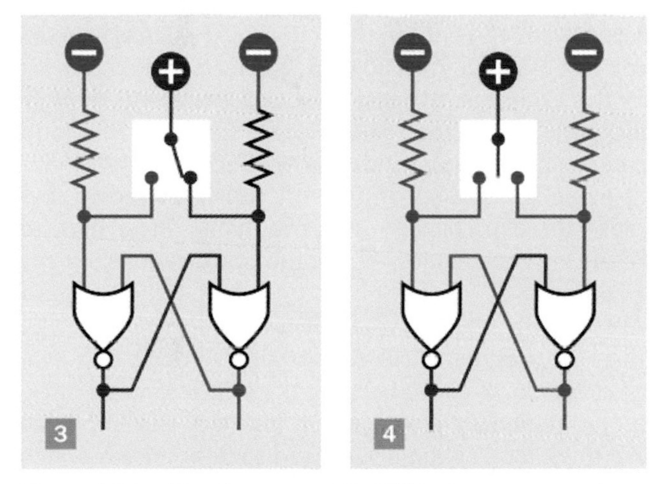

Figure 4-132 *After the states of the NOR gates are reversed, they remain that way when the switch moves back to its neutral center position.*

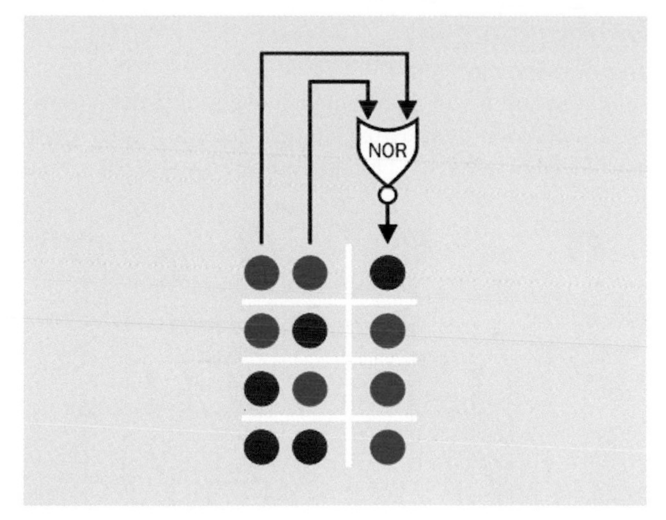

Figure 4-133 *A reminder of the truth table for a NOR gate.*

Now comes the clever part. In Step 2, suppose that you move the switch so that it doesn't touch either of its contacts. (Or suppose that the switch contacts are bouncing, and failing to make a good contact. Or suppose you disconnect the switch entirely.) Without a positive supply from the switch, the lefthand input of the left NOR gate goes from positive to negative, as a result of the pulldown resistor. But the righthand input of this gate is still positive, and one positive is all it takes to make the NOR maintain its negative output, so nothing changes. In other words, the circuit has "flopped" in this state, regardless of whether the switch is disconnected.

Referring to Figure 4-133, if the switch turns fully to the right and supplies positive power to the righthand pin of the right NOR gate, that NOR recognizes that it now has a positive logical input, so it changes its logical output to negative. That goes across to the other NOR gate, which now has two negative inputs, so its output goes positive, and runs back to the right NOR.

In this way, the output states of the two NOR gates change places. They flip, and then flop there, even if the switch breaks contact or is disconnected again, as in Step 4.

If a switch bounces so severely that the pole connection fluctuates all the way between one contact and the other, this circuit won't work. It only works if the output alternates between one connection, and no connection at all. But this is generally the case with a SPDT switch.

Debouncing with NANDs

The drawings in Figure 4-134 and Figure 4-135 show a similar sequence of events if you use a negatively powered switch with two NAND gates. To refresh your memory of NAND behavior, I am including Figure 4-136.

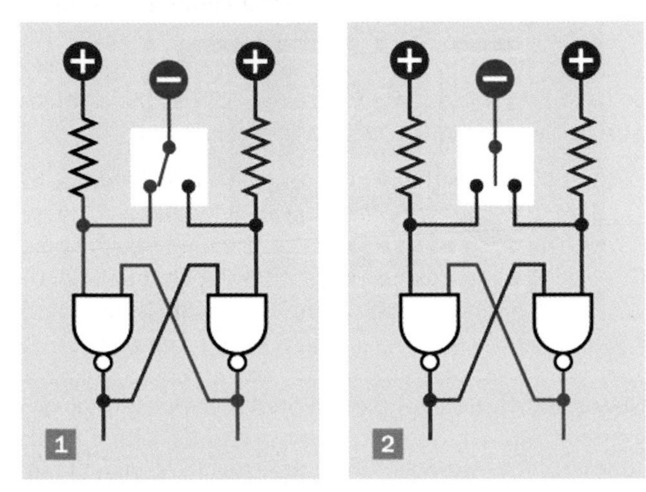

Figure 4-134 *Two NAND gates can be used as a flip-flop with pullup resistors and a switch providing negative power.*

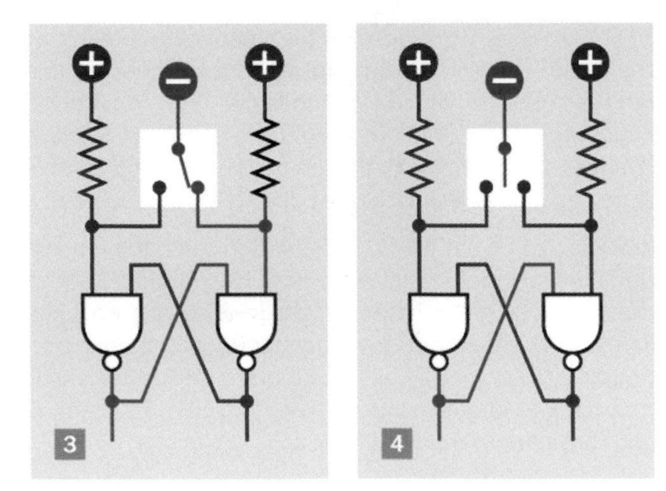

Figure 4-135 *Once again, the gate states remain unchanged when the switch is unconnected with either of them.*

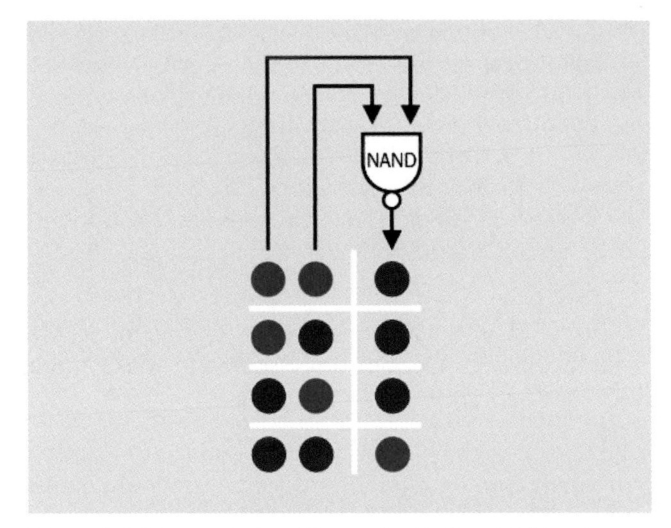

Figure 4-136 *A reminder of the truth table for a NAND gate.*

If you want to verify the function of the NAND circuit, you can use your 74HC00 chip, specified in the parts list for this experiment, to test it yourself. Be careful, though: the gates inside the NOR chip are upside-down compared with the gates inside the NAND chip. You will have to move some wires around on your breadboard, because the two chips are not swappable. Refer back to Figure 4-83 and Figure 4-95 for the specifications.

Jamming Versus Clocking

The NOR and the NAND circuits are examples of a *jam-type flip-flop*, so called because the switch forces it to respond immediately, and jams it into that state. You can use this circuit anytime you need to debounce a switch (as long as it's a double-throw switch).

A more sophisticated version is a *clocked flip-flop*, which requires you to set the state of each input first and then supply a clock pulse to make the flip-flop respond. The pulse has to be clean and precise, which means that if you supply it from a switch, the switch must be debounced—probably by using another jam-type flip-flop! Considerations of this type have made me reluctant to use clocked flip-flops in this book. They add a layer of complexity that I prefer to avoid in an introductory text. If you want to know more about flip-flops, I explore them in greater detail in *Make: More Electronics*. It's not a simple topic.

What if you want to debounce a single-throw button or switch? Well, you have a problem! One solution is to buy a special-purpose chip such as the 4490 "bounce eliminator," which contains digital delay circuitry. A specific part number is the MC14490 from On Semiconductor. This contains six circuits for six separate inputs, each with an internal pullup resistor. It's relatively expensive, however—more than 10 times the price of a 74HC02 containing NOR gates. Really, you will have an easier time if you avoid single-throw switches and use double-throw switches (or pushbuttons) that are more easily debounced.

Or, you could use a 555 timer wired in flip-flop mode. My preference for that option now seems to make more sense.

Experiment 24: Nice Dice

Electronic circuits to simulate the throw of one or two dice have been around for decades. However, new configurations still exist, and this project provides an opportunity to learn more about logic while ending up with something useful. I especially want to introduce you to binary code, the universal language among digital chips.

What You Will Need

- Breadboard, hookup wire, wire cutters, wire strippers, multimeter
- 9-volt power supply (battery or AC adapter)
- 555 timer (1)
- 74HC08 logic chip (1), 74HC27 logic chip (1), 74HC32 logic chip (1)
- 74HC393 binary counter (1)
- Tactile switch (1)
- SPDT slide switches (2)
- Resistors: 100 ohms (6), 150 ohms (6), 220 ohms (7), 330 ohms (2), 680 ohms (4), 2.2K (1), 10K (2), 1M (1)
- Capacitors: 0.01μF (2), 0.1μF (2), 0.33μF (1), 1μF (1), 22μF (1)
- LM7805 voltage regulator (1)
- Low-current LEDs (15)
- Generic LED (1)

A Binary Counter

At the heart of every electronic dice circuit that I have ever seen is a counter chip of some kind. Often it's a *decade counter* with 10 "decoded" output pins that are energized one at a time, in sequence. A die only has six faces, but if you tie the seventh pin of the counter back to its reset pin, the counter will restart after it reaches six. (Note that "dice" is really a plural word that should only be used for two dice or more, while "die" is the singular word, although many people don't realize this.)

I always like to do things differently, so I decided not to use a decade counter, partly because I wanted a binary counter to satisfy my desire to demonstrate binary code. This added a little complexity to the circuit, but will enrich the learning process—and when all is said and done, you'll have a circuit that runs two dice (not just one) with a modest chip count, while still fitting on a breadboard.

The counter chip that I've chosen is widely used: the 74HC393. It actually contains two counters, but the second one can be ignored for the time being. The pinouts are shown in Figure 4-137.

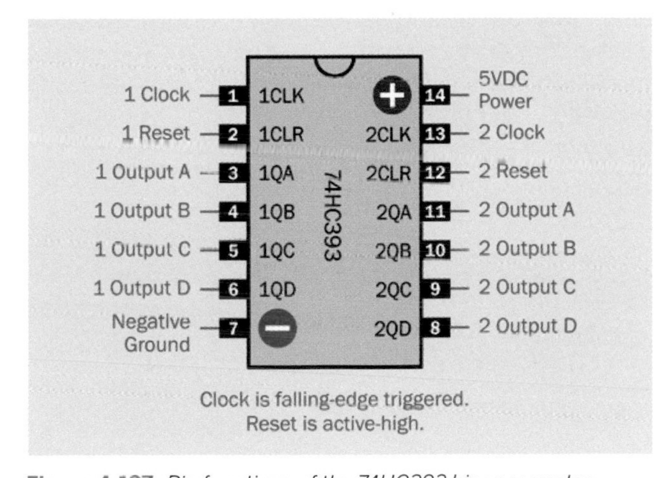

Figure 4-137 *Pin functions of the 74HC393 binary counter.*

Manufacturers have an odd habit of identifying the pin functions of digital chips using as few letters as possible. These cryptic abbreviations can be difficult to understand. To give you an example, in Figure 4-137 the pin labels inside the outline of the chip are the ones I found in a datasheet from Texas Instruments. (To make things even more confusing, other manufacturers use different abbreviations of their own. There is no standardization.)

Outside the outline of the counter I have restated the pin functions in relatively plain English, in green lettering. The number preceding each pin function refers to counter #1 or counter #2, packaged separately in the chip.

Counter Testing

The best way to understand this chip is to bench-test it. Figure 4-138 shows the schematic, and Figure 4-139 shows it on a breadboard. Figure 4-140 shows the breadboarded component values.

Remember:

- This is a 5V logic chip. Don't leave out the voltage regulator.
- Note there is a 0.1µF capacitor between the power-supply pin of the timer, and ground. This is to suppress little voltage spikes that the timer tends to generate. They can confuse the counter if they are uncontrolled.

The capacitor and resistors that I have specified with the timer will run at about 0.75Hz. In other words, the beginning of one pulse, and the beginning of the next pulse, will be slightly more than one second apart. You can see this by watching the yellow LED on the timer output. (If the yellow LED does not behave this way, you made a wiring error somewhere.)

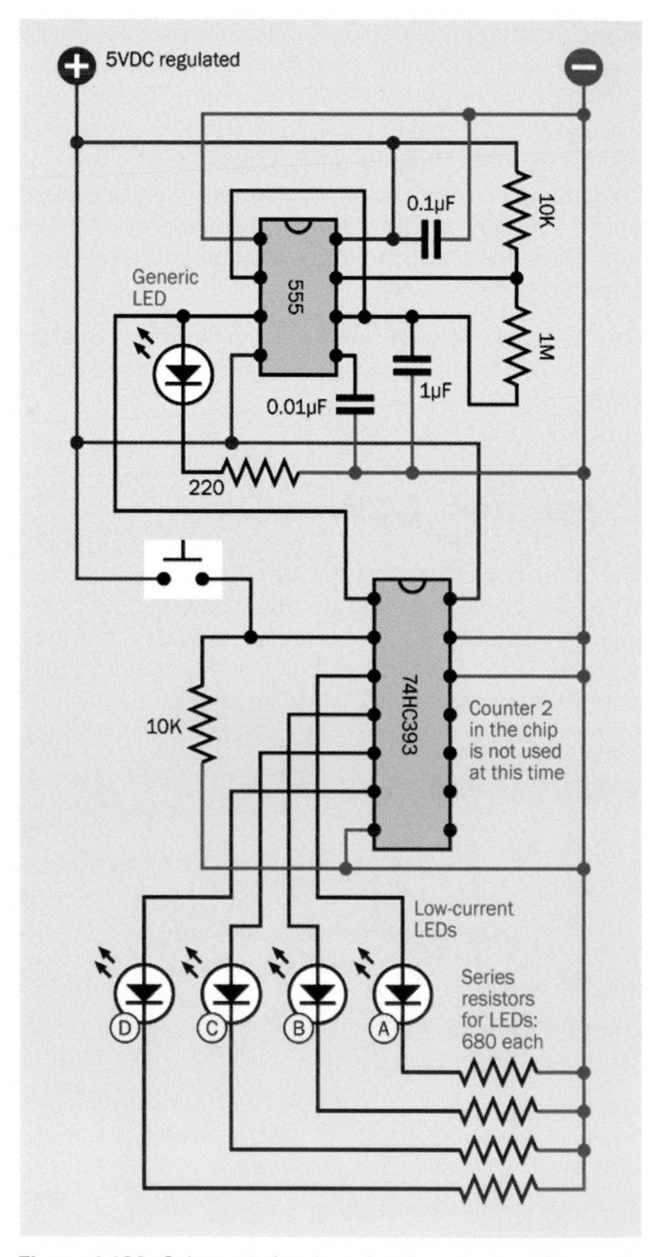

Figure 4-138 *Schematic for observing the output and the reset function of the 74HC393 decade counter.*

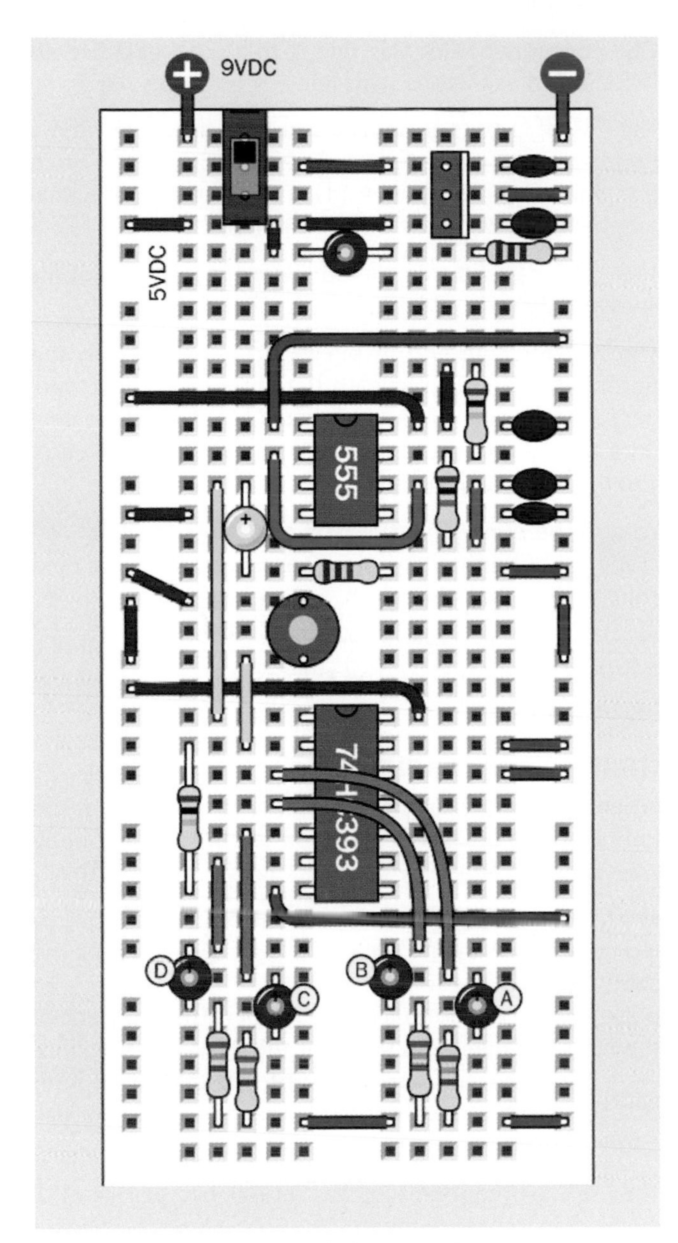

Figure 4-139 *The breadboarded test circuit.*

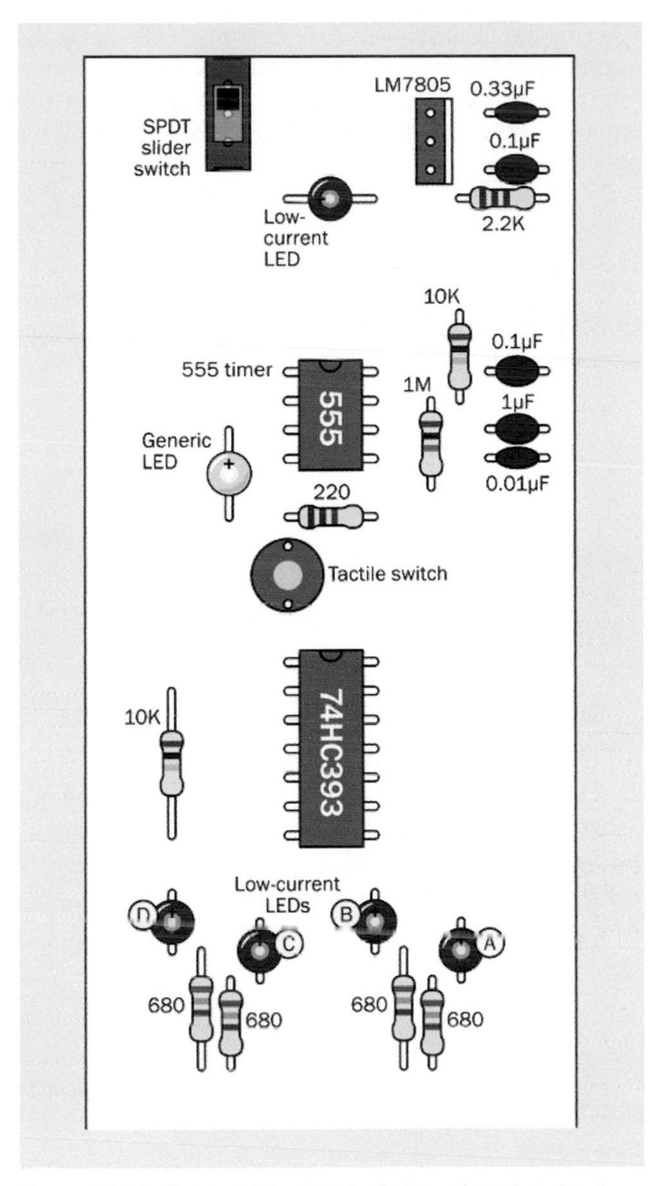

Figure 4-140 *Component values in the breadboarded circuit.*

The four red LEDs labeled A, B, C, and D will display the output states of the counter. If your connections are correct, they will run through the sequence shown in Figure 4-141, where a black circle indicates that an LED is not lit, and a red circle indicates that it is lit.

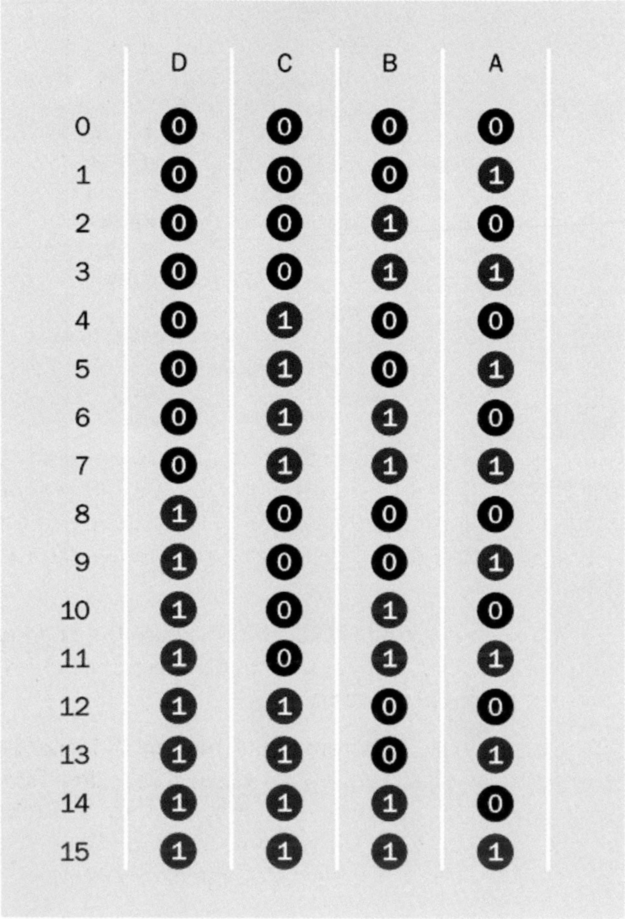

	D	C	B	A
0	0	0	0	0
1	0	0	0	1
2	0	0	1	0
3	0	0	1	1
4	0	1	0	0
5	0	1	0	1
6	0	1	1	0
7	0	1	1	1
8	1	0	0	0
9	1	0	0	1
10	1	0	1	0
11	1	0	1	1
12	1	1	0	0
13	1	1	0	1
14	1	1	1	0
15	1	1	1	1

Figure 4-141 *The full sequence of outputs from a binary counter.*

I'm going to tell you some more, now, about binary and decimal arithmetic. Do you really need to know this? Yes, it's useful. A variety of chips such as decoders, encoders, multiplexers, and shift registers use binary arithmetic, and of course it is absolutely fundamental in almost every digital computer that has ever been made.

Fundamentals: Binary Code

As you can see in Figure 4-141, every time the LED in column A goes out, the LED in column B reverses its state—from on to off, or off to on. Every time the LED in column B goes out, it reverses the LED in column C, and so on.

One consequence of this rule is that each LED flashes twice as fast as the one to its left.

The row of LEDs represents a *binary number*, meaning a number written in only two digits: 0 and 1, as shown in the white font in Figure 4-141. The equivalent decimal number is shown in the black font on the left.

The LEDs can be considered *binary digits*, commonly known as *bits*.

The rule for counting in binary is very simple. In the rightmost column, start with 0, then add 1—and then, because you can only count in ones and zeroes, the next time you want to add 1, you have to revert to 0 and carry 1 to the next column to the left.

What if the numeral in the next column to the left is already a 1? Change it back to 0 and carry 1 to the next column after that. And so on.

The rightmost LED represents the *least significant bit* of a four-bit binary number. The leftmost LED is showing us the *most significant bit*.

Rising Edge, Falling Edge

When you run the test, notice that each transition of the rightmost red LED (either from on to off, or off to on) always occurs when the yellow LED goes out. Why is this?

Most counters are *edge triggered*, meaning that the rising edge or the falling edge of a high pulse nudges the counter along to the next value in its series, when the pulse is applied to the clock input pin. The behavior of the LEDs clearly shows you that the 74HC393 is falling-edge triggered. In Experiment 19, we used a counter that was rising-edge triggered. The type that you use depends on your application.

The 74HC393 counter also has a reset pin, just like the 4026B chip from Experiment 19.

- Some datasheets describe a reset pin as a "master reset" pin, which may be abbreviated MR.
- Some manufacturers call a "reset" pin a "clear" pin, which may be abbreviated CLR on a datasheet.

Whatever it's called, the reset pin will always have the same end result. It forces all the outputs of the counter to go low—which in this case means 0000 binary.

A reset pin requires a separate pulse. But does the reset occur when the pulse begins, or when it ends?

Let's find out. If you built the circuit carefully, the reset pin is held in a low state through a 10K resistor. But there is also a tactile switch that can connect the reset pin directly to the positive bus. This overwhelms the 10K resistor, and forces the reset pin to go high.

As soon as you press the tactile switch, all the outputs go dark, and they stay dark until you let go of the tactile switch. Evidently, the reset function of the 74HC393 is triggered and held by a high state.

The Modulus

Switch off the power, disconnect the pullup resistor and the tactile switch from the reset pin (pin 2), and substitute a wire as shown in Figure 4-142. All the previous connections have been grayed out. The new wire, in solid black, connects the fourth digit, from output D, to the reset pin. Figure 4-143 shows the revision on the breadboard; the new connection is colored green.

What do you think will happen?

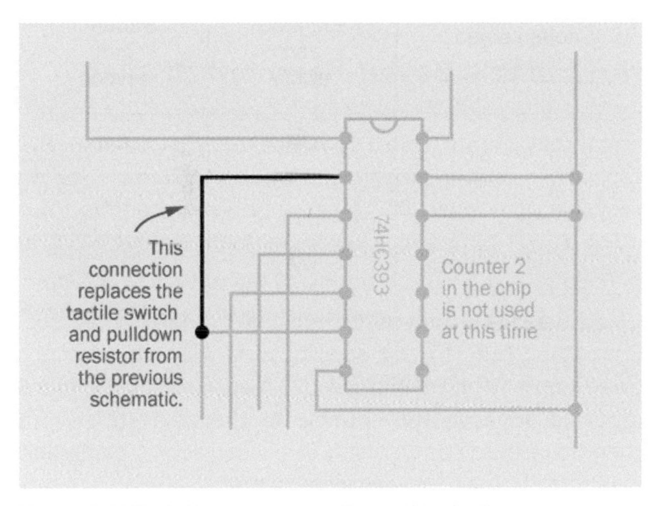

Figure 4-142 *Adding an automatic reset to the timer.*

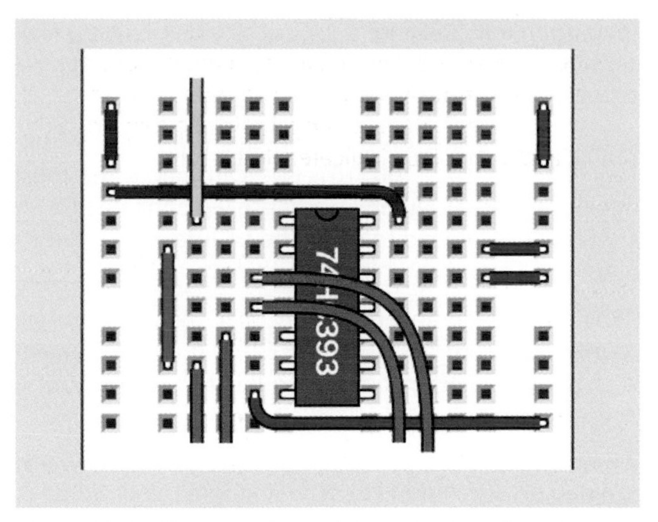

Figure 4-143 *Close-up of the revision on the breadboard. The pulldown resistor, the tactile switch, and associated connections have been removed. The green wire has been inserted.*

Run the counter again. It counts from 0000 up to 0111. The very next binary output should be 1000, but as soon as the fourth digit transitions from 0 to 1, the high state is sensed by the reset pin, forcing the counter back to 0000.

Can you see the leftmost LED flicker, before the counter resets? I doubt it, because the counter responds in less than a millionth of a second.

The counter now runs from 0000 through 0111 before automatically repeating itself. Because counting from 0000 through 0111 binary is equivalent to counting from 0 to 7 in decimal, we now have a *divide-by-8* counter. (Previously, it was a divide-by-16 counter.)

Suppose you move your reset wire from the fourth digit to the third digit. Now you have a divide-by-4 counter.

- You can easily wire almost any 4-bit binary counter so that it resets after 2, 4, or 8 input pulses.

The number of states in the output of a counter, before it repeats, is known as the *modulus*, often abbreviated as "mod." A mod-8 counter repeats after eight pulses (which are numbered from 0 through 7).

Converting to Modulus 6

What about the project that we are supposed to be working on here, to generate electronic dice patterns?

I'm getting to it. Because a die has six sides, I have a feeling that we may need to rewire the counter so that it repeats after six states.

In binary code, the output sequence will look like this: 000, 001, 010, 011, 100, 101. (We can ignore the most significant bit, in column D, because we don't need it for just six states.) I need to make the counter reset after its output of 5 decimal, which is 101 binary.

(Why 5 decimal, and not 6 decimal? Because we are counting upward from 0. It would be more convenient in this project if the counter would start from 1, but it doesn't do that.)

What's the next output after 101 binary? The answer is 110 binary.

Is there something distinctive about 110? If you study the sequence, you'll see that 110 is the first in the series that begins with two high bits.

How can we tell the counter, "When you have a 1 in column B, and a 1 in column C, reset to 0000?" The word "and" in the last sentence should give you a clue. An AND gate has a high output when, and only when, its two inputs are high. Just what we need.

Can we drop it right in? Absolutely, because all the members of the 74HCxx chip family are designed to talk to each other. In Figure 4-144 you will see that I have added an AND gate. Of course, to use it on your breadboard, you'll have to add an appropriate chip, which is a 74HC08. It contains four AND gates, of which we only need one. Therefore, in addition to power connections, its unused inputs must be grounded. This is a bit of a hassle, but I'll show you how to do it, as soon as I make a few more additions and modifications. (The unused outputs must be left unconnected.)

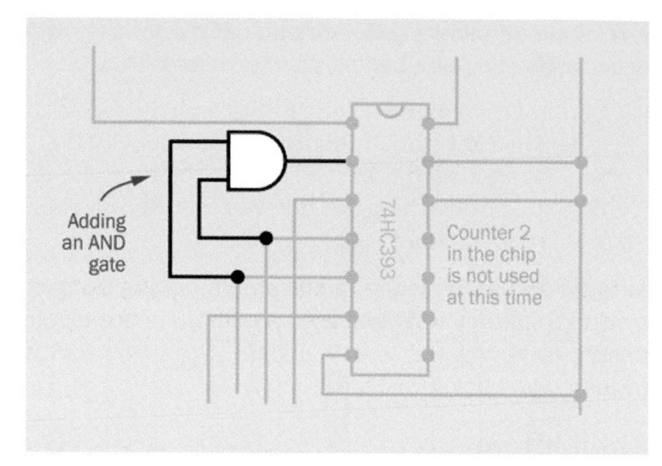

Figure 4-144 *An AND gate has been added, to adjust the counter so that it only cycles through six output states instead of its usual 16.*

Meanwhile, please remember this take-home message:

- You can use logic chip(s) with a counter to change the modulus of the counter by looking for a distinctive pattern in the output states, and feeding back a signal to the reset pin.

Not Using a Seven-Segment Display

For the dice display, I could just use a seven-segment numeral that counts from 1 to 6. But this is a problem, because the counter runs from 0 through 5. I don't see an easy way to convert 000 binary to a seven-segment numeral 1, 001 binary to a seven-segment numeral 2, and so on.

Could we make the counter skip 000 binary somehow? Er, maybe, but I'm not sure how. Perhaps by using a three-input OR gate that would feed back to the clock input to advance the counter to the next state—but then this would conflict with the usual clock signals, and it all sounds like a lot of trouble to me.

In any case, I'm not very excited by using a seven-segment numeral in this project, because it's not visually appealing. Why not use LEDs that emulate the pattern of spots on an actual die? The sequence is shown in Figure 4-145.

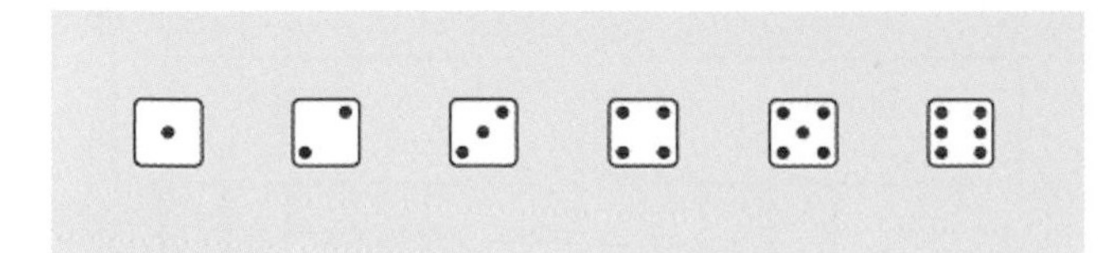

Figure 4-145 *The series of spot patterns to be reproduced by LEDs.*

Can you figure out a way to convert the binary output from the counter to illuminate the LEDs in those patterns?

Choosing the Gates

I'll start with the easiest option. If I connect output A from the counter (see Figure 4-138) to the LED that represents the center spot of the die, that will work out well, because the center spot is only illuminated for patterns 1, 3, and 5, and goes out for patterns 2, 4, and 6. This is exactly the way that output A behaves.

Then things get a little tricky. I need to illuminate a diagonal pair of spots for patterns 4, 5, and 6, and the other diagonal pair for 2, 3, 4, 5, and 6. But how?

Figure 4-146 shows my answer to the problem. You'll see I have added a couple more logic gates: a three-input NOR, and a two-input OR. Alongside I have shown the sequence of binary numbers, and the spot patterns that each number creates on the die.

To make things work, I had to begin with the die showing a 6 pattern when the counter starts at 000 binary. The sequence of patterns really doesn't matter, so long as they are all represented. They're going to be selected at random anyway.

Figure 4-147 shows how the counter outputs light up the different patterns of spots. And just in case this still isn't entirely clear, I have created a snapshot sequence showing high and low states in the circuit as the counter increments from 000 through 101. The snapshots are shown in Figure 4-148, Figure 4-149, and Figure 4-150.

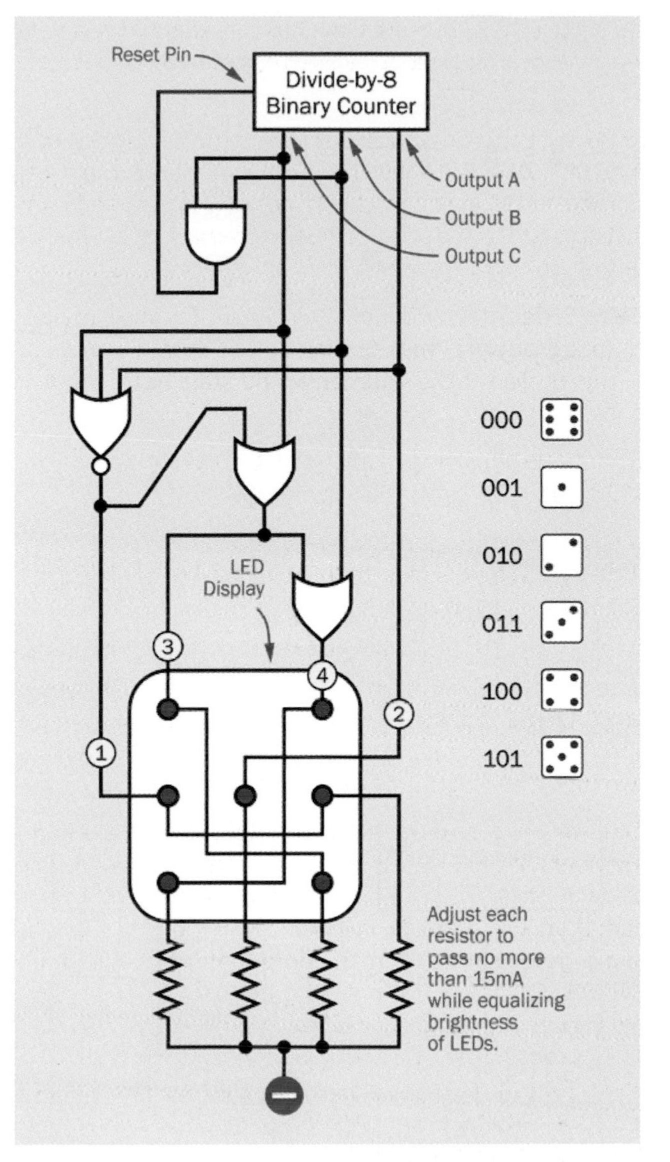

Figure 4-146 *A logic network to create the sequence of spots on a die.*

I squeezed the snapshots so that I could fit two into the width of a column, and I omitted the AND gate, because it doesn't do anything during the counting range of 000 through 101. It responds only when the counter tries to advance to 110—at which point, the AND sets it back to 000.

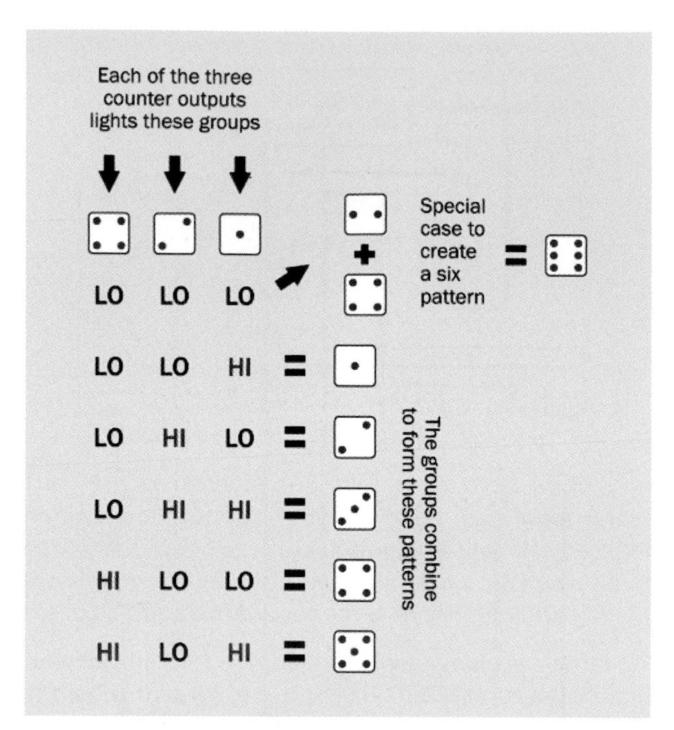

Figure 4-147 *How the outputs from the binary counter are used to light the patterns of spots.*

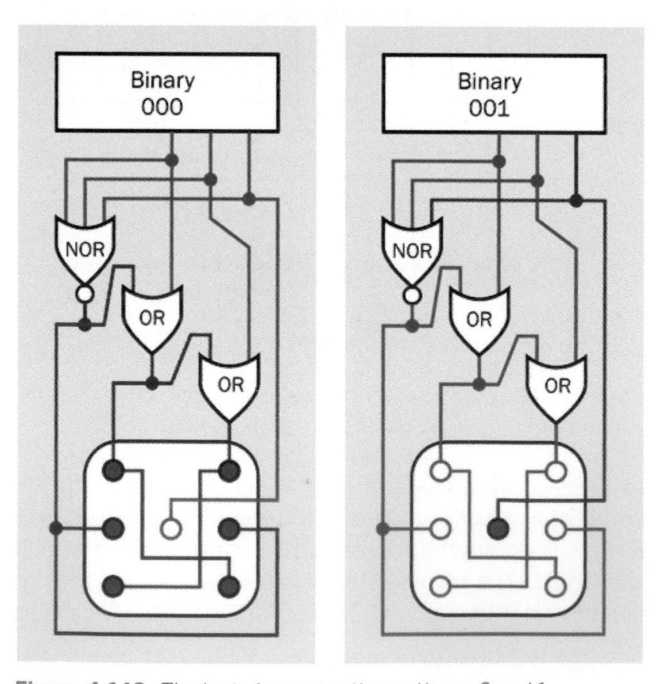

Figure 4-148 *The logic for generating patterns 6 and 1.*

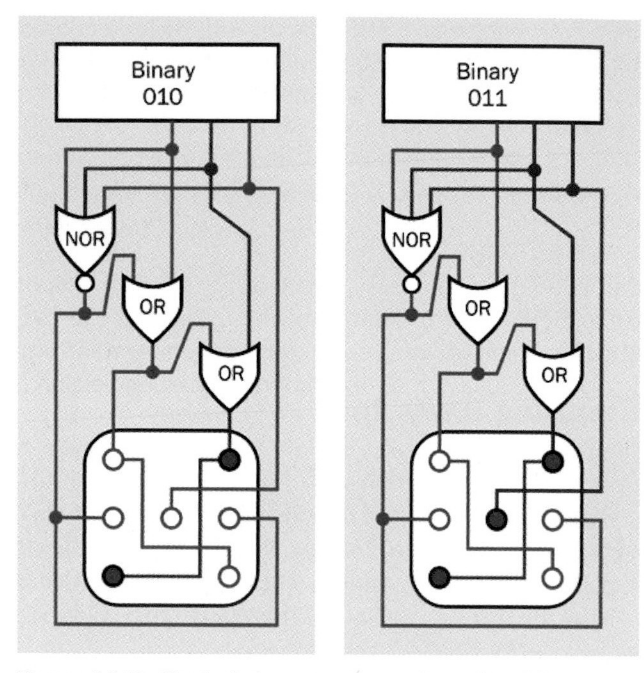

Figure 4-149 *The logic for generating patterns 2 and 3.*

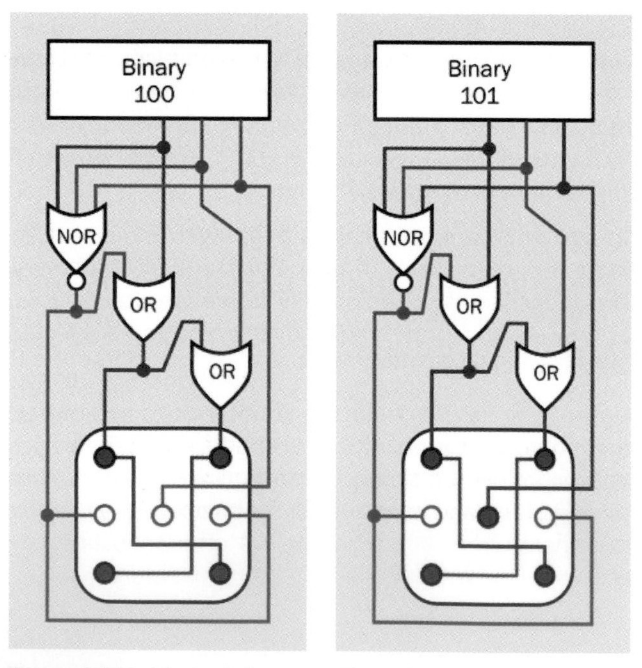

Figure 4-150 *The logic for generating patterns 4 and 5.*

If you're wondering how I came up with this selection of logic gates to translate the counter output into the dice patterns, I'm not sure I know how to tell you. There's a certain amount of trial and error, and some intuitive guesses involved in creating this kind of logic diagram. At least, that's the way it is for me. There are more rigorous and formal ways of doing it, but personally I don't find them so easy.

The Finished Circuit

The schematic in Figure 4-151 was derived from the logic diagram in Figure 4-146. The breadboarded version is shown in Figure 4-152.

The component values are shown in Figure 4-153. Note that I changed the timing resistor and timing capacitor for the 555 timer, so that it now runs around 5kHz. The idea of this circuit is that you will stop the timer at an arbitrary moment, after it has gone through hundreds of cycles. This way, you should end up with a random number.

I have added a switchable 22µF capacitor that can make the timer run slowly (about 2Hz) if you want to demonstrate how the counting works, to anyone who is skeptical.

I haven't bothered to show the values for the lower half of the breadboard, because the only components there are the chips. That's a nice aspect of building circuits based on logic: you don't have to worry about squeezing in resistors and capacitors. Chips and wires do most of the work.

The numbered outputs at the bottom of the circuit in Figure 4-151 and Figure 4-152 correspond with the inputs to the pattern of LEDs shown in Figure 4-154. There isn't room on the breadboard to add the LEDs, so you will either need a second breadboard, or you can drill some holes to mount the LEDs in a piece of plywood or plastic.

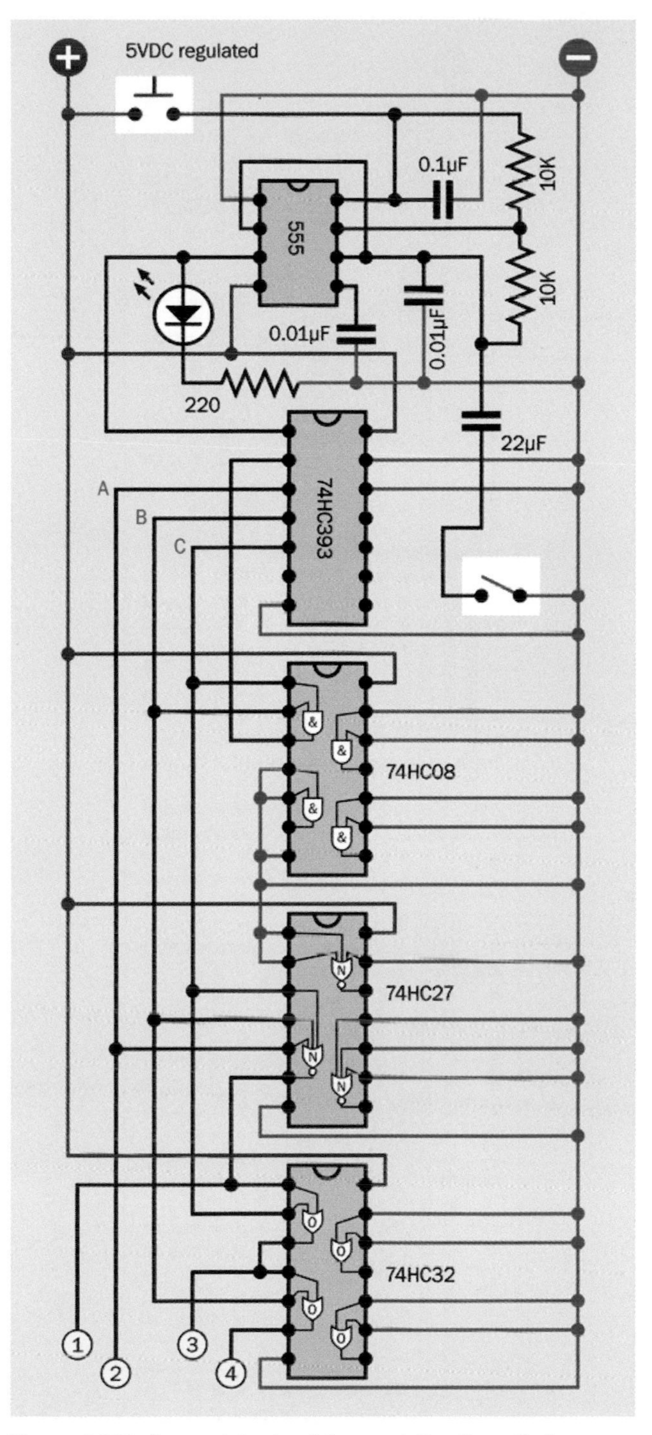

Figure 4-151 *A complete circuit for emulating the roll of a single die.*

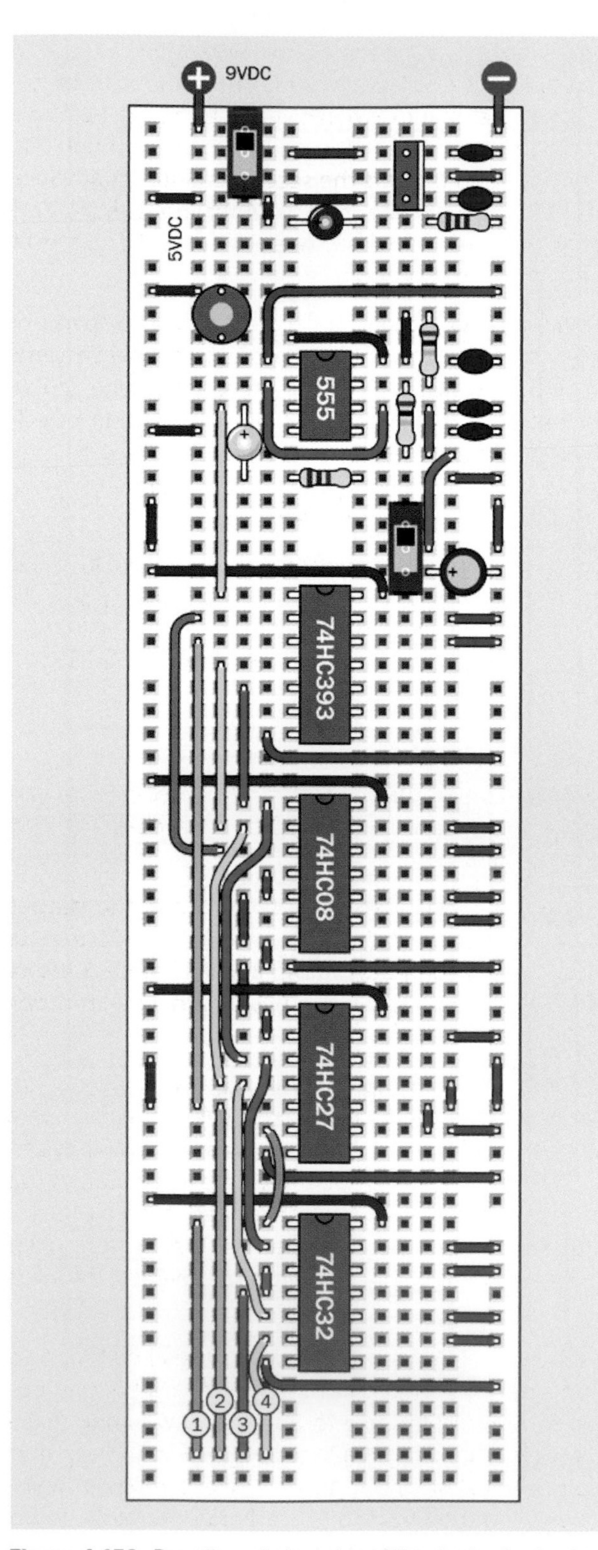

Figure 4-152 *Breadboarded version of the single-die circuit.*

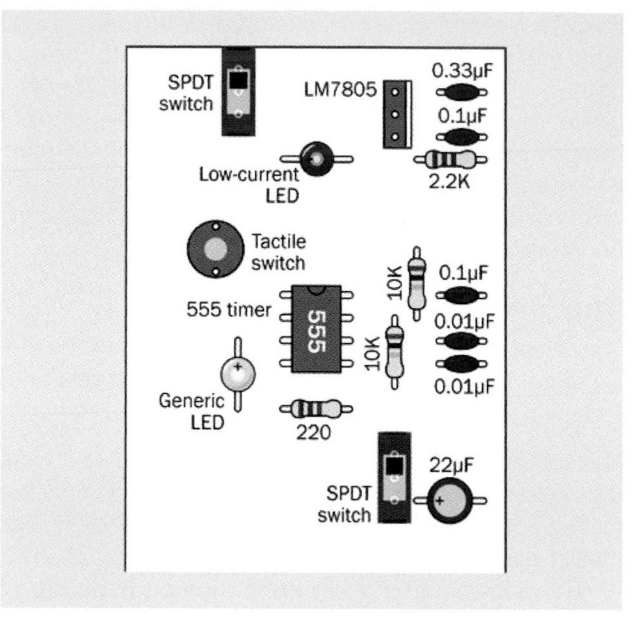

Figure 4-153 *Values for the components in the control section of the dice simulation.*

Three pairs of the LEDs are wired in series, because a logic chip isn't powerful enough to drive a pair of LEDs in parallel. Putting them in series will require you to use a lower-value resistor than usual. The way to do this is to apply 5VDC to one of the pairs of LEDs through your meter, set to measure milliamps. Try a 220-ohm series resistor, and see how much current you measure. If you aim for a maximum of 15mA, that will be within the specifications of HC chip outputs. You may need a resistor with a value of 150 ohms or 100 ohms, depending on the characteristics of the LEDs that you are using.

Finally apply 5VDC through a 330-ohm resistor to the center LED, and compare its brightness to the LEDs wired in pairs. You may have to increase the resistor value to make the center LED visually equivalent to the others.

Hook up the LEDs to the logic circuit, hold down the button, release it, and you have your dice value (or, to be semantically correct, your die value).

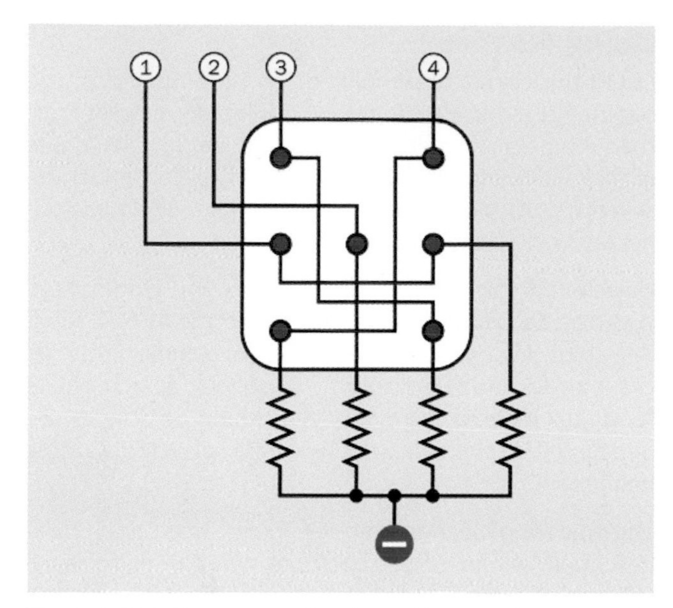

Figure 4-154 *Wiring seven LEDs (six of them in pairs, in series) to display the spot patterns of a single die.*

How do you know that it gives you a really random result? Really, the only way to be sure is to use it repeatedly and note how many times each number comes up. You might need to run it maybe 1,000 times to get decent verification. Because the circuit relies on the behavior of a human being pressing a button, there is no way to automate the testing process. All I can say is, the result really *should* be random.

Good News

There are more chips in this circuit than you've used in previous circuits in the book, but in the immortal words of Professor Farnsworth, in one of my favorite TV shows, *Futurama*: "Good news, everyone!"

The good news is that you can modify this circuit to simulate two dice instead of one, simply by adding some more wires and LEDs. You don't need any more chips at all.

We have a lot of unused logic gates in the AND, NOR, and OR chips. Three ANDs are left over, two NORs, and two ORs. In addition, there is also an entirely separate counter in the 74HC393 chip. This is exactly what you need.

The question is, how to create a second series of random numbers, different from the first. Maybe add another 555 timer, running at a different speed?

I don't like this idea, because the two timers will go in and out of phase with each other, and some pairs of values may come up more than others. I think it would be a better idea if the first counter runs from 000 through 101 binary, and then triggers the second counter to advance from 000 to 001. The first counter goes from 000 through 101 again, and triggers the second counter to move on to 010. And so on.

The second counter will run at one-sixth the speed of the first, but if you drive them fast enough, the patterns will still be too fast to see. The great advantage of this arrangement is that every possible combination of values will be displayed an equal number of times, so they should all have an almost-equal chance of coming up, just as if you were using two real dice.

Why do I say "almost equal"? Because, bear in mind there is a tiny delay when a counter resets itself from 101 binary to 000 binary. But if the lefthand counter is running at around 5kHz, a delay of less than one-millionth of a second seems trivial.

Chained Counters

The last remaining question is how the first counter can advance the second counter when it reaches 101 and flips back to 000.

This is easy. Consider what happens when the output from the first counter changes from 011 to 101 to 110, the last value lasting only for an instant before it resets to 000. After the C output reaches a high state, it drops low.

What does the clock input of the second counter need, to make it advance by one count? You already know that. It needs a high state that drops low. All you have to do is to connect the C output, from the first timer, to the clock input of the second timer. Indeed, the chip is designed to work this way, so that the falling state from one timer acts as a "carry" signal to advance the next timer.

The schematic in Figure 4-155 shows the circuit for two dice. I'm not including another breadboard picture, because you should be able to add the new wiring yourself. It is almost exactly a mirror image of the wiring that you already installed, but don't forget to move it down by one row of holes on the breadboard, to allow room for the positive power supply to each chip.

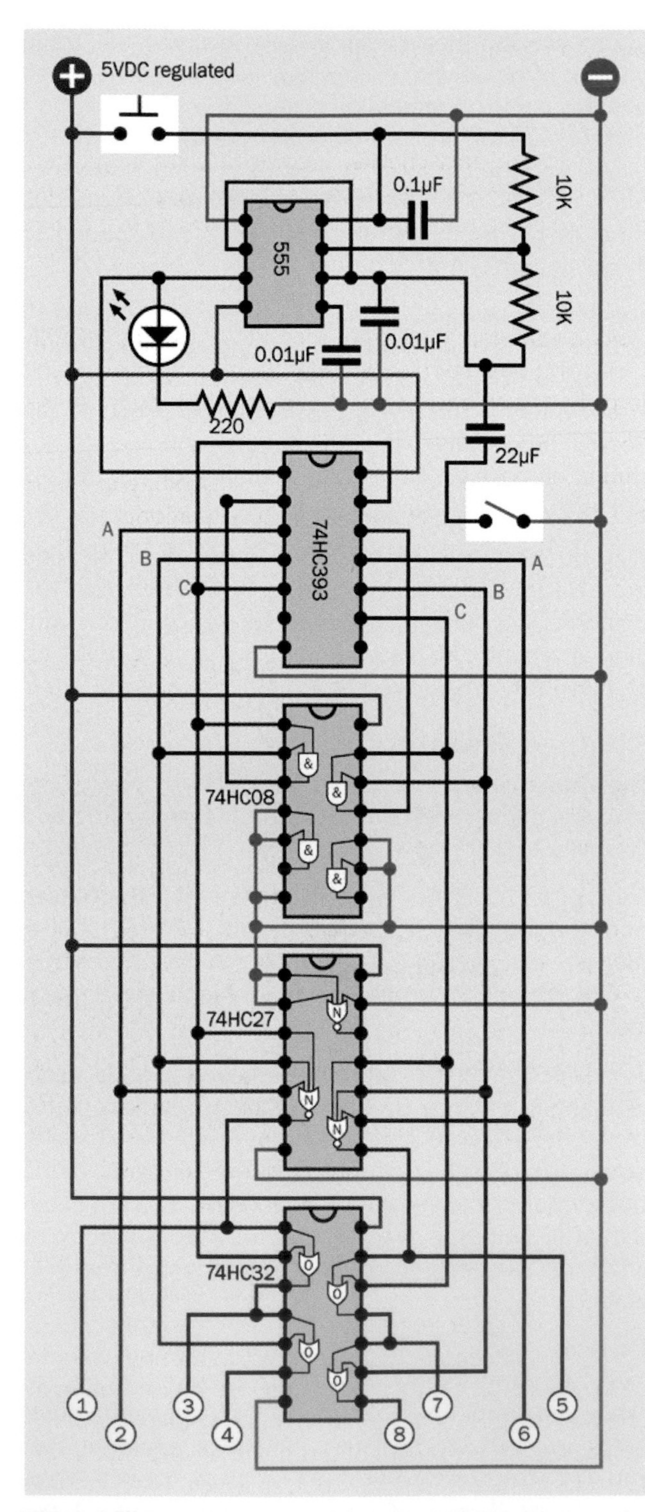

Figure 4-155 *Complete circuit to run two LED dice.*

Going Further

Could the circuit be simplified? As I mentioned at the beginning, a decade counter would require simpler logic than a binary counter. You wouldn't need an AND gate to make it count with a modulus of six, because just the seventh output pin on the decade counter could be connected back to the reset.

However, if you want to run two dice, you would need two decade counters, and that would mean two separate chips. Also, you would still need two chips to handle the logic for the two displays. To see why, search online for digital dice. At this point, you should be able to understand the schematics that will be displayed in Google Images.

The only simplification I can see, in the circuit that I have described, would be to substitute two diodes for each OR gate. The circuits that you'll find online often do this, but you will end up with a signal passing through two diodes in succession, which will drop the voltage below a level that I feel is acceptable.

The Slowdown Problem

The version of this project that I included in the first edition of *Make: Electronics* included a nice extra feature. When you took your finger off the "run" button, the die pattern display gradually slowed before it stopped. This increased the suspense of waiting to see what the final number would be.

The feature was enabled by splitting the power supply to the 555 timer. The timer was "always on" but voltage to its RC network was shut off when the player stopped pressing the "run" button. At that point, a large capacitor slowly discharged into the network, and the timer slowed as the voltage diminished.

A reader named Jasmin Patry sent email telling me that when he used the circuit, the value 1 came up disproportionately often, and he suspected that this had something to do with the slowdown feature.

Jasmin turned out to be a video game designer who understands much more about randomicity than I do. He had the polite, patient style of a man who really knew what he was talking about, and he seemed interested in helping to fix the problem that he had identified.

After he sent me graphs showing the relative frequency of each number in the simulation, I had to agree that the problem existed. I suggested many possible explanations, all of which turned out to be wrong. In the end Jasmin successfully proved that the lower power consumption of a single LED, relative to the higher power consumption of six LEDs, allowed the timer to keep running for a little longer when the voltage was marginal. This increased the odds that it would stop during that period.

Eventually Jasmin suggested a substitute circuit, in which a second 555 timer was added, and the outputs from the two timers were merged through an XOR gate. He successfully proved that this eliminated the bias toward one number. I was delighted that one of my readers had learned so much from reading my book, he was able to identify and fix a mistake that he found in it.

In this new edition, I have omitted the slowdown capacitor that caused the trouble in the first edition. But I haven't adopted Jasmin's circuit, because it was quite complicated. A single die would have needed its own pair of 555 timers, plus the XOR gate that he wanted. He also used diodes that I would have replaced with OR gates, and there wasn't nearly enough room on the breadboard.

With his permission, I will send his circuit as a freebie to anyone who registers with me (using the procedure described in the Preface—see "Me Informing You" on page 10). I cannot easily reprint it here, because I would have to redraw it completely to fit the two-column format.

Slowdown Alternatives

You'd think there must be a simpler way to make the display slow down, without affecting the randomicity. When I looked online, I found that someone had used an NPN transistor with its emitter connected to pin 7 of the timer, and a capacitor between its base and its collector, so that when power was disconnected, the transistor output would gradually diminish. Several other people had done the same thing in their dice circuits. However, I suspect that this configuration may be subject to the same problems that Jasmin found.

I have also seen circuits using exactly the same slowdown capacitor configuration that I used (for example, on the Doctronics website). I think they are almost certainly susceptible to the problem that I have described.

My final answer on this topic may be dissatisfying: I don't know how to achieve a slowing effect without adding more components that will greatly complicate the circuit.

However—just before the text of this book was finalized, my friend and fact-checker Fredrik Jansson suggested powering the 555 timer from a separate voltage regulator, to isolate it from power fluctuations in the rest of the circuit. I like this idea, but there wasn't time to try it before the book went into production.

I built a completely different dice circuit using a PICAXE microcontroller, but discovered that this had its own randomicity issues because of the imperfect random-number generator built into the chip.

In Experiment 34 (the last one in this book), you'll find that I have created yet another dice simulation, using an Arduino. But here again, I had to depend on the built-in random-number generator, and I am skeptical that it creates an evenly distributed range of numbers.

Figure 4-156 *This electronic dice display uses 10mm LEDs embedded in a box of sanded polycarbonate plastic.*

The issue of randomness is really not simple at all. I became so interested in it after my emails with Jasmin Patry, I explored it at length in *Make: More Electronics*, and also wrote a column about it in *Make:* magazine (volume 45) in collaboration with Aaron Logue, who runs his own little website describing projects that he builds. He introduced me to the concept of using a reverse-biased transistor to generate random noise, which is then processed with a clever algorithm attributed to the great computer scientist, John von Neumann. This, I think, is as close to a perfect random-number generator as you can get—but the chip count is not trivial.

All of these enhancements go beyond the range of an introductory-level book. If any readers have a *really* simple enhancement to the dice circuit presented here, to add the slowdown effect, my email inbox is always open, and yes, I do read the messages.

Meanwhile, I'm including a couple of photographs of finished electronic-dice projects. The one in Figure 4-156 was included in the first edition of this book, in 2009. Figure 4-157 shows one that I built around 1975, after Don Lancaster's amazing *TTL Cookbook* told me how I could use 74xx logic chips. Forty years later, the LEDs still light up randomly. (At least, I think they're random.)

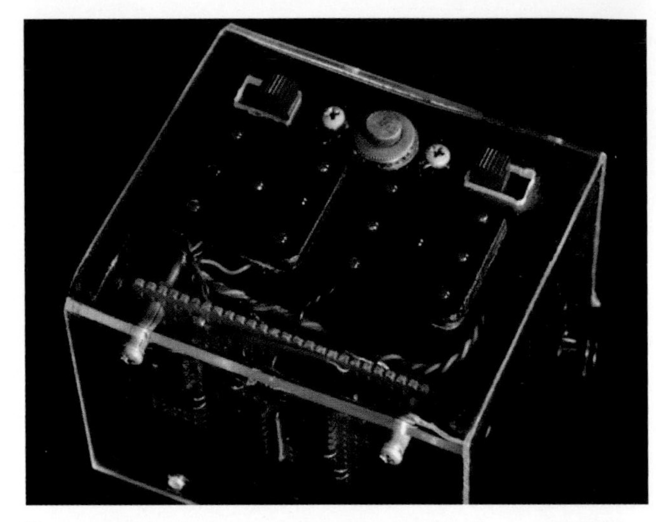

Figure 4-157 *Electronic dice designed and built around 1975, in a box of Lucite and black-painted plywood.*

What Next?

5

At this point, we can branch out in numerous directions. Here are some possibilities:

Audio: This is a large field including hobby projects such as amplifiers and "stomp boxes," to modify guitar sound.

Electromagnetism: This is a topic that I haven't even mentioned yet, but it has some fascinating applications.

Radio-frequency devices: Anything that receives or transmits radio waves, from an ultra-simple AM radio onward.

Programmable microcontrollers: These are tiny computers on a single chip. You write a little program on your desktop computer, and load it into the chip. The program tells the chip to follow a sequence of procedures, such as receiving input from a sensor, waiting for a fixed period, and sending output to a motor. Popular controllers include the Arduino, PICAXE, BASIC Stamp, and many more.

I don't have space to develop all of these topics fully, so what I'm going to do is introduce you to them by describing just a few projects in each category. You can decide which interests you the most, and then proceed beyond this book by reading other guides that specialize in that interest.

I'm also going to make some suggestions about setting up a productive work area, reading relevant books, catalogs, and other printed sources, and generally proceeding further into hobby electronics.

Tools, Equipment, Components, and Supplies

No additional tools or equipment are needed for this final chapter of the book. For a summary of all the components, see Figure 6-8. See "Supplies" on page 316 for a list of additional supplies (primarily wire, for the coils in Experiments 25, 26, 28, 29, and 31).

Customizing Your Work Area

At this point, if you're getting hooked on the fun of creating hardware but haven't allocated a permanent corner to your new hobby, I have some suggestions. Having tried many different options over the years, my main piece of advice is this: don't build a workbench!

Many hobby electronics books want you to go shopping for 2 × 4s and plywood, as if a workbench has to be custom-fabricated to satisfy strict criteria about size and shape. I find this puzzling. To me, the size and shape is not very important. I think the most important issue is storage.

I want tools and parts to be easily accessible, whether they're tiny transistors or big spools of wire. I certainly don't want to go digging around on shelves that require me to get up and walk across the room.

This leads me to two conclusions:

- You need storage around the workbench.
- You need storage below the workbench.

Many DIY workbench projects allow little or no storage underneath. Or, they suggest open shelves, which will be vulnerable to dust. My minimum configuration would be a pair of two-drawer file cabinets with a slab of 3/4-inch plywood or a Formica-clad kitchen countertop placed across them. File cabinets are ideal for storing all kinds of objects, not just files, and you can often find cheap ones at yard sales and thrift stores.

Of all the workbenches I've used, the one I liked best was an old-fashioned steel office desk—the kind of monster that dates back to the 1950s. They're difficult to move (because of their weight) and don't look beautiful, but you can buy them cheaply from used office furniture dealers, they're generous in size, they withstand abuse, and they last forever. The drawers are deep and usually slide in and out smoothly, like good file-cabinet drawers. Best of all, the desk has so much steel in it that you can use it to ground yourself before touching components that are sensitive to static electricity. If you use an anti-static wrist strap, you can simply attach it to a sheet-metal screw that you drive into one corner of the desk.

What will you put in the deep drawers of your desk or file cabinets? Some paperwork may be useful, perhaps including the following documents:

- Product datasheets
- Parts catalogs
- Sketches and plans that you draw yourself

The remaining capacity of each drawer can be filled with plastic storage boxes. The boxes can contain tools that you don't use so often (such as a heat gun or a high-capacity soldering iron), and larger-sized components (such as loudspeakers, AC adapters, project boxes, and circuit boards). You should look for storage boxes that measure around 11 inches long, 8 inches wide, and 5 inches deep, with straight sides. Boxes that you can buy at Walmart will be cheaper, but they often have tapering sides, which are not space-efficient.

The boxes that I like best are Akro-Grids, made by Akro-Mils (see Figure 5-1 and Figure 5-2). These are very rugged, with optional transparent snap-on lids. Perspective in the photographs make the boxes look as if they taper downward, but they don't. You can download the full Akro-Mills catalog online and then search online for retail suppliers. You'll find that Akro-Mils also sells an incredible variety of parts bins, but I don't like open bins because their contents are vulnerable to dust and dirt.

Figure 5-1 *Akro-Grid boxes contain grooves allowing them to be partitioned into numerous compartments for convenient parts storage. The height of the box in this photograph allows three to be stacked in a typical file-cabinet drawer.*

Figure 5-2 *Lids are sold separately for Akro-Grid boxes to keep the contents dust-free. The taller box shown here allows two to be stacked in a file-cabinet drawer.*

For medium-size components, such as potentiometers, power connectors, control knobs, and toggle switches, I like storage containers measuring about 11 inches long, 8 inches wide, and 2 inches deep, divided into four to six sections. You can buy these from Michaels (the craft store), but I prefer to shop online for the Plano brand, as they seem more durably constructed. The Plano products that are most suitable for medium-size electronic parts are classified as fishing-tackle boxes.

For undivided, flat-format storage boxes, the Prolatch 23600-00 is ideally sized to fit a file-cabinet drawer, and the latches are sufficiently secure that you could stack a series of them on their long edges. See Figure 5-3.

Figure 5-3 *This Plano brand box is undivided, making it useful for storing spools of wire or medium-size tools. When stacked upright on its long edge, three will fit precisely in a file-cabinet drawer.*

Plano also sells some really nicely designed toolboxes, one of which you can place on your desktop. It will have small drawers for easy access to screwdrivers, pliers, and other basics. Because you need a work area that's only about three feet square for most electronics projects, surrendering some desk space to a toolbox is not a big sacrifice.

If you have a steel desk with relatively shallow drawers, one of them can be allocated for printed catalogs. Don't underrate the usefulness of hardcopy, just because you can buy everything online. The Mouser catalog, for instance, has an index, which is more useful in some respects than their online search feature, and the catalog is divided into helpful categories. Many times I've found useful parts that I never knew existed, just by browsing, which is much quicker than flipping through PDF pages online, even with a broadband connection. Currently, Mouser is still quite generous about sending out their catalogs, which contain over 2,000 pages. McMaster-Carr will also send you a catalog, but only after you've ordered from them, and only once a year. It is probably the most comprehensive, most wonderful catalog of tools and hardware in the world.

Now, the big question: how to store all the dinky little parts, such as resistors, capacitors, and chips? I've tried various solutions to this problem. The most obvious is to buy a case of small drawers, each of which is removable, so you can place it on your desk while you access its contents. But I don't like this system, for two reasons. First, for very small components, you need to subdivide the drawers, and the dividers are never secure. And second, the removability of the drawers creates the risk of accidentally emptying the contents on the floor. Maybe you're too careful to allow this to happen, but I'm not. In fact, on one occasion I tipped the entire case of drawers onto the floor.

My personal preference is to use Darice Mini-Storage boxes, shown in Figure 5-4. You can find these at Michaels in small quantities, or buy them more economically in bulk online—just search for:

```
darice mini storage box
```

The blue boxes are subdivided into five compartments that are exactly the right size and shape for resistors. The yellow boxes are subdivided into 10 compartments, which are ideal for semiconductors. The purple boxes aren't divided at all, and the red boxes have a mix of divisions. All the boxes share the same stock number: 2505-12.

The dividers are molded into the boxes, so you don't have the annoyance associated with removable dividers that slip out of position, allowing components to mix together. The box lids fit tightly, so that even if you drop one of the boxes, it probably won't open. The lids have metal hinges, and a ridge around the edge that makes the boxes securely stackable.

Figure 5-4 *Darice Mini-Storage boxes are ideal for components such as resistors, capacitors, and semiconductors. The boxes can be stacked stably, stored on shelves, or grouped in larger boxes. The brand sticker is easily removed after being warmed with a heat gun.*

After considerable searching, I found cheap plastic bins, with lids, measuring about 8 inches by 13 inches by 5 inches deep. Each bin will hold nine Darice parts boxes. The bins can then be categorized and stored on shelves.

Labeling

No matter which way you choose to store your parts, labeling them is essential. Any ink-jet printer will produce neat-looking labels, and if you use peelable (nonpermanent) labels, you'll be able to reorganize your parts in the future, as always seems to become necessary. I use color-coded labels for my collection of resistors, so that I can compare the stripes on a resistor with the code on the label, and see immediately if the resistor has been put in the wrong place. See Figure 5-5.

Even more important: you need to place a second (nonadhesive) label inside each compartment with the components. This label tells you the manufacturer's part number and the source, so that reordering is easy. I buy a lot of items from Mouser, and whenever I open their little plastic bags of parts, I snip out the section of the bag that has the identifying label on it, and slide it into the compartment of my parts box before I put the parts on top of it. This saves frustration later.

Figure 5-5 *To check that resistors are not placed in the wrong compartments, print the color code on each label.*

If I were *really* well organized, I would also keep a database on my computer listing everything that I buy, including the date, the source, the type of component, and the quantity. But I'm not that well organized.

On the Bench

Some items are so essential that they should sit on the bench or desktop on a permanent basis. These include your soldering iron(s), Helping Hand with magnifier, desk lamp, breadboard, power strip, and power supply. For a desk lamp, I prefer to use one that contains an LED bulb, for reasons explained in Experiment 14.

A power supply for your projects is a matter of personal preference. If you're serious about electronics, you can buy a unit that delivers properly smoothed current at a variety of properly regulated and calibrated voltages. Your little wall-plug AC adapter cannot do this, and its output may vary depending on how heavily you load it. Still, as you've seen, it is sufficient for basic experiments, and when you're working with logic chips, you need to mount a 5-volt regulator on your breadboard anyway. Overall, I consider a good power supply optional.

Another optional item is an oscilloscope. This will show you, graphically, the electrical fluctuations inside your wires and components, and by applying probes at different points, you can track down errors in your circuit. It's a neat gadget to own, but it will cost a few hundred dollars, and for our tasks so far, it has not been necessary. If you plan to get seriously into audio circuits, an oscilloscope becomes far more important, because you'll want to see the shapes of the waveforms that you generate.

You can try to economize on an oscilloscope by buying a unit that plugs into the USB port of your computer and uses your computer monitor to display the signal. I have tried one of these, and was not entirely happy with the results. It worked, but did not seem accurate or reliable for low-frequency signals. Maybe I was unlucky, but I decided not to try any other brands.

The surface of your desk or workbench will undoubtedly become scarred by random scuffs, cut marks, and drops of molten solder. I use a piece of half-inch plywood, two feet square, to protect my primary work area, and I clamp a miniature vise to its edge. In the past I used to cover the plywood with a square of conductive foam, to reduce the risk of static discharge from me to sensitive components. Over the years, though, I realized that my particular combination of carpet, chair, and shoes does not cause me to suffer from static. This is a matter for you to determine by experience. If you see a tiny little spark sometimes when you touch a metal object, and you feel a little zap of electricity, you need to consider grounding yourself and perhaps using anti-static foam (or a piece of metal) on your working surface.

Inevitably, during your work you'll create a mess. Little pieces of bent wire, stray screws, fasteners, and fragments of stripped insulation tend to accumulate, and can be a liability. If metal parts or fragments get into a project that you're building, they can cause short circuits. So you need a trash container. But it has to be easy to use. I use a full-size garbage pail, because it's so big that I can't miss it when I throw something toward it, and I can never forget that it's there.

Last, but most essential: a computer. Now that all datasheets are available online, and all components can be ordered online, and many sample circuits are placed online by hobbyists and educators, I don't think anyone can work efficiently without quick Internet access. To avoid wasting space, you can place a tower computer on the floor and mount its monitor on the wall—or use a tablet, or a small, cheap laptop that has a minimal footprint.

A possible workbench configuration, using a steel desk, is shown in Figure 5-6. A more space-efficient configuration is suggested in Figure 5-7.

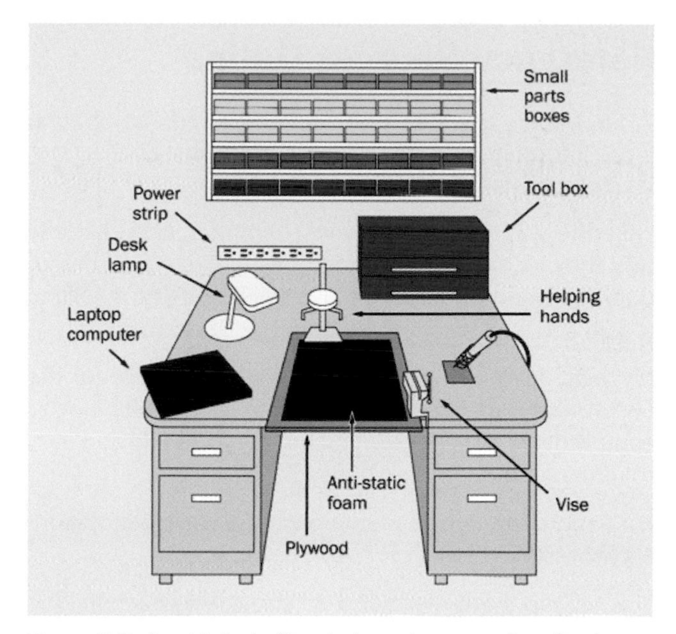

Figure 5-6 *An old steel office desk can be as good as, if not better than, a conventional workbench when building small electronics projects. It provides a large work area and ample storage, and has sufficient mass for you to ground yourself when dealing with components that are sensitive to static electricity.*

Figure 5-7 *For maximum utilization of available space, consider walling yourself in.*

Reference Sources Online

When people ask me to recommend a website offering basic information at entry level, I recommend Doctronics (*http://www.doctronics.co.uk*).

I like the way they draw their schematics, and I like the way they include many illustrations of circuits on breadboards, as I do myself. They also sell kits, if you're willing to pay and wait for shipping from the UK.

My next favorite hobby site is also British-based: the Electronics Club (*http://electronicsclub.info*). It's not as comprehensive as Doctronics, but very friendly and easy to understand.

For a more theory-based approach, try ElectronicsTutorials (*http://www.electronics-tutorials.ws*).

This will go a little farther than the theory sections I've included here.

For an idiosyncratic selection of electronics topics, try Don Lancaster's Guru's Lair (*http://www.tinaja.com*).

Lancaster wrote *The TTL Cookbook* more than 30 years ago, which opened up electronics to at least two generations of hobbyists and experimenters. He knows what he's talking about, and isn't afraid of getting into some fairly ambitious areas such as writing his own PostScript printer drivers and creating his own serial-port connections. You'll find a lot of ideas there.

Books

Yes, you do need books. A few of the ones I use are stacked up in Figure 5-8.

As you're already reading this one, I won't recommend other beginners' guides. Instead, I'll suggest some titles that will take you farther in various directions, and can be used for reference.

Make: More Electronics is the sequel that I wrote to this book, which includes all the topics (such as op-amps) for which I did not have space here. Some of the circuits are more ambitious. If you read that book, and this book, you will cover most aspects of electronics that are accessible to an individual on a moderate budget.

Figure 5-8 *A sun-damaged copy of Don Lancaster's classic guide to TTL chips remains at the top of my stack of reference books. It opened up a whole new era of hobby electronics more than 40 years ago. Much of its information is still useful, and copies are available from secondhand sources, including Amazon.*

The Encyclopedia of Electronic Components is a project that I began before I realized quite how demanding it would be. Consequently, its three volumes were repeatedly delayed. Volumes 1 and 2 are in print as I write this. By the time you read this, maybe Volume 3 will be in print too. The idea of these books is that they are ideal for quick reference. They can remind you of what you forgot, and they go into a lot of detail. By comparison, *Make: Electronics* is a teaching guide full of hands-on tutorials, in which I try not to get bogged down in detail.

Now let me list the books that I consider most important, written by other people:

Practical Electronics for Inventors, by Paul Scherz with Simon Monk (McGraw-Hill, Second Edition, 2013): This is a massive, comprehensive book, well worth the $40 cover price. Despite its title, you won't need to invent anything to find it useful. It's my primary reference source, covering a wide range of concepts, from the basic properties of resistors and capacitors all the way to some fairly highend math.

Getting Started with Arduino, by Massimo Banzi and Michael Shiloh (Make, 2014): This is the simplest introduction around, and will help to familiarize you with the processing language used in Arduino (similar to the C language, if you know anything about that).

Making Things Talk, by Tom Igoe (Make: Books, 2011): This ambitious and comprehensive volume shows how to make the most of the Arduino's ability to communicate with its environment, even getting it to access sites on the Internet.

TTL Cookbook, by Don Lancaster (Howard W. Sams & Co, 1974): The 1974 copyright date is not a misprint! You may be able to find some later editions, but whichever one you buy, it will be secondhand. Lancaster wrote his guide before the 74xx series of chips was emulated on a pin-for-pin basis by CMOS versions such as the 74HCxx series, but it's still a good reference, because the concepts and part numbers haven't changed, and his writing is so accurate and concise. Just bear in mind that his information about high and low logic voltages is no longer accurate.

CMOS Sourcebook, by Newton C. Braga (Sams Technical Publishing, 2001): This book is entirely devoted to the 4000 series of CMOS chips, not the 74HCxx series that I've dealt with primarily here. The 4000 series is older and must be handled more carefully, because it's more vulnerable to static electricity than the generations that came later. Still, the chips remain widely available, and their great advantage is their willingness to tolerate a wide voltage range, typically from 5 to 15 volts. This means you can set up a 12-volt circuit that drives a 555 timer, and use output from the timer to go straight into CMOS chips (for example). The book is well organized in three sections: CMOS basics, functional diagrams (showing pinouts for all the main chips), and simple circuits showing how to make the chips perform basic functions.

The Encyclopedia of Electronic Circuits, by Rudolf F. Graf (Tab Books, 1985): A totally miscellaneous collection of schematics, with minimal explanations. This is a useful book to have around if you have an idea and want to see how someone else approached the problem. Examples are often more valuable than general explanations, and this book is a massive compendium of examples. Many additional volumes in the series have been published, but start with this one, and you may find it has everything you need.

The Circuit Designer's Companion, by Tim Williams (Newnes, Second Edition, 2005): Much useful information about making things work in practical applications, but the style is dry and fairly technical. May be useful if you're interested in moving your electronics projects into the real world.

The Art of Electronics, by Paul Horowitz and Winfield Hill (Cambridge University Press, Second Edition, 1989): The fact that this book has been through 20 printings tells you two things: (1) many people regard it as a fundamental resource; (2) secondhand copies should be widely available, which is an important consideration, as the list price is over $100. It's written by two academics, and has a more technical approach than *Practical Electronics for Inventors*, but I find it useful when I'm looking for backup information.

Getting Started in Electronics, by Forrest M. Mims III (Master Publishing, Fourth Edition, 2007): Although the original dates back to 1983, this is still a fun book to have. I think I have covered many of its topics here, but you may benefit by reading explanations and advice from a completely different source, and it goes a little farther than I have into some electrical theory, on an easy-to-understand basis, with cute drawings. Be warned that it's a brief book with eclectic coverage. Don't expect it to have all the answers.

Experiment 25: Magnetism

Now that I have surveyed your future options, let me deal with a very important topic that has been waiting in the background: the relationship between electricity and magnetism. Quickly this will lead us into audio reproduction and radio, and I'll describe the fundamentals of self-inductance, which is the third and final basic property of passive components (resistance and capacitance being the other two). I left self-inductance until last because it has limited application to DC circuits. But as soon as we start dealing with analog signals that fluctuate, it becomes fundamental.

Fundamentals: A Two-Way Relationship

Electricity can create magnetism:

- When electricity flows through a wire, the electricity creates a magnetic force around the wire.

This principle is used in almost every electric motor in the world.

Magnetism can create electricity:

- When a wire moves through a magnetic field, the field creates a flow of electricity in the wire.

This principle is used in power generation. A diesel engine, or a water-powered turbine, or a windmill, or some other source of energy can turn coils of wire through a powerful magnetic field. Electricity is induced in the coils. With the exception of solar panels, all practical sources of electric power use magnets and coils of wire.

In the next experiment, you'll see a dramatic mini-demo of this effect. It should be a part of any school science class, but even if you've done it in the past, I suggest that you do it again, because setting it up takes only a matter of moments.

What You Will Need

- Large screwdriver (1)
- 22-gauge wire, or thinner (no more than 6 feet)
- 9-volt battery (1)
- Paper clip (1)

Procedure

This couldn't be simpler. Wind the wire around the shaft of the screwdriver, near its tip. The turns should be neat and tight and closely spaced, and you'll need to make 100 of them, within a distance of no more than 2 inches. To fit them into this space, you'll have to make turns on top of previous turns. If the final turn tends to unwind itself, secure it with a piece of tape.

Now apply the 9-volt battery. At first sight, this looks like a very bad idea, because you're going to short out your battery just as you did in Experiment 2. But when you pass current through a wire that's coiled instead of straight, the flow of current is inhibited (in a way that I will explain shortly), and the current does some work (such as, it can move a paper clip).

Put a small paper clip near the screwdriver blade, as shown in Figure 5-9.

Figure 5-9 *This most basic electromagnet is just strong enough to attract a paper clip.*

The surface should be smooth, so that the paper clip can slide across it easily. Because many screwdrivers are already magnetic, you may find that the paper clip is naturally attracted to the tip of the blade. If this happens, move the clip just outside the range of attraction. Now apply the 9 volts to the circuit, and the clip should jump to the tip of the screwdriver.

Congratulations: you just made an electromagnet. The schematic is shown in Figure 5-10.

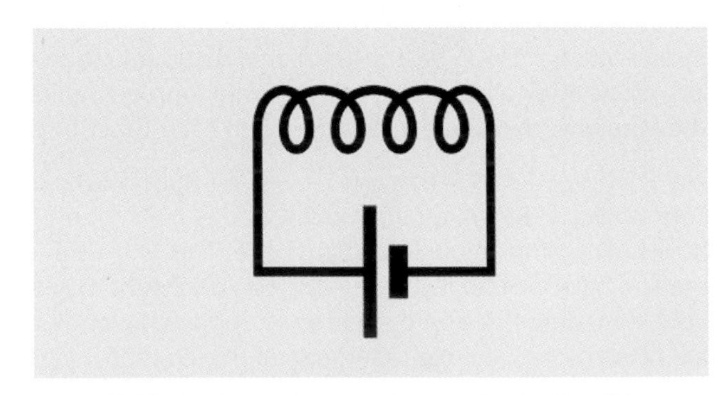

Figure 5-10 *A schematic can't get much simpler than this.*

Theory: Inductance

When electricity flows through a wire, it creates a magnetic field around the wire. Because the electricity "induces" this effect, it is known as *inductance*. This is illustrated in Figure 5-11.

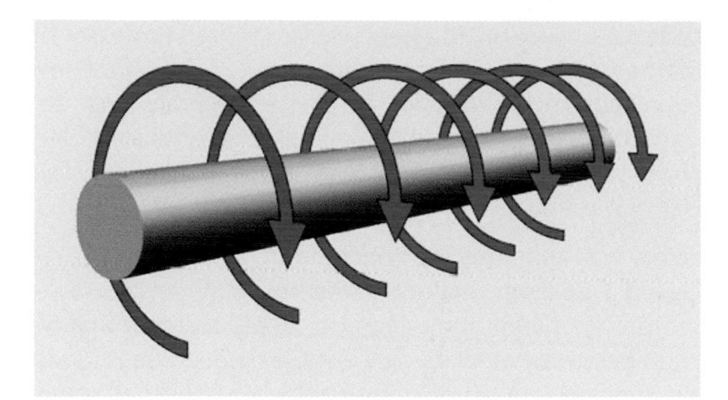

Figure 5-11 *When the flow of electricity is from left to right along this conductor, it induces a magnetic force shown by the green arrows.*

The field around a straight wire is very weak, but if we bend the wire into a circle, the magnetic force starts to accumulate, pointing through the center of the circle, as shown in Figure 5-12. If we add more circles, to form a coil, the force accumulates even more. And if we put a steel or iron object (such as a screwdriver) in the center of the coil, the effectiveness increases further.

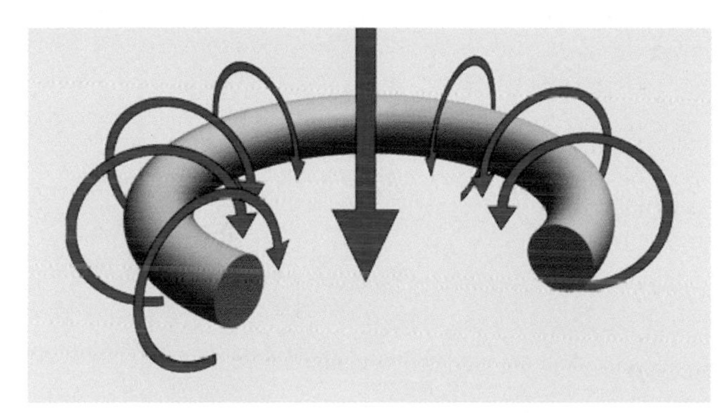

Figure 5-12 *When the conductor is bent to form a circle, the cumulative magnetic force acts through the center of the circle, as shown by the large arrow.*

Figure 5-13 shows this graphically, along with a formula known as "Wheeler's approximation," which allows you to calculate the inductance of a coil approximately, assuming you know the inner radius, the outer radius, the width, and the number of turns. (The dimensions must be in inches, not metric.) The basic unit of inductance is the henry, named after American electrical pioneer Joseph Henry. Because this is a large unit (like the farad), the formula expresses inductance in microhenries.

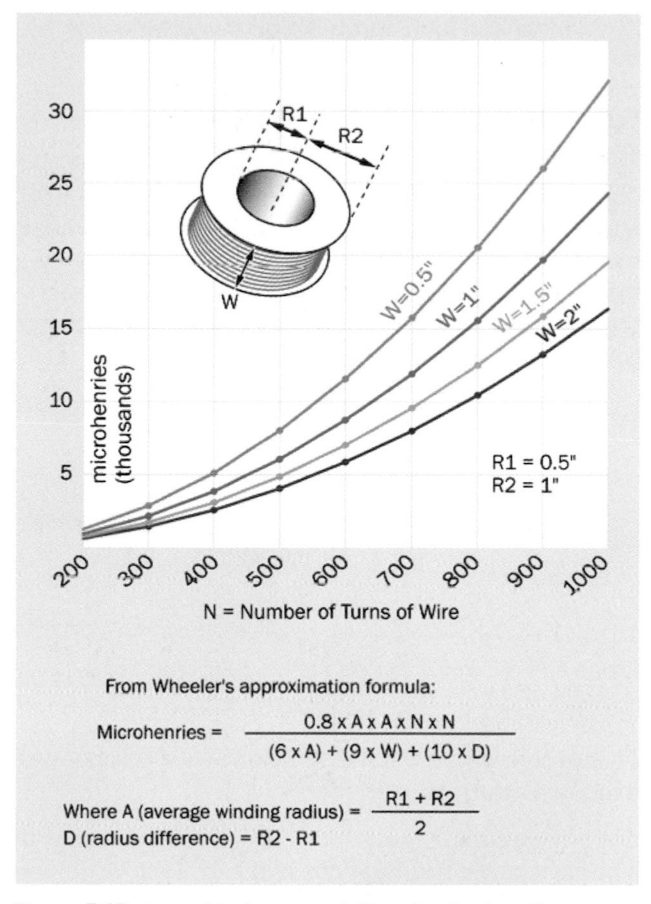

Figure 5-13 *A graphical representation showing how the dimensions and number of turns in a coil affect its inductance, calculated approximately with a simple formula.*

You'll see from the graphs that if you keep the basic size of a coil the same, and double the number of turns (by using thinner wire or wire with thinner insulation), the reactance of the coil increases by a factor of four. This is because the formula includes the factor N x N at the top. Here are some take-home messages:

- Inductance increases with the diameter of the coil.
- Inductance increases approximately with the square of the number of turns. (In other words, three times as many turns create nine times the inductance.)
- If the number of turns remains the same, inductance is lower if you wind the coil so that it's

slender and long, but is higher if you wind it so that it's fat and short.

Fundamentals: Coil Schematics and Basics

Check the schematic symbols for coils in Figure 5-14. Moving from left to right, either of the first two symbols represents a coil with an air core (the first symbol is older than the second). The third and fourth symbols indicate that the coil is wound around a solid iron core, or a core composed of iron particles or ferrite, respectively.

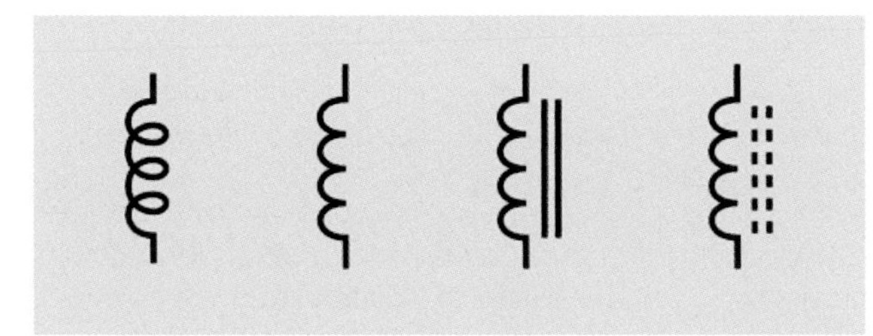

Figure 5-14 *Schematic symbols to represent coils. See text for details.*

An iron core will add to the inductance of a coil, because it increases the magnetic effect.

If you measure the magnetic field created by a coil with a positive power source at one end and negative ground at the other end, the field will reverse if you reverse the polarity of the power supply.

Perhaps the most widespread application of coils is in transformers, where alternating current in one coil induces alternating current in another, often sharing the same iron core. If the primary (input) coil has half as many turns as the secondary (output) coil, the voltage will be doubled, at half the current—assuming hypothetically that the transformer is 100% efficient.

Background: Joseph Henry

Born in 1797, Joseph Henry was the first to develop and demonstrate powerful electromagnets. He also originated the concept of "self-inductance," meaning the "electrical inertia" that is a property of a coil of wire.

Henry started out as the son of a day laborer in Albany, New York. He worked in a general store before being apprenticed to a watchmaker, and was interested in becoming an actor. Friends persuaded him to enroll at the Albany Academy, where he turned out to have an aptitude for science. In 1826, he was appointed Professor of Mathematics and Natural Philosophy at the Academy, even though he was not a college graduate and described himself as being "principally self-educated." Michael Faraday was doing similar work in England, but Henry was unaware of it.

Henry was appointed to Princeton in 1832, where he received $1,000 per year and a free house. When Morse attempted to patent the telegraph, Henry testified that he was already aware of its concept, and indeed had rigged a system on similar principles to signal his wife, at home, when he was working in his laboratory at the Philosophical Hall.

Henry taught chemistry, astronomy, and architecture, in addition to physical science, and because science was not divided into strict specialties as it is now, he investigated phenomena such as phosphorescence, sound, capillary action, and ballistics. In 1846, he headed the newly founded Smithsonian Institution as its secretary. His photograph appears in Figure 5-15.

Figure 5-15 *Joseph Henry was an American experimenter who pioneered the investigation of electromagnetism. This photograph is archived in Wikimedia Commons.*

Experiment 26: Tabletop Power Generation

In Experiment 5, you saw that chemical reactions can generate electricity. Now it's time to see electricity generated by a magnet.

What You Will Need

- Wire cutters, wire strippers, test leads, multimeter
- Cylindrical neodymium magnet, 3/16" diameter by 1.5" long, axially magnetized (1)
- Hookup wire, 26-gauge, 24-gauge, or 22-gauge, total 200 feet
- Low-current LED (1)
- Capacitor, 1,000µF (1)
- Switching diode, 1N4001 or similar (1)

Optional extras:

- Cylindrical neodymium magnet, 3/4" diameter by 1" long, axially magnetized (1)
- Half-inch diameter wooden dowel, 6" long (minimum)
- Steel screw, #6 size with flat head
- PVC water pipe, 3/4" internal diameter, 6" long (minimum)
- Two pieces of 1/4" plywood, each about 4" x 4" (you will need a 1" hole saw or Forstner bit to drill a hole through the plywood)
- Spool of magnet wire, quarter-pound, 26-gauge, about 350 feet (1)

Procedure

First, you need a magnet. Neodymium magnets are the strongest available, and are fairly cheap if you choose the small cylindrical type. A magnet of just 3/16" diameter and 1.5" long will be sufficient. Wrap about 10 turns of 22-gauge wire tightly around it, as shown in Figure 5-16. Now allow the wire to loosen slightly, so that the magnet can slide through the coils.

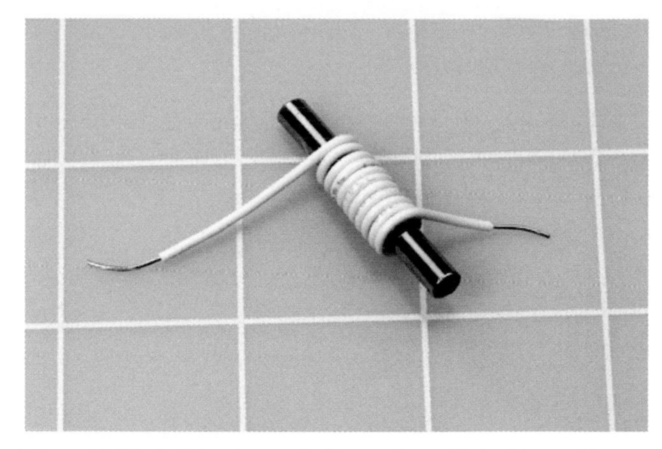

Figure 5-16 *Just ten turns of wire can be sufficient to create a small electrical potential when a magnet moves through them.*

Set your meter to measure millivolts AC (not DC, because we're going to be dealing with alternating pulses of electricity). Strip a little insulation from each end of the coil, and use alligator test leads to attach the meter. Grasp the magnet between finger and thumb, and shuttle it quickly to and fro inside the coil. I'm guessing you'll see a value of 3mV to 5mV on your meter. Yes, this small magnet, and 10 turns of wire, can generate a few millivolts.

Try winding a bigger coil, with the layers overlapping, as shown in Figure 5-17. Move the magnet quickly again. You should find that you are generating more voltage.

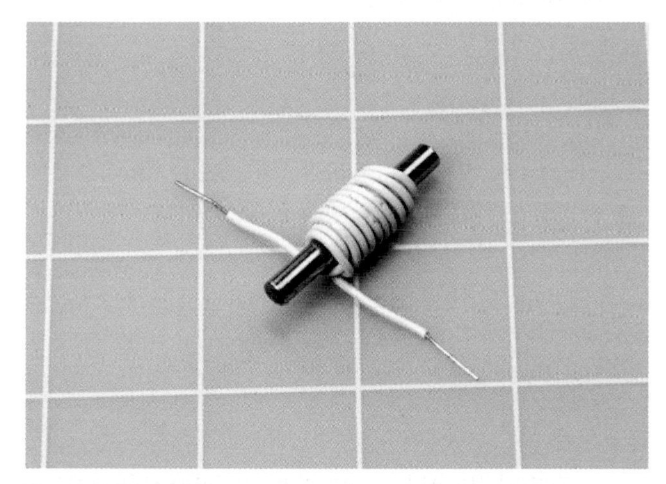

Figure 5-17 *Adding more turns of wire will increase the measured voltage when the magnet moves through them.*

Remember the formula from the previous experiment, in which I showed how passing electricity through more coils of wire would induce a stronger magnetic field. It works both ways:

- More coils of wire will generally induce a higher voltage, when a magnet moves through the coils.

This leads me to wonder—if we had a bigger, stronger magnet and a *lot* of turns of wire, could we generate enough electricity to power something, such as, maybe, an LED?

Lighting an LED

I'm going to use 22-gauge wire, because you already have it for the other experiments. The trouble is, it is relatively thick and has thick insulation. Two hundred turns of this wire really begin to bulk up. This is why we should be using *magnet wire*, which is pure copper wire with an ultra-thin coating of insulation made from shellac or plastic film. Magnet wire is designed to pack as densely as possible.

However, you may not feel like spending the money for a spool of magnet wire, bearing in mind that you are unlikely to find any other use for it. So, I decided to see if 22-gauge hookup wire would be workable in this experiment. The answer is yes, but only just.

You do need 200 feet. That will cost a bit of money, but you can always reuse the wire for normal purposes, such as creating jumpers for breadboards.

You can join two or more pieces of wire together when you are winding a spool, and so long as you twist the stripped ends tightly, you don't need to solder them.

You also need a more powerful magnet. The smallest that worked for me is cylindrical, measures 1″ long and 3/4″ in diameter, and is axially magnetized, meaning that the north and south poles are at opposite ends of its axis. (The axis is an imaginary line that runs through the center of the cylinder, parallel with its curved sides. You can imagine the cylinder like a shaft rotating around its axis.)

The equipment that I ended up with is shown in Figure 5-18. The magnet is at the right-hand side. The spool I made from 1/4″ plywood, and it is just over four inches in diameter. A piece of 3/4″ PVC water pipe runs through the center, and its inside diameter is just a fraction wider than the diameter of the magnet, so that the magnet can slide through it freely.

Figure 5-18 *Two hundred turns of 22-gauge wire on a homemade spool, with a magnet that attaches itself to a screw in a wooden dowel.*

Push-fit the plywood circles onto the pipe, to create a spool. Now you need to wind 200 feet of wire onto the spool—taking care to leave access to the inside end of the wire. I drilled a small hole in one of the plywood circles, near the center, and poked the wire out through the hole.

The width of the coil that you are going to wind should be about the same as the length of the magnet, and the magnet inside the tube should be able to emerge completely on either side of the coil. The cross-section of the spool in Figure 5-19 shows what I mean.

To hold the magnet conveniently, I drilled a hole in one end of a piece of 1/2″ wooden dowel, and inserted a #6 1″ flat-headed screw. I was then able to hold the dowel like a handle, while the magnet attracted itself firmly to the screw.

Now for the big moment. Use a couple of alligator test leads to attach the ends of your coil to the inputs of your meter, and set the meter to AC volts, as you did before. This time, though, set it to measure up to 2 volts.

With the magnet attached to the dowel, you can push it as quickly as possible in and out of the PVC pipe. Alternatively, remove the magnet from the dowel, drop it into the pipe, and rattle it up and down with your finger and thumb over the ends of the pipe. If you really work hard, your meter should show a voltage of around 0.8V.

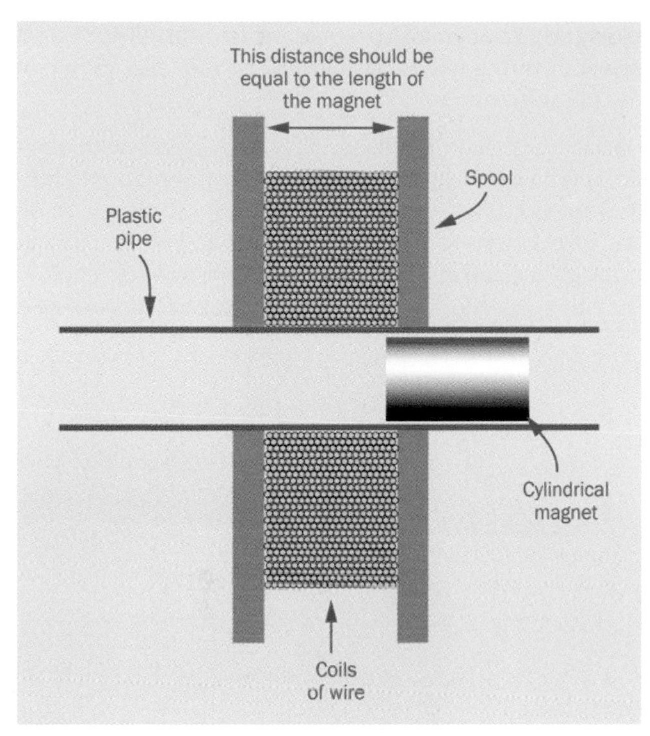

Figure 5-19 *Setup for generating just enough power to light an LED.*

You took all that trouble, and you got less than a volt?

Ah, but your meter is *averaging* the current. Each pulse is probably peaking at a higher voltage.

Disconnect your test leads from the meter, and attach them to a low-current LED. Clamp the LED so that it doesn't jostle around. Now when you move the magnet vigorously, I think you will see the LED flicker. If it doesn't, reverse the orientation of the magnet in the tube and try again. You really need a low-current LED to make this work.

Optional Extensions

If you are willing to spend a little more money, you can get much more impressive results.

First, use a bigger magnet. I get excellent results from one that is 2" long and 5/8" in diameter. Of course, you'll need a larger diameter of PVC pipe to accommodate the magnet.

Second, buy a spool of proper magnet wire. I used about 500 feet of 26-gauge wire. It's easy to buy online; there are dozens of suppliers.

If you're fortunate, your magnet wire will be supplied on a plastic spool with a hole in the middle just a little bit bigger than the diameter of your magnet. And, better still, the spool of magnet wire will allow you access to the "tail" of the wire sticking out in the center of the spool, as shown circled in red in Figure 5-20.

Figure 5-20 *A spool of magnet wire with the inside end accessible, circled in red.*

To remove the thin film of insulation from the ends of the magnet wire, you can scrape them very gently with a knife blade, or rub them with fine sandpaper. Check with a magnifying glass to make sure that some insulation has been removed. You can also apply your meter to check the resistance, which should be less than 100 ohms.

Now you can attach an LED to each end of the magnet wire on the spool, and generate voltage by pushing your magnet in and out of the center of the spool, as shown in Figure 5-21.

If the spool is the wrong size, or if the tail of the wire isn't accessible, you'll just have to rewind the wire from one spool to another. Suppose you have 500 feet; that will entail rewinding about 2,000 turns. If you can make four turns per second, you'll require 500 seconds—a little less than 10 minutes, which I think is tolerable.

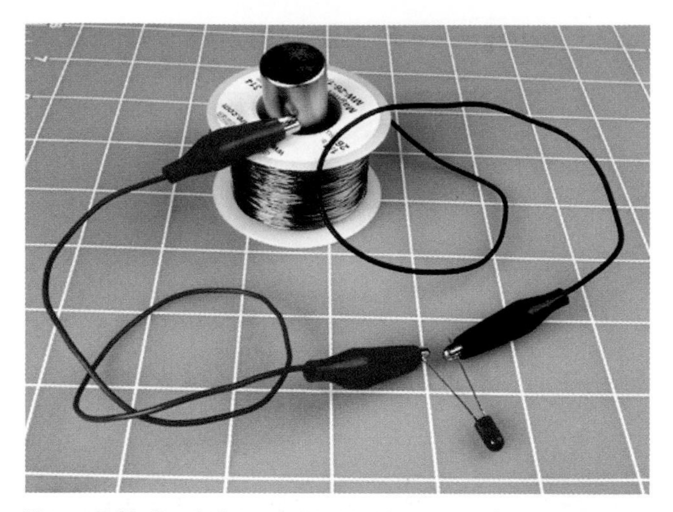

Figure 5-21 *Ready for power generation, on a rather small scale.*

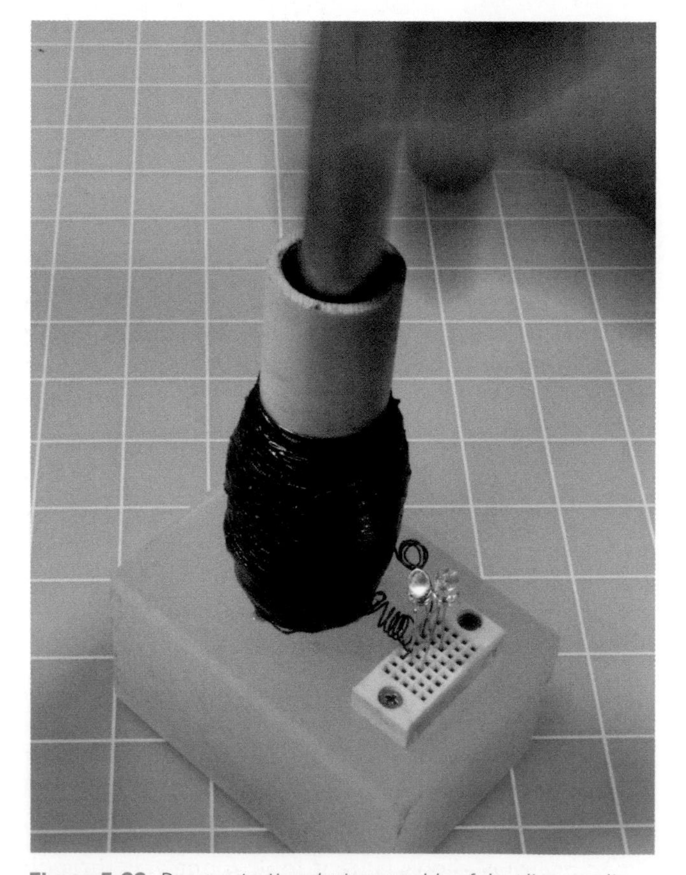

Figure 5-22 *Demonstration device capable of dazzling results.*

Figure 5-22 shows a larger-scale device that I built for demonstration purposes. The coil of magnet wire is coated with epoxy glue, so that it won't unravel, and I mounted the pipe in a block of plastic that holds it securely. My neodymium magnet attaches itself to a steel screw in to the end of an aluminum rod, also visible in the photograph.

I added two high-intensity LEDs to the coil, with their polarity in opposite directions. When the magnet shuttles up and down, the LEDs light up the whole room. Also, their opposite polarity shows that voltage travels through the coil in one direction on the upstroke, and in the other direction on the downstroke. See Figure 5-23.

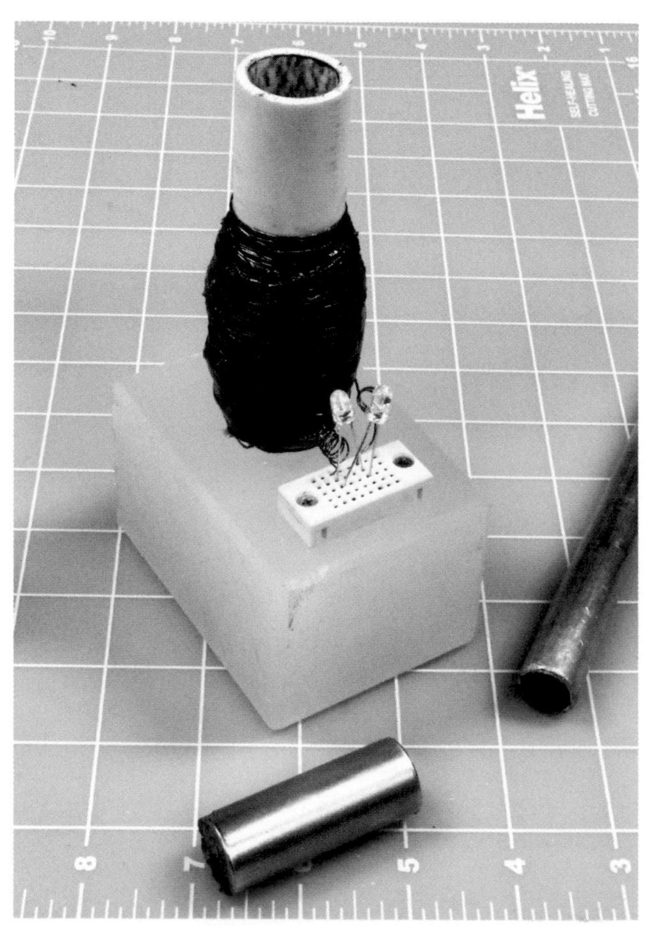

Figure 5-23 *The LED generator in action.*

Caution: Blood Blisters and Dead Media

Beware of the spooky capabilities of neodymium.

Neodymium magnets are breakable. They're brittle and can shatter if they slam against a piece of magnetic metal (or another magnet). For this reason, many manufacturers advise you to wear eye protection.

You can easily pinch your skin and get *blood blisters* (or worse). Because a magnet pulls with increasing force as the distance between it and another object gets smaller, it closes the final gap very suddenly and powerfully. Ouch!

Magnets never sleep. In the world of electronics, we tend to assume that if something is switched off, we don't have to worry about it. Magnets don't work that way. They are always sensing the world around them, and if they notice a magnetic object, they want it, *now*. Results may be unpleasant, especially if the object has sharp edges and your hands are in the way. When using a magnet, create a clear area on a nonmagnetic surface, and watch out for magnetic objects underneath the surface. For example, my magnet sensed a steel screw embedded in the underside of a kitchen countertop, and slammed itself into contact with the countertop unexpectedly.

It's hard to take this seriously until it happens to you. But, seriously, neodymium magnets don't fool around. Proceed with caution.

Also, remember that *magnets create magnets.* When a magnetic field passes across an iron or steel object, the object picks up some magnetism of its own. If you wear a watch, be careful not to magnetize it. If you use a smartphone, keep it away from magnets. Likewise, any computer or disk drive is vulnerable. The magnetic stripe on a credit card is easily erased. Also keep magnets well away from TV screens and video monitors (especially cathode-ray tubes). Last but not least, powerful magnets can interfere with the normal operation of cardiac pacemakers.

Charging a Capacitor

Here's another thing to try. Disconnect the LED from whatever coil of wire you created, and connect a 1,000µF electrolytic capacitor in series with a 1N4001 signal diode, as shown in Figure 5-24. Attach your meter, measuring DC volts (not AC, this time), across the capacitor.

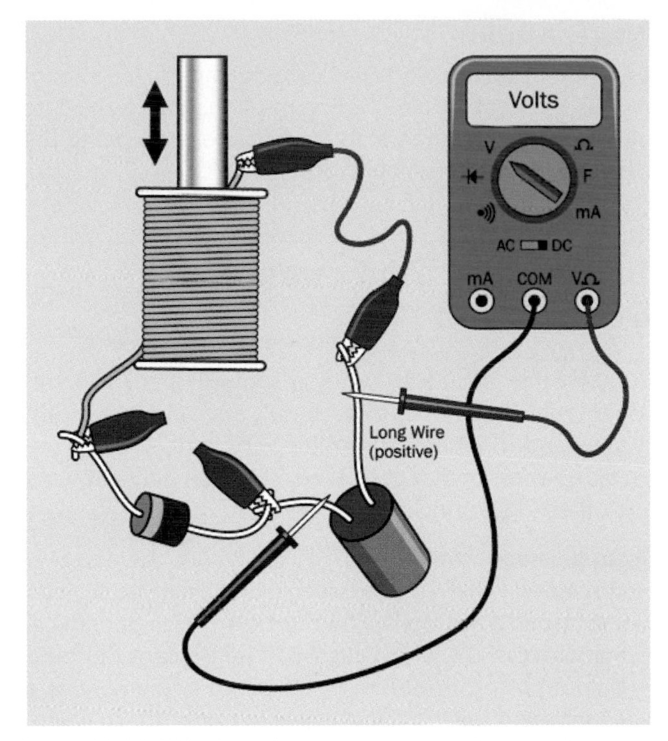

Figure 5-24 *A diode enables you to accumulate voltage from your coil in a capacitor.*

If your meter has a manual setting for its range, set it to at least 2VDC. Make sure the positive (unmarked) side of the diode is attached to the negative (marked) side of the capacitor, so that positive voltage will pass through the capacitor and then through the diode.

Now move the magnet vigorously up and down in the coil. The meter should show that the capacitor is accumulating charge. When you stop moving the magnet, the voltage reading may decline very slowly, mostly because the capacitor discharges itself through the internal resistance of your meter.

This experiment is more important than it looks. Bear in mind that when you push the magnet into the coil, it induces current in one direction, and when you pull it back out again, it induces current in the opposite direction. You are actually generating alternating current.

The diode only allows current to flow one way through the circuit. It blocks the opposite flow, which is how the capacitor accumulates its charge. If you jump to the conclusion that diodes can be used to convert alternating current to direct current, you're absolutely correct. We say that the diode is "rectifying" the AC power.

Next: Audio

Experiment 25 showed that voltage can create a magnet. Experiment 26 has shown that a magnet can create voltage. We're now ready to apply these concepts to the detection and reproduction of sound.

Experiment 27: Loudspeaker Destruction

You saw that electricity running through a coil can create enough magnetic force to pull a small metallic object toward it. What if the coil is very light, and the object is heavier? In that case, the coil can be pulled toward the object. This principle is at the heart of a loudspeaker.

To understand how a loudspeaker works, there's really no better way than to disassemble it. Maybe you'd prefer not to spend a few dollars on this destructive but educational process—in which case, you might consider picking up a piece of nonfunctional audio equipment at a yard sale, and pulling a loudspeaker out of that. Or simply take a look at my photographs illustrating the process step by step.

What You Will Need

- Cheapest possible loudspeaker, 2″ minimum (1)
- Utility knife (1)

Procedure

Figure 5-25 shows a small loudspeaker seen from the rear. A magnet is hidden in the sealed cylindrical section.

Figure 5-25 *The back of a small loudspeaker.*

Turn the loudspeaker face-up, as shown in Figure 5-26. Cut around the perimeter of its cone with a sharp utility knife or X-Acto blade. Then cut around the circular center and remove the O-shaped circle of black paper that you've created.

Figure 5-26 *A two-inch loudspeaker ready for its fate.*

The unconed speaker is shown in Figure 5-27. The yellow weave at the center is the flexible section that normally allows the cone to move in and out, while preventing it from deviating from side to side.

Figure 5-27 *The loudspeaker with its cone removed.*

Cut around the outside edge of the yellow weave, and you should be able to pull up a hidden paper cylinder, which has a copper coil wound around it, as shown in Figure 5-28. In the photograph, I've turned it over so that it is easily visible.

Figure 5-28 *The copper coil is normally hidden inside the groove of the magnet, below.*

The two ends of this copper coil normally receive power through flexible wires from two terminals at the back of the speaker. When the coil sits in the groove visible in the magnet, the coil reacts to voltage fluctuations by exerting an up-and-down force in reaction to the magnetic field. This vibrates the cone of the loudspeaker and creates sound waves.

Large loudspeakers in your stereo system work exactly the same way. They just have bigger magnets and coils that can handle more power (typically, as much as 100 watts).

Whenever I open up a small component like this, I'm impressed by the precision and delicacy of its parts, and the way it can be mass-produced for such a low cost. I imagine how astonished Faraday, Henry and the other pioneers of electrical research would be, if they could see the components that we take for granted today. Henry spent days winding coils by hand to create electromagnets that were far less efficient than this cheap little loudspeaker.

Background: Origins of Loudspeakers

As I mentioned at the beginning of this experiment, a coil will move if its magnetic field interacts with a heavy or fixed object. If the object is a permanent magnet, the coil will interact with it more strongly, creating more vigorous motion. This is how a loudspeaker works.

The idea was introduced in 1874 by Ernst Siemens, a prolific German inventor. (He also built the world's first electrically powered elevator in 1880.) Today, Siemens AG is one of the largest electronics companies in the world.

When Alexander Graham Bell patented the telephone in 1876, he used Siemens' concept to create audible frequencies in the earpiece. From that point on, sound-reproduction devices gradually increased in quality and power, until Chester Rice and Edward Kellogg at General Electric published a paper in 1925 establishing basic principles that are still used in loudspeaker design today.

At sites such as Radiola Guy (*http://bit.ly/radiolaguy*), you'll find photographs of very beautiful early loudspeakers, which used a horn design to maximize efficiency, as shown in Figure 5-29. As sound amplifiers became more powerful, speaker efficiency became less important compared with quality reproduction and low manufacturing costs. Today's loudspeakers convert only about 1% of electrical energy into acoustical energy.

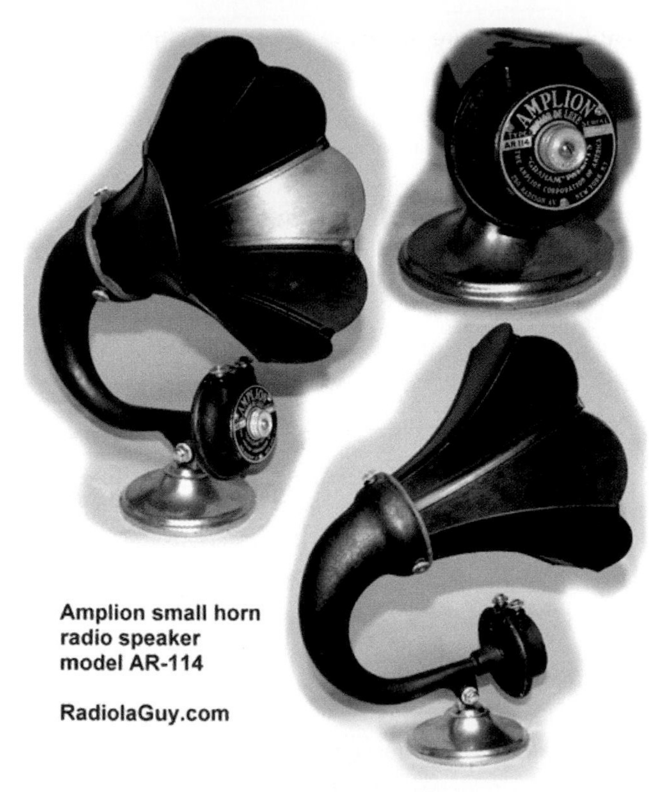

Amplion small horn
radio speaker
model AR-114

RadiolaGuy.com

Figure 5-29 *This beautiful Amplion AR-114x illustrates the efforts of early designers to maximize efficiency in an era when the power of audio amplifiers was very limited. Photos by "Sonny, the RadiolaGuy." Many early speakers are illustrated at http://www.radiolaguy.com. Some are for sale.*

Theory: Sound, Slectricity, and Sound

Time now to establish a more specific idea of how sound is transformed into electricity and back into sound again.

Suppose someone bangs a gong with a stick, as shown in Figure 5-30. The flat metal face of the gong vibrates in and out, creating pressure waves that the human ear perceives as sound. Each wave of high air pressure is followed by a trough of lower air pressure, and the wavelength of the sound is the distance (usually ranging from meters to millimeters) between one peak of pressure and the next.

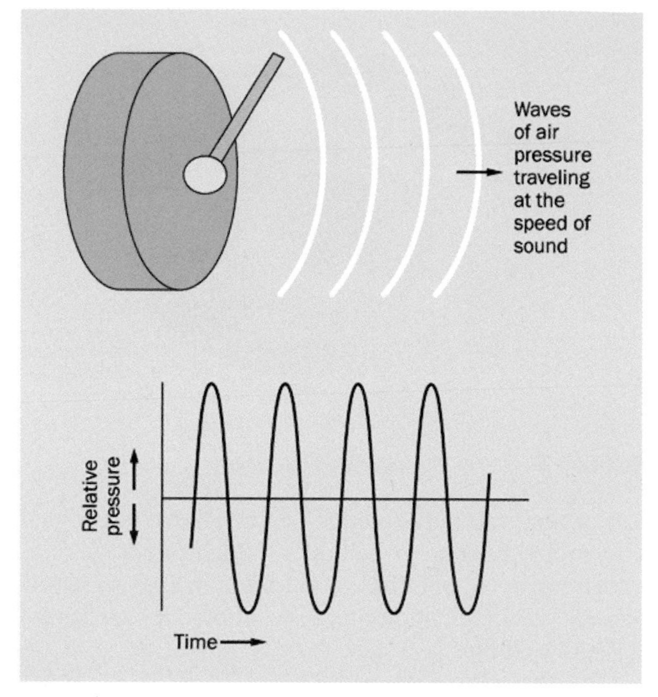

Figure 5-30 *Striking a going makes its flat surface vibrate. The vibrations create waves of pressure in the air.*

The frequency of the sound is the number of waves per second, usually expressed as hertz.

Suppose we put a very sensitive little membrane of thin plastic in the path of the pressure waves. The plastic will flutter in response to the waves, like a leaf fluttering in the wind. Suppose we attach a tiny coil of very thin wire to the back of the membrane so that it moves with the membrane. And let's position a stationary magnet inside the coil of wire. This configuration is like a tiny, ultra-sensitive loudspeaker, except that instead of electricity producing sound, the sound will produce electricity. Pressure waves make the membrane oscillate along the axis of the magnet, and the magnetic field creates a fluctuating voltage in the wire. The principle is illustrated in Figure 5-31.

This is known as a *moving-coil* microphone. There are other ways to build a microphone, but this is the configuration that is easiest to understand. Of course, the voltage that it generates is very small, but we can amplify it using a transistor, or a series of transistors, as suggested in Figure 5-32.

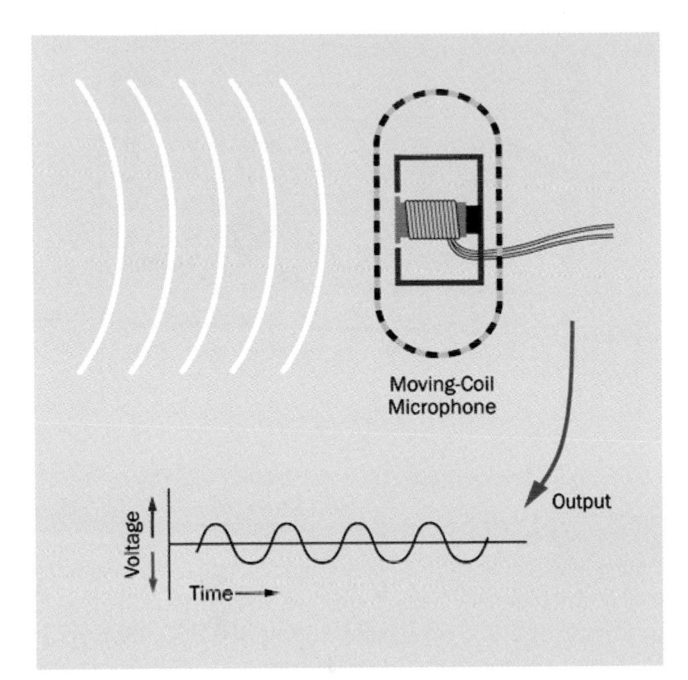

Figure 5-31 *Sound waves entering a moving-coil microphone make a membrane vibrate. The membrane is attached to a coil on a sleeve around a magnet. Motion of the coil induces small currents.*

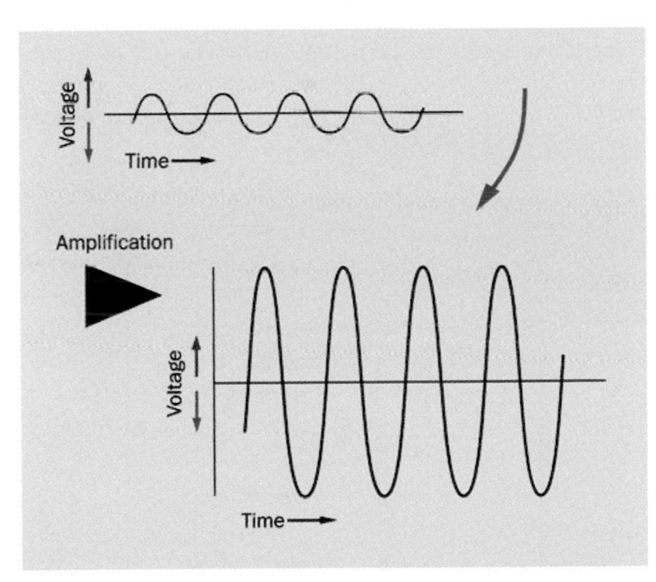

Figure 5-32 *Tiny signals from the microphone pass through an amplifier, which enlarges their amplitude while retaining their frequency and the shape of their waveform.*

Then we can feed the output through the coil around the neck of a loudspeaker, and the loudspeaker will re-create the pressure waves in the air, as shown in Figure 5-33.

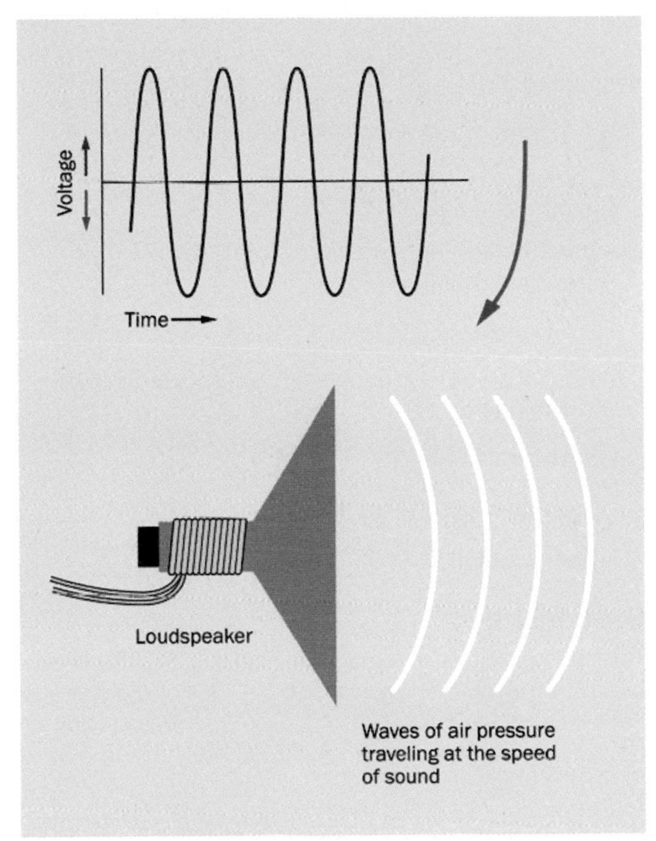

Figure 5-33 *The amplified electrical signal is passed through a coil around the neck of a loudspeaker cone. The magnetic field induced by the current causes the cone to vibrate, reproducing the original sound.*

Somewhere along the way, we may want to record the sound and then replay it. But the principle remains the same. The hard part is designing the microphone, the amplifier, and the loudspeaker so that they reproduce the waveforms *accurately* at each step. It's a significant challenge, which is why accurate sound reproduction can be elusive.

Experiment 28: Making a Coil React

You've seen that when you pass current through a coil, the current creates a magnetic field. When you disconnect the current, what happens to the field that it created?

The energy in the field is converted back into a brief pulse of electricity. We say that this happens when the field *collapses*.

This experiment will enable you to see it for yourself.

What You Will Need

- Breadboard, hookup wire, wire cutters, wire strippers, multimeter
- Low-current LEDs (2)
- Hookup wire, 22 gauge (26 gauge preferred), 100 feet (1 spool)
- Resistor, 47 ohms (1)
- Capacitor, 1,000 µF or larger (1)
- Tactile switch (1)

Procedure

Take a look at the schematic in Figure 5-34. The breadboarded version is shown in Figure 5-35. For the coil, you can use a spool of 100 feet of 22-gauge hookup wire. Alternatively, if you created your own coil of 200 feet of wire in Experiment 26, you can use that; and if you splurged on a spool of magnet wire, that will be even better.

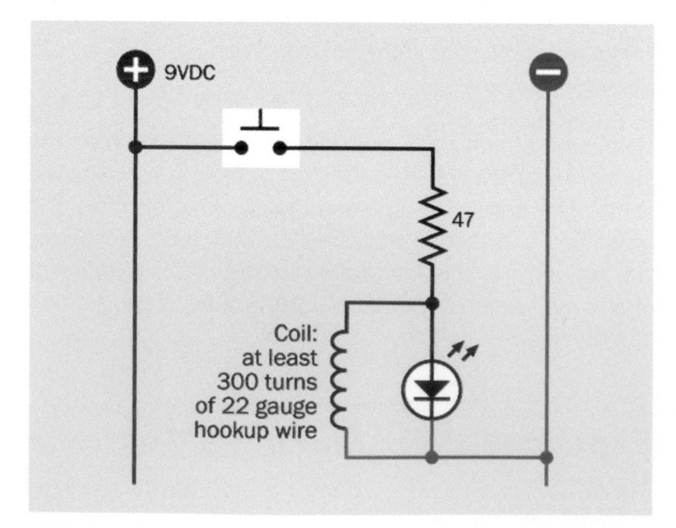

Figure 5-34 *A simple circuit to demonstrate the self-inductance of a coil.*

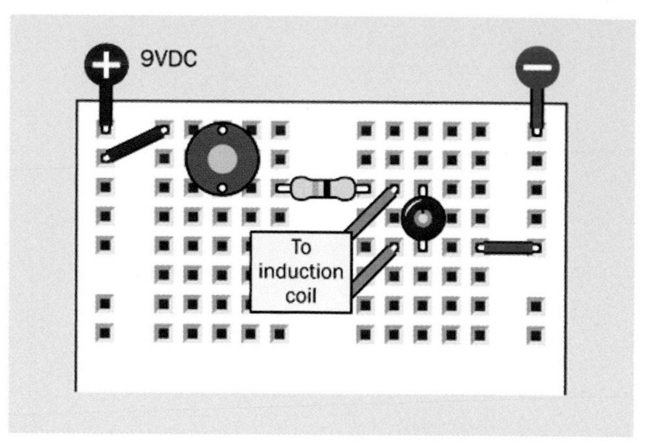

Figure 5-35 *Breadboarded version of the self-inductance experiment.*

When you look at the schematic, it doesn't seem to make much sense. The 47-ohm resistor seems too small to protect the LED—but why should the LED light up at all, when the electricity can go around it through the coil?

Now test the circuit, and I think you'll be surprised. Each time you press the button, the LED blinks briefly. Can you imagine why that should be?

Try adding a second LED, the other way up, as in Figure 5-36 and Figure 5-37. Press the button again, and the first LED flashes, as before. But now when you release the button, the second LED flashes.

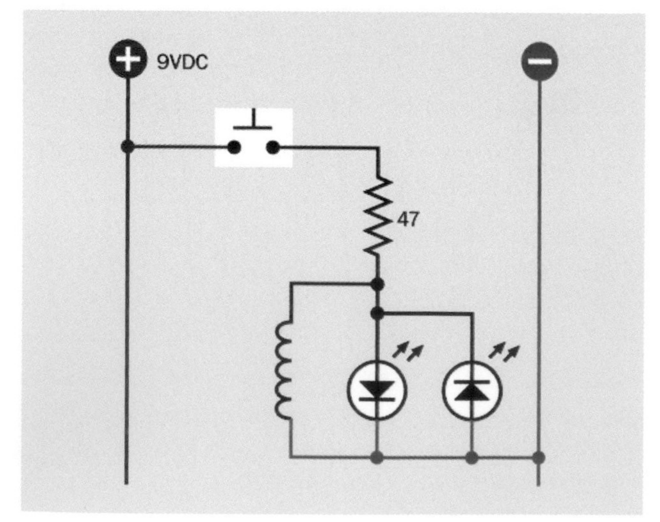

Figure 5-36 *One LED flashes when a magnetic field is created; the other flashes when the field collapses.*

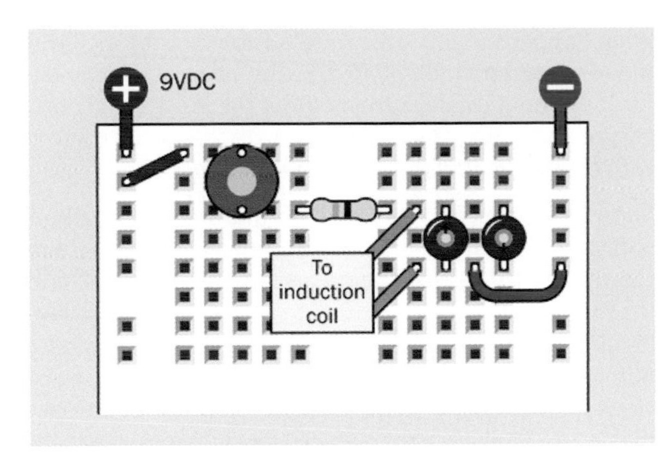

Figure 5-37 *Breadboarded version of the two-LED demonstration circuit.*

A Collapsing Field

Here's what happened during this experiment. At first, the coil required a brief amount of time to build up a magnetic field. This took a moment, and during that moment the coil blocked some of the flow of current. As a result, some of the current detoured through the first LED. Once the magnetic field was established, current flowed through the coil more normally.

This response of the coil is known as *self-inductance.* Sometimes people use the term *inductive reactance*, or just *reactance*, but since self-inductance is the correct term, I'll be using it here.

When you disconnected the power, the magnetic field collapsed, and the energy from the field was converted back into electricity in a short, brief pulse. This caused the second LED to flash when you let go of the button.

Naturally, different sizes of coil store and release different amounts of energy.

Perhaps you remember in Experiment 15, I advised you to tie a diode across the coil of a relay to absorb the surge that occurs when a relay coil is switched on and off. You have now seen this effect for yourself.

Resistors, Capacitors, and Coils

The three primary types of passive components in electronics are resistors, capacitors, and coils. We can now list and compare their properties.

A *resistor* constrains current flow, and drops voltage.

A *capacitor* allows a pulse of current to flow initially, but blocks direct current.

A *coil* (often referred to as an *inductor*) blocks DC current initially, but allows a continuing flow of direct current.

In the circuit that I just showed you, I didn't use a higher-value resistor because I knew the coil would allow only a very brief pulse. The blinking LEDs would have been less easily visible if I had used a more usual 330-ohm or 470-ohm resistor.

Don't try to run the circuit without the coil of wire included. You will quickly burn out one or both of the LEDs. The coil may look as if it isn't doing anything, but it is.

Here's one last variation on this experiment, to test your memory and understanding of electrical fundamentals. Build a new circuit shown in Figure 5-38 and Figure 5-39 using a 1,000µF capacitor instead of a coil (be careful to get its polarity the right way around, with the positive lead at the top.) Also, use a 470-ohm resistor, because the coil isn't there to block and divert current anymore.

First hold down button B for a second or two to make sure that the capacitor is discharged. Now, what will you see when you press button A? Maybe you can guess. Remember, a capacitor will pass an initial pulse of electricity. Consequently, the bottom LED lights up—and then gradually fades out, because the capacitor accumulates a positive charge on its upper plate and a negative charge on its lower plate. As this occurs, the potential across the lower LED diminishes to zero.

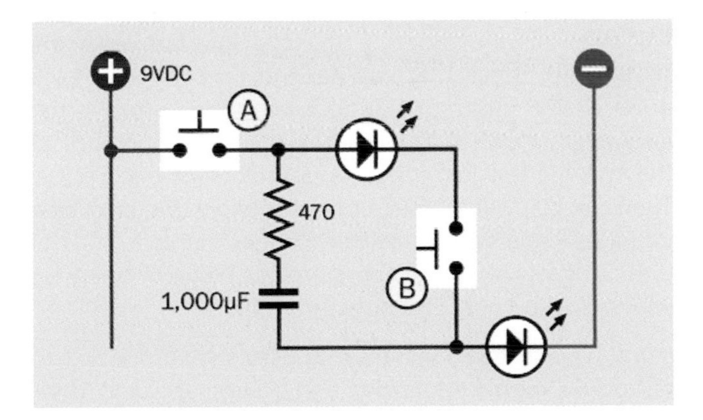

Figure 5-38 *In many ways, the behavior of a capacitor is opposite to the behavior of a coil.*

Figure 5-39 *Breadboarded version of the capacitor demo.*

The capacitor is now charged. Press the right-hand button, and the capacitor discharges through the upper LED. You can see this as the equivalent of the experiment in Figure 5-37, but using a capacitor instead of a coil.

Capacitors and inductors both store power. You were able to see this more obviously with the capacitor, because a high-value capacitor is much smaller than a high-value coil.

Theory: Alternating Current Concepts

Here's a simple thought experiment. Suppose you set up a 555 timer to send a stream of pulses through a coil. This will be a primitive form of alternating current.

Will the self-inductance of the coil interfere with the stream of pulses? That will depend on how long each pulse is, and how much inductance the coil has. If the frequency of pulses is just right, the self-inductance of the coil will last just long enough to block each pulse. Then the coil will recover in time to block the next one. In conjunction with a resistor (or just the resistance of the loudspeaker) a coil can suppress some frequencies while allowing others to pass through.

If you have a stereo system that uses a small speaker to reproduce high frequencies and a large speaker to reproduce low frequencies, almost certainly there is a coil somewhere in the speaker cabinet, stopping the higher frequencies from reaching the large speaker.

What happens if you substitute a capacitor for a coil? If the AC pulses are long relative to the value of the capacitor, it will tend to block them. But if the pulses are shorter, the capacitor can charge and discharge in rhythm with the pulses, and will allow them through.

I don't have space in this book to get deeply into alternating current. It's a vast and complicated field where electricity behaves in strange and wonderful ways, and the mathematics that describe it can become quite challenging, involving differential equations and imaginary numbers. (What is an imaginary number? The most obvious example is the square root of minus-one. How can that exist? Well—it can't, which is why we say that it's imaginary. Yet it crops up in electrical theory. If that sounds interesting, you may want to check it out.)

But I haven't finished with coils yet. The next experiment will demonstrate the audio effects that I just described above.

Experiment 29: Filtering Frequencies

In this experiment, you'll change the sound of sound. Using coils and capacitors, you can filter sections of the audible spectrum to create a rich variety of effects.

What You Will Need

- Breadboard, hookup wire, wire cutters, wire strippers, test leads, multimeter
- 9-volt power supply (battery or AC adapter)
- Loudspeaker, 8 ohms impedance, minimum 4 inches diameter (1)
- Audio amplifier chip, LM386 (1)
- 22-gauge hookup wire, 100 feet
- Small plastic storage bin as a loudspeaker enclosure (1)
- 555 timer (1)
- Resistors, 10K (2)
- Capacitors: 0.01µF (3), 2.2µF (1), 100µF (1), 220µF (3)
- Trimmer potentiometers: 10K (1), 1M (1)

- SPDT slide switches (4)
- Tactile switch (1)

A Home for Your Speaker

The little speaker that I recommended for previous projects was adequate when all you needed was a few beeps, but small speakers have a limited capability to reproduce bass notes. Since I want you to be able to hear how electronic components can affect those notes, it's time to consider a larger speaker, such as the one in Figure 5-40, which has a cone four inches in diameter.

Figure 5-40 *A loudspeaker that would be suitable for this project.*

Bearing in mind my previous comments on the need to suppress out-of-phase sound waves from the back of the speaker, you'll need a box to contain it. The box will also boost the sound by resonating, in the same way that the body of an acoustic guitar resonates with the vibrations of its strings.

If you have time to make a plywood box, that would be ideal, but the simplest and cheapest enclosure is probably a plastic storage bin with a snap-on lid. Figure 5-41 shows the speaker bolted into the bottom of a bin. Drilling holes neatly in thin plastic is quite a challenge, and —well, I didn't try too hard.

Figure 5-41 *A resonant enclosure is necessary if you want to hear some bass (lower frequencies) from your speaker. A cheap plastic storage bin is sufficient for demo purposes.*

To improve the attributes of the plastic bin, you can put some soft, heavy fabric inside it before you snap the lid on. A hand towel or some socks should be sufficient to absorb some of the vibration.

A Single Chip

Back in the 1950s, you needed vacuum tubes, transformers, and other power-hungry, heavyweight components to build an audio amplifier. Today, you can buy a chip for about $1 that will do the job, if you add a few capacitors around it, and a volume control.

One of the simplest, cheapest, and easiest to use is the LM386, which is available from multiple manufacturers, each of which prefaces or appends some extra identifying letters and numbers to it. The LM386N-1, LM386N, and LM386M-1 are all basically the same for our purposes. Just make sure you buy the through-hole version, not the surface-mount version. Pinouts for this amplifier are shown in Figure 5-42.

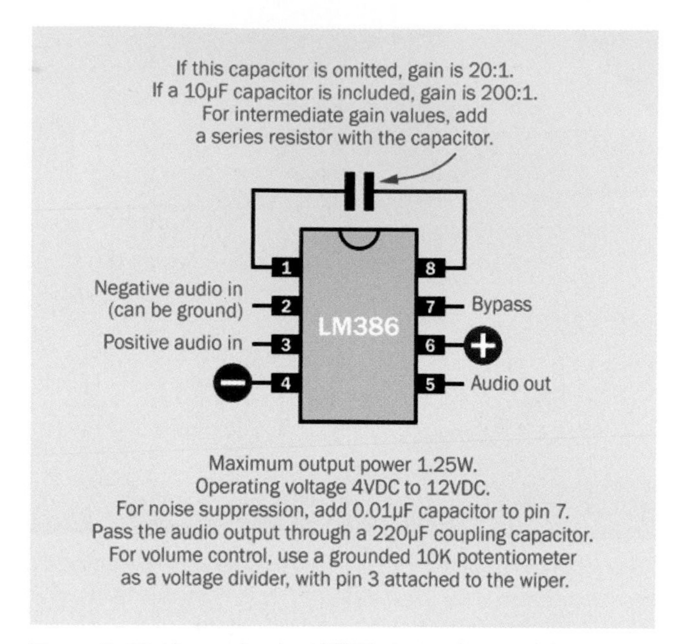

Figure 5-42 *Pinouts for the LM386 single-chip amplifier.*

This little chip works with a power supply ranging from 4VDC to 12VDC, and although it is rated for just 1.25 watts, you'll be surprised how loud it can sound. It has a nominal amplification ratio of 20:1.

Test, 1-2-3

For testing purposes, I want a source of frequencies covering a large range of the audible spectrum. A simple way to achieve this is with a 555 timer. The schematic in Figure 5-43 shows the timer at the top, with component values that can deliver a range from around 70Hz to 5KHz when you twiddle the 1M trimmer potentiometer. Unfortunately you won't hear this as a linear response, by which I mean that a small rotation of the trimmer will have a much greater audible effect on high frequencies than on low frequences. But, it will be good enough for demo purposes, and the lower frequencies provide a more dramatic demo of audio filtering anyway.

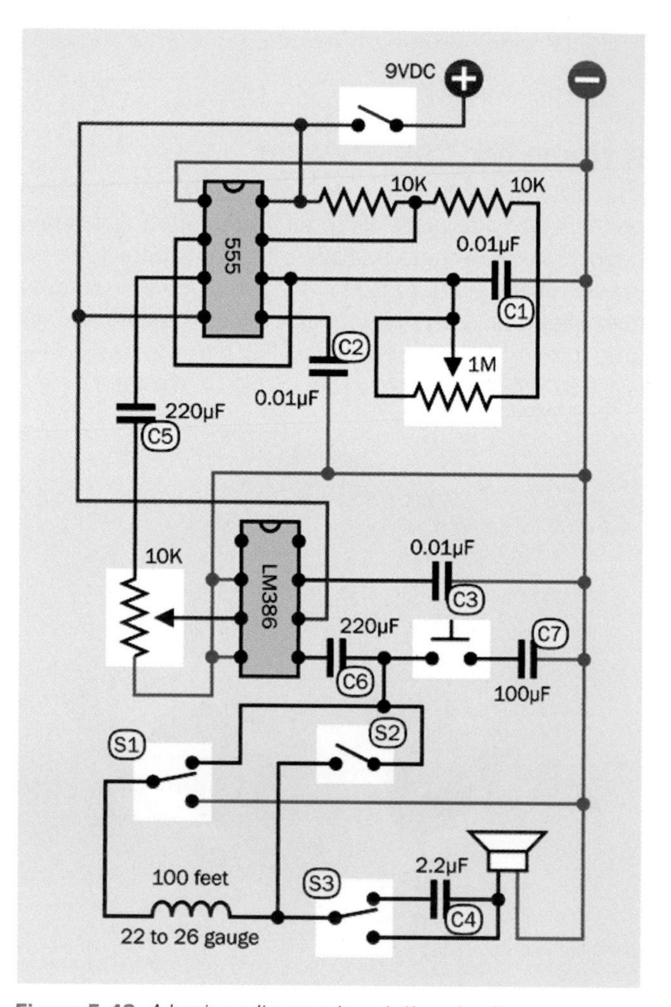

Figure 5-43 *A basic audio experimentation circuit.*

The breadboarded version of the circuit is shown in Figure 5-44 and the component values are shown in Figure 5-45.

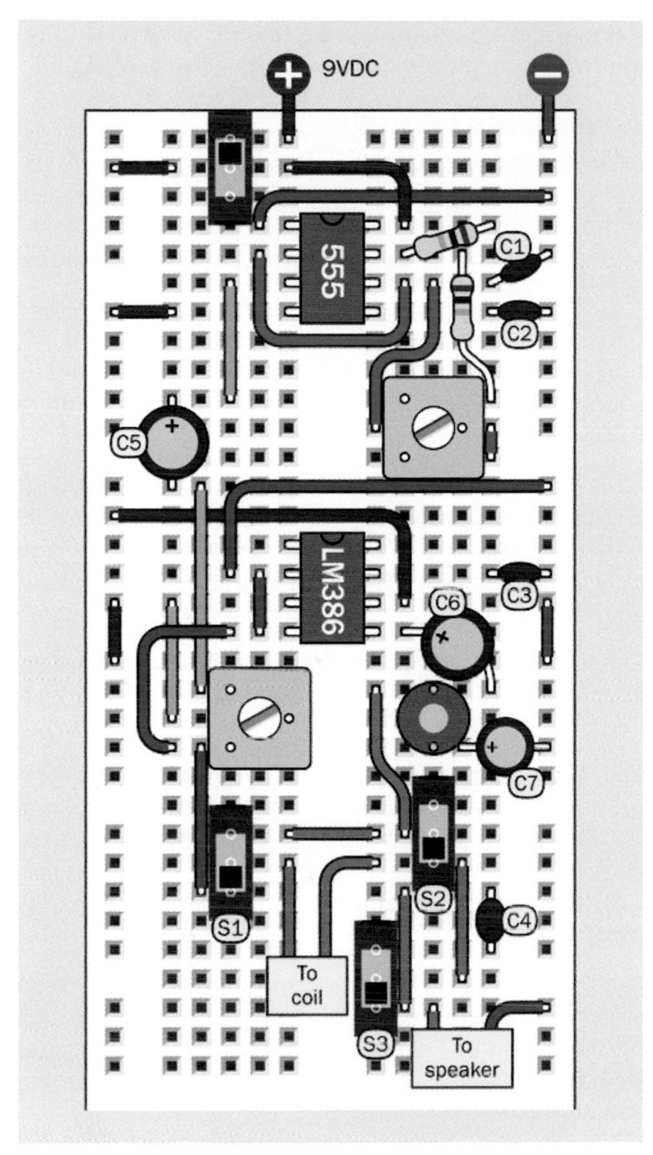

Figure 5-44 *Breadboarded version of the audio experimentation circuit.*

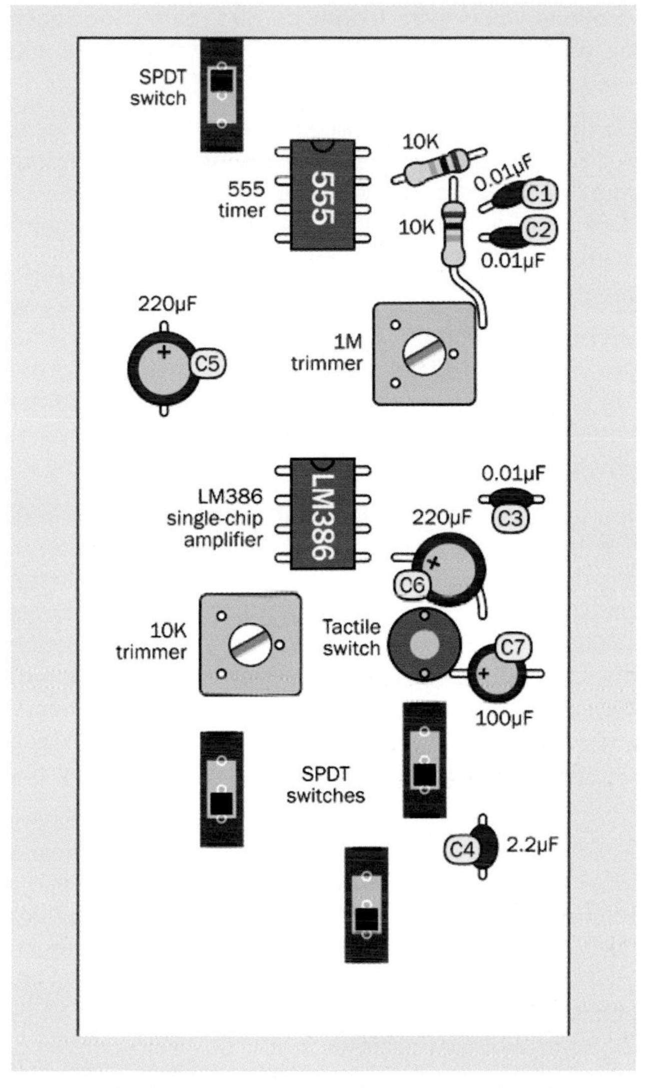

Figure 5-45 *Component values in the audio experimentation circuit.*

When you build this circuit, I have to warn you that amplifiers are sensitive to all electrical fluctuations, not just the ones you want to hear. Any electrical interference will be reproduced as a mess of scratchy, buzzing sounds, and this problem will be much worse if you use unnecessarily long pieces of wire to connect components.

The kind of jumper wires with little plugs on the end are especially undesirable in an amplifier circuit, as they behave like radio antennas. I have tried to limit the lengths of all the wires in the breadboard layout shown in Figure 5-44, and I encourage you to do the same. The only locations where wire length doesn't matter too much are on the power-output side of the chip, where you need to attach wires to your loudspeaker and to a coil.

For the coil, magnet wire of 22 gauge or thinner is ideal, but you will get some audible results from a 100-foot spool of 22-gauge hookup wire, and the 200 feet of hookup wire that I suggested in the previous experiment would be better.

Now, before powering up your breadboard, please pay attention to the three slide switches near the bottom of the circuit, and make sure that all of them are in the "down" position. In other words, slide them toward the bottom of the breadboard. Also turn the two trimmer potentiometers about halfway through their range.

You can power this project with an AC adapter or a 9-volt battery, with no regulation necessary. However, if you use an adapter, it may introduce some hum into the circuit. You can reduce this by placing a capacitor of 1,000μF or more between the two buses of the breadboard. If you use a battery, the power consumption of the amplifier will limit battery life to two or three hours, and some of the sound filters will pull down the voltage slightly, affecting the audio frequency created by the 555 timer.

As soon as you switch on the power, you should hear a tone. If you don't, your first troubleshooting strategy should be to disconnect the upper lead of the 220μF capacitor from the output pin of the 555 timer, and touch your speaker wires very briefly between that pin and the negative bus. If you don't hear anything, you made a wiring error around the timer. If you do hear something, then your error is related to the LM386 amplifier chip.

Make sure you connected power to the correct pins of the LM386. The positive and negative supply pins are not in the same positions as on logic chips.

Still no sound? Detach the top end of the short vertical piece of blue wire above the 10K trimmer. Touch the end of this piece of wire with your finger, and you should hear some whistling and buzzing sounds, because this is attached to the input pin of the amplifier (pin 4). *Still* nothing? Try attaching your speaker between the negative side of capacitor C6, and the negative power supply bus. C6 is a coupling capacitor connecting you directly with the output pin of the LM386.

If none of these attempts is successful, you'll have to go around the circuit with your meter, checking voltages.

Adventures in Audio

Assuming that your circuit is now up and running, I'll explain the functions of the components before I suggest some things to try. I'm going to refer to the labels that I applied to components in the breadboard layout shown in Figure 5-44.

Capacitor C1 sets the frequency of the timer, in conjunction with the 1M trimmer. Just in case you want to hear a sound higher in pitch than 5KHz, you can substitute a 0.0068μF (6.8nF) capacitor.

C5 is a coupling capacitor. It has a large value so that it will be transparent to a wide range of frequencies. Its purpose is to block DC from the 555 timer, because you only want to amplify the fluctuations, not the basic voltage.

Capacitor C6 is another coupling capacitor, protecting your loudspeaker from the DC coming out of the amplifier.

Capacitor C7 couples the amplifier output to negative ground when you press the button beside it. The value of C7 is chosen so that it takes away the higher frequencies, shunting them to ground. Without those frequencies, the sound that goes to the loudspeaker sounds more mellow.

Capacitor C4 is switched in and out of the circuit by slide switch S3. When the slide is up, sound from the 555 passes through C4 on its way to the amplifier. Because C4 has a small value, it blocks low frequencies, leaving you with a thin, tinny sound.

The complicated part of the circuit relates to the coil. I wanted you to hear the difference when a coil is connected in parallel with the loudspeaker, and when it is in series with the loudspeaker. Switches S1 and S2 give you those options, as shown in Figure 5-46 and Figure 5-47. When the coil is in parallel with the speaker, this is sometimes described as bypassing the speaker.

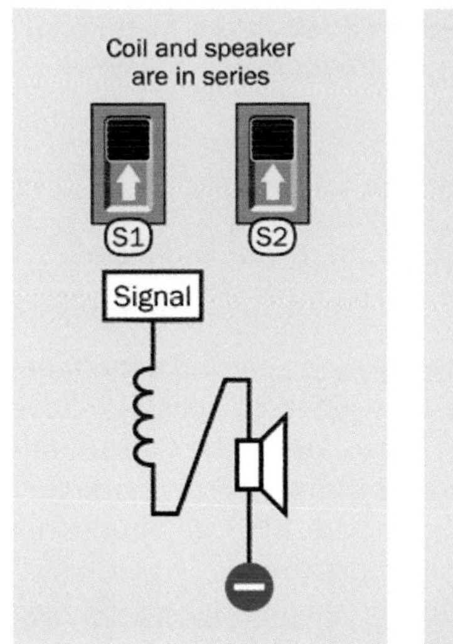

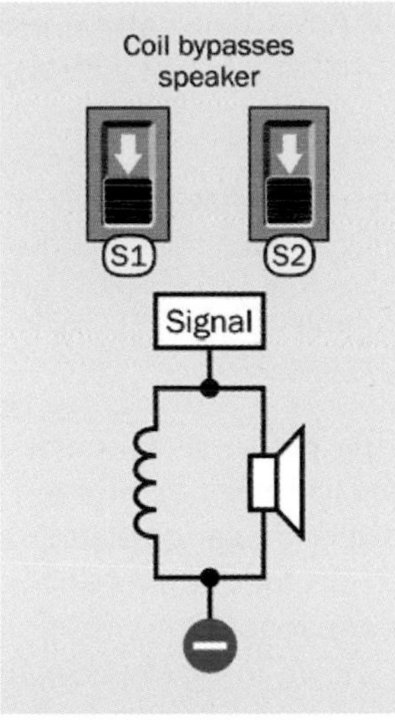

Figure 5-46 *Switches S1 and S2 (identified in the breadboard diagram for the circuit) allow you to feed audio to the speaker in parallel with an external coil, or in series with it.*

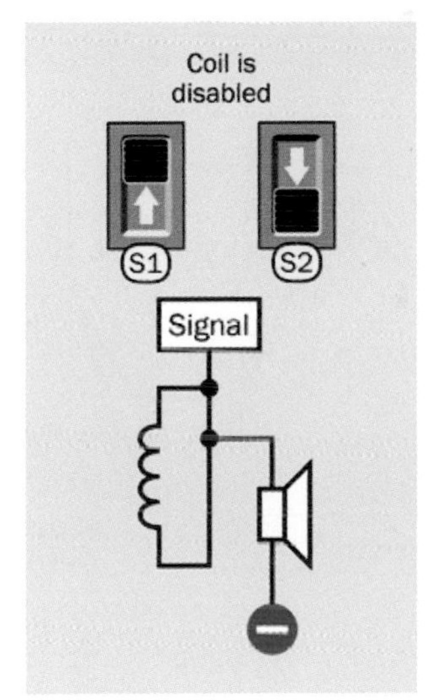

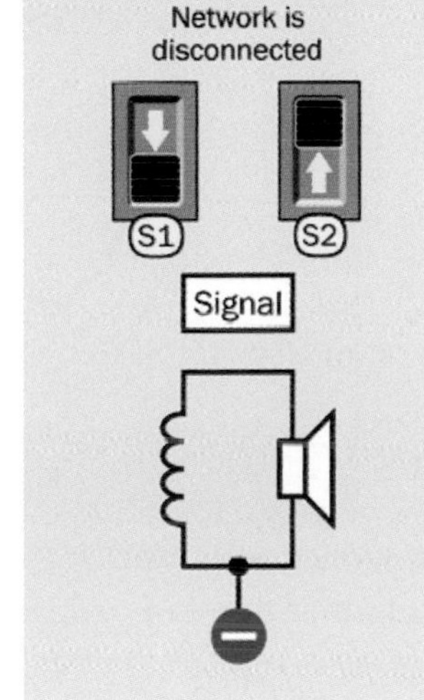

Figure 5-47 *Two other configurations of S1 and S2 allow you either to bypass the coil, or mute the output from the amplifier.*

You have quite a lot to play with here, especially bearing in mind that you can adjust the frequency and volume of the sound while you test the various filters. You can also test the effect of using two filters simultaneously. For instance, press the button to activate the bypass capacitor C7, which cuts the high frequencies, and switch C4 into the circuit at the same time, to cut the low frequencies. Now you have a *bandpass filter*, so called because you are passing just a narrow band of frequencies in the midrange.

The trimmer at bottom-left functions as a volume control, but you'll find that it only works properly in the middle part of its range. If you push it too high or too low, the circuit starts oscillating. This is a problem with amplifier circuits. The solution tends to involve adding small and large capacitors in various locations. I decided not to bother, because the midrange of the trimmer is usable.

The capacitors and coils in this circuit are all operating on a *passive* basis. They block some frequencies, but they don't boost any frequencies. A more sophisticated audio filtering system uses transistors to provide *active* filtering, but requires a lot more electronics.

Theory: Waveforms

If you blow across the top of a bottle, the mellow sound that you hear is caused by the air vibrating inside the bottle. If you could make a graph of the pressure waves, they would have a rounded profile.

If you could slow down time and draw a graph of the alternating voltage in any power outlet in your house, it would have the same profile.

If you could measure the speed of a pendulum swinging slowly to and fro in a vacuum, and draw a graph of the speed relative to time, once again it would have the same profile.

That profile is a *sine wave*, so called because you can derive it from basic trigonometry. In a right-angled triangle, suppose one of the sides adjacent to the right angle is called "a." If you divide the length of "a" by the length of the sloping side of the triangle (the hypoteneuse), the result is the *sine* of the angle opposite side "a."

To make this simpler, imagine a ball on a string rotating around a center point, as shown in Figure 5-48. Ignore the force of gravity, the resistance of air, and other annoying variables. Just measure the vertical height of the ball and divide it by the length of the string, at regular instants of time, as the ball moves around the circular path at a constant speed. Plot the result as a graph, and there's your sine wave, shown in Figure 5-49. Note that when the ball circles below its horizontal starting line, we consider its distance negative, so the sine wave becomes negative, too.

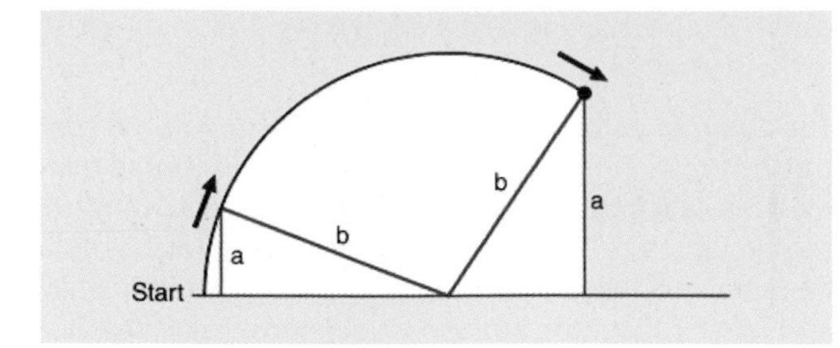

Figure 5-48 *You can draw a sine wave by beginning with simple geometry.*

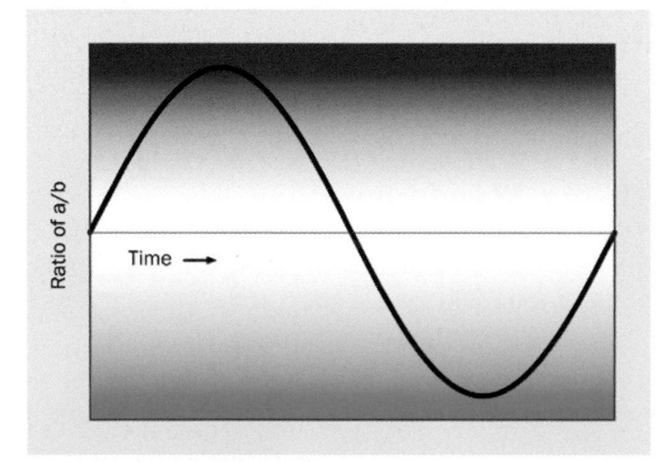

Figure 5-49 *An audio sine wave is generated by any instrument that uses a vibrating column of air—such as a flute. It is a gentle, harmonious sound.*

Why should this particular curve turn up in so many places and so many ways in nature? There are reasons for this rooted in physics, but I'll leave you to dig into that topic if it interests you. Getting back to the subject of audio reproduction, what matters is this:

- The static pressure in the air around you is called *ambient pressure*. It results from air being pulled down by the force of gravity. (Yes, air does have weight.)
- Almost any sound consists of a wave that is higher than ambient pressure, followed by a wave that is lower than ambient pressure—just like waves in the ocean.
- We can represent the higher and lower waves of pressure by voltages that are relatively high and relatively low, which is why I used the red and blue background in Figure 5-49.
- Any sound can be broken down into a mixture of sine waves of varying frequency and amplitude.

Or, conversely:

- If you put together the right mix of audio sine waves, you can create *any sound at all*.

Suppose that there are two sounds playing simultaneously. Figure 5-50 shows one sound as a purple curve and the other as a green curve. When the two sounds travel either as pressure waves through air or as alternating electric currents through a wire, their amplitudes are added together to make the more complex curve, which is shown in black. Now try to imagine dozens or even hundreds of different frequencies being added together, and you have an idea of the complex waveform of a piece of music.

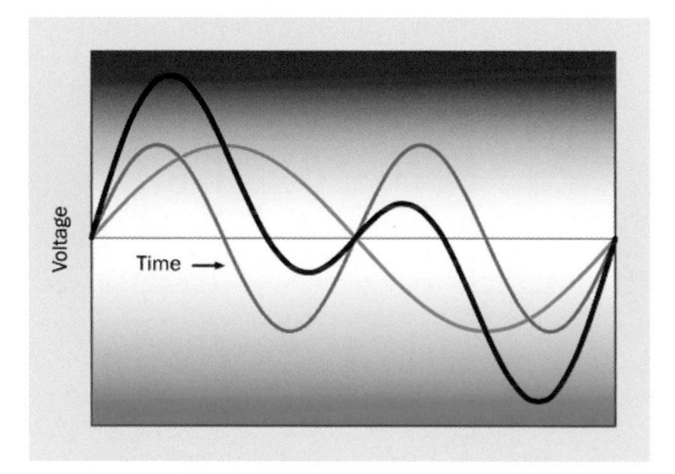

Figure 5-50 *When two sine waves are generated at the same time (for instance, by two musicians, each playing a flute), the combined sound creates a compound curve. The purple sine wave is twice the frequency of the green sine wave. The compound curve (black line) is the sum of the distances of the sine waves from the baseline of the graph.*

An astable 555 timer circuit creates a *square wave*. This is because the output from the timer switches abruptly from low to high and back again. The result is shown in Figure 5-51. A sine wave sounds gentle and melodious, as it varies smoothly. A square wave tends to sound harsh, and has a "buzz" to it. That buzz is really composed of *harmonics*, meaning frequencies that are two or more times the basic frequency.

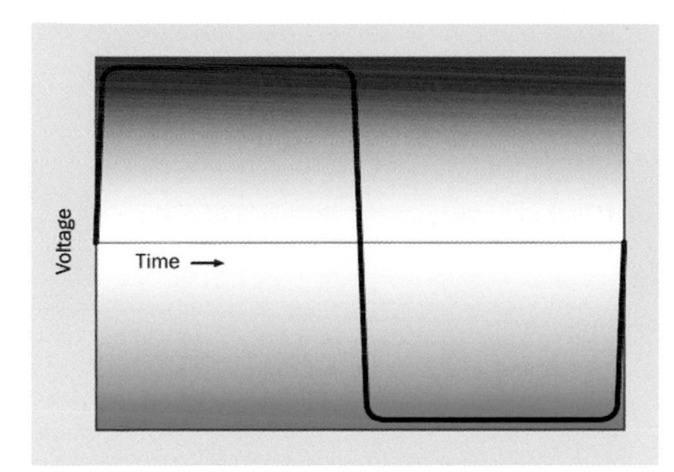

Figure 5-51 *The square wave that you might obtain from a source such as a 555 timer, which switches its output abruptly on and off.*

Because a square wave contains high-frequency harmonics, it's a good choice for testing audio filters. A low-pass filter, which only allows lower frequencies, will remove the buzz by rounding off the corners of the square wave.

Mangling Some Music

You may be wondering, if the LM386 is an audio amplifier, can it amplify music? Yes, and in fact, that's what it is designed to do. You can test this for yourself using any audio device with a headphone output.

Bear in mind that the LM386 is only a mono amplifier, so you won't be able to hear both audio channels from your music player. To obtain just one of them, use a cable with a miniature audio jack at each end. Cut off one of them, strip the insulation from the wire, and you will find a mesh of fine wires that are the shielding in the cable, to be connected with negative ground. Inside the shielding will be two conductors, carrying the signals for the left channel and the right channel. Snip off one of them (it doesn't matter which one), and throw it away—but don't allow the conductor in the residual stub of wire to short-circuit with the shielding.

Strip insulation from the remaining conductor. The wires inside are very thin, and you will be able to deal with them more easily in this experiment if you add a little solder. The desired result is shown in Figure 5-52.

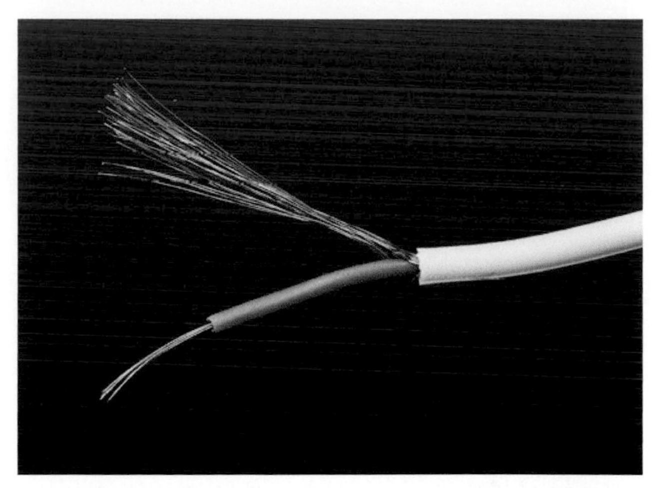

Figure 5-52 *An audio cable stripped to expose its shielding and one conductor. The shielding connects with negative ground.*

Make sure the power to your amplifier circuit is off, and push all the slide switches to the down positions. Remove the orange piece of wire that connects pin 3 of the 555 timer with the 220µF capacitor below it. You are taking the 555 timer out of the circuit, and using the positive end of capacitor C6 as your input point.

Use one of your alligator test leads to grab the positive capacitor lead, and attach the other end of the test lead to the audio conductor in your cable. Use another test lead to connect the shielding from the cable to negative ground in your circuit. It's essential that your music player must share negative ground with your amplifier circuit.

Switch on your circuit, then switch on your music player, and you should hear music. If it's too loud and distorted, you may need to insert a 1K or 10K resistor between the audio wire from the music player and the positive end of the capacitor.

Once you have the volume right, you can play with your high-pass and low-pass filters to see how they affect the music. They won't make it sound good, but they will make it sound different.

Background: Crossover Networks

In a traditional audio system, each speaker cabinet contains two drivers—one of them a small speaker called a *tweeter*, which reproduces high frequencies, the other a large speaker known as a *woofer*, which reproduces low frequencies. (Modern systems often remove the woofer and place it in a separate box of its own that can be positioned almost anywhere, because the human ear has difficulty sensing the direction of low-frequency sounds. In this system, the woofer may be referred to as a *subwoofer*, because it is capable of reproducing very low frequencies.)

Audio frequencies are divided between a tweeter and a woofer by filtering them, so that the tweeter doesn't try to deal with any low frequencies, and the woofer is protected from high frequencies. The circuit that takes care of this is called a "crossover network," and truly hardcore audiophiles have been known to make their own (especially for use in car systems) to go with speakers of their choice in cabinets that they design and build themselves.

If you want to make a crossover network, you should use high-quality polyester capacitors (which have no polarity, last longer than electrolytics, and are better made) and a coil that has the right number of turns of wire and is the right size, to cut high frequencies at the appropriate point. Figure 5-53 shows a polyester capacitor, while Figure 5-54 shows an audio crossover coil that I bought on eBay for $6. I was curious to find out what was inside it, so I took it apart.

Figure 5-53 *Some nonelectrolytic capacitors have no polarity, such as this high-quality polyester film capacitor.*

Figure 5-54 *What exotic components may we find inside this high-end audio component?*

First I peeled away the black vinyl tape that enclosed the coil. Inside was some typical magnet wire—copper wire thinly coated with shellac or semitransparent plastic, as shown in Figure 5-55. I unwound the wire, and as I did so, I counted the number of turns.

Figure 5-55 *The black tape is removed, revealing a coil of magnet wire.*

Figure 5-56 shows the wire and the spool that it was wound around.

Figure 5-56 *The audio crossover coil consists of a plastic spool and some wire. Nothing more.*

So here's the specification for this particular coil in an audio crossover network. Forty feet of 20-gauge copper magnet wire, wrapped in 200 turns around a small plastic spool.

Conclusion: there's a lot of mystique attached to audio components. They are frequently overpriced, and you can make your own coil if you start with these parameters and adjust them to suit yourself.

Suppose you want to put some thumping bass speakers into your car. Could you build your own filter so that they only reproduce the low frequencies? Absolutely—you just need to wind a coil, adding more turns until it cuts as much of the high frequencies as you choose. Just make sure the wire is heavy enough so that it won't overheat when you push 100 or more audio watts through it.

Here's another project to think about: a color organ. You can tap into the output from your stereo and use filters to divide audio frequencies into three sections, each of which drives a separate set of colored LEDs. The red LEDs will flash in response to bass tones, yellow LEDs in response to the midrange, and green LEDs in response to high frequencies (or whatever colors you prefer). You can put signal diodes in series with the LEDs to rectify the alternating current, and series resistors to limit the voltage across the LEDs to, say, 2.5 volts (when the music volume is turned all the way up). You'll use your meter to check the current passing through each resistor, and multiply that number by the voltage drop across the resistor, to find the wattage that it's handling, to make sure the resistor is capable of dissipating that much power without burning out.

Audio is a field offering all kinds of possibilities if you enjoy designing and building your own electronics.

Experiment 30: Make It Fuzzy

Let's try one more variation on the circuit in Experiment 29. This will demonstrate another fundamental audio attribute: distortion.

What You Will Need

- The breadboarded circuit from Experiment 29, plus:
- 2N2222 transistor (1)
- Resistors: 330 ohms (1), 10K (1)
- Capacitors: 1µF (2) and 10µF (1)

Making the Mods

The modifications to the circuit are very minor. You need to add a transistor, two resistors, and three capacitors. Figure 5-57 shows the new components at the top end of the breadboard, with the preexisting components grayed out.

Figure 5-58 shows the same components, and their values, in the relevant section of the schematic, with other components omitted.

The 2N2222 transistor overloads the input of the LM386, while the 1µF capacitors, C8 and C9, limit low frequencies to emphasize the fuzzy effect.

The purpose of C10 is to boost the LM386. This is a feature of the chip: if you add a capacitor between pin 1 and pin 8, the power of the amplifier can increase from 20:1 to 200:1.

Thus, in two separate ways this unfortunate little amplifier chip is being forced to do more than its designers intended. Naturally enough, it will complain about this cruel treatment.

Make the modifications, and switch on the power. The output previously had some buzz in it, because it was basically a square wave. But now if you experiment with the 10K trimmer and the 1M trimmer, you can make the output scream, Hendrix-style.

If the result is too extreme, you can remove the 330-ohm resistor and substitute one with a slightly higher value. And what exactly is going on here?

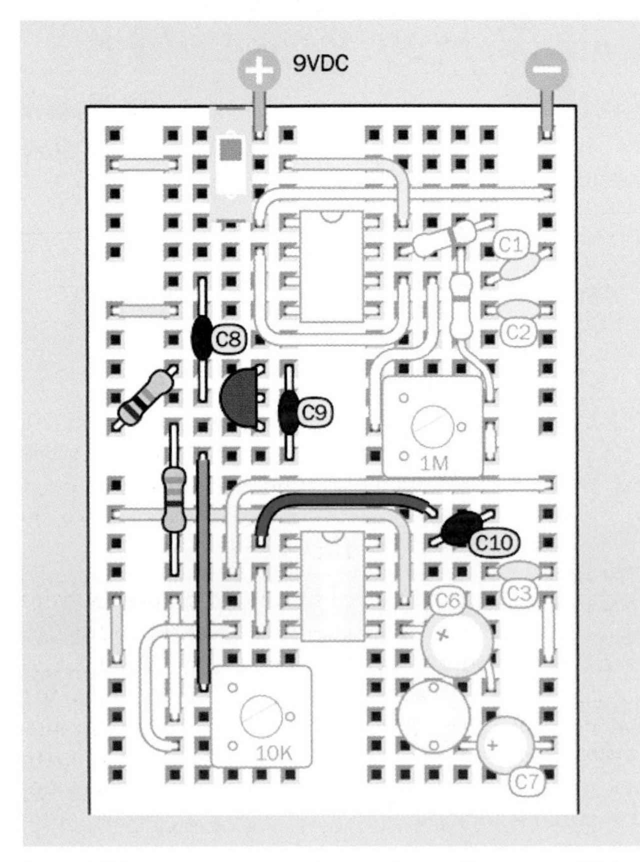

Figure 5-57 *Modifications to the circuit from Experiment 29, to add more distortion.*

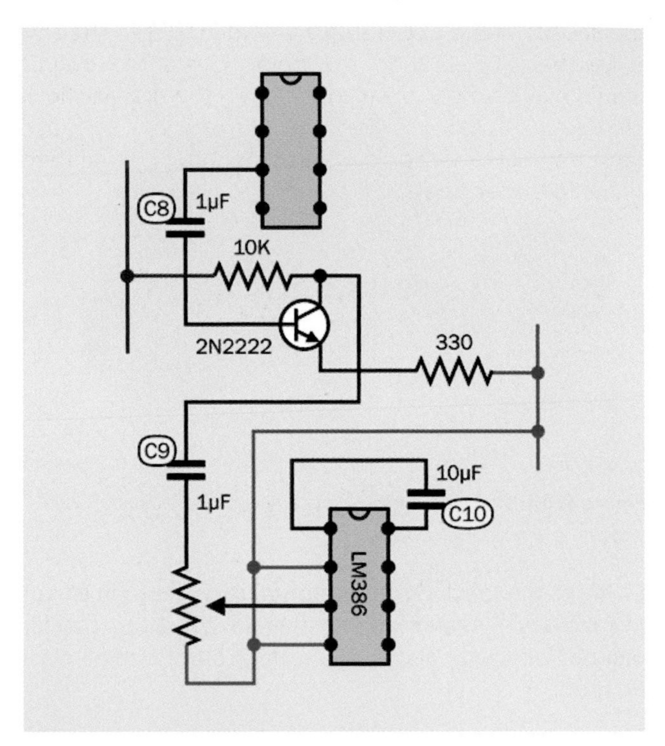

Figure 5-58 *The additional components, with values.*

Background: Clipping

In the early days of "hi-fi" (high-fidelity) sound, engineers labored mightily to perfect the process of audio reproduction. They wanted the waveform at the output end of the amplifier to look identical with the waveform at the input end, the only difference being that it should be bigger, so that it would be powerful enough to drive loudspeakers. The slightest distortion of the waveform was unacceptable.

Little did they realize that their beautifully designed tube amplifiers would be abused by a new generation of rock guitarists whose intention was to create as much distortion as possible.

If you push a vacuum tube—or a transistor—to amplify a sine wave beyond the component's capabilities, it runs out of power and "clips" the top and bottom of the curve. This makes it look more like a square wave, and as I explained Experiment 29, a square wave has a harsh, buzzing quality. For rock guitarists trying to add an edge to their music, the harshness is a desirable feature.

The sequence in Figure 5-59 shows what happens. So long as the output stays within the voltage limits of the amplifier, the signal can be faithfully reproduced. But in the second frame of the sequence, the input to the amplifier has increased to the point where the output would exceed the limits (suggested by the gray sections of the curve). Because the amplifier only has so much power available, it clips the signal, as shown in the third frame in the sequence.

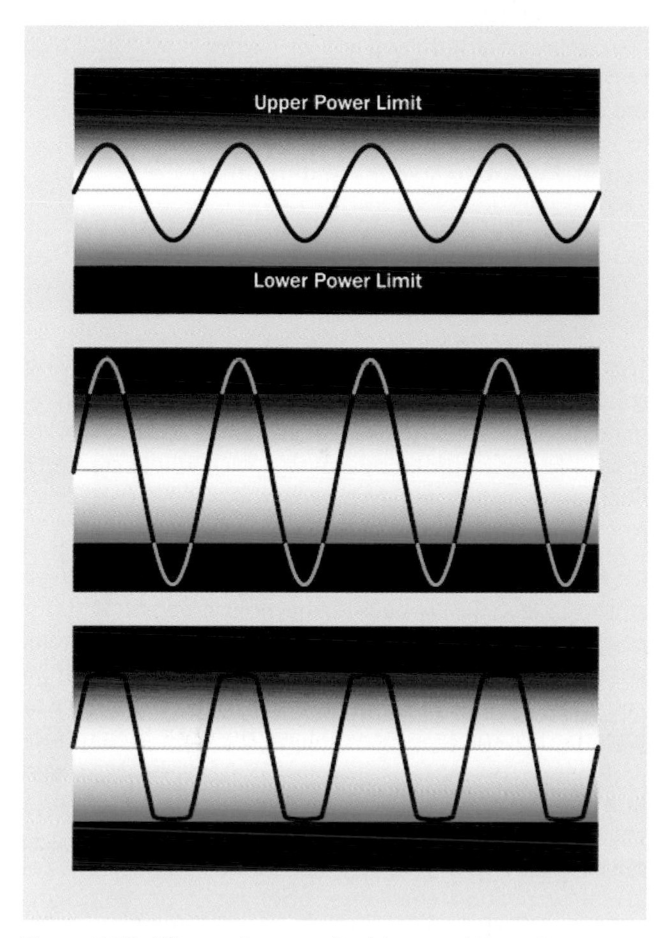

Figure 5-59 *When a sine wave (top) is passed through an amplifier that is turned up beyond the limit of its components, the amplifier chops the wave (bottom).*

For rock guitarists, clipping sounded good, and "stomp boxes" were introduced to create the effect. A very early example is shown in Figure 5-60

Figure 5-60 *This Vox Wow-Fuzz pedal was one of the early stomp boxes, which deliberately induced the kind of distortion that audio engineers had been trying to get rid of for decades.*

Background: Stomp-Box Origins

The Ventures recorded the first single to use a fuzz box, titled "The 2,000 Pound Bee," in 1962. Truly one of the most awful instrumentals ever made, it used distortion as a gimmick and must have made other musicians conclude that this was a sound destined for oblivion.

Then Ray Davies of the Kinks started experimenting with distortion, initially by plugging the output from one amp into the input of another, supposedly during the recording of his hit "You Really Got Me." This overloaded the input and created clipping that sounded more musically acceptable. From there it was a short step to Keith Richards using a Gibson Maestro Fuzz-Tone when the Rolling Stones recorded "(I Can't Get No) Satisfaction" In 1965.

Today, you can find thousands of advocates promoting as many different mythologies about "ideal" distortion. In Figure 5-61, I've included a schematic from Flavio Dellepiane (*http://www.redcircuits.com*), a circuit designer in Italy who gives away his work (with a little help from Google AdSense).

Flavio is a self-taught Maker, having gained much of his knowledge from electronics magazines such as the old British publication, *Wireless World*. In the fuzz circuit that I'm including here, he uses a very high-gain amplifier consisting of three field-effect transistors (FETs), which closely imitate the rounded square wave typical of an overdriven tube amp.

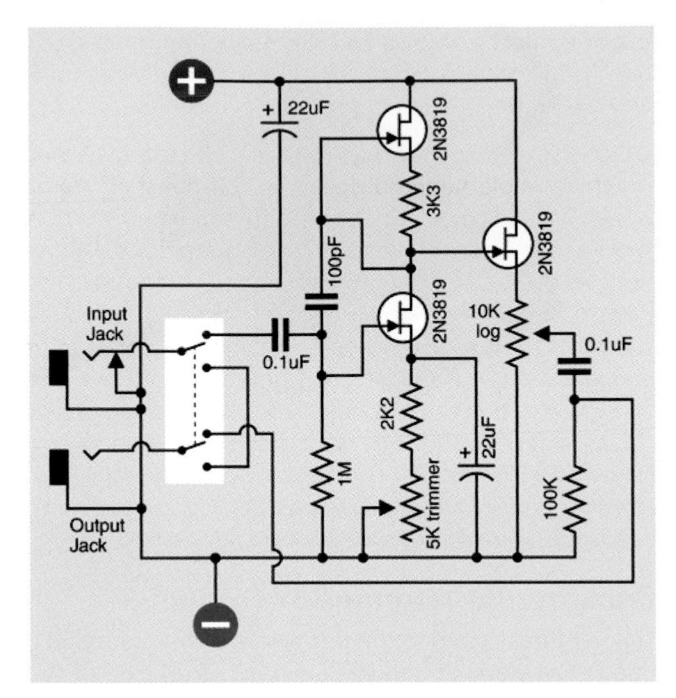

Figure 5-61 *This circuit designed by Flavio Dellepiane uses three transistors to simulate the kind of distortion that used to be created by overloading the input of a tube amplifier.*

Flavio offers dozens more schematics on his site, developed and tested with a dual-trace oscilloscope, low-distortion sinewave oscillator (so that he can give audio devices a "clean" input before he abuses it), distortion meter, and precision audio voltmeter. This last item, and the oscillator, were built from his own designs, and he gives away their schematics, too. Thus his site provides one-stop shopping for home-audio electronics hobbyists in search of a self-administered education.

Before fuzz, there was tremolo. A lot of people confuse this with vibrato, so let's clarify that distinction right now:

- Vibrato applied to a note makes the frequency waver up and down, as if a guitarist is bending the strings.
- Tremolo applied to a note makes its volume fluctuate, as if someone is turning the volume control of an electric guitar up and down very quickly.

Harry DeArmond sold the first tremolo box, which he named the Trem-Trol. It looked like an antique portable radio, with two dials on the front and a carrying handle on top. Perhaps in an effort to cut costs, DeArmond didn't use any electronic components. His steam-punk-ish Trem-Trol contained a motor fitted with a tapered shaft, with a rubber wheel pressing against it. The speed of the wheel varied when you turned a knob to reposition the wheel up and down the shaft. The wheel, in turn, cranked a little capsule of "hydro-fluid," in which two wires were immersed, carrying the audio signal. As the capsule rocked to and fro, the fluid sloshed from side to side, and the resistance between the electrodes fluctuated. This modulated the audio output.

Today, Trem-Trols are an antique collectible. Johann Burkard has posted an MP3 of his DeArmond Trem-Trol online (*http://bit.ly/tremolo-clip*) so you can actually hear it.

The idea of using a mechanical source for electronic sound mods didn't end there. The original Hammond organs derived their unique, rich sound from a set of toothed wheels turned by a motor. Each wheel created a fluctuating inductance in a sensor like the record head from a cassette player.

It's fun to think of other possibilities for motor-driven stomp boxes. Going back to tremolo: imagine a transparent disc masked with black paint, except for a circular stripe that tapers at each end. While the disc rotates, if you shine a bright LED through the transparent stripe toward a phototransistor, you have the basis for a tremolo device. You could even create never-before-heard tremolo effects by keeping a library of discs with different stripe patterns. Figure 5-62 shows what I have in mind, while Figure 5-63 suggests some disc patterns. For a real fabrication challenge, how about an automatic disc changer?

In the world of solid-state electronics, today's guitarists can choose from a smorgasbord of effects, all of which can be home-built using plans available online. For reference, try these special-interest books:

Analog Man's Guide to Vintage Effects by Tom Hughes (For Musicians Only Publishing, 2004). This is a guide to every vintage stomp box and pedal you can imagine.

Figure 5-62 *A hypothetical neo-electromechanical tremolo generator.*

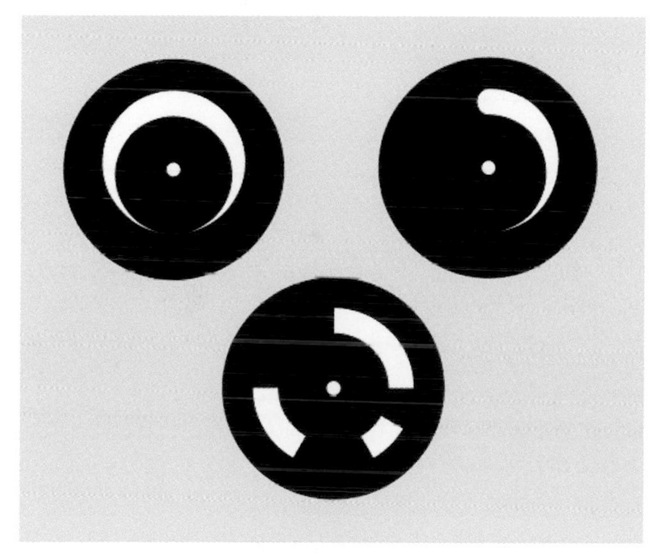

Figure 5-63 *Different stripe patterns could be used to create various tremolo effects.*

How to Modify Effect Pedals for Guitar and Bass by Brian Wampler (Custom Books Publishing, 2007). This is an extremely detailed guide for beginners with little or no prior knowledge. Currently it is available only by download, from sites such as Open Library (*http://www.openlibrary.org*).

but you may be able to find the previously printed edition from secondhand sellers, if you search for the title and the author.

Of course, you can always take a shortcut by laying down a couple hundred dollars for off-the-shelf stomp boxes that use digital processing to emulate distortion, metal, fuzz, chorus, phaser, flanger, tremolo, delay, reverb, and several more, all in one convenient package. Purists, of course, will claim that it "doesn't quite sound the same," but maybe that's not the point. Some of us simply can't get no satisfaction until we build our own stomp box and then tweak it, in search of a sound that doesn't come off-the-shelf and is wholly our own.

Experiment 31: One Radio, No Solder, No Power

Getting back to the principle of inductance, I want to show you how it can enable a simple circuit that receives AM radio signals without a power supply. This is often known as a crystal radio, because the earliest examples used a natural mineral crystal that functioned as a semiconductor. The idea originated at the dawn of telecommunications, but if you've never tried it, you've missed an experience that is truly magical.

What You Will Need

- Rigid cylindrical object, about three inches in diameter, such as a vitamin bottle or water bottle (1)
- Hookup wire, 22 gauge, 60 feet minimum
- Heavier wire, 16 gauge preferred, 50 to 100 feet (this wire may be stranded, and you can try a thinner gauge to reduce the cost, although your radio may not pull in so many stations)
- Polypropylene rope ("poly rope") or nylon rope, 10 feet
- Germanium diode (1)
- High-impedance earphone (1)
- Test lead (1)
- Alligator clips (3) or use extra test leads

Optional:

- 9-volt power supply (battery or AC adapter)
- LM386 single-chip amplifier
- Small loudspeaker (2" acceptable)

The diode and headphone can be ordered from the Scitoys Catalog (*http://www.scitoyscatalog.com*). A high-impedance earphone is also available from amazon.com.

Step 1: The Coil

You need to create a coil that will resonate with radio transmissions in the AM waveband. The coil will consist of 65 turns of 22-gauge hookup wire, measuring approximately 60 feet.

You can wind the coil around any empty glass or plastic container, so long as it has parallel sides providing a constant diameter close to 3". A water bottle will do, if it isn't the type made of extremely thin plastic that will be easily squashed or deformed under pressure.

I just happened to have a vitamin bottle that was exactly the right size. In the photographs, you'll notice that it has no label. I softened its adhesive with a heat gun (lightly, to avoid melting the bottle) and then peeled it off. Some remaining adhesive residue was removed with a little xylene.

After you prepare a clean, rigid bottle, use a sharp object such as an awl or a nail to punch two pairs of holes in it, as shown in Figure 5-64. The holes will be used to anchor the ends of the coil.

Strip some insulation from the end of your hookup wire, and anchor it in one pair of holes, as shown in Figure 5-65. Now wrap five turns of wire around the bottle, and keep it from unwinding itself by applying a small, temporary piece of tape. Duct tape is ideal, or regular Scotch tape will do. "Magic" tape isn't strong enough and will be difficult to remove.

Figure 5-64 *The holes will anchor wire wrapped around the bottle.*

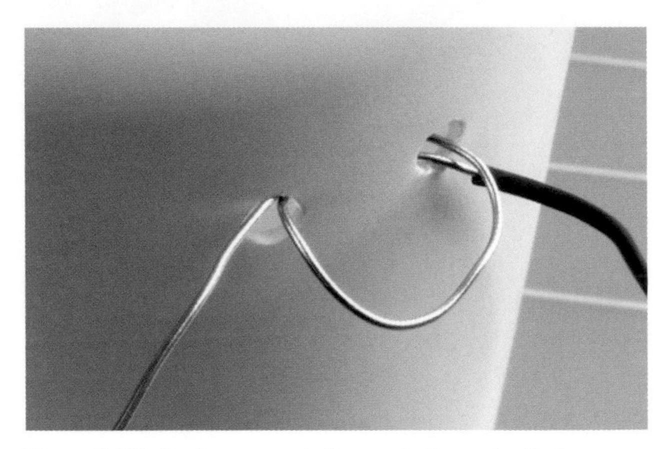

Figure 5-65 *Anchor one end of your wire in a pair of holes.*

Now you need to strip away about half an inch of insulation from the wire. The idea is that you should be able to tap into the coil at this point. Using your wire strippers, make an incision in the insulation and then pull the plastic coating away from your incision. See Figure 5-66.

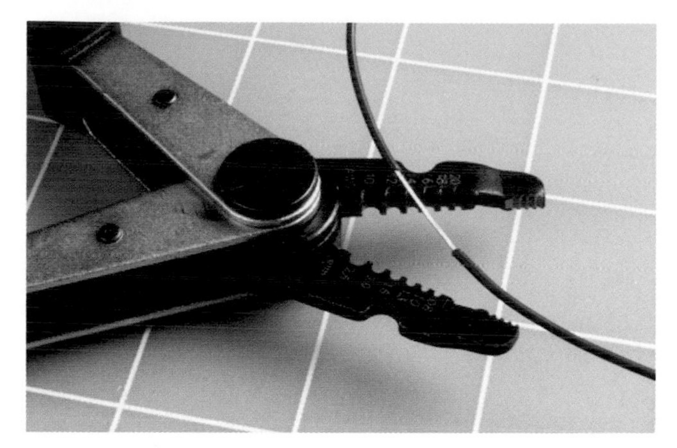

Figure 5-66 *Use your wire strippers and your thumbnails to pull back about half an inch of insulation.*

The next step is to twist the exposed wire into a loop, to make it easily accessible and prevent the insulation from closing up. See Figure 5-67.

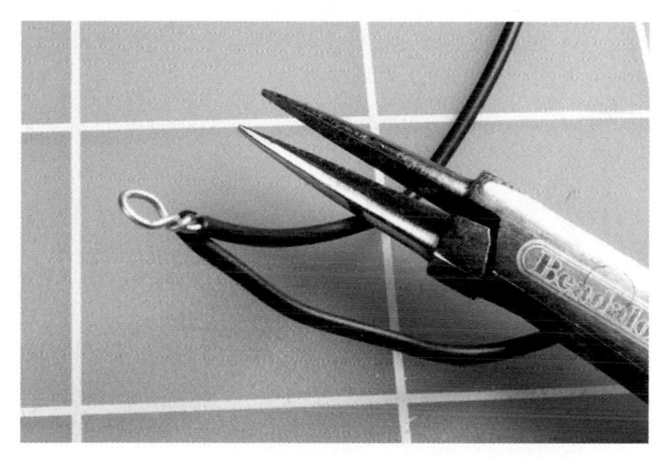

Figure 5-67 *Create a loop in the section of wire that you exposed.*

You just created a *tap* on your coil. Remove the piece of tape that you used to hold your first five turns temporarily, and wind another five turns around the bottle. Apply the tape again, and create another tap. You'll need a total of 12 of them, altogether. It doesn't matter if they don't line up with each other precisely. When you have made the last tap, wind five more turns around the bottle and then cut the wire. Bend the end into a U shape about a half-inch in diameter, so that you can hook it through the pair of holes that you drilled at the far end of the bottle. Pull the wire through, then loop it around again to make a secure anchor point.

My coil wrapped around a vitamin bottle is shown in Figure 5-68.

Figure 5-68 *The completed coil, wrapped tightly around the bottle.*

Your next step is to set up an antenna, which will be a section of wire that is as thick as possible and as long as possible. If you live in a house with a yard outside, this is easy: just open a window, toss out a reel of 16-gauge wire while holding the free end, then go outside and string up your antenna by using polypropylene rope ("poly rope") or nylon rope, available from any hardware store, to hang the wire from any available trees, gutters, or poles. The total length of the wire should be 50 to 100 feet. Where it comes in through the window, suspend it on another length of poly rope. The idea is to keep your antenna wire as far away from the ground or from any grounded objects as possible.

If you don't have an accessible yard, you can string up your antenna indoors, hanging it with poly rope or nylon rope from window treatments, door knobs, or anything else that will keep it off the floor. The antenna doesn't have to be in a straight line; in fact you can run it all around the room.

Caution: High Voltage!

The world around us is full of electricity. Normally we're unaware of it, but a thunderstorm is a sudden reminder that there's a huge electrical potential between the ground below and the clouds above.

If you put up an outdoor antenna, never use it if there is any chance of a lightning strike. This can be extremely dangerous. Disconnect the indoor end of your antenna, drag it outside, and push the end of the wire into the ground to make it safe.

Antenna and Ground

Use an alligator test lead to connect the end of your antenna wire with the top end of the coil that you made.

Next you need to establish a ground wire. This literally has to connect with the ground outside. Ideally, you should bury a couple of feet of bare wire in soft, moist earth—although this may be problematic if you live in a desert area, as I do. If you use a grounding stake of the type sold by wholesale electrical supply houses to ground welding equipment, be careful where you hammer it into the ground. You don't want to hit any hidden conduits.

A cold-water pipe is often suggested as a good connection with the ground, but (duh!) this will work only if the pipe is made of metal. Even if your home is plumbed with copper pipes, a section may have been repaired and replaced with plastic at some time in the past.

Probably the most reliable option is to attach the wire to the screw in the cover plate of an electrical outlet, as the electrical system in your house will be ultimately grounded. But be sure to anchor the wire securely, so there is absolutely no risk of it touching the sockets in the outlet. I would prefer not to insert the ground wire in the ground socket of the outlet, because there is the risk of poking it into the live socket by mistake.

Now you need a couple of slightly hard-to-find items: a germanium diode, which functions like a silicon-based diode but is better suited to the tiny voltages and currents that you'll be dealing with, and a high-impedance earphone. The kind of earphones or ear buds that you use with a media player will not work here; this has to be an old-school item, like the one shown in Figure 5-69. If it has a plug on the end, you'll have to snip it off and then carefully strip insulation from the tip of each wire.

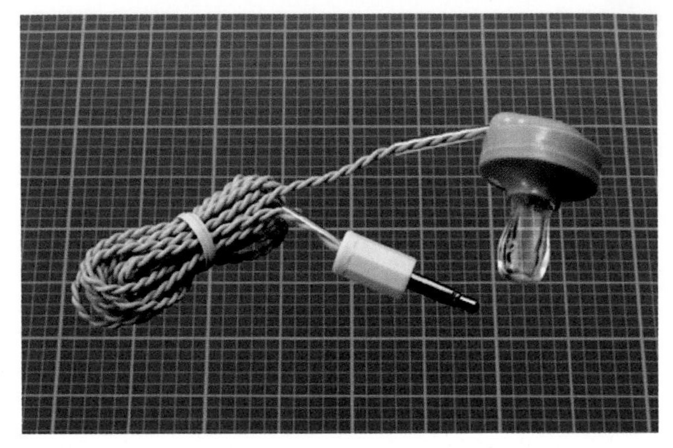

Figure 5-69 *This is the type of earphone you need for your no-power radio.*

The parts are assembled with test leads and alligator clips, as shown in Figure 5-70. The real-world version that I built isn't as neat as the diagram, but the connections are still the same, as shown in Figure 5-71. Notice that the test lead at the bottom can latch on to any of the taps on your coil. This is how you will be tuning your radio.

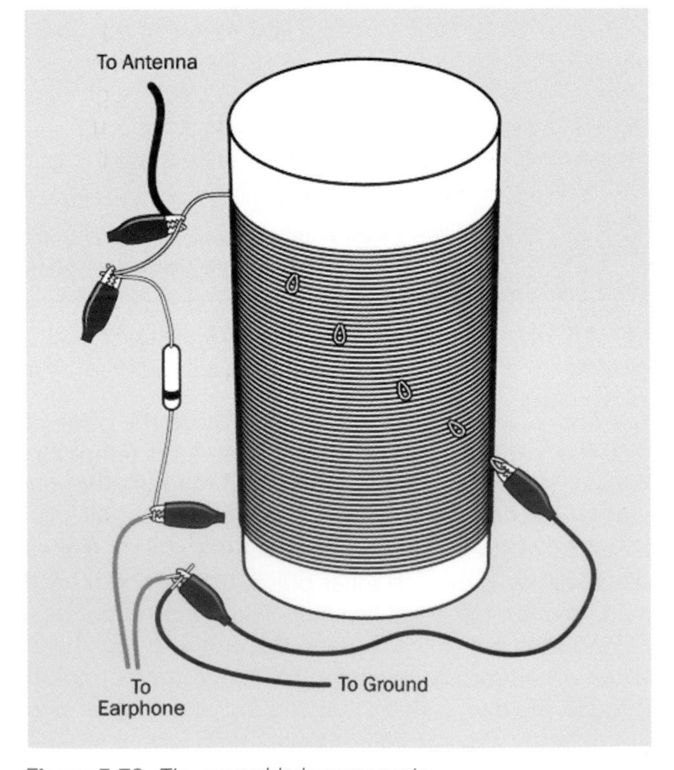

Figure 5-70 *The assembled components.*

Figure 5-71 *The real-world version.*

If you followed the instructions, and you live within 20 or 30 miles of an AM radio station, and your hearing is reasonably good, you will be able to listen to the faint sounds of radio on your earphone—even though you are not applying any power to the circuit that you built. This project is many decades old, but can still be a source of surprise and wonder. (See Figure 5-72.)

If you live too far from a radio station, or you can't put up a very long antenna, or your ground connection isn't very good, you may not hear anything. Don't give up; wait till sunset. AM radio reception changes radically when the sun is no longer exciting the atmosphere with its radiation.

To choose among radio stations, move the alligator clip at the end of your test lead from one tap to another on your coil. Depending on where you live, you may pick up just one station, or several, playing individually or simultaneously.

Figure 5-72 *The simple pleasure of picking up a radio signal with ultra-simple components and no additional power.*

It may seem that you're getting something for nothing here, but really you are taking energy from a source of power—the transmitter located at a radio station. A transmitter pumps power into a broadcasting tower, modulating a fixed frequency. When the combination of your coil and antenna resonates with that frequency, you're sucking in just enough voltage and current to energize a high-impedance headphone.

The reason you had to make a good ground connection is that power will only flow through your coil if it has somewhere to go. You can think of the ground as being an almost infinite power sink, with a reference voltage of zero. The transmitter at an AM radio station is also likely to have a potential relative to ground. See Figure 5-73.

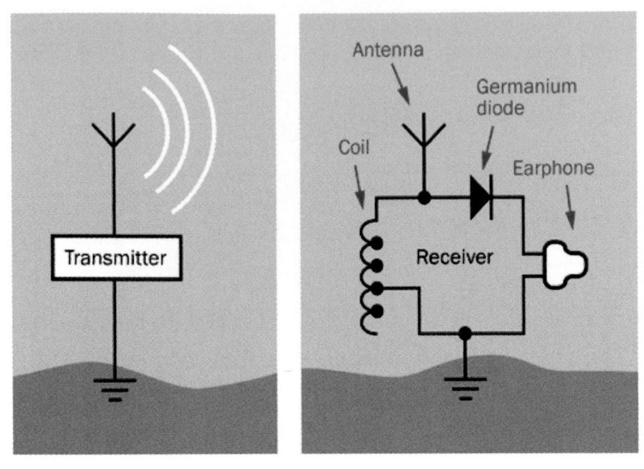

Figure 5-73 *Your no-power radio takes just enough energy from a distant transmitter to create a barely audible sound in your earphone.*

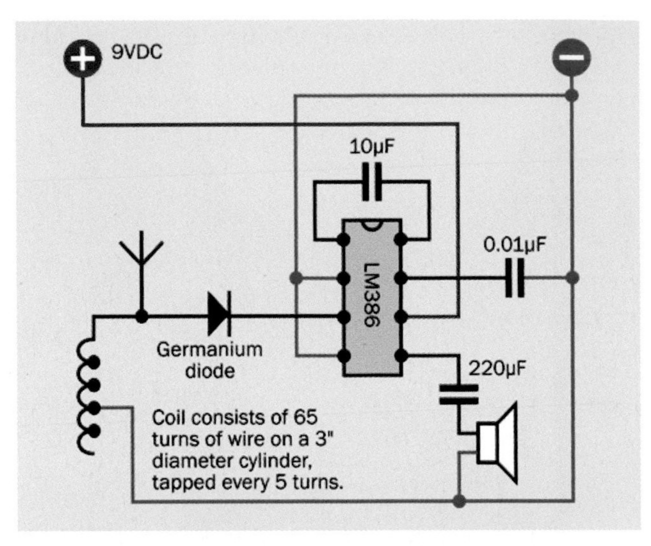

Figure 5-74 *The LM386 single-chip amplifier can make your crystal-set radio audible through a loudspeaker.*

Enhancements

If you have difficulty hearing anything through your earphone, try substituting a piezoelectric transducer, also known as a piezo beeper. You need the type that does not have an oscillator built in, and functions passively, like a loudspeaker. Press it tightly against your ear, and you may find that it works as well as an earphone, or better.

You can also try amplifying the signal. Ideally you should use an op-amp for the first stage, because it has a very high impedance. However, I decided to put op-amps in *Make: More Electronics*, where I had room to explore the topic more thoroughly. As a substitute, you can feed the signal directly into the same LM386 single-chip amplifier from Experiment 29.

Figure 5-74 shows how simple the circuit can be. The germanium diode can connect directly with the LM386 input, as I don't think you'll need a volume control. Be sure to include the 10µF capacitor between pins 1 and 8, to increase the amplification of the chip to its maximum value. Even where I live, about 120 miles from Phoenix, Arizona, I was able to pick up a station broadcasting from the Phoenix area.

If you want to improve the selectivity of your radio, you can add a variable capacitor to tune the resonance of your circuit more precisely. Variable capacitors are uncommon today, but you can find one at the same specialty source that I recommended for the earphone and the germanium diode: the Scitoys Catalog (*http://www.scitoyscatalog.com*).

This source is run by a smart man named Simon Quellan Field, whose site suggests many fun projects that you can pursue at home. One of his clever ideas is to remove the germanium diode from your radio circuit and substitute a low-current LED in series with a 1.5-volt battery. This didn't work for me, because I live in a remote location; but if you're close to a transmitter, you may be able to see the LED varying in intensity as the broadcast power runs through it.

Theory: How Radio Works

High-frequency electromagnetic radiation can travel for miles. To make a radio transmitter, I could use a 555 timer chip running at, say, 850 kHz (850,000 cycles per second), and would pass this stream of pulses through an extremely powerful amplifier to a transmission tower—or maybe just a long piece of wire. If you had some way to block out all the other electromagnetic activity in the air, you could detect my signal and amplify it.

This was more or less what Guglielmo Marconi did when he performed a groundbreaking experiment in 1901, except that he had to use a primitive spark gap, rather than a 555 timer, to create the oscillations. His transmissions were of limited use, because they had only two states: on or off. You could send Morse code messages, and that was all.

Marconi is pictured in Figure 5-75.

Figure 5-75 *Guglielmo Marconi, the great pioneer of radio (photograph from Wikimedia Commons).*

Five years later, the first true audio signal was transmitted by imposing lower audio frequencies on the high-frequency carrier wave. In other words, the audio signal was "added" to the carrier frequency, so that the power of the carrier varied with the peaks and valleys of the audio. This is shown in Figure 5-76.

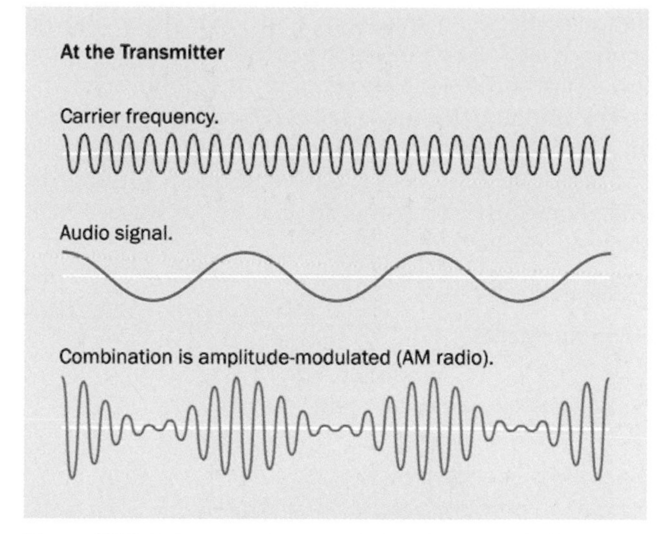

Figure 5-76 *Using a carrier wave of fixed frequency to transmit an audio signal*

At the receiving end, a very simple combination of a capacitor and a coil detected the carrier frequency out of all the other noise in the electromagnetic spectrum. The values of the capacitor and the coil were chosen so that their circuit would resonate at the same frequency as the carrier wave. The basic circuit is shown in Figure 5-77, where the variable capacitor is represented by a capacitor symbol with an arrow through it.

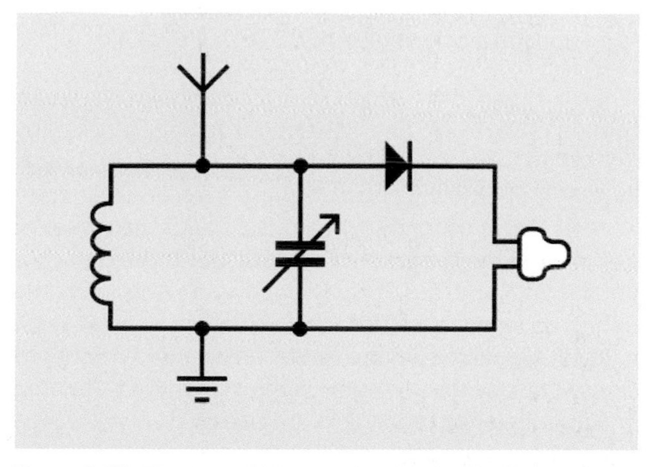

Figure 5-77 *When a variable capacitor is added to the previous circuit, it enables better discrimination among different signals sharing the spectrum.*

The carrier wave fluctuates up and down so rapidly, an earphone cannot possibly keep up with the positive-negative variations. It will remain hesitating at the midpoint between the highs and lows, producing no sound at all. A diode solves this problem by blocking the lower half of the signal, leaving just the positive voltage spikes. Although these are still very small and rapid, they are now all pushing the diaphragm of the earphone in the same direction, so that it averages them out, approximately reconstructing the original sound wave. This is shown in Figure 5-78.

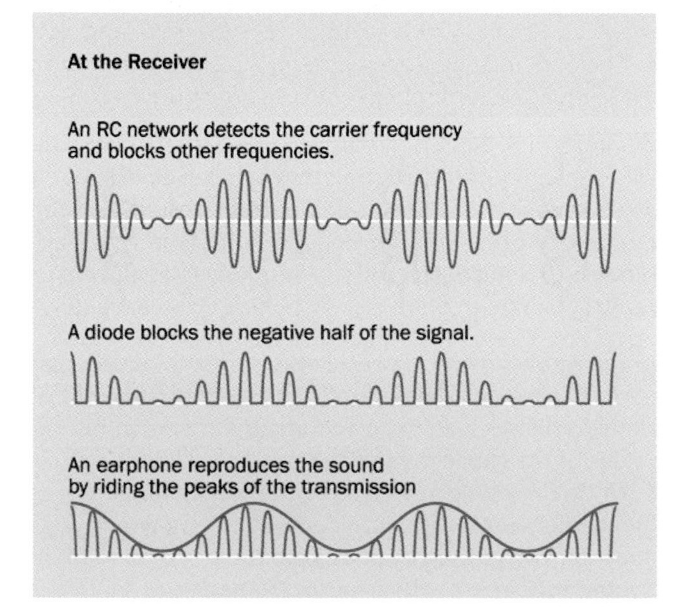

At the Receiver
An RC network detects the carrier frequency
and blocks other frequencies.
A diode blocks the negative half of the signal.
An earphone reproduces the sound
by riding the peaks of the transmission

Figure 5-78 *How a simple AM radio receiver decodes the signal and reproduces it on an earphone.*

When a capacitor is added to a receiving circuit, an incoming pulse from the transmitter is initially blocked by the self-inductance of the coil, while it charges the capacitor. If an equally negative pulse is received after an interval that is properly synchronized with the values of the coil and the capacitor, it coincides with the capacitor discharging and the coil conducting. In this way, the right frequency of carrier wave makes the circuit resonate in sympathy. At the same time, audio-frequency fluctuations in the strength of the signal are translated into fluctuations in voltage in the circuit.

If you are wondering what happens to other frequencies pulled in by the antenna, the lower ones pass through the coil to ground, while the higher ones pass through the capacitor to ground. They are just "thrown away."

The waveband allocated to AM radio ranges from 300kHz to 3MHz in carrier frequencies. Many other frequencies are allocated for special purposes, such as ham radio. It's not so difficult to pass the ham radio exam, and with appropriate equipment and a well-situated antenna, you can talk directly with people in widely scattered locations—without relying on any communications network to connect you.

Experiment 32: Hardware Meets Software

I'm guessing that many readers of this book—perhaps most readers—have heard of the Arduino. This experiment, and the next two, will show you how you can set up an Arduino and then write programs for it, instead of just downloading applications that you find online.

What You Will Need

- Arduino Uno board, or compatible clone (1)
- USB cable with A-type and B-type connector at opposite ends (1)
- A desktop or laptop computer with an available USB port (1)
- Generic LED (1)

Definitions

A microcontroller is a chip that works like a little computer. You write a program consisting of instructions that the microcontroller can understand, and copy them into some memory in the chip. The memory is nonvolatile, meaning that it will preserve its contents even when the power is switched off.

Normally I would suggest jumping right in and writing a program, but learning to use a microcontroller entails a bigger investment of time and mental energy than was required by the components that I have dealt with previously. How do you know if you want to get into this, until you know some more details? Therefore, I must begin with an explanation and orientation. After that, this experiment will take you through the process of setting up an Arduino and performing the most basic test. Experiments 33 and 34 will take you further into programming the Arduino and using other components in conjunction with it.

The setup-and-test procedure may take an hour or two. You should allocate the time when you will be able to follow the instructions without distraction. Once you have gone through this initial process, everything will be much easier.

Real-World Applications

A typical microcontroller application might run like this:

- Receive an input from a rotational encoder that is functioning as a volume control on a car stereo.
- Figure out which way the encoder is rotating.
- Count the pulses from the encoder.
- Tell a programmable resistor how many equivalent steps it should take, to adjust the volume of the stereo up or down.
- Wait for more inputs.

A microcontroller could also handle a much larger application, such as all the inputs, outputs, and decisions associated with the intrusion alarm system in Experiment 15. It would scan sensors, activate a siren through a relay after a delay period, receive and verify a keypad sequence when you want to shut off the alarm—and much more.

All modern cars contain microcontrollers, which deal with complex tasks such as timing the ignition in the engine—and simple matters such as sounding a chime if you don't fasten your seat belt.

A microcontroller can handle small but important tasks that I discussed in previous experiments, such as debouncing a pushbutton or generating an audio frequency.

Bearing in mind that one little chip can do so many different things, why don't we use it for everything?

The Right Tool for the Job?

A microcontroller is versatile and powerful, but is more suitable in some situations than others. It is ideal for applying logical operations along the lines, "If this happens, do that, but if that happens, do something else," but it adds cost and complexity to a project, and of course it entails a major learning process: you have to master the computer language that tells the microcontroller what to do.

If you don't want to learn the language, you can download and use programs that other people have written. Many Makers use this option, because it provides immediate results. You can find thousands of Arduino programs in libraries online, and they cost nothing.

But a program probably won't do exactly what you want. Inevitably, you will need to modify it—and you will be back in a situation where you need to understand the language in order to get full use out of the chip.

Writing a program for the Arduino can be relatively simple, depending on the application. However, it is not a one-step operation. The code has to be tested, and the processes of revision and debugging can be time-consuming. One little error will create unexpected results, or can stop anything from happening at all. You have to reread your code, find the mistakes, and try again.

After you have everything working, the results can be impressive. For this reason, personally I think that programming a microcontroller is worthwhile, so long as your expectations are realistic.

You will have to try it for yourself to find out if it's something you want to pursue actively.

One Board, Many Chips

I'll begin with the most basic question. What is an Arduino?

If you think that it's a chip, you are not quite correct. Each product with the Arduino name on it consists of a small circuit board designed by Arduino, containing a microcontroller chip made by a completely different company. The Arduino Uno board uses an Atmel ATmega 328P-PU microcontroller. The board also includes a voltage regulator, some sockets that enable you to plug in wires or LEDs, a crystal oscillator, a power-supply connector, and a USB adapter that enables your computer to communicate with the board. See Figure 5-79 for a photograph of the board with some of the components identified.

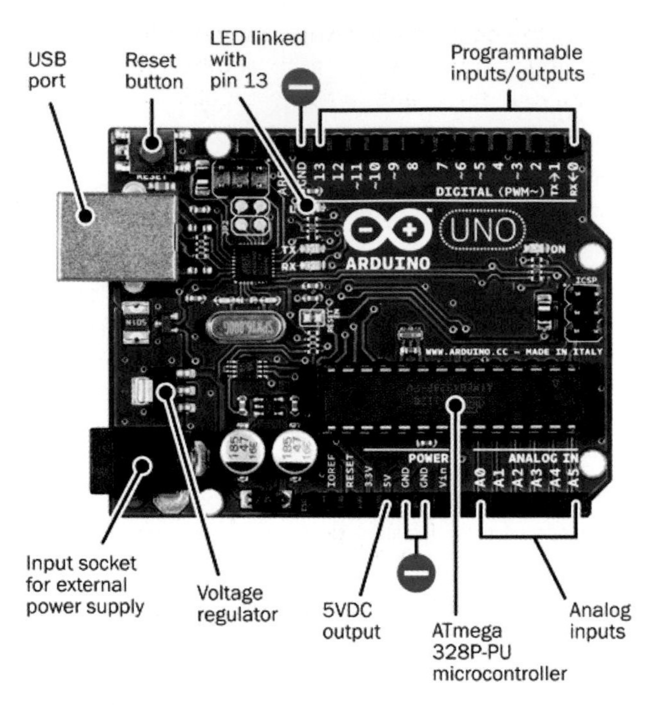

Figure 5-79 *An Arduino Uno board, featuring the ATmega 328P-PU microcontroller made by Atmel.*

If you buy an ATmega 328P-PU on its own from a component supplier, it will cost you less than one-sixth the retail price of an Arduino board containing the chip. Why should you have to pay so much more, just to get that little circuit board? The answer is that development of the board and the clever software that enables you to use it was not a trivial matter.

The clever software is called an IDE, which is an acronym for *integrated development environment*. After you install it on your computer, the IDE is a user-friendly place in which to write a program and then *compile* it, meaning that it converts the C language instructions (which human beings can understand) into machine code (which the Atmel chip will understand). You then copy it to the ATmega chip.

In case this is confusing, I'll summarize the story so far.

- An Arduino is a circuit board into which is plugged an Atmel microcontroller.
- IDE software written by the Arduino company enables you to write a program on your computer.
- After you write your program, the IDE software compiles it, to create code that the chip understands.
- The IDE software sends the code to the Atmel chip, which stores it.

Once your code is in the chip, the chip doesn't really need the Arduino board anymore. Theoretically, you could unplug the ATmega328 and use it somewhere else —in a breadboard, or in a circuit that you solder around it. The chip will still do what you programmed it to do, because the code remains stored in the chip.

In reality there are some little snags associated with this, but you can learn about them by reading a very good book titled *Make: AVR Programming* by Elliot Williams. It tells you exactly how to transplant the ATmega chip.

If you learn how to do this, the implications are significant. You will only need one Arduino board, while you can buy a lot of Atmel chips very cheaply. Put one chip into the board, program it, remove it, and use it in a standalone project. Put another chip into the board, transfer a different program into it, and use it in a different project.

This is relatively easy if you have the version of the Arduino Uno where the microcontroller is a through-hole chip mounted in a socket. You can pry it out with a miniature screwdriver, and substitute another chip by pushing it in with your finger and thumb. (Another version of the Uno comes with a surface-mounted microcontroller, soldered in. This version does not allow you to relocate the chip.)

Beware of Imitations?

Now that I have explained the basics, I'll explain how to set everything up.

While various models of Arduino exist, the one I'm going to deal with here is the Arduino Uno, and my instructions will apply to version R3 or later.

You can buy the Arduino board from many sources, because it was designed and marketed as an "open source" product, allowing anyone to create a copy of it, just as any manufacturer can make a 555 timer (although for slightly different reasons).

Mouser, Digikey, Maker Shed, Sparkfun, and Adafruit all sell genuine Arduino products. On eBay, however, you can find *unlicensed* copies of the Arduino board for one-third of the price. You can see that they are unlicensed, because they don't have the Arduino logo printed on them. To help you in distinguishing the real boards from the imitations, you can see the logo at Figure 5-80.

Figure 5-80 *Only the boards manufactured or licensed by Arduino are supposed to have this logo on them.*

The unlicensed boards are completely legal. This is not like buying pirated software or music. The only feature that Arduino chose to control is its trademark, which may not be used by other manufacturers. (Actually, some scammers use the logo illegally, but you'll know they are not genuine Arduino boards, because they're so cheap.) Just to make things more complicated, because of a dispute between Arduino and its former manufacturer, authentic Arduino boards are marketed under the name Genuino outside of the United States.

If you buy an imitation board, will it be reliable? I would trust the ones from Adafruit, Sparkfun, Solarbotics, Evil Mad Scientist, and a few others. I can't buy and test them all, so you have to make up your own mind about this, based on feedback from other buyers and your general impression of a supplier. But remember, you only have to buy one board, after which you can use it to program multiple Atmel chips, using the plan I described above. So maybe it's not such a big deal to pay a little extra for the genuine Arduino product. By doing so, you help the company to continue its development of new products in the future.

Personally, I bought a genuine Arduino board.

Setup

I will assume, now, that you have acquired an Arduino Uno, or an imitation that you believe is trustworthy. You will also need a standard USB cable with a type A plug on one end and a type B plug on the other end, as shown in Figure 5-81. This is not usually supplied when you buy the board on its own. If you don't have a spare cable, you can borrow one from another device while you do your setup and initial test. Cables are cheaply available online from sites such as eBay.

Figure 5-81 *You will need this type of USB cable to attach your Arduino board to a USB port on your computer.*

You have the board and the cable, so now you need the IDE software. Go to the Arduino website (*http://www.arduino.cc*) and click the "download" tab. Then choose the IDE software that is appropriate for your computer. It is currently available in variants for Mac OS, or Linux, or Windows. I will be using version 1.6.3, but my instructions should also apply to later versions. You can download the IDE software for free.

Note that you need a computer running Windows XP or later, or running Mac OS X 10.7 or later, or Linux 32-bit, or Linux 64-bit. (These requirements are valid at the time of writing. Arduino may change its requirements in the future.)

The procedure for setup under the three different operating systems is described below, primarily based on instructions in the nice little introductory guide titled *Getting Started with Arduino* by Massimo Banzi and Michael Shiloh. You can also find installation instructions on sites such as SparkFun (*http://www.sparkfun.com*) and Adafruit (*http://www.adafruit.com*). Lastly, there are instructions on the Arduino website (*http://www.arduino.cc*) as well.

Unfortunately, all these instructions are slightly different. For instance, Arduino's website tells me to plug in my board before running the installer, but the *Getting Started* book tells me to run the installer before plugging in my board. This makes my task difficult, because both the Arduino website and the *Getting Started* book were written in collaboration with Arduino developers.

Below I will give you my best guess about what will work on each system.

Linux Installation

This is likely to be the most challenging, as there are so many variants of the operating system. I will have to refer you to the Arduino site (*http://playground.arduino.cc/learning/linux*) for guidance.

I regret that I can't help you, myself, with Linux.

Windows Installation

I am inclined to use the procedure recommended in *Getting Started with Arduino*. This is also recommended at the website maintained by Sparkfun.

Do not plug in the board yet. First, identify the IDE installation program that you downloaded. Its name may be something like *arduino-1.6.3-windows.exe*, although by the time you read this, the version numbers almost certainly will have changed. The *.exe* at the end of the file name may not be visible, depending on the system settings on your computer.

- Some guides refer to the installation download as being a zipped file, which has to be unzipped. So far as I can determine, file zipping has been discontinued by Arduino. You can run it as is.

Double-click the icon, and you should see an installation sequence that is familiar to you from installing other software from other vendors.

You have to agree to the terms of a licensing agreement (if you don't agree, you can't run the software).

You are asked if you want shortcuts on your desktop and in your Start menu. Allow the shortcut to be added to your desktop. The Start menu is up to you.

You will be asked to approve the installation folder for the IDE software, and you should accept the default that is suggested.

If you are a dinosaur like me, still using Windows XP (and actually, there are quite a few million of us), you may see a warning such as the one shown in Figure 5-82. The exact look of this will depend on your version of Windows. Ignore the warning by choosing to "continue anyway." If you are asked permission to install device drivers, say "yes."

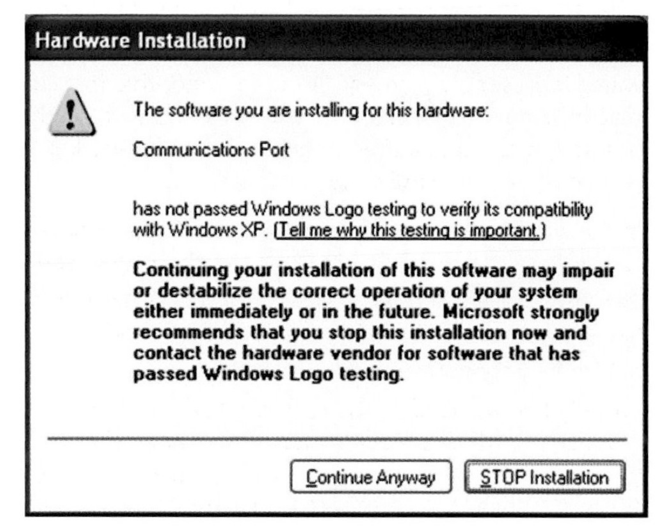

Figure 5-82 *Users of the old WinXP operating system can ignore this warning.*

- On Windows 8, a security feature prevents you from installing unsigned device drivers. This should not be an issue with modern versions of the Arduino IDE installer, but if you somehow obtain older versions from somewhere else, you can Google:

  ```
  sparkfun disable driver signing
  ```

  which should take you to the section of the Sparkfun site that contains helpful advice on the topic.

After installing the IDE software, plug your Arduino into your computer with the USB cable.

You don't need to use the circular power input jack on the Arduino board, so long as the board is connected with a USB port on your computer. The board receives power through the USB cable. Bear in mind that a shorter, thicker cable will minimize voltage drop. Also, if you are using a laptop computer, especially if it is an older one, it may limit the power that you can draw from a USB port to 250mA. Even on a desktop computer, which is supposed to deliver 500mA via USB, that power may be shared across three or four USB ports. A device such as an external hard drive can take a significant amount.

Watch the board while it identifies itself to the computer. You should see a green LED that lights up steadily, and a yellow LED that blinks. A couple more LEDs nearby on the board, labeled TX and RX, should flicker briefly. They show that data is being transmitted and received.

On your computer, go to the shortcut for the IDE software. It is simply named "Arduino," and the installer placed it on your desktop. If you don't want it on the desktop, drag it to a different location. Double-click it to launch the Arduino IDE software.

In the window that opens, pull down the Tools menu, go to the Ports submenu (which is currently named Serial Ports in the Mac version), and you should see a list of serial ports on your computer. They will be named COM1, COM2, and upward.

What is a serial port? In the early days of Windows (and before that, MS-DOS) computers didn't have USB connectors. They used "serial protocol" through D-shaped connectors, and the computer kept track of the connector that it was using by assigning it a "port number." This system is still buried in Windows, even though decades have passed since it was established and the protocol has become rare for domestic applications.

All you need to know is if the Arduino IDE software and Windows have agreed on the port number that has been assigned to your Uno board. Ideally, when you choose Tools > Ports in the IDE software, you will see the Uno in the list with a check mark beside it, and everything is fine. If so, skip the troubleshooting section that follows, and go directly to the Old Arduino Blink Test, below (see "The Old Arduino Blink Test" on page 284).

Windows Troubleshooting

There are two bad possibilities that can affect your port assignment.

- In the Port submenu of the Arduino IDE software, you may see the Arduino Uno listed, but it is not checkmarked. Another port may be checkmarked instead. Try checkmarking the correct port. You may see a warning if the IDE software doesn't approve of the Uno board that you are using. Ignore the warning, click the box saying "Don't show me this again," and proceed to the Old Arduino Blink Test (see "The Old Arduino Blink Test" on page 284).
- You may see no port in the list labeled Arduino Uno. In that case, make a note of the COM ports that are listed. Close the IDE menus. Unplug the Uno board. Wait five seconds. Open the IDE Port submenu again, and see which COM port has disappeared. Close the submenu. Plug the Uno back in. Reopen the submenu. Click the port that has reappeared, to checkmark it. Proceed to the Old Arduino Blink Test (see "The Old Arduino Blink Test" on page 284).

Windows allows you an option to verify your port settings. Click your Start menu and choose the Help and Support service. In the window that opens, type the words "Device Manager" as your search term. It should be the first hit in the list of search results. Open Device Manager. If you are using Windows XP, Device Manager shows you a Ports list. On versions of Windows later than XP, when you go to Device Manager you may have to select View > Show Hidden Devices to reveal the Ports list.

You should see your Arduino Uno listed there. If there is a yellow circle beside it, or an exclamation point, right-click it to find out what's wrong.

If Windows complains that it cannot find the device driver for your board, tell it to look in the Arduino folder containing all the files that were extracted by the installer.

One known issue with the Arduino Uno and Windows ports is that the IDE software can get confused if you have more than nine ports already assigned. This is unusual, but if you have this problem, try unassigning some ports, or manually assign an unused port that has a single-digit number.

If you are still having difficulties, go to "If All Else Fails" on page 284 below.

Mac Installation

After completing the IDE installer download, locate the icon that the computer has created, double-click it, and you'll see a disk image that contains the Arduino IDE software. You can drag this into your Applications folder.

Now plug your Arduino into your computer with the USB cable.

- You don't need to use the circular power input jack on the Arduino board, so long as the board is connected with your computer. The board takes its power through the USB cable.

Watch the board while it identifies itself to the computer, and you should see a green LED that lights up steadily, and a yellow LED that blinks. A couple more LEDs nearby on the board, labelled TX and RX, should flicker briefly. They show that data is being transmitted and received.

If a window opens telling you that a "new network interface" has been detected, click Network Preferences, and then Apply. It doesn't matter if the Uno is described as "not configured." Close the window.

Double-click the program icon for the Arduino IDE that you dragged into your Applications folder. You need to select the correct port for communication with the Uno board. In the Tools menu of the IDE software, click the Serial Port option, and select `/dev/cu.usbmodemfa141` (or a similarly named port) from the list that pops up.

If everything has proceeded as described here, you can continue with the Old Arduino Blink Test (see "The Old Arduino Blink Test" on page 284).

If All Else Fails

This book may stay in print for a while. At least, I hope it does! Software, on the other hand, changes frequently. My installation instructions for the Arduino IDE may be out of date by the time you read this.

I will try to revise my instructions in each new printing and each new ebook version, so that they will be as accurate as possible. But of course you may happen to be reading an older printing or an older ebook.

What to do? Your best bet is to go to the Arduino site, or the Sparkfun site, and follow installation procedures listed there. A website is more easily and quickly updated than a book.

The Old Arduino Blink Test

I am assuming that you have launched the IDE software. Its main window should look something like the screenshot in Figure 5-83, although subsequent versions may change somewhat.

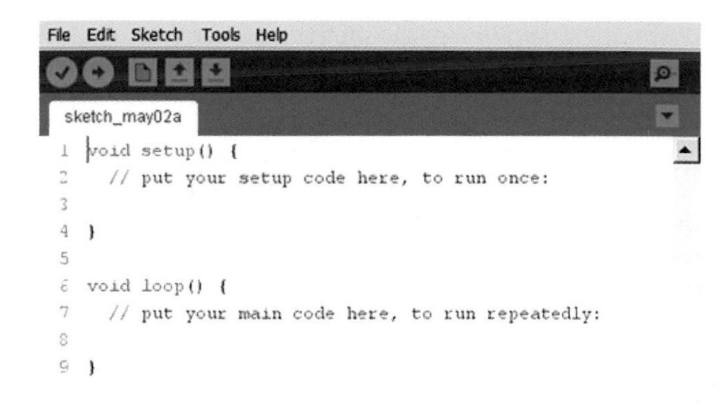

Figure 5-83 *Default window that opens when the Arduino IDE is launched.*

Before you can make your Arduino do something, you have to check that the IDE software has correctly identified the version of the board that you connected with your computer.

In the IDE main window, pull down the Tools menu, open the Boards submenu, and verify that the Arduino Uno has a bullet point beside it, as in Figure 5-84. If it doesn't, click it to select it.

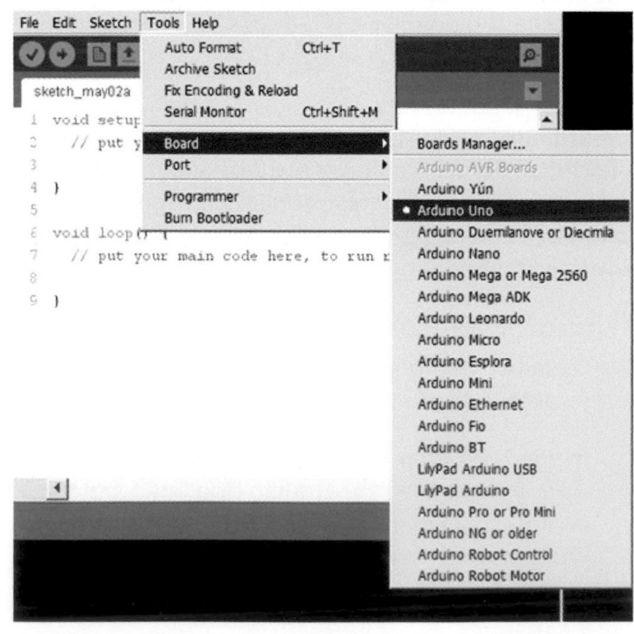

Figure 5-84 *When you are using the Arduino Uno, it should have a bullet point beside it in the Boards Manager submenu.*

Now you are ready to give your Arduino some instructions. At the top of the workspace in the main IDE window, you'll see the word "sketch" followed by today's date and the letter "a." What is this "sketch"? Is it a picture that you are going to draw?

No, in the Arduino world, "sketch" means the same thing as "program." Perhaps this is because the developers didn't want to make people feel intimidated by the idea that they were going to program a computer. Similarly, while he was still alive, Steve Jobs seemed to feel that users of handheld devices would be more comfortable if he referred to programs as "apps." Jobs was probably right, but where Makers are concerned, I don't think they are so easily intimidated. In fact I think they *want* to program computers. Otherwise, why would you be reading this?

"Sketch" means "program" for the Arduino, but I will continue to use the word "program," because that is what it really is, and I feel silly calling it a "sketch." When you read source materials online, people use the word "program" at least as often as they use the word "sketch."

Now I need to remind you of the sequence of events that we are going to follow. First, you *write a program* in the IDE window. Then you *compile* it by selecting a menu option, to convert it into instructions that a microcontroller can understand. Then you *upload* it to the Arduino board, and then the board automatically *runs the program*.

The IDE window on my computer contains some default text shown in Figure 5-83. Future versions of the IDE may do things a little differently, but the principle will be the same. You will see some lines beginning with two slash marks, like this:

```
// put your setup code here, to run once.
```

This is known as a *comment line*. It is for human interest, to explain what's going on.

- When the program that you write is compiled for the microcontroller, the compiler will ignore all lines that begin with // marks.

The next line reads:

```
void setup() {
```

This is a line of *program code*, for the compiler and the microcontroller to understand. But you need to know what it means, because the setup routine has to be at the beginning of every Arduino program, and I'm hoping you will start to write your own programs in the future.

The word `void` tells the compiler that this procedure won't generate any numerical result or output.

The term `setup()` says that the following procedure is something that has to be done once only, right at the beginning.

Notice that there is a { mark following `setup()`.

- Every complete function in C language should be contained within a { mark and a } mark.

Because a { mark must always be followed by a } mark, there must be a } mark somewhere on the opening screen. Yes, it's a couple of lines farther down. There's nothing between the { mark and the } mark, so there are no instructions in this procedure. This is because you are going to write them.

- It doesn't matter if { and } are on separate lines. The compiler for the Arduino ignores line breaks and all whitespace larger than a single space between words.
- The { and } are properly known as *braces*.

It's time to type something on the empty line beneath "put your setup code here." Try this:

```
pinMode(13, OUTPUT);
```

You have to type it exactly. The compiler won't tolerate typographical errors. Also, because the C language is *case sensitive*, you must distinguish between uppercase and lowercase letters. `pinMode` must be `pinMode`, not `pin mode` or `Pinmode`. `OUTPUT` must be `OUTPUT`, not `output` or `Output`.

The word `pinMode` is a command to the Uno, telling it how to use one of its pins. The pin can receive data as an input, or can send data as an output. `13` is the pin number, and if you check your board, you'll find one of the little connectors is identified as 13, right beside the yellow LED. I have selected 13 as an arbitrary choice.

A semicolon marks the end of the instruction.

- A semicolon must be included at the end of each instruction. Always. Don't forget!

Now move on down to the empty line below the message that says:

```
// put your main code here, to run repeatedly.
```

You can tell from the double slash marks that this is another comment line for your information. The compiler will ignore it. Below it, carefully type these instructions:

```
void loop() {
digitalWrite(13, HIGH);
delay(100);
digitalWrite(13, LOW);
delay(100);
}
```

If you have any prior familiarity with an Arduino, you'll be groaning, now, as you think to yourself, "Oh no, it's the old blink test!" Yes indeed, and that's why I subtitled this section, "The Old Arduino Blink Test." This is the program that almost everyone uses as a preliminary test (although I have changed the delay values, for reasons that will become apparent). Please humor me by typing the program into the IDE. I will be getting to some more challenging projects soon enough.

Also, you may be just a little sketchy on what some of the statements actually mean.

`void` means the same as before.

`loop()` is an instruction telling the Arduino to do something over and over again. What does it have to do? It has to obey the procedure between the braces.

`digitalWrite` is a command to send something out of a pin. Which pin? I am specifying `13` because its mode was defined previously.

- You can't use a digital pin unless you have previously specified what its mode will be.

What should the pin do? Go to a HIGH state.

Don't forget the semicolon at the end of the instruction.

`delay` tells the Arduino to wait for a while. How long? `100` means 100 milliseconds. There are 1,000 milliseconds in a second, so the Arduino is going to wait for one-tenth of a second. During that time, pin 13 will stay high.

I think you can figure out what the next two lines mean.

In a moment, you can activate the program. But first, go to your board and insert the leads of an LED between connector 13 and the connector labelled GND, right beside it.

- Make sure the short lead of the LED is in the GND connector.
- The LED does not need a series resistor, as this is built in to connector 13.

The little yellow LED on my board was already blinking, by default, as soon as I plugged in the board. The LED that I just inserted also starts blinking, because the yellow surface-mount LED built into the board is wired in parallel with pin 13.

In previous versions of the Uno, the onboard LED did not start blinking as soon as you plugged in the board. In future versions, Arduino may disable this "blinking by default." Either way, it doesn't matter, because your program is going to change the blink rate.

Verify and Compile

Next you must check if you made any typographical errors. Pull down the Sketch menu and choose Verify/Compile, as shown in Figure 5-85. The IDE examines your code, and if it sees any problems, it complains about them.

Figure 5-85 *Choose the Verify/Compile option before sending your program to the Arduino.*

You can test this. In your program listing, change `pinMode` to `piMode`, then Verify/Compile, and see what happens.

You get an error message in the black space at the bottom of the IDE window. You can stretch this black space by dragging its upper border with your mouse, so you can see more than two lines at a time, without scrolling them. The error message that I get says, "*piMode* was not declared in this scope."

You see, there are *reserved words* and *defined functions* in the C language, which have special meanings. You have used a couple of them already, such as `digitalWrite` and `delay`.

But `piMode` doesn't exist as a reserved word or defined function, so the compiler complains that you have not declared it, to say what it is.

Fix your program text till you can Compile/Verify without any errors.

Upload and Run

Now pull down the File menu, and choose Upload. Personally I always imagine that I am downloading from my big computer into the little Arduino, but everyone calls it uploading, so I guess that's what it is.

If the upload is successful, you'll see the message "Done Uploading" just above the black error window.

If it is not successful, and the uploading process never quite ends—this is not good. It means you still have some communications problems, probably because the COM port assignments don't match. Go back to the troubleshooting section for your type of computer, above. But save your program first. Pull down the File menu, choose Save, and give your program a name. After you fix your COM issues, you will be able to reload the program, if necessary, and try again.

If everything worked as planned, the onboard yellow LED, and your LED, are now blinking rapidly—on for 1/10th of a second, and off for 1/10th of a second, in accordance with the instructions in your program.

You may feel that this is a small achievement after taking so many steps, but we have to start somewhere, and a blinking LED is usually where microcontroller programming begins. In the next experiment, you will write a new program that will do something much more useful.

Here's a summary of what you may have learned so far, and what you generally have to do, to program the Arduino:

- Start a new program (or "sketch," as Arduino prefers to call it).
- Select the New option from the File menu if necessary.
- Every program must begin with a `setup` function, which will run once only.
- You must declare the number of a digital pin, and its mode, by using the `pinMode` command, before you can do something with the pin later.
- The mode of a pin can be INPUT or OUTPUT.
- Some pin numbers are not valid. Look at your Uno board to see the numbering system that is used.
- A pair of braces must enclose every function or block in a program. But they can be on separate lines.
- The compiler ignores line breaks and extra whitespace.
- Every instruction in a function or block must end in a semicolon.
- Every Arduino program must contain a `loop` function (after the `setup` function), which will run repeatedly.
- `digitalWrite` is a command to make a pin that is set for output have a state that is specified as HIGH or LOW.
- `delay` makes the Arduino do nothing for a specified number of milliseconds (thousandths of a second).
- The numbers in parentheses after a command are *parameters* telling the Arduino how the command must be applied.
- Use the Verify/Compile option in the Sketch menu to check your program before you try to upload it to the Arduino.
- You must fix any errors found in the Verify/Compile operation.
- Reserved words are a vocabulary of commands that the Arduino understands. You have to spell them correctly. Uppercase is considered different from lowercase.
- After you upload your program, it will start running automatically, and will continue to do so

until you disconnect power from the board or upload a new program.

- There is a Reset button (a tactile switch) beside the USB connector on the Uno board. When you press it, the Arduino restarts your program at the beginning, with all the values reset.

Caution: Lost Code

If you modify your program and upload it to the microcontroller, the new version will *overwrite* the old version. In other words, the old version will be erased. If you didn't save it on your computer under a different filename, it may be gone forever. Be very careful when uploading revised programs. Saving each version on your computer, under a new name, is a sensible precaution.

After program instructions have been uploaded into the microcontroller, there is no way to read them back out again.

Programming Entails Detail

I don't know if you noticed, but the summary of points to remember, in this experiment, has been much longer than the summaries in any experiments where you used individual components. Writing a program entails a lot of detail, and you have to get everything exactly right. Personally I enjoy this, because once you get it right, it will always be right, and it will always work the same way. Programs never wear out. If you save them in a suitable medium, they can last forever. The software that I wrote in the 1980s will still run 30 years later in a DOS window on my desktop computer.

Some people don't enjoy detail work, or they tend to make typing errors, or they don't like the way in which a computer language makes inflexible demands (such as *always* insisting that you begin a program with a `Setup` function, even if you don't have a setup). Different kinds of people enjoy different aspects of electronics, and that's the way it should be. If everyone wanted to write programs, and no one ever wanted to touch a piece of hardware, we wouldn't have any computers. It's up to you to decide what activity suits you.

Personally, I'm going to continue with another experiment that uses the Arduino in a more interesting way. I want to show you how a microcontroller can do things more easily than individual components in some cases.

Before I end this experiment, though, you may be wondering what happens if you disconnect the Arduino from your computer.

- The Arduino needs power to *run* your program.
- The Arduino does not need power to *store* your program. The program is stored automatically in the microcontroller, like data in a flash drive.
- If you want to run your program when the board is not connected with your computer, you have to provide power in the round black socket next to the USB socket on the board.
- The power supply can range from 7VDC to 12VDC. It does not have to be regulated, because the Arduino board contains its own regulator, which changes your power input to 5VDC on the board. (Some Arduinos use 3.3VDC, but not the Uno.)
- The power supply jack is 2.1mm in diameter, with center pin positive. You can buy a 9V AC-to-DC adapter with that kind of plug on its output wire.
- If you connect external power while the Arduino is also connected with a USB cable, the Arduino automatically uses the external power.
- You can disconnect your Arduino from the serial cable anytime, without bothering to use the "Safely Remove Hardware" option that exists in some versions of Windows.

Background: Origins and Options Among Programmable Chips

In factories and laboratories, many procedures are repetitive. A flow sensor may have to control a heating element. A motion sensor may have to adjust the speed of a motor. Microcontrollers are perfect for this kind of routine task.

A company named General Instrument introduced an early line of microcontrollers in 1976, and called them PICs, meaning Programmable Intelligent Computer—or Programmable Interface Controller, depending on which historical source you believe. General Instrument sold the PIC brand to another company named Microchip Technology, which owns it today.

Arduino uses Atmel microcontrollers, but PICs are still an alternative, and are used as the basis for a hobby-education version licensed by a British company named Revolution Education Ltd. They call their range of chips the PICAXE, for no apparent rational reason other than they must think that it sounds cool. (I'm not so sure that they're right about that.)

The PICAXE comes with its own IDE, which uses a different computer language named BASIC. In some ways this is a simpler language than C. Another range of microcontrollers, the BASIC Stamp, also uses BASIC, with additional, more powerful commands.

If you search for PICAXE on Wikipedia, you'll find an excellent introduction to all the various features. In fact, I think it's a clearer overview than you'll get from the PICAXE website.

Unlike the Arduino, you don't need to buy a special board to program PICAXE chips. A customized USB cable is all you need—in addition to the appropriate IDE software, which you can download for free.

The first edition of this book contained some introductory information about PICAXE products. You can probably find that edition secondhand, if you're interested.

Fundamentals: Advantages and Disadvantages

Now that you have learned some basics, I need to discuss issues that may affect your decision about whether to use a microcontroller in a project.

Longevity

The flash memory that stores a program in an ATmega328 is guaranteed by the manufacturer for 10,000 read-write operations, with provision for automatically locking out memory locations that go bad. This would seem to be ample, and we may hope that a microcontroller will last almost indefinitely. However, we don't know yet if it will really have the same longevity as an old-school logic chip, some of which are still working 40 years after they were manufactured. Does this matter? You have to make up your own mind about that.

Obsolescence

Microcontrollers are maturing rapidly as a technology. When I wrote the first edition of this book, the Arduino was relatively new, and its future was uncertain. The Arduino now dominates the hobby-electronics field, but what will the situation be another five years from now? No one knows. A product such as the Raspberry Pi is an entire computer on a chip. No one can predict whether it, or something else like it, will displace the Arduino.

Even if the Arduino remains the microcontroller system of choice, we have already seen new versions of the hardware, and updates to the IDE software that must be used when you program the chip. One way or another, you may have to keep yourself informed about developments in the field as they occur, and you may even have to abandon one brand of microcontroller and switch to a different brand.

By comparison, in most cases, individual components in through-hole format have reached the end of their development cycle. Some relatively recent innovations have been introduced, such as rotational encoders or small dot-matrix LED and LCD displays. However, most of these new products are designed for use with microcontrollers. In the simple world of transistors, diodes, capacitors, logic chips, and single-chip amplifiers, the knowledge that you acquire today should be valid 10 years from now.

Hybrid Circuits

Last, and perhaps most important, microcontrollers cannot be used alone. They always require some other component, even if it's just a switch, a resistor, or an LED, and the other components must be properly compatible with the inputs and outputs of the microcontroller.

Therefore, to make practical use of a microcontroller, you still have to be familiar with electronics generally. You need to understand basic concepts such as voltage, current, resistance, capacitance, and inductance. You should probably know about transistors, diodes, alphanumeric displays, Boolean logic, and other topics that I have covered so far in this book. And if you are going to build prototypes, you will still need to know how to use a breadboard or make solder joints.

Bearing all this in mind, I can summarize the pros and cons.

Individual Components: Advantages

Simplicity.

Instant results.

No programming language needed.

Cheap, for small circuits.

Today's knowledge will be valid tomorrow.

Better for analog applications such as audio.

Still necessary when using microcontrollers.

Individual Components: Disadvantages

Capable of performing one function only.

Circuit design is challenging for applications involving digital logic.

Not easily scalable. Large circuits are difficult to build.

Revisions to a circuit may be difficult or even impossible.

More components in a circuit generally require more power.

Microcontrollers: Advantages

Extremely versatile, able to perform many functions.

Additions or revisions to a circuit can be easy (just rewrite the program code).

Huge online libraries of applications, freely available.

Ideal for applications involving complex logic.

Microcontrollers: Disadvantages

Relatively expensive for small circuits.

Significant programming skills required.

Time-consuming development process: write code, install code, test, revise-and-debug, reinstall—in addition to troubleshooting the circuit hardware.

Rapidly evolving technology requires a continuing learning process.

Each microcontroller has individual quirks and features, requiring study and memorization.

Greater complexity means more things that can go wrong.

Requires a desktop or laptop computer, and data storage for programs. Data may be lost accidentally.

Requires a regulated power supply (usually 5VDC or 3.3VDC), like a logic chip. Limited output of 40mA per pin, or less. Cannot drive a relay or a loudspeaker, as a 555 timer can. You must buy a separate driver chip if you need to deliver more power.

Summing Up

Now I am ready to answer the question, "Should I use microcontrollers or individual components?"

My answer is that you need both. That is why I am including microcontrollers in a book that is primarily about individual components.

In the next experiment, I'll show how a sensor and a microcontroller can work together.

Experiment 33: Checking the Real World

A switch is either "on" or "off," but most of the inputs we receive from the real world tend to vary between those extremes. A thermistor, for instance, is a sensor that varies its electrical resistance over a wide range, depending on its temperature.

A microcontroller would be very useful if it could process that kind of input. For instance, it could receive the input from the thermistor, and then it could function like a thermostat, turning on a heater if the temperature falls below a minimum value, and turning off the heater when the room is warm enough.

The ATmega328 that is used in the Arduino Uno can do this, because six of its pins are classified as "analog inputs," meaning that they don't just evaluate an input as "logic-high" or "logic-low" on a digital basis. They convert it internally using what is called an *analog-to-digital converter*, or ADC.

On the 5-volt version of the Arduino, an analog input must range from 0VDC to 5VDC. (Actually, the upper limit can be modified, but that introduces some complexity that I will leave until later.) A thermistor doesn't generate any voltage; it just varies its resistance. So, I'll have to think of a way in which a change in resistance can deliver a change in voltage.

Once that problem has been dealt with, an ADC inside the microcontroller will convert the voltage on the analog pin into a digital value ranging from 0 through 1023. Why that numeric range? Because it can be expressed in

10 binary digits, and the ADC isn't accurate enough to justify a wider range with smaller increments.

After the ADC has supplied a number, your program can compare it with a target value, and can take appropriate action—such as changing the state of an output pin, which can supply voltage to a solid-state relay, which can activate a room heater.

The sequence beginning with the thermistor and ending with the digital value is shown visually in Figure 5-86.

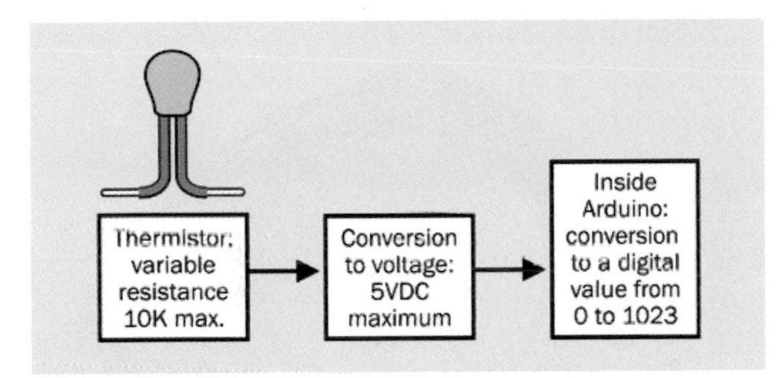

Figure 5-86 *A simplified view of the plan for processing the status of a thermistor.*

The following experiment will show you how to do this.

What You Will Need

- Breadboard, hookup wire, wire cutters, wire strippers, test leads, multimeter
- Thermistor, 10K, 1% or 5% accuracy (1) (This must be the NTC type, meaning that its resistance drops as the temperature increases. A PTC thermistor behaves oppositely)
- Arduino Uno board (1)
- Laptop or desktop computer with an available USB port (1)
- USB cable with A-type and B-type connector at opposite ends (1)
- 6.8K resistor (1)

Using a Thermistor

The first step is to get to know your thermistor. It has very thin leads, because they must not conduct heat into or out of the tip where the temperature-measuring junction is located. The leads are probably too thin to plug into your breadboard reliably, so I suggest you grab them with a couple of alligator test leads, and use the leads to grab the probes of your meter, as shown in Figure 5-87.

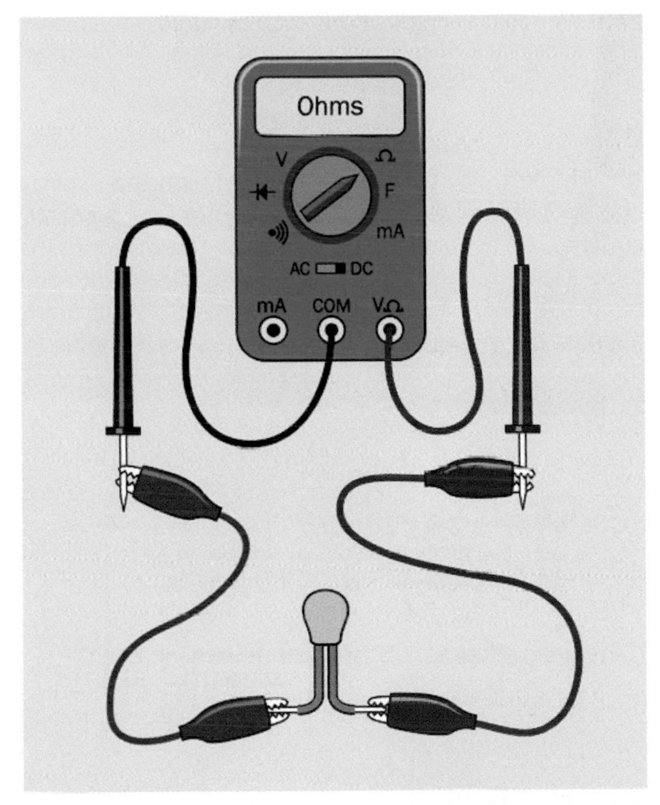

Figure 5-87 *Testing a thermistor.*

The thermistor that I am recommending is rated at 10K. That's its maximum resistance when it gets really cold. Its resistance doesn't change much as the temperature rises until around 25 degrees Celsius (77 degrees Fahrenheit). After that, the resistance declines more rapidly.

You can test it with your meter. At room temperature, the thermistor should have a resistance around 9.5K. Now grip it between your finger and thumb. As it absorbs your body heat, its resistance goes down. At body temperature (arbitrarily agreed to be 37 degrees Celsius or 98.6 degrees Fahrenheit) its resistance is around 6.5K.

How can you convert this resistance range to the 0V to 5V required by the microcontroller?

First bear in mind that the maximum value corresponding with room temperature should actually be lower than 5V. The real world is an unpredictable place. What if your thermistor gets much hotter than you expected, for some surprising reason? Maybe you place your soldering iron beside it, or maybe you rest it on a warm piece of electronic equipment.

Here we have the first lesson in analog–digital conversion: allow for unexpected, extreme values when measuring the everyday world.

Range Conversion

The simplest way to convert the thermistor's resistance into a voltage value is to choose a resistor approximately equal to the average resistance of the thermistor in the temperature range that interests us. Put the resistor and the thermistor in series to create a voltage divider, apply 5VDC at one end and 0VDC at the other end, then take the voltage at the midpoint between the components, as shown in Figure 5-88.

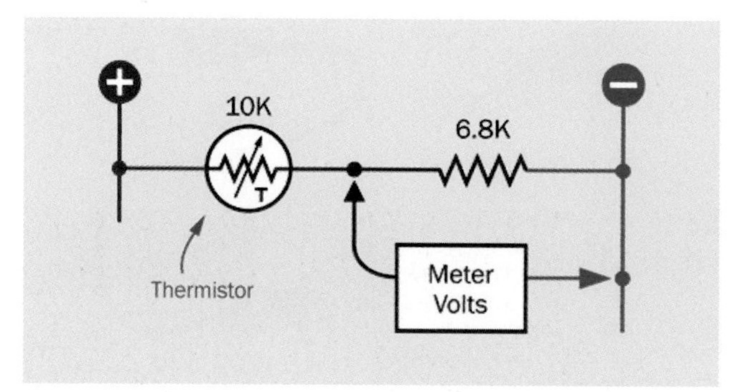

Figure 5-88 *The simplest circuit for deriving a voltage from the changing resistance of a thermistor.*

Normally, to set up this circuit, you would need to install a voltage regulator to provide the 5VDC. However, the Arduino has its own voltage regulator, and conveniently provides a 5VDC output (see Figure 5-79). You can tap into this output and take it across to your breadboard with a jumper wire. You'll also need to tap one of the ground outputs from the Arduino, and take that to your breadboard as well.

When I tried that, and varied the thermistor temperature from around 25 to 37 Celsius, my meter measured a voltage from 2.1V to 2.5V. You should try that yourself, to check my numbers.

Obviously we run no risk of endangering the microcontroller with those voltages. But now I see a different problem: the range isn't wide enough to achieve optimal accuracy.

Figure 5-89 shows the conversion between input voltage and internal digital equivalents. The range from 2.1V to 2.5V is defined by the darker blue rectangle. It will be converted into digital values from around 430 to 512, which is a spread of 82—only a small fraction of the complete range from 0 through 1,023.

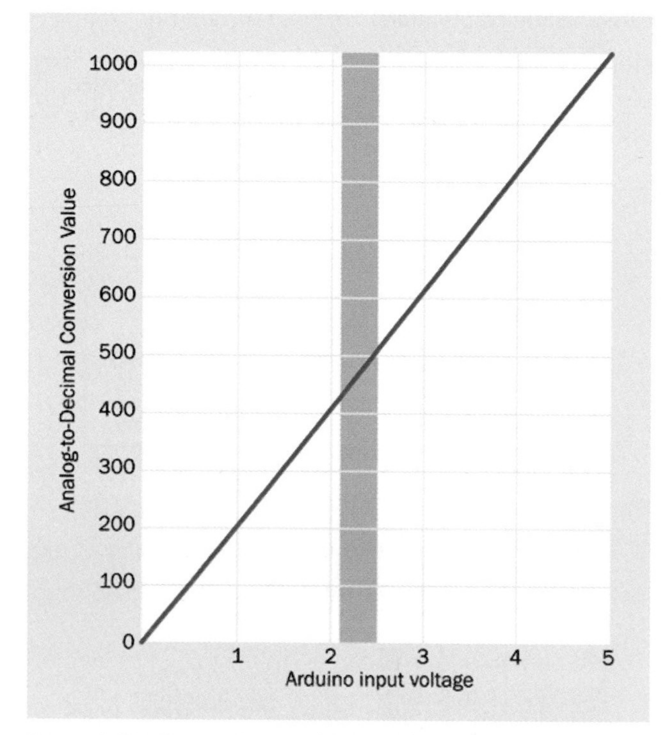

Figure 5-89 *Conversion graph from Arduino input voltage to ADC values. The blue rectangle identifies approximate voltages derived from a 10K thermistor in series with a 6.8K resistor, for temperatures between 75 and 95 degrees Fahrenheit.*

Using this limited range will be like using a small number of pixels from a high-resolution photograph. Inevitably, there will be a lack of detail. Wouldn't it be nice if we could somehow convert our voltages into a digital range covering maybe 500 values instead of 82 values?

One way to do that would be by amplifying the voltage, but that will entail using an additional component, such as an op-amp. This is possible, but then we will need resistors to control feedback, and everything will get complicated. The whole idea of the microcontroller was to keep it simple!

There is another option, which is to use a feature of the Arduino that sets a lower maximum voltage for the range. But this requires me to supply a sample of the new maximum voltage to one of the pins. To create that voltage, I would need to use another voltage divider, and then calculate a new conversion from voltage inputs to ADC values. Really, I want to avoid that kind of thing, at least until I have a simple program running.

After thinking about this a bit more—maybe I can live with a range of 82 values to represent a range from approximately 75 to 95 degrees Fahrenheit. That's an accuracy of about 1/4 degree for each digital step created by the ADC. It's not good enough for a clinical thermometer, but perfectly adequate for room temperature.

Connections

So let's give it a try. But wait—are we still going to use a separate breadboard with that funky-looking Uno board on which the microcontroller is mounted?

Yes, that's the plan. There are three ways to hook everything up:

- You can buy a gadget called a *protoshield*, which is like a mini-breadboard that sits on top of the Uno board and plugs into the connectors. I am not a fan of this device, because it leads us away from an eventual finished circuit on a regular breadboard.
- You can pry the microcontroller off the Uno board and plug it into a breadboard, where you can connect components with its pins in a normal manner. But if you do that, you have no way to load a program into the microcontroller, and you will need a crystal oscillator to make the microcontroller run at the same speed as when it was on the Arduino board.
- You can mount your thermistor and resistor on a regular breadboard, and then run the voltage across from the thermistor circuit to the Uno board on a jumper wire, in the same way that you are already supplying positive voltage and ground from the Arduino to the breadboard. It's messy, but it is what most people seem to do. If you finish a program and mount it permanently on the microcontroller, then maybe you can move the chip to a more convenient location.

Figure 5-90 shows the arrangement. Figure 5-91 shows how it looks in a photograph. I have to admit, this is one occasion where those little wires with plugs at each end are convenient, although I still don't entirely trust them.

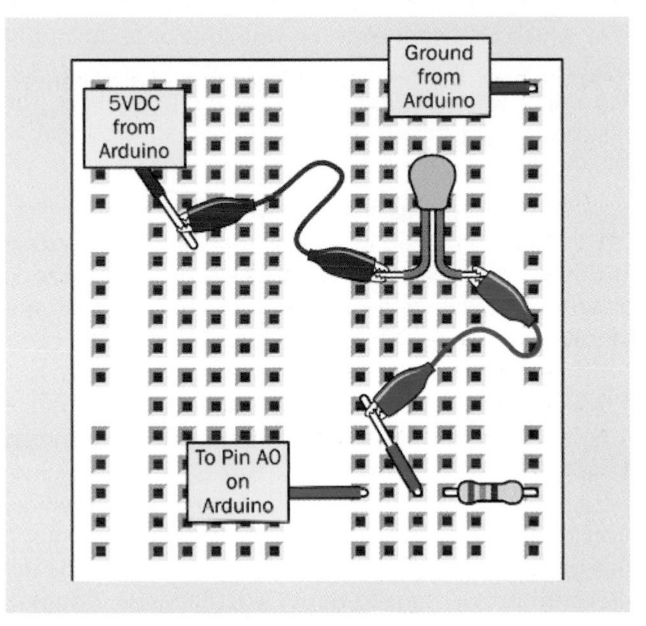

Figure 5-90 *Wiring the thermistor circuit to the Arduino.*

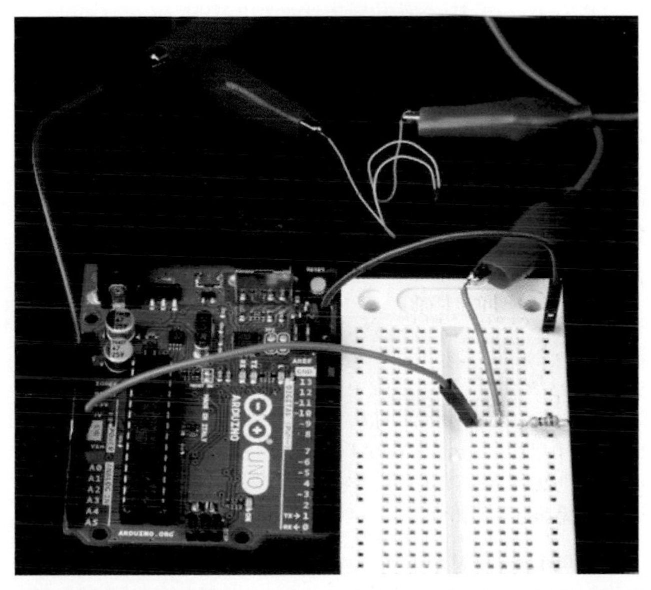

Figure 5-91 *Using jumper wires to connect the thermistor circuit with the Arduino.*

What, No Output?

Now you're all set to convert an analog input to an internal digital value. But wait—something is missing, here. You don't have an output!

In an ideal world, the Uno board would be sold with a nice little alphanumeric display, so that you could use it like a real computer. In fact, you can obtain displays like that which will work with the Uno, but once again this would entail some complexity. Like almost everything in the microcontroller world, it is not plug-and-play. The microcontroller has to be programmed to send text to the display.

So, I will keep things simple. I will use the little yellow LED on the Uno board as an indicator. I will pretend that the indicator represents a room heater, which comes on if the temperature is low and switches off when the temperature is high.

Hysteresis

Suppose we're heating a greenhouse where we want the temperature to be a toasty 85 degrees Fahrenheit. Suppose the voltage from the resistor-thermistor combination is 2.3V at that temperature. Look it up on the graph in Figure 5-89 and you'll see that the ADC inside the microcontroller will convert this to a digital value of around 470.

So, 470 is our threshold. If the number goes down to 469, we put on the heat (or simulate it, by lighting the LED). If the number goes up to 471, we turn off the heat.

But wait. Does this make sense? It means a fractional increase in temperature sensed by the thermistor will trigger the LED, and a fractional decrease will switch it off. The system will be forever fluctuating.

A real thermostat doesn't respond to every little temperature variation when someone opens or closes a door. When it switches on, it stays on, until the temperature is a bit above the target. Then when the heat is off, it stays off, until the temperature is a bit below the target.

This behavior is known as *hysteresis*, and I discuss it in more detail, in relation to a component called a *comparator*, in my sequel to this book, *Make: More Electronics*.

How can we implement hysteresis in a microcontroller program? We need a broader range of values than 469 through 471. The program could say something like, "If the LED is on, keep it on until the temperature value exceeds 490. Then switch if off." And, "If the LED is off, keep it off until the temperature value drops below 460. Then switch it on."

Can we do this? Yes, quite easily. The program listed in Figure 5-92 uses this logic. I derived it from a screen capture of the Arduino IDE, so I have good reason to think that it works.

```
// Heater Control Simulation
// by Charles Platt

int digitemp = 0;
// digitemp is a variable to store
// a digitized temperature value.

int ledstate = 0;
// Will be 0 if LED is currrently off.
// Will be 1 if LED is currently on.

void setup()
{
     pinMode (13, OUTPUT);
     // Onboard LED shows the output.

     // (No need to set the analog pin
     // which is input by default.)
}

void loop()
{
     digitemp = analogRead (0);
     // Thermistor is on analog input A0.

     if (ledstate == 1 && digitemp > 490)
          {
          ledstate = 0;
          digitalWrite (13, LOW);
          }

     if (ledstate == 0 && digitemp < 460)
          {
          ledstate = 1;
          digitalWrite (13, HIGH);
          }

     delay (100);
}
```

Figure 5-92 *Program listing for control of an imaginary heating device.*

The program also introduces some new concepts—but first, type it into the IDE. You don't need to include all the comment lines, which I just added as explanations. You can type the much shorter version in Figure 5-93 where the comment lines have been omitted.

```
int digitemp = 0;
int ledstate = 0;

void setup()
{
     pinMode (13, OUTPUT);
}

void loop()
{
     digitemp = analogRead (0);
     if (ledstate == 1 && digitemp > 490)
          {
          ledstate = 0;
          digitalWrite (13, LOW);
          }
     if (ledstate == 0 && digitemp < 460)
          {
          ledstate = 1;
          digitalWrite (13, HIGH);
          }
     delay (100);
}
```

Figure 5-93 *The same program as before, with the comment lines omitted.*

Verify/Compile your program, and fix any typing errors (probably there will be a semicolon missing in one place or another—that's the most common error).

Plug in your Arduino, upload the program, and if the temperature of your thermistor is below 85 degrees Fahrenheit, the yellow LED should switch on.

Hold the thermistor between your finger and thumb to make the thermistor think that the room temperature has increased. After a few seconds, the LED goes out. Now let go of the thermistor, and it cools down—but the LED stays on for a while, because the hysteresis in the system is telling it to wait until the temperature is sufficiently low. Eventually, the LED switches on again. Success!

But how does the program work?

Line by Line

This program introduces the concept of a *variable*. This is a little space in the microcontroller's memory where a digital value can be stored. You can think of it as a "memory box." On the outside of the box is a label that has the name of the variable. On the inside is a numeric value.

`int digitemp = 0;` means I have invented a variable named `digitemp`. It is an *integer* (a whole number), and it starts with a value of zero.

`int ledstate = 0;` means I have invented another integer variable, to keep track of whether the LED on the board is on or off. There's no easy way I can get the microcontroller to look at the LED and tell me what its state is, so I will have to keep track of it myself.

The `setup` of the program just tells the microcontroller to use pin 13 as an output. I don't need to tell it to use pin A0 as an input, because analog pins are inputs by default.

Now comes the heart of the program, in the loop. First, I use the command `analogRead` to tell the microcontroller to read the state of an analog port. Which one? I specify `0`, which means analog port A0. This is where the connection from my breadboard is plugged in.

What do I do with the information from the ADC after it reads the port? There is only one sensible place to put it: in the variable named `digitemp` that I created for this purpose.

Now that `digitemp` contains a value, I can examine it. First I want to say, if the heat is on (the LED is illuminated) and the temperature value is greater than 490, it's time to turn the heat off. The "if" part is tested like this:

```
if (ledstate == 1 && digitemp > 490)
```

The double `==` means, "make a comparison and see if these two items are the same." A single equals sign means, "assign this value to a variable," which is a different operation.

The double `&&` is a "logical AND." Yes, we have Boolean logic here, just as if we were using an AND logic gate. But instead of wiring up a chip, we just type a line of code.

The > symbol means, "is greater than."

The "if" test is contained in parentheses. If the statement in the parentheses is true, the microcontroller performs the following procedure, between the braces. In that procedure, `ledstate = 0` records the fact that the LED is going to be switched off. `digitalWrite (13, LOW);` actually switches it off.

The second "if" test is very similar, except it only applies if the LED is off, and the temperature has fallen much lower. Then, we switch the LED on.

Finally there is a delay of 1/10th of a second because we don't need to check the temperature more often than that.

And that's it.

Additional Details

I have thrown in some pieces of syntax, such as the "if" test, and the double equals sign, and the && logical operator, without giving you a whole list of all the syntax that you can use in C language. You can find that kind of list online. I don't have space for it here.

Notice a couple of things about the program:

- Lines are indented to clarify the logical structure. The compiler ignores extra white space, so you can feel free to add as much of it as you like.
- The IDE uses color to help you see if you made any typing errors.
- When you make up a name for a variable, you can use any combination of letters, numerals, and underscore characters—so long as the combination is not the same as a word that already has a special meaning in the C language. For instance, you can't have a variable named `void`.
- Some people like to begin variable names with a capital letter, and some don't. It's your choice.
- Each variable should be declared at the beginning of the program, so that the compiler knows what to expect.
- An integer (declared with the `int` term) can have a value ranging from -32,768 to +32,767. The C language on this microcontroller allows you other types of variables that have a wider range of values, or that can have fractional values. But I won't need to use a larger numeric value until Experiment 34.

For a beginning language reference, go to the main Arduino website, click the Learning tab, and choose Reference from the menu that drops down. You can also pull down the Help menu from the Arduino IDE and choose the Reference submenu.

Enhancements

This program does the job that I set out to do, but it's very limited. The biggest limitation is that it uses specific numbers for the minimum and maximum temperature values. That's like having a thermostat that is glued into just one position and can never be adjusted. How could the program be enhanced so that the user can adjust the threshold temperature for the heat to go on and off?

I think the way to do it would be to add a potentiometer. The ends of the potentiometer's track would be connected to 5V and 0V, and the wiper would be connected to another analog input on the microcontroller. This way, the potentiometer functions as a voltage divider and supplies the full range from 0VDC to 5VDC.

Then I would add another procedure in the loop where the microcontroller would check the potentiometer setting, and digitize it.

The result would be a number in the full range from 0 through 1,023. I would have to convert this to a number compatible with the likely range of values for the `digitemp` variable. Then I would put the result in a new variable named, perhaps, `usertemp`. Then I would see if the actual room temperature, measured by the thermistor, is significantly above or below `usertemp`.

Notice that I skipped over one little detail: exactly how I would convert the potentiometer input to the range suitable for `usertemp`. All right, I'll deal with that now.

If the likely range of values for the thermistor is 430 through 512 as I estimated previously, that range can be thought of as having a middle value of 471, plus or minus 41. The potentiometer has a middle value of 512, plus or minus 512 for its full range. Therefore:

```
usertemp = 471 + ( (potentiometer - 512) * .08)
```

where `potentiometer` is the value from the potentiometer input, and the asterisk symbol is used by the C language as a multiplication sign. That would be close enough.

Yes, arithmetic does tend to be involved in programming, sooner or later, somewhere or other. There is no way to pretend otherwise. But it seldom goes beyond high-school math.

In the enhanced version of the program, I would still have to take care of hysteresis. The first "if" statement would have to be converted so that it says something like,

```
if (ledstate == 1 && digitemp > (usertemp + 10) )
```

then switch off the LED. But

```
if (ledstate == 0 && digitemp < (usertemp - 10) )
```

then switch on the LED. That would give me a hysteresis range of plus-or-minus 10, using the values from the ADC.

Now that I've described this modification, maybe you can consider making it yourself. Just remember to declare each new variable before you use it in the body of the program.

Experiment 34: Nicer Dice

In this last experiment, I'm going to revisit Experiment 24, which used logic chips to create dice patterns. Instead of logic chips, we can now use "if" statements with logical operators in a microcontroller program. We can replace several pieces of hardware with a dozen lines of computer code, and instead of a 555 timer, a counter, and three logic chips, we will need just one microcontroller. This is a great example of an appropriate application. (Of course, it does still require some LEDs and series resistors.)

What You Will Need:

- Breadboard, hookup wire, wire cutters, wire strippers, test leads, multimeter.
- Generic LED (7).
- Series resistor, 330 ohms (7).
- Arduino Uno board (1).
- Laptop or desktop computer with an available USB port (1).
- USB cable with A-type and B-type connector at opposite ends (1).

The Limits to Learning by Discovery

Learning by discovery works well when you're getting to know an electronic component. You can put it on a breadboard, apply power, and see what happens. Even when you're designing a circuit, you can use some trial and error and make modifications as you go along.

Writing programs is different. You have to be disciplined and logical; otherwise, you will tend to write buggy program code that doesn't work reliably. Also, you have to plan ahead; otherwise, you will waste a lot of time redoing previous work or throwing it away entirely.

I don't enjoy planning, but I dislike wasting time even more. So, I do plan, and in this final project, I'm going to describe the planning process. I'm sorry that you won't get some immediate pleasure from just putting some parts together and seeing what happens, but if I don't explain the software development process, I will be creating a misleading impression by making programming seem simpler than it really is.

Randomicity

The first question seems obvious: "What do I really want this program to do?" The question is necessary because if you don't define your goals clearly, there is no way the microcontroller can figure them out for you. This is similar to the process that I described in Experiment 15, of making a "wish list" for the intrusion alarm; but for a microcontroller, greater detail is required.

The basic requirement is very simple. I want a program that will choose a random number and display it in some LEDs that resemble the spots on a die (the word "die" being the singular of "dice," as I mentioned once before).

Because choosing a random number is fundamental to this program, you need to be properly informed on the subject. So let's check the Arduino website, where there is a language reference section. It is not as comprehensive as I would like, but is a good starting point.

To find it, go to the Arduino home page, click the Learning tab, and select Reference, and you will find a section titled Random Numbers. When you click that, you find there is a function specifically created for the Arduino, named `random()`.

This is not entirely surprising, because almost all high-level computer languages have some kind of random function built in, and it always works by using mathematical trickery to generate a stream of numbers that are unpredictable to a human observer and continue for a very long time before the sequence repeats. The only problem is that because they are created mathematically, the sequence will start from the same place each time you run the program.

What if you want the sequence to start at a different place? There is another Arduino function named `randomSeed()` that initializes the number generator by looking at the state of a pin on the microcontroller that is not connected with anything. As I have mentioned before, a floating logic pin picks up any electromagnetic radiation that may be around, and you never know what to expect from it. So `randomSeed()` can be genuinely random, and sounds like a good idea—although we would have to remember not to use that unconnected, floating pin for anything else.

Setting aside the issue of seeding the random number generator for the time being, let's suppose I use the Arduino's `random()` function to choose a number for the output of the dice program. How should this actually work?

I think the player will press a button, at which point the randomly selected pattern of dice spots can be displayed. Job done! Then if you need to "throw the die" a second time, you just press the button again, and another randomly selected pattern of dice spots will be created.

That sounds very convenient—but I don't think it will look very interesting. It may not seem very plausible, either. People may wonder if that number is really random. I think the problem is that control of the procedure has been taken away from the user.

Going back to the hardware version of this project, I liked the way it displayed a rapid blur of patterns, and I liked that the player could press a button to stop the sequence arbitrarily.

Maybe the program should emulate this, instead of using the `random()` function. It can count through numbers 1 through 6 over and over again, very fast—like the counter chip in the hardware version of Nice Dice.

But this worries me in a different way. When the program counts from 1 to 6, and then repeats, I think the microcontroller will take an extra few microseconds to go back to the start of the loop. So, a 6 will always be displayed for a fraction longer than the other numbers.

Maybe I can combine the two concepts. I can use the random number generator to create a series of numbers, and I will display them in very quick succession, until the player presses a button to stop at an arbitrary moment.

I like that plan. But then what? I could have another button to restart the rapid number display. But, no, that's not necessary: the same button can do it. Press to stop, press to restart.

You see, now I am getting a clearer idea of what I want this program to do. This will help me to take the next step, figuring out the instructions to get the microcontroller to do it.

Pseudocode

I like to write *pseudocode,* which is a series of statements in English that will be easily converted into computer language. Here is my pseudocode plan for the program that I am naming Nicer Dice. Bear in mind that these instructions will be executed extremely fast, so that the numbers are a blur.

Main loop:

- Step 1. Choose a random number.
- Step 2. Convert it to a pattern of dice spots, and light up the appropriate LEDs.
- Step 3. Check to see if a button has been pressed.
- Step 4. If a button has not been pressed, go back to Step 1 and choose another random number, so that the sequence will repeat rapidly. Otherwise...
- Step 5. Freeze the display.
- Step 6. Wait for the player to press the button a second time. Then go back to Step 1 and repeat.

Do you see any problems with this sequence? Try to visualize it from the microcontroller's point of view. If you were receiving the instructions in the program, would you have everything you need to get the job done?

No, you wouldn't, because some instructions are missing. Step 2 says, "light up the appropriate LEDs," but—there is no instruction anywhere to switch them off!

You always have to remember:

- The computer *only* does what you tell it to do.

If you want the illuminated LEDs to be switched off before a new number is displayed, you have to include an instruction to do that.

Where should I put it? Well, I need to blank the display immediately before each new number is chosen and displayed. Therefore, the right place to zero the display is at the beginning of the main loop. I'll include it as Step 0.

- Step 0. Switch off all the LEDs.

But wait. Depending on the number that was displayed in the previous cycle, some of the LEDs in the pattern will be on, and some will be off. If we switch off all the LEDs to clear the display, this will include some of the LEDs that were off already. The microcontroller won't care, but it will waste some time executing that instruction. Maybe it would be more efficient just to switch off the LEDs that were previously on, and ignore the ones that are already off.

That will entail more complexity in the programming, and it may not be necessary. In the very early days of computing, people had to *optimize* a program to save processor cycles, but I think even a microcontroller is fast enough, now, for us not to care about the time that will be wasted by switching off two or three LEDs that were off already. I'll use an all-purpose routine to switch off all the LEDs, regardless of their current states.

Button Inputs

What else is missing from the list of pseudocode instructions?

There is a button issue.

Once again, I need to visualize what I want the program to do. The rapid display is cycling through numbers very quickly. The player presses a button to stop it. The display freezes, showing the current number. In Step 6 the microcontroller waits indefinitely until the player presses the button again, to create a new rapid display.

Wait a minute. How can the player press the button "again" without releasing it first?

With the pseudocode in its current form, this is what the microcontroller will actually do, bearing in mind that it performs tasks very, very quickly:

- Program tells the microcontroller to check the button.
- Microcontroller finds that the button is pressed.
- The display freezes. Microcontroller waits for the button to be pressed again.
- But it finds that the button is still being pressed, because the player hasn't had time to stop pressing it.
- Microcontroller says, "Oh, the button is pressed, so I should resume the rapid display."

Consequently the frozen display will only last for an instant.

Here's the solution to the problem. An extra step in the sequence:

- Step 5A. Wait for the player to release the button.

This will stop the computer from running ahead and displaying more numbers, until the player is ready.

Is this it? Are we done, yet?

No, I'm afraid not. You may think this is getting a bit laborious, but in that case I have to say, I'm sorry, but this is what programming is like. If someone tells you that you can just throw some instructions together and see them work, I'm afraid this is often not the case.

One more button problem still exists. Step 6 says, wait for the button to be pressed again to restart the rapid display. OK, the player presses the button, the display resumes—but the microcontroller is so fast, it will zip through the process of zeroing the current display and displaying a new die pattern, and once again it will recheck the button before the player has time to stop pressing it. Consequently, when the microcontroller gets to Step 4, it will find that the button is still being pressed, so it will freeze the display again.

Now what? Maybe I should add a new Step 7, telling the microcontroller to wait for the button to be released before resuming the rapid display.

That's counter-intuitive. I don't think anyone will understand that you have to press the button, *and release it*, before the rapid display resumes. We could just say, "Oh, you have to do that, because the program requires it." But that is a *very bad way of thinking*.

- A program has to do what the user expects. We should never force the user to do things to satisfy the program.

In any case, the idea of waiting for the button to be released before the rapid display resumes still won't work. There's another issue: contact bounce. This happens when the button is pressed, *and* when the button is released. Consequently, if someone lets go of a button and the process resumes, and the program checks the button again a millisecond later, the contacts may still be vibrating, and they may appear to be open or closed, on an unpredictable basis.

This is the kind of thing we run into when a microcontroller interacts with the physical world. The microcontroller wants everything to be precise and stable, but the physical world is imprecise and unstable.

I had to think carefully about this particular problem before I decided how I wanted to solve it.

One solution would be to go back to having two buttons, one to start the rapid display, and one to stop it. This way, as soon as the "start" button is pressed, the microcontroller can ignore its status and its contact bounce, and just wait for the "stop" button to be pressed.

But from the player's point of view, I like the simplicity of just having one button. Surely, there's a way to make it work?

I went back to describing what I wanted the program to do, as clearly as possible. I said to myself: "I want the program to resume showing the rapid sequence as soon as the button is pressed for a second time. But after that, the program should ignore the button until after it is released, *and* until the contacts have stopped bouncing."

Why not simply lock out the button for a second or two? Actually this will be a good idea, because the random display should run for a little while before the player can stop it again. This will make the display look "more random," as it runs through all those numbers.

Suppose I lock out the button for, say, two seconds after the rapid display begins. Step 4 should be rewritten like this:

- Step 4. If a button has not been pressed, OR the rapid display has been running for less than two seconds, go back to the top and choose another random number. Otherwise…

Notice the OR word. Those Boolean operators really come in handy.

The System Clock

I think I have resolved all the button problems, but now I have a new problem. I need to measure two seconds.

Is there a system clock inside the microcontroller? Probably there is. Maybe the C language can access it, and ask it to measure a time interval.

Check the language reference. Yes, there is a function named `millis()` that measures milliseconds. It runs like a clock, starting from zero each time a program begins. This function is capable of reaching such a high number, it doesn't reach its limit and restart from zero until about 50 days have passed. That should certainly be enough.

But—no, there is one more little snag. The Arduino doesn't allow my program to reset the system clock on demand. The clock starts counting like a stopwatch when the program starts running, but unlike a stopwatch, I can't stop it.

How can I deal with this? I will have to use the system clock in the same way that I would use the clock on my kitchen wall, in the real world. When I want to cook a hard-boiled egg, I make a mental note of the time when the water starts boiling. Let's suppose it is 5:02pm and I want to boil the egg for seven minutes. I say to myself, "5:02 plus seven minutes is 5:09, so I'll take the egg out at 5:09."

What I am doing, inside my head, is comparing the clock, which keeps running, with my memorized time limit of 5:09. I am saying to myself, "Has the clock reached 5:09 yet?" When the time on the clock is equal to 5:09, or greater, my egg is cooked.

The way to do this in the dice program is to invent a variable that will work like my memory of the time at the start of the egg-boiling process. Immediately before the rapid display begins, I store the current value of the system clock in the variable, plus two seconds. Then I can get the program to say, "Has the system clock reached the value of my variable yet?" until it finally gets there.

Suppose I name the variable "ignore," because it will tell me at what time in the future the program should stop ignoring the button. Then Step 4 can ask the microcontroller, "Has the system clock exceeded the *ignore* variable yet?" and if it has, the program can resume paying attention to the button.

I cannot reset the system clock, but I can reset the "ignore" variable to match the current value of `millis()`, plus two seconds, each time a new rapid display session begins.

Final Draft of Pseudocode

Bearing all these issues in mind, here is the revised and, I hope, final sequence of events for the program:

- Before the loop begins, establish input and output of the logic pins, and set the "ignore" variable to the current time, plus two seconds.
- Step 0. Switch off all the LEDs.
- Step 1. Choose a random number.
- Step 2. Convert it to a pattern of dice spots, and light up the appropriate LEDs.
- Step 3. Check to see if the button has been pressed.
- Step 4. Check to see if the system clock has caught up with the "ignore" variable yet.
- Step 4a. If the button has not been pressed, OR the system clock has not caught up with the "ignore" variable yet, go back to Step 0. Otherwise…
- Step 5. Freeze the display.
- Step 5A. Wait for the player to release the button.
- Step 6. Wait as long as it takes for the player to press the button once more, to restart the display.
- Step 7. Reset the "ignore" variable to the system clock, plus two seconds.
- Go back to Step 0.

Do you think it will work? Let's find out.

Hardware Setup

Figure 5-94 shows seven LEDs wired on a breadboard to display the spots on a die. The concept is the same as in Figure 4-146, except that the Arduino can deliver 40mA from each output pin, and therefore I don't have to drive pairs of LEDs in series. A single output pin can easily drive a pair of LEDs in parallel, and for each generic LED, a 330-ohm resistor is ample.

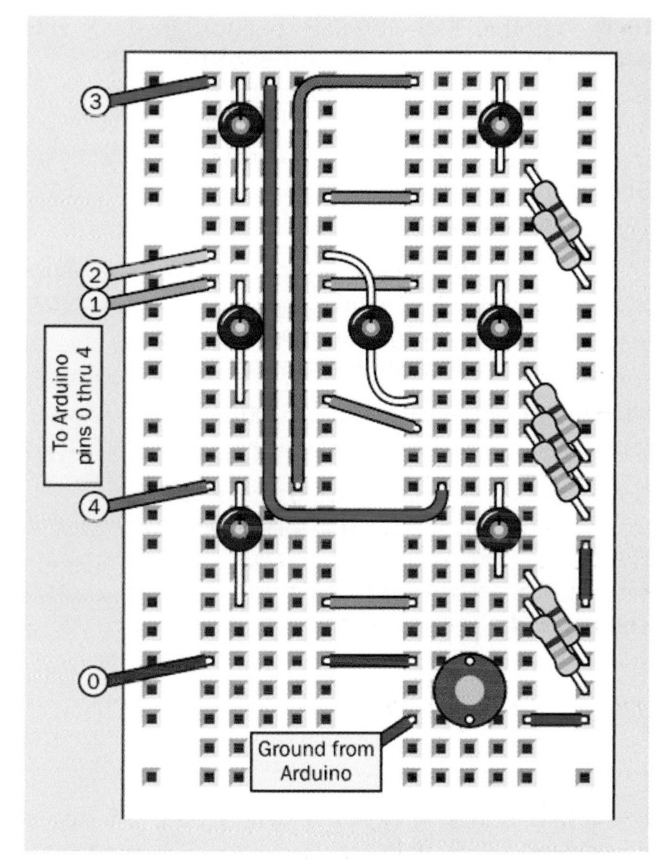

Figure 5-94 *Seven LEDs wired on a breadboard to display the spot patterns of a die.*

The numbers on the wires are the same as the numbering system in Figure 4-146. They have nothing to do with the die values. It's just an arbitrary way to identify each wire. Also, I can plug the wires numbered 1 through 4 into digital outputs numbered 1 through 4 on the Uno board. This will help to make everything clear.

I'll use digital connection 0, on the Uno board, as an input that checks the status of the pushbutton. However, note that the Uno uses digital pins 0 and 1 when receiving USB data. If you have any problem uploading your program, disconnect the wire from digital input 0 temporarily.

Don't connect the ground wire from the breadboard to the Uno board just yet. It's safer to upload the program first, because the program will tell the microcontroller which pins are outputs and which pins are inputs. A previous program may have configured them differently, and so long as the Arduino board is plugged in, it will want to run whatever program is in its memory. This may not be safe for the Arduino outputs, because:

- You must be very careful not to apply any voltage to a digital pin that is configured as an output.

Now, the Program

Figure 5-95 shows the program that I wrote to match the pseudocode. The same program is listed in Figure 5-96 with the comments removed, so that you'll be able to copy-type it more quickly. Please type it in the IDE editing window.

```
// Nicer Dice
// by Charles Platt

int spots = 0;       // How many spots to display.
int outpin = 0;      // The number of an output pin.
long ignore = 0;     // When to stop ignoring the button.

void setup()
{
  pinMode(0, INPUT_PULLUP);
  pinMode(1, OUTPUT);
  pinMode(2, OUTPUT);
  pinMode(3, OUTPUT);
  pinMode(4, OUTPUT);
  ignore = 2000 + millis();
}

void loop()
{
  // First, we must blank the display.
  for (outpin = 1; outpin < 5; outpin++)
  { digitalWrite (outpin, LOW); }

  // Now pick a random number from 1 through 6.
  spots = random (1, 7);

  // Now display the appropriate spot pattern.
  if (spots == 6)
  { digitalWrite (1, HIGH); }   // Side pair of spots

  if (spots == 1 || spots == 3 || spots == 5)
  { digitalWrite (2, HIGH); }   // Center spot

  if (spots > 3)
  { digitalWrite (3, HIGH); }   // Diagonal spots, left

  if (spots > 1)
  { digitalWrite (4, HIGH); }   // Diagonal spots, right

  // Add a small delay for a pleasing display speed.
  delay (20);

  // After 2 seconds have passed, stop ignoring the button.
  // If the button is pressed, call the checkbutton function.
  if ( millis() > ignore && digitalRead(0) == LOW )
  { checkbutton(); }
}

// This function waits for the button to be released,
// then waits for it to be pressed to start the next run.
void checkbutton()
{
  delay (50);                        // Button pressed; debounce.
  while (digitalRead(0) == LOW)      // While button is pressed,
    { }                              // do nothing while waiting.
  delay (50);                        // Button released, debounce.
  while (digitalRead(0) == HIGH)     // While button is released,
    { }                              // do nothing while waiting.
  ignore = 2000 + millis();          // Set the new ignore time,
}                                    // and return to the main loop.
```

Figure 5-95 *The Nicer Dice program listing.*

```
int spots = 0;
int outpin = 0;
long ignore = 0;

void setup()
{
  pinMode(0, INPUT_PULLUP);
  pinMode(1, OUTPUT);
  pinMode(2, OUTPUT);
  pinMode(3, OUTPUT);
  pinMode(4, OUTPUT);
  ignore = 2000 + millis();
}

void loop()
{
  for (outpin = 1; outpin < 5; outpin++)
  { digitalWrite (outpin, LOW); }

  spots = random (1, 7);

  if (spots == 6)
  { digitalWrite (1, HIGH); }

  if (spots == 1 || spots == 3 || spots == 5)
  { digitalWrite (2, HIGH); }

  if (spots > 3)
  { digitalWrite (3, HIGH); }

  if (spots > 1)
  { digitalWrite (4, HIGH); }

  delay (20);

  if ( millis() > ignore && digitalRead(0) == LOW )
  { checkbutton(); }
}

void checkbutton()
{
  delay (50);
  while (digitalRead(0) == LOW)
    { }
  delay (50);
  while (digitalRead(0) == HIGH)
    { }
  ignore = 2000 + millis();
}
```

Figure 5-96 *The same program listing with comments removed.*

While you are typing it, you will find that the second "if" statement contains a character that you haven't seen before. In fact, you may have never typed it, ever, on your keyboard. It is a vertical line, sometimes called the "pipe" symbol. On a Windows keyboard, you are likely to find it above the Enter key. You create it by pressing Shift and the backslash \ character. The listing uses two pairs of pipe symbols in the second "if" statement, and I will explain them when I go through the program line by line.

When you finish, choose the Sketch > Verify/Compile option in the IDE to see if you made any errors.

Some of the error messages can be difficult to understand, and they refer to line numbers. But line numbers are not displayed with the program! This seems like a cruel joke, telling you which line your error is on, but not showing you the number. Maybe there is a way to switch on line numbering? If you check the Help, and search for "line numbers," I don't think you'll find anything. Check the Arduino forums, and you will find a lot of people complaining that they can't display line numbers.

Ah, but the forums show the oldest posts first. If you scroll all the way down to the newest posts, you find that the problem finally was solved. Arduino just didn't get around to documenting it. Go to File > Preferences, and you will see an option to switch on the line numbers.

Of course the error messages can be a bit difficult to understand, but here are the most common causes of errors, which you should check before you try to fix anything.

- You may have forgotten to put a semicolon at the end of an instruction.
- You may have left out a closing brace. Remember, braces { must always } be in pairs.
- Although a command word often contains a mix of uppercase and lowercase letters, as in `pinMode`, you may have typed it in all lowercase letters. The IDE should display the command words in red, when they are correctly typed. If you see one that is in black, it contains a typing error.
- You may have omitted parentheses from a function name such as `void loop()`.
- You may have used a single = sign where you should have two == signs. Remember, = means "assign a value" while == means "compare a value."
- You may have typed only one | symbol or only one & symbol where they should be used in pairs.

When your Verify/Compile operation finds no additional error messages, upload the program. Now plug in the ground wire connecting your breadboard with the Uno board, and the LEDs should start flashing. Wait a couple of seconds, then press the button, and the display stops, displaying a random spot pattern. Press the button again, and the display resumes. Hold down the button, and after the two-second "ignore" period, the display stops again.

The pseudocode has been successfully implemented!

Now, how does the program work?

Short and Long Integers

The program includes some words that you haven't seen before, and one really important new concept.

One new word is `long`. Until now, you have typed `int` (meaning "integer") before each variable name. But the value of `int` is limited between -32,768 to +32,767. When you need to store a bigger number than that, a long integer can be used, allowing a range of values from -2,147,483,648 to 2,147,483,647.

Why not use long integers for everything? Then we don't have to worry about the limit for a regular integer. True, but a long integer takes twice as long to process (or more), and takes twice as much memory. We don't have a whole lot of memory on the Atmel microcontroller.

The system clock uses the `millis()` function to count milliseconds. If we only allowed it to count up to 32,767, that would be enough for just over half a minute. We may need more than that, so the function stores its value in a long integer. (How do I know this? I read it in the language reference. You have to read the documentation to use a computer language.)

When I created my "ignore" variable to memorize the current value of the system clock, the variable had to be defined so that it is compatible with the clock; so, it has to be defined as a long integer, by using the word `long`.

What happens if you try to store a number outside of the permitted range in an integer (or long integer) variable? Your program will produce unexpected results. It's up to you to make sure that this never happens.

Setup

The setup section of the program is fairly straightforward. You haven't used those `pinMode()` instructions before, but they are easy to understand.

The first one has a parameter, `INPUT_PULLUP`, which is very useful: it activates a pullup resistor built into the microcontroller, so you don't have to add a pullup resistor yourself. But, bear in mind, it is a *pullup* resistor, not a *pulldown* resistor. Therefore, the input state of the pin is normally high, and when you use a button, it must ground the pin of the chip, to make it low. Remember:

- When the button is pressed, the `digitalRead()` function returns a `LOW` value.
- When the button is released, the `digitalRead()` function returns a `HIGH` value.

The "for" Loop

At the beginning of the `void loop()` function, there is a different kind of loop. It is called a "for" loop because it begins with the word `for`. This is a very basic and convenient way to make the microcontroller count through a series of numbers, storing each new number in a variable, and throwing out the previous value. The syntax works like this:

- The reserved word `for` is followed by three parameters in parentheses.
- Each parameter is separated from the next by a semicolon.
- The first parameter is the first value that will be stored in the specified variable. (It is properly known as the initialization code.) In this program, the first value is 1, stored in a variable which I created, named `outpin`.
- The second parameter is the value where the loop stops counting (properly known as the stop condition). Because the loop stops at that point, the final value in the variable will actually be one less than the limit. In this program, the limit is `< 5`, meaning "less than 5." So the loop will count from 1 through 4, using the `outpin` variable.
- The third parameter is the amount that the loop adds to the variable in each cycle (properly known as the iteration expression). In this case, we are counting in ones, and the C language allows me to specify that by using two ++ symbols. So, `outpin++` means, "add 1 to the outpin variable in each cycle."

"For" loops allow you to specify all kinds of conditions. They are extremely flexible. You should read up on them in the language reference. This "for" loop just counts from 1 through 4, but it could count as easily from 100 to 400, or to any range you like, limited by the type of the integer used in the loop (`int` or `long`).

During each cycle of the loop, the microcontroller is told to do something. The procedure it performs is listed in braces after the loop is defined. Like any procedure, it can contain numerous operations, each ending in a semicolon. There is only one operation in this procedure: write a `LOW` state to the pin specified by the variable `outpin`. Because `outpin` is going to count from 1 through 4, the "for" loop is going to create a low output on digital pins 1 through 4.

Ohhhh, *now* you can see what this is all about. The loop is switching off all the LEDs.

Isn't there a simpler way to do this? Sure, you could use these four commands:

```
digitalWrite (1, LOW);
digitalWrite (2, LOW);
digitalWrite (3, LOW);
digitalWrite (4, LOW);
```

But I wanted to introduce you to the concept of a "for" loop, because it's basic and important. Also, what if you wanted to turn off nine LEDs? Or what if you wanted the microcontroller to flash an LED 100 times? A "for" loop is often the best way to make a procedure efficient when repetition is involved.

The Random Function

After the "for" loop has zeroed the die display, we get to the `random()` function, which chooses a number between the limits in the parentheses. We want a die value from 1 through 6, so why is the range listed from 1 through 7? Because actually the function is choosing fractional values from something like 1.00000001 to 6.99999999 and throwing away the numbers which follow the decimal point. So, 7 is a limit that is never actually reached, and the output will be from 1 through 6.

Whatever the random number is, it is stored in another variable that I made up, named `spots`, meaning the number of spots on the face of a die.

The "if" Statement

Now it's time to see what the value of `spots` is, this time around, and light up the appropriate LEDs.

The first "if" statement is simple enough. If we have 6 spots, this will be the only occasion where we write a high value through output pin 1, which is connected with the left and right LEDs.

Why don't we switch on all the diagonal LEDs, too? The answer is, they will be switched on for other die values also, and it's more efficient to minimize the number of "if" tests. You'll soon see how this works.

The next "if" uses the pipe symbol that I mentioned previously. A pair of || symbols means OR in the C language. So, the function says that if we have a die value of 1, OR 3, OR 5, we light up the center LED, by putting a high state on pin 2.

The third "if" says that if the `spots` value is greater than 3, we need to light two of the diagonally placed LEDs. These will be required to display the patterns for a 4, 5, or 6.

The last "if" says that if the `spots` value is greater than 1, the other diagonally placed LEDs must be illuminated, too.

You can test the logic of these "if" functions by looking back at the spot patterns in Figure 4-146. The logic gates in that figure were chosen to fit the binary output from the counter chip, so they're different from the logical operations in the "if" functions of the program. Still, the LEDs are paired in the same way.

Flash Speed

After the "if" functions, I inserted a delay of 20 milliseconds, because I think it makes the display more interesting. Without this delay, the LEDs flash so rapidly, they are just a blur. With the delay, you can see them flashing, but they're still too fast for you to stop at a number that you want—although you can try!

You may wish to adjust the delay value to a number higher or lower than 20.

Creating a New Function

Now comes the important part. In the pseudocode that I wrote, we have reached Steps 3, 4, and 4a. To refresh your memory:

- Step 3. Check to see if the button has been pressed.
- Step 4. Check to see if the system clock has caught up with the "ignore" variable yet.
- Step 4a. If the button has not been pressed, OR the system clock has not caught up with the "ignore" variable yet, go back to Step 0. Otherwise...

These steps can be combined in one "if" function. In pseudocode, it would be like this:

- If (the button is not pressed OR the system clock is less than the "ignore" value), go back to step 0.

But there is a problem with this. The term "go back to..." suggests that I want to direct the microcontroller to a specific part of the program. This may seem a natural thing to do, but when you write in C, you should try to avoid transferring control from one section of a program to another.

The reason is that a lot of "go here" and "go there" instructions make the program difficult to understand—not just for other people, but for you, too, when you take another look at it six months from now, and you can't remember what you had in mind.

The concept of C is that each part of a program is contained in a separate block, and the program runs by *calling* them when you want them. Think of each block of instructions as an obedient servant who only does one thing, such as washing the dishes or taking the garbage out. When you need that task to be executed, you just call the servant by name.

The blocks are properly known as *functions*, which is confusing, because we have been dealing with functions such as `setup()` and `loop()` already. But in fact you can write your own function, which works in basically the same way.

I decided that the correct way to write this program is to split off the button-checking function into—well, a function. I have called it `checkbutton()`, but I could have called it anything at all, so long as the word wasn't already being used for some other purpose.

You see the `checkbutton()` function at the bottom of the listing, preceded by the word `void`, because this function doesn't send any value back to the rest of the program.

`void checkbutton()` is the *header* for the function, after which the procedure is contained within braces, as usual. All this function does is:

- Wait 50ms for the contact to stop bouncing.
- Wait for the button to be released.
- Wait another 50ms for the end of the contact bounce created by releasing the button.
- Wait for the button to be pressed again (in other words, wait for the released state to end).
- Reset the `ignore` variable.

When the microcontroller gets to the end of the function, where does it go? Simple: back to the line immediately following the one that called the function. Where is that? Immediately below the "if" function, above. That is how you call a function: you just state its name (including the parentheses, which sometimes have parameters inside them, although not in this case).

You can, and should, create as many functions as you like in a program, using each one to perform a separate task. To learn about this, I suggest you read any general reference on the C language. The Arduino documentation doesn't go into much detail on functions, because —well, they're a little difficult to understand, if they start to pass values to and fro. Still, they are at the heart of the C language.

Structure

The line that begins `if ( millis() > ignore` has the same purpose as Step 4 in my pseudocode, except that it now works the other way around. Instead of deciding whether to send the microcontroller back to the beginning, it determines whether to call my `checkbutton()` function. Previously I summarized the logic as, "if (the button is not pressed OR the system clock is less than the "ignore" value), go back to step 0." The revised version says, "if the button-ignore period is over, AND the button is pressed, make a detour to the checkbutton() function."

After the microcontroller does that and returns, it reaches the end of the main `loop` function, and the `loop` function always repeats automatically.

Really, this program does only one thing. It selects random numbers and displays them as spot patterns, over and over again. If the button is pressed, it pauses and waits, but when the button is pressed again, the program resumes doing what it was doing before. The button-checking routine is just a momentary interruption.

Therefore, the natural structure for this program is to have a main loop that just selects and displays numbers, and then if the button is pressed, the microcontroller is sent on a brief detour to the `checkbutton()` function and back again.

The Arduino documentation doesn't say anything about structure, because it wants to get you started in making things happen as quickly as possible. So the Arduino simply forces you to use the mandatory `setup` function, followed by the `loop` function, and that's it.

But as soon as a program begins to grow in size, you really need to divide it into your own functions, to keep it from becoming a complicated mess. A standard C-language tutorial will explain this in more detail.

Of course, if you just want to use the Arduino to do one simple thing, such as switch on a heater when a room gets cold, you can put all the procedures in the main `loop` function, and that will be all you need. But this is a waste of the microcontroller's capabilities. It can do so much more. The trouble is, when you try to do something more ambitious—such as simulate the throw of a die—the instructions accumulate. Structuring them helps to keep everything clear.

There is another advantage to dividing a program into functions. You can save the functions separately, and reuse them in other programs later. The `checkbutton()` function could be reused in any game where you want to stop the action by pressing a button and restart it by pressing the button a second time.

Likewise, you can use other people's functions in your own programs, provided the authors don't restrict you from doing so by controlling their copyright. A vast number of functions in the C language are available freely online, many of them specifically written for the Arduino. Functions exist to control almost any alphanumeric display, for instance. This leads to a very important but often ignored recommendation for programmers:

- Don't reinvent the wheel.

You don't need to waste time writing your own function if someone else will let you use theirs.

This is another reason why the concept of functions is so important in C.

But Is It Too Difficult?

The more you write programs, the easier it gets. The learning curve is steep at the beginning, but after some practice you'll write a "for" loop without thinking much about it. Everything will seem obvious.

That's what programmers like to say. Is it true?

Sometimes it is, and sometimes it isn't. In the maker movement we tend to assume that anyone can take control of the techno-world around us. In fact I subscribe to this belief myself—but computer programming pushes this philosophy to the limit.

I used to teach an introductory programming class, and I noticed a very wide range of aptitudes among the students. Some of them found programming a very natural thought process, while others found it extremely difficult, and this didn't always have much to do with intelligence.

At one end of the scale, at the end of my 12-week, 36-hour programming course, one student wrote an entire simulation of a slot machine, including graphics showing the spinning wheels, and money that came tumbling out.

At the other end of the scale, I had a student who was a pharmacist. He was a very smart, well-educated man, but no matter how hard he tried, he couldn't get the syntax right, even in simple "if" statements. "This is really annoying me," he said, "because it makes me feel stupid. And I know I'm not stupid."

He was right, he wasn't stupid, but I came to the conclusion that I couldn't help him, because I had learned a fundamental fact.

- To be good at writing programs, you must be able to think like a computer.

For whatever reason, the pharmacist couldn't do that. His brain just didn't work that way. He could describe to me the pharmacology of a medication, its molecular structure, and a lot more; but that was of no help to him in writing programs.

When the Arduino was marketed, evangelists described it as a device for creative people and others who didn't think of themselves as programmers. Supposedly, it would be so simple, everyone could use it.

The trouble is, I'm old enough to remember when HTML was introduced with the same idea—that it would be so easy, everyone would be coding their own web pages. Well, a few people did, but not "everyone." Today, only a tiny minority will hand-code HTML (I am one of them, but I'm eccentric for doing so).

Going back even farther, to the dawn of computing as we know it, the BASIC computer language was created with the idea that "everyone" could use it. In the 1980s, with the advent of desktop computers, evangelists predicted that people would be writing little programs in BASIC to balance their checkbooks or store recipes. Well, a lot of people gave it a try, but how many people today still do so?

My purpose in emphasizing this is to reassure you that if you are one of the people who finds it difficult, there's no stigma attached to this. I am sure you have other skills that you can pursue instead. In fact, building things with individual components could be one of them, as I think it requires different thought processes. Personally, I find writing programs much easier than designing circuits, but for someone else, the reverse could be equally true.

Upgrading the Nicer Dice Program

As in the version of this program in Experiment 24, the obvious upgrade is to add a second die display. This can be done very easily with the Arduino board, because it has additional digital outputs that can drive a second set of LEDs. You simply need to duplicate the section of the program that begins with zeroing the display and ends with the `delay(20);` function. Substitute the new pin numbers for your additional LEDs in the `digitalWrite()` functions, and job done!

Other Microcontrollers

I already mentioned the PICAXE. Its documentation is good, the tech support is excellent, and the language is easier to learn than C. So why didn't the PICAXE capture everyone's imagination? I don't know; maybe because it has a wacky name. I think you should check it out. Start by looking at its entry in Wikipedia.

The BASIC Stamp has a larger vocabulary of commands than the PICAXE, and a bigger range of add-on devices (including displays with graphical capability, and a little keyboard that is specifically designed for use with the controller). You can buy it in the form of surface-mount components squeezed onto a tiny board that will plug into a breadboard, as shown in Figure 5-97. A very nice design.

Figure 5-97 *The BASIC Stamp microcontroller consists of surface-mounted components on a platform that has pins spaced at 1/10-inch intervals, for insertion in a breadboard or perforated board.*

On the downside, you'll find that everything associated with the BASIC Stamp is a bit more expensive than in the PICAXE world, and the download procedure isn't quite as simple.

New products such as the Raspberry Pi extend the functionality of a microcontroller to the point where it becomes a real computer. By the time you read this, still more alternatives will be emerging in this turbulent field. Before you commit yourself to learning one of them in detail, I think it's a good idea to spend a day or two studying the online documentation and forums.

When I am thinking about learning something new, I do Google searches such as:

```
microcontroller problems OR difficulties
```

(I would substitute the name of an actual product for "microcontroller" in this search phrase.)

This is not because I am negative by nature. I just don't want to spend a lot of time on a product that turns out to have unresolved issues.

Unexplored Territory

It's time, now, for me to do some general summing up.

If you've taken the time to complete most of the projects in this book with your own hands, you have gained a very rapid introduction to the most fundamental areas of electronics.

What have you missed along the way? Here are some topics that remain wide open for you to explore. Naturally you should search online if they interest you.

The informal approach of Learning by Discovery that I have used in this book tends to be light on theory. I've avoided most of the math that you'd be expected to learn in a more rigorous course on the subject. If you have mathematical aptitude, you can use it to gain a much deeper insight into the way in which circuits work.

We didn't go very far into binary code, and you didn't build a half-adder, which is a great way to learn how computers function on the most fundamental level. But in *Make: More Electronics*, I show you how to do this.

I avoided going deeply into the fascinating and mysterious properties of alternating current. Here again, some math is involved, and just the behavior of current at high frequencies is an interesting topic in itself.

For reasons already stated, I avoided surface-mount components—but you can go into this area yourself for a relatively small investment, if you like the idea of creating fascinatingly tiny devices.

Vacuum tubes were not mentioned, because at this point, they are mainly of historical interest. But there's something very special and beautiful about tubes, especially if you can enclose them in fancy cabinetwork. In the hands of a skilled craftsperson, tube amplifiers and radios become art objects.

I didn't show you how to etch your own printed circuit boards. This is a task that appeals only to certain people, and the preparation for it requires you to make drawings or use computer software for that purpose. If you happen to have those resources, you might want to do your own etching. It could be a first step toward mass-producing your own products.

I didn't cover static electricity at all. High-voltage sparks don't have any practical applications, and they entail some safety issues—but they are stunningly impressive, and you can easily obtain the necessary information to build the equipment. Maybe you should try.

Op-amps and higher-level digital logic are other topics that I haven't touched here. However, they are included in *Make: More Electronics*.

In Closing

I believe that the purpose of an introductory book is to give you a taste of a wide range of possibilities, leaving you to decide for yourself what you want to explore next. Electronics is ideal for those of us who like to do things ourselves, because almost any application—from robotics, to radio-controlled aircraft, to telecommunications, to computing hardware—can be pursued by just one person, working at home, with limited resources.

As you delve deeper into the areas of electronics that interest you most, I trust you'll have a satisfying learning experience. But most of all, I hope you have a lot of fun.

Tools, Equipment, Components, and Supplies

5

This chapter is divided into five parts.

Kits. Various kits have been prepared, containing components and supplies that you can use to complete the projects in the book. See "Kits" on page 311 just below, for details.

Searching and Shopping Online. Instead of buying kits, you may prefer to do your own shopping. I have compiled some tips to help you. See "Searching and Shopping Online" on page 311 just below, for details.

In **Checklists of Supplies and Components** you will find an itemization of everything that you need. The lists of supplies begin at "Supplies" on page 316, and the lists of components begin at "Components" on page 317.

Buying Tools and Equipment. I have listed all the tools that are discussed at the beginning of each chapter of the book, and I have some suggestions about where to find them. See "Buying Tools and Equipment" on page 324.

Suppliers is a list of sources. Abbreviations in this list are used in the buying guides. See "Suppliers" on page 326.

Kits

Kits of components for the experiments in this book are still being finalized as the book goes to press. One kit should contain all the parts that you need for Chapters One, Two, and Three of the book. An additional kit should offer parts for Chapter Four. A separate soldering kit may also be available.

For information, please visit *www.plattkits.com*.

The page will be updated as more options become available. Please note that kits may be offered by independent suppliers who are not affiliated with Maker Media in any way.

Searching and Shopping Online

I'm including some general advice about searching for parts because many readers of the first edition of this book seemed to have difficulty getting the results they were looking for. I'll start with the most basic considerations, and work up. Even some seasoned shoppers may find a couple of tips here that are useful.

For a comprehensive list of all suggested suppliers, see "Suppliers" on page 326. Here are the ones that I consider primary:

Electronic components are available from large retail vendors online, most of which don't impose minimum quantities. Mouser, Digikey, and Newark are the obvious choices, maintaining huge inventories. Find them at:

Mouser Electronics (*http://www.mouser.com*) ships from Texas

Digi-Key (*http://www.digikey.com*) ships from Minnesota

Newark element 14 (*http://www.newark.com*) ships from Arizona

In addition, don't forget eBay, where prices are often lower than from other sources, especially if you use

Asian sellers. eBay is not so useful for parts for which there is less demand, such as logic chips.

Tools and equipment are available from eBay, Amazon, and Sears, but if you want a truly monumental selection, McMaster-Carr (*http://www.mcmaster.com*) is unbeatable.

They also have excellent tutorials—for instance, on the properties of different types of plastic, or the relative advantages of different drill bits.

The Art of Searching

The easiest search is for a specific part number, if you have one. You can enter it into the search field on a site such as mouser.com, where the algorithm is smart enough to allow some flexibility. For example, suppose you want a 7402 logic chip. Mouser will helpfully suggest that a SN7402N from Texas Instruments may be what you want, bearing in mind that Texas Instruments adds SN to the front and N to the end of the basic chip type.

However, the search will not be helpful if there is an additional code in the middle of a part number. When you search for a 7402, Mouser won't think of showing you any chips in the 74HC02 family, because HC is added in the middle.

Try the Chat Option

Suppose you only have an incomplete part number, or you don't know if the part is obsolete, or you generally need a bit of help. Don't overlook the option to make a voice call. A large distributor will have sales representatives who can assist you. It doesn't matter that you are an individual buying small quantities.

Better still, open a chat window. This will allow you to copy-paste a part number into the window and get a fairly quick answer advising you of similar options if the part is unavailable.

Parts on Google

If you want to do comparison shopping, use a generic search engine. I'll assume that Google is your default, because I feel it's the most appropriate for our purposes.

If a part number is long and complicated, you will have a better chance of finding what you want, and not being offered things that you don't want. Searching for a 7402 on Google will generate results including a Pantone ink color and an Institutes of Health standard. Searching for 74HC02 will narrow the results to logic chips.

Unfortunately, you are now likely to get hits from a lot of datasheet resellers. These companies harvest datasheets from electronics manufacturers and repackage them for you with ads that pay for this "service." It wouldn't matter, except that the reseller often shows you just one datasheet page at a time, because each page will have a new set of ads, all of which make money for the reseller. Waiting for each page to display is a waste of time, so I often use a hyphen as a minus sign to block datasheets when I search for components on Google, like this:

```
74HC02 -datasheet
```

Note that when you specify a part number, the search engine is less likely to compensate for any little errors that you make. Google understands that if you type "compoments" you probably meant "components," but it will not know that an 84HC02 chip should really be a 74HC02.

Datasheets

What if you actually want to see a datasheet, because you need to check the specification of a component before buying it? Go to one of the big distributors, find the part that you want, and you will see the option to click a datasheet icon. This will link you with a printable multipage document (almost always in PDF format) maintained by the component manufacturer itself. For me, this is a lot quicker than dealing with datasheet resellers on Google.

General Search Techniques

If you are looking for a component type, a search term that is brief and vague is usually inadequate. Suppose you search for:

```
switch
```

In my location, the first hit from this search was for light switches, the second hit was for a local wine bar, and I was then offered a variety of network switches (which are like routers). I also found a company named Switch that helps people to find a new job. How can you avoid these irrelevant hits?

As a first step, add a word to define your area of interest. For example, this may help:

```
switch electronic
```

Better still, if you want a DPDT toggle switch rated for 1 amp, just say so:

```
"toggle switch" dpdt 1a
```

Note the use of quote marks to nail down a specific phrase, discouraging Google from showing near-miss search results that are not quite what you asked for. Also note that search terms are not case-sensitive; there's no advantage in putting a term such as dpdt in caps.

You can narrow your search even further by naming a source, such as:

```
"toggle switch" dpdt 1a amazon
```

Why mention Amazon, if you can go to amazon.com and do your searching there? Because the search capability at amazon.com has fewer features than Google. In this example, it would not recognize the use of quote marks.

Fortunately Amazon allows Google to crawl all over its site and index everything, so that a search from Google can jump you straight to an Amazon list of toggle switches.

Exclusions

Use the minus option to avoid items that you don't want. For instance, if you are only interested in a full-size toggle switch, you could try this:

```
"toggle switch" dpdt 1a amazon -miniature
```

Note that the minus sign is another piece of syntax that Amazon's search feature doesn't understand.

Alternatives

Don't forget the AND and OR logical operators. If a single-pole, double-throw switch will work just as well for you as a double-pole, double-throw switch, you could try this on Google:

```
"toggle switch" dpdt OR spdt 1a -miniature
```

But even this may cause some problems, because naming conventions in electronics can be inconsistent. Some people refer to a DPDT switch as a 2P2T switch. Some call a SPDT switch a 1P2T switch. You'll need a lot of ORs to cover these alternatives.

Too Much Typing?

Personally I find that a carefully constructed, detailed search will save time by avoiding subsequent searches. However, if you don't want the chore of typing an elaborate search term, you have other options. One is to click the word "Images" which Google displays immediately above each set of search results, adjacent to the word "Web." Google Images will show you pictures of every conceivable kind of switch, and because our brains are well equipped to recognize images quickly, scrolling through a lot of pictures can be a more efficient way to find what you want than scrolling through a lot of text.

Alternatively, you can click the "Shop" option above Google's search results. This will give you the ability to list items in order of price from dozens or hundreds of different vendors. Some vendors will not be included, however.

Vendor Categories

Another option is to go to a vendor's site and use their system of categories. At mouser.com, digikey.com, and newark.com, if you search for "switch" you'll be shown a list of different kinds of switches. Click the type you want, and you'll be offered additional options, to narrow your search one step at a time.

Ultimately on mouser.com and other large-vendor sites you are likely to see little windows listing attributes such as voltages, amperages, and other values. This can be frustrating, because the lists are not intelligently managed. Some switches rated for half an amp, for instance, are grouped under 0.5A, while others are grouped separately as 500mA. These ratings are identical, but the people who create the listings just seem to copy the specifications from datasheets, some of which use amps while others use milliamps.

What to do? Use your Control-click option (Command-click on a Mac). Holding down the Ctrl key (or Command key) while you click the additional selection allows you to select the 0.5A switches *and* the 500mA switches *and* any others that might be suitable—including 1A switches, as a higher current rating will work fine at lower currents.

Which to Click First?

When using categories on a vendor site, it's useful to begin by choosing the attributes that you absolutely, positively need. For instance, if you are shopping for logic chips, begin by selecting through-hole versions, because you definitely won't want the tiny surface-mount versions. But note that a "DIP" package (meaning, dual-inline pin) is almost the same thing as a "PDIP" (plastic DIP) package, which is the same as a "through-hole" version.

Conversely, any chip format identified by an acronym beginning with S is almost always a surface-mount version, which you do not want. SMT, in particular, means surface-mount.

A Real-Life Search

Here's an example of an actual search that I performed for a part that I used in this book. I knew what I wanted it to do, but I didn't know the part number.

I wanted a counter with a 3-bit output for use in the "Nice Dice" circuit (see "Experiment 23: Flipping and Bouncing" on page 218). So I went to Mouser Electronics and I started by searching for:

```
counter
```

While I was typing my search term, Mouser suggested an autocomplete:

```
Counter ICs
```

An IC is an integrated circuit, which is the same thing as a chip. So I clicked the autocomplete suggestion, which took me to a page suggesting 821 matches. Little scrolling windows would allow me to narrow the search by manufacturer, counter type, logic family, and much more. How should I proceed?

I scrolled horizontally to the window allowing me to choose the mounting style. Only two options: SMD/SMT (which are surface mount chips) and through-hole (the chips that plug into breadboards, not requiring a magnifying glass). I clicked the through-hole option and then clicked the Apply Filters button. That gave me 177 matches.

All the logic chips in this book are HC type in the 7400 family, so I went to the Logic Family window, and clicked 74HC. But, not so fast! I know that Mouser often lists the same thing under different names, so I scrolled through the other options. Sure enough I found HC listed separately from 74HC. I Control-clicked it to select both terms.

Now I had 52 options to choose from. As Counter Type, I selected Binary, because I wanted a binary output. This left me with 33 remaining matches.

There were no 3-bit chips, but I could use a 4-bit chip and ignore the highest bit. I saw two options in Number of Bits: 4 and 4-bit. I Control-clicked to select them both.

Counting sequence could be Up, or Up/Down. I only wanted Up, so I clicked that. Now only nine matches left! Time to inspect the results. I wanted to use the most commonly available chip, which I determined by seeing how many of each one were in stock. I saw more than 7,000 of the SN74HC393N by Texas Instruments.

I clicked the datasheet link to make sure it would do what I wanted. A 14-pin chip providing maximum continuous output current of plus-or-minus 25mA with a nominal 5-volt supply ("nominal" means "typically used"). Yes, this was a standard logic chip in the 74HCxx family. Actually it contained two 4-bit counters, and I only needed one, but I wasn't going to quibble over that, and in fact I realized I could make use of the second counter in the chip, if I enlarged the scope of my project.

The 74HC393 would cost me about 50 cents. Might as well put six of them in the shopping cart. That's only $3, so maybe I should look for something else, reasonably small and light, so that I could add it without paying any additional shipping charge. But first I printed the datasheet for the 74HC393 and added it to my paper-based file-folder system.

You can see that this process entailed a lot of clicking. But it took me less than 10 minutes, and I found exactly what I wanted.

I could have followed a different path. Because I knew I wanted a chip in the 74xx family, I could have gone to this URL, which I keep bookmarked for easy reference:

www.wikipedia.org/wiki/List_of_7400_series_integrated_circuits

This includes all the 74xx logic chips that have ever been made. If you go to this page, you can press Control-F to search the text, and then type in:

```
4-bit binary counter
```

It has to be an exact match, which means you must type 4-bit, not 4 bit. The search yields 13 hits, and you can compare the features of the chips. After you choose one, you can copy its number and paste it into the search field of a site like Mouser, which takes you to that one component.

The only problem is, the Wikipedia page doesn't tell me which chips are old and almost out of production, and which are still popular. For my purposes, writing a book that I hope will be around for a while, I have to stick with the most popular components. This may be a good idea from your point of view, too, because if you build a circuit around an old chip, you are locking yourself into the past.

I could have used yet another approach, doing a Google search for people discussing and advising each other about counter chips. But you get the general idea. You don't need a part number, to find what you want.

eBay Options

I buy a lot of parts through eBay because I find bargains there, and also because most companies selling through eBay are extremely quick and reliable. To minimize your time and trouble, you need to know a few search basics that are specific to eBay.

First, don't hesitate to click the little "Advanced" option just to the right of the Search button on eBay's home page. This will allow you to specify attributes such as the country of origin (if you want to avoid overseas suppliers), and can limit your search to Buy It Now items. You can also specify a minimum price, which can be useful to eliminate stuff that is too cheap to be any good. Then, before starting the actual search, I usually click the display option for Price + Shipping: Lowest First.

Once you find what you want, it's time to check the seller's feedback. For sellers within the United States, I want 99.8% or better. I've never had a problem with sellers rated 99.9%, but I have been disappointed sometimes with service from sellers rated 99.7%.

If a supplier is in an Asian nation such as China, Hong Kong, Thailand, or others, you can be less fussy about feedback, because a lot of buyers give bad feedback when they don't receive something as quickly as they expect. Overseas sellers will warn you that a small packet will take 10 to 14 days in transit, but buyers complain anyway, and this drags down the feedback rating. In reality, in my experience, every item that I have ordered from overseas sources has always turned up, and has always been what I wanted. You just need to exercise a little patience.

After you find what you want on eBay, you may want to click the Add to Cart button, rather than the Buy it Now button, because you can look for additional items from the same seller, and you'll save time by grouping them into one shipment. This should also reduce shipping costs.

Click the Visit Store option in the Seller Information window, or if the seller doesn't have an eBay store, click to See Other Items. You then have the option to search within that seller's list of products. After you add as many as you want to your cart, it's checkout time.

You can make direct contact with overseas suppliers, instead of finding them through eBay. Tayda Electronics in Thailand (abbreviated tay in my list—see "Suppliers" on page 326) is a popular source.

Amazon

I don't think amazon.com is very useful for components, but it can be a good source for tools and for supplies such as wire or solder. The only problem I have is that Amazon doesn't like to show me the cheapest stuff first. You have to choose that option repeatedly after every search, and if the products are scattered among different store categories, the option to resequence the results will not exist. Even when you can prioritize the list by cheapness, Amazon (unlike eBay) is not smart enough to factor in the shipping cost. Pliers with a price of $4.95 and $6 shipping will be listed as being cheaper than pliers selling for $5.50 with $3 shipping. On the other hand, Amazon ships fast and if you buy a bunch of items at one time, all of them warehoused by Amazon themselves, you may be able to spend enough to get free shipping.

Killing Autocomplete

One last tip regarding Google. The default mode of the search engine will prompt you with a pop-up list of similar terms while you are still trying to type your search string. I get really annoyed with the interference of this autocomplete option, so I have disabled it—and you can, too.

In the address bar of your browser, use this URL to launch Google:

http://www.google.com/webhp?complete=0

Now save it as one of your favorites, and when you click the favorite, Google won't try to tell you what it thinks you are looking for. It will wait quietly for you to finish typing.

You can also use the URL as the default page that will open every time you launch your browser.

Is Searching Worth the Trouble?

You may feel that you don't want to memorize all these searching techniques. OK, that's why Maker Shed and myself are offering kits to go with this book. Buy a kit, and you should have all the components that you need, with no searching necessary.

But what are you going to do if you get interested in projects outside of this book? Suppose you see a circuit online—or suppose you want to modify a circuit, or design your own. At that point, I think you'll have to buy your own parts, and even if you try to obtain them all from one source, search techniques can be valuable.

Checklists of Supplies and Components

Photographs and general information are provided at the beginning of each chapter of this book. See "Necessary Items for Chapter One" on page 1, "Necessary Items for Chapter Two" on page 41, "Necessary Items for Chapter Three" on page 99, and "Necessary Items for Chapter Four" on page 142.

Below you will find listings of all the components and supplies. But I need to clarify the distinction between these two words.

Supplies are items such as solder or wire that I suggest you buy in a one-time purchase, sufficient for all the experiments. It doesn't make sense to consider how many inches of wire you will need for each project.

Components become an integral part of a project. You may be able to reuse these items, but only if you remove them from a previous project. Therefore, to take one example, a breadboard is included with components.

Supplies

The following supplies will be sufficient for all the projects. See "Suppliers" on page 326 for a list of sources from which you can buy these supplies, and abbreviations that I will use to refer to them.

Hookup Wire

You need 22 gauge, solid conductor, in at least two colors (red and blue), and preferably two more colors (your choice). Automotive wire is acceptable, so long as it is solid-core. Search eBay or Google for

```
solid wire 22 gauge OR awg
```

or check discount suppliers such as all, elg, and jam, or hobby suppliers such as ada and spk. (AWG is an acronym for American Wire Gauge.)

Quantities? If you want to do the experiments in Experiments 26, 28, 29, 30, and 31 that explore the world of inductance, you really need 200 feet of wire. Different colors can be joined together temporarily when winding coils. The wire can be unwound afterward and reused for other purposes.

If you are willing to skip the inductance experiments, I suggest you buy three spools of 25 feet each. You can find smaller lengths than 25 feet, but the price per foot goes up rapidly.

Jumpers

Personally I prefer not to use precut jumper wires, but if you choose them, one box should be sufficient. In addition you will need 25 feet of raw hookup wire, to make connections that are longer than the longest precut jumper. To find precut jumpers, you have to use the right search term. On Google, search for:

```
jumper wire box
```

The word "box" is the key to finding what you want. It automatically eliminates the type of undesirable flexible jumpers with plugs at each end, which are generally sold in bundles, not boxes. I don't think they are a good idea.

Stranded Wire

This is an optional addition, for situations where flexibility is important. One 25-foot spool will be enough.

Solder

This is usually sold by weight. See "Essential: Solder" on page 105 for the pros and cons of solder containing lead. Either way, be sure you buy *electronic solder* with a *rosin core*. The thickness may range from 0.02" to 0.04" (0.5mm to 1mm). If you only want to solder a couple of projects, three feet of solder will be sufficient, and some sources on eBay will sell very small quantities. Otherwise, try all, elg, jam, ada, amz, and spk.

Heat-Shrink Tubing

This is optional, but useful. One assortment of three or four (small) sizes will be enough. Because it has automotive applications, you can find it from hardware suppliers such as hom, har, and nor as well as hobbyist sources.

Perforated Board (unplated)

Only required in Experiment 14, although you can use it to build a permanent version of any projects in the book if you are willing to solder point-to-point wiring. A small piece, perhaps 4" x 8", will be enough for three average projects. Finding unplated board can be difficult, as most of it has copper or nickel solder pads. I think these are undesirable when doing point-to-point wiring, as they increase the risk of short circuits. Search for:

```
perforated board bare -copper
```

Also try searching for "prototyping board," or "proto board," or "phenolic board." Note that unplated board is also known as "unclad" in some places. At the time of writing, Keystone Electronics makes very small, cheap pieces of unplated perforated board, available through mou and dgk. You can also find unplated board at jam.

Perforated Board (plated)

This type of board is used for the finished version of Experiment 18, but of course you can use it for other projects where you want to make a permanent version. For convenience, use the type that has copper traces in the same pattern as connections inside a breadboard. This can be hard to find, because there are many different patterns, and the pattern that you want does not have a generally agreed name.

The BusBoard SB830 describes it as a "solderable breadboard" and it is currently available through amz. At ada, you find something similar named "Perma-Proto." GC Electronics 22-508 is another option, available through jam.

The Schmartboard 201-0016-31 (available through mou) is a two-part package consisting of a breadboard and a matching perforated board. The manufacturer suggests placing the perforated board over the breadboard and inserting components through both of them while developing and testing the circuit. Then lift up the perforated board, and the components are already in position, ready for soldering. Unfortunately this may not work for components with very short leads.

Machine Screws (bolts)

These and nuts with nylon inserts are available from hardware stores, but probably not in the small sizes necessary for attaching perforated board to the inside of a project box, or similar tasks. I suggest you buy #4 size flat-headed bolts in lengths of 3/8" and 1/2". My favorite source for this kind of hardware is McMaster-Carr.

Project Boxes

These vary a lot in price. Those made of ABS plastic are usually the cheapest. Try discount suppliers such as all, elg, and jam, or hobby suppliers such as ada and spk.

Components

Quantities and specifications for resistors, capacitors, and other components are listed below. See "Suppliers" on page 326 for a list of sources, and abbreviations that I will use to refer to them. The biggest suppliers are dgk, eby, mou, and nwk. You may find lower prices at all, elg, jam, and spk, but the selections will be smaller, and you should compare the shipping costs of buying from multiple suppliers with the cost of buying all your components in one shipment from a source where the prices are slightly higher.

Resistors

Any manufacturer is acceptable. Lead length is usually unimportant. All projects in this book can use a quarter-watt power rating (the most common value). A tolerance of 10% is acceptable, and the color bands on 10% resistors are easier to read than the bands on 5% or 1% resistors. However, you can buy 5% or 1% resistors if you wish.

The total number of resistors used in each section of the book is shown in Figure 6-1, but because resistors and capacitors are cheaply available in quantity, I don't think it makes sense to buy specific numbers for individual experiments. You will save time and money by buying packaged assortments.

- To buy sufficient resistors for *all* the projects in this book (with some to spare), get at least 10 of each of these values: 47 ohms, 220 ohms, 330 ohms, 1K, 2.2K, 4.7K, 6.8K, 10K, 47K, 100K, 220K, 330K, 470K, 680K, 1M. Also, get 20 of 470 ohms. Prepackaged assortments are your best bet. The quantities I have specified assume you will reuse some resistors after they serve their purpose in simple demonstration experiments.

Resistors	Book Chapters					Total
	1	2	3	4	5	
47 ohms				2	1	3
100 ohms				6		6
150 ohms				6		6
220 ohms				8		8
330 ohms				3	8	11
470 ohms	2	6	4	12		24
680 ohms				10		10
1K	2	2	1	4		9
2.2K	1			5		6
4.7K		4	2			6
6.8K					1	1
10K		1	1	41	4	47
47K				1		1
100K		2	1	4		7
220K		2				2
330K				1		1
470K		4	2			6
1M		1		4		5

Figure 6-1 *The number of resistors used in experiments in each chapter of the book.*

Capacitors

Find them at the same sources listed for resistors, above. Any manufacturer is acceptable. Radial leads are preferred, meaning that both leads emerge from the same end of the capacitor, instead of a lead at each end. A working voltage of at least 16VDC is recommended for power supplies up to 12VDC. You can substitute capacitors with higher working voltages, but the components will be physically larger. Other ratings such as temperature and impedance are not important for our purposes.

Ceramic capacitors are likely to last for many decades, while the longevity of electrolytic capacitors is a subject of some debate. For larger values, you have to use electrolytics, because the cost of ceramics becomes prohibitive. Personally I would use ceramics for values below 10µF and electrolytics for 10µF and upward, but you'll probably save money if you use electrolytics for values of 1µF and upward.

If you want to know the exact number of capacitors required in each chapter of the book, see Figure 6-2.

- To buy sufficient capacitors for *all* the projects in this book (with some to spare), get at least five of each of these values: 0.022µF, 0.047µF, 0.33µF, 1µF, 2.2µF, 3.3µF, 10µF, 100µF, 220µF. Also get at least 10 of values 0.01µF and 10µF. You only need two of each of these values: 15µF, 22µF, 68µF, 1,000µF. The quantities I have specified assume you will reuse some capacitors after they serve their purpose in simple demonstration experiments.

Capacitors	Book Chapters					Total
	1	2	3	4	5	
0.01µF		2		18	3	23
0.022µF				1		1
0.047µF				1		1
0.1µF		3		9		12
0.33µF		2		5		7
1µF		2		4	2	8
2.2µF					1	1
3.3µF		2	2	3		7
10µF		1		8	1	10
15µF				1		1
22µF				2		2
33µF		1				1
68µF				2		2
100µF		2		5	1	8
220µF		1	1		3	5
1,000µF		2			2	4

Figure 6-2 *The total number of capacitors used in experiments in each chapter of the book.*

Components other than Resistors or Capacitors	Book Chapters 1, 2, and 3	Extra for Chapter 4
LED (generic)	4	2
LED (low current)	1	15
Battery 9V	1	
Battery 9V connector	1	
Battery 1.5V	2	
Battery 1.5V carrier	1	
Breadboard	1	
Trimmer 500K	1	
Trimmer 100K		1
Trimmer 20K or 25K		1
Transistor 2N2222	6	
Speaker (small)	1	
Toggle switch	2	
Tactile switch	2	6
SPDT slide switch		2
Relay 9VDC DPDT	2	
AC-DC Adapter	1	
Diode 1N4001	1	
Perforated board 3" x 6"	1	
Fuse 3A	2	
Potentiometer 1K	2	
Lemons (or lemon juice)	2	
Galvanized brackets 1"	4	
Diode 1N4148		3
Timer 555 TTL type		4
7-seg LED display		3
4026B counter		3
74HC00 2-input NAND		1
74HC08 2-input AND		1
LM7805 regulator		1
74HC32 2-input OR		1
74HC02 2-input NOR		1
74HC27 3-input NOR		1
74HC393 counter		1

Figure 6-3 *Minimum numbers of components, assuming you will reuse items from each experiment in subsequent experiments. Items for Chapter Four are additional to items listed for Chapters One, Two, and Three.*

Other Components

For components other than resistors and capacitors, the minimum necessary quantities to build all the projects in Chapter One, Chapter Two, and Chapter Three of the book are shown in Figure 6-3. These quantities assume you will reuse all the components from each experiment in subsequent experiments. Components for Chapter Four are additional to those for previous sections of the book. Components for Chapter Five are not listed here, as the experiments are so diverse; see the beginning of each experiment in Chapter Five for a summary of options.

If you are concerned about burning out chips or transistors, which are vulnerable to damage, please add at least 1 to each quantity in Figure 6-3.

What if you may want to keep some of the projects that you build, instead of reusing the components for subsequent projects? In that case, please consult the tables for individual experiments that follow, and add the numbers of components for those experiment(s) which interest you.

The information that you need for finding and buying components is also provided below.

See "Suppliers" on page 326 for a list of suppliers that are referred to by abbreviations of their names. For most electronic components, go to all, eby, elg, jam, and spk for special deals, or dgk, mou, and nwk for one-stop shopping where everything will be available.

Components for Chapter One

Components for Chapter One, other than resistors and capacitors, are listed in Figure 6-4.

Components for Chapter One	Experiments 1	2	3	4	5	Total
LED (generic)			1	2		3
LED (low current)					1	1
Battery 9V	1		1	1		3
Battery 1.5V		2				2
Battery 1.5V carrier		1				1
Fuse 3A		2				2
Lemons (or lemon juice)					2	2
Galvanized brackets 1"					4	4
Potentiometer 10K				2		2
Deionized water (1 glass)					1	1

Figure 6-4 *Components other than resistors and capacitors, used in the first chapter of this book.*

Generic LEDs

The Lumex SLX-LX5093ID or Lite-On LTL-10223W are examples, but generic LEDs can be from any manufacturer. Probably 5mm LEDs are easier to handle, but 3mm LEDs can fit more easily into a crowded breadboard.

A typical forward current would be 20mA, typical forward voltage around 2VDC (blue and white LEDs will require a higher voltage). If you find a bunch of LEDs bundled for a low price on a site such as eBay, they can be considered generic.

Low-Current LEDs

These should be rated for 3.5mA forward current or less. The Kingbright WP710A10LID is an example, although the manufacturer, physical size, and color are unimportant. You could actually use this type of LED in all the experiments, but if you do, you should double the values of all the series resistors to protect it, as its maximum rating may be as low as 6mA.

Batteries

Nine-volt batteries can be the everyday alkaline type, available from supermarkets and convenience stores. Rechargable 9-volt batteries are an acceptable alternative.

The AA-size 1.5-volt batteries used in Experiment 2 must be alkaline. Do not use any type of rechargeable battery in this experiment.

Battery Connectors and Carriers

Just one carrier for a 1.5-volt battery will be necessary and sufficient. Note that a battery carrier may also be described as a *battery holder* or *battery receiver*. Make sure you get the type that holds only one AA battery (not two, three, or four). The Eagle 12BH311A-GR is an example.

You should buy at least three connectors for 9-volt batteries, because you may want to leave them attached to circuits that you build. Nine-volt connectors are sometimes described as *snap connectors* or *battery snaps*. Typical examples are the Keystone model 235 or Jameco Reliapro BC6-R. Buy whatever is cheapest, but make sure it terminates in wire leads.

Fuse

The 3A fuse in Experiment 2 ideally should be the automotive type, as its blades are easy to grip with alligator clips. Any automotive parts source will stock this type of fuse. The physical size is unimportant. Alternatively buy a 2AG size cartridge fuse, which is the smallest cartridge size, from an electronics supplier. It should be the fast-blow type, not a delay-fuse or "slow-blo" type. The voltage rating is not important. Littelfuse 0208003.MXP is an example.

Potentiometer

The full-size 1K potentiometer required for Experiment 4 should ideally be 1" in diameter, but sizes as small as half an inch are acceptable. Power rating, voltage rating, tolerance, shaft type, shaft diameter, and shaft length are unimportant. Select a potentiometer that has linear taper, and makes one turn, with panel mounting style, and solder lug terminals. Buy two. The Alpha RV24AF-10-15R1-B1K-3 and Bourns PDB181-E420K-102B are examples.

Juice and Brackets

If you use lemon juice in a squeeze bottle for Experiment 5, make sure it is undiluted and unsweetened. Vinegar is an acceptable substitute.

The 1" brackets for Experiment 5 must be galvanized. Pipe straps and hanger straps, to mount conduits and pipes, are an acceptable substitute. Any hardware source will have them cheaply available.

Deionized Water

This is often known as distilled water. Your local supermarket should have this, but make sure it is not "purified" and is not "spring water." It must have zero mineral content.

Components for Chapter Two

Components for Chapter Two, other than resistors and capacitors, are listed in Figure 6-5.

Components for Chapter Two	Experiments						Total
	6	7	8	9	10	11	
LED (generic)	1		2	1	1	1	6
Battery 9V	1	1	1	1	1	1	6
Battery 9V connector			1	1	1	1	4
Breadboard			1	1	1	1	4
Trimmer 500K					1		1
Transistor 2N2222					1	6	7
Speaker (small)						1	1
Toggle switch SPDT	2						2
Tactile switch		1	1	2			4
Relay 9VDC DPDT		2	1				3

Figure 6-5 *Components other than resistors and capacitors used in the second chapter of this book.*

Breadboards

A breadboard is classified here as a component, because it cannot be separated from the circuit; on the contrary, it is the foundation for the circuit. You have to decide how many circuits you may want to keep on their breadboards, and how many you are likely to disassemble so that the breadboard can be reused. Ideally, each breadboard should have a single bus on each side, and 700 connection points, as shown in Figure 2-10. Search Google or eBay for

```
solderless breadboard 700
```

However, if you prefer, you can use a dual-bus breadboard and ignore the extra lines of holes.

Trimmer Potentiometer

Trimmers of the recommended type are shown at left and at right in Figure 2-22, and a discussion of other types accompanies that photograph. The power rating is not important. The preferred type is single-turn, and it must terminate in pins that are spaced in multiples of 0.1" (2.54mm or 2.5mm). The Vishay T73YP504KT20 is a low-cost 500K trimmer.

Transistors

Before purchasing any 2N2222 transistors, see "Essential: Transistors" on page 49 for an important cautionary note.

Toggle Switch

This should be panel-mount type, ideally with screw terminals, although pins or solder lugs will be acceptable. It can be SPDT or DPDT. Voltage and current ratings are unimportant for the experiments in this book. The NKK S302T is an example, but you can find cheaper switches on eBay.

Tactile Switch

The type of tactile switch shown in Figure 2-19 is very strongly recommended, with two pins 0.2" apart, ideal for insertion in a breadboard. Avoid buying the more common tactile switches that have four pins or leads. The Alps SKRGAFD-010 is preferred (currently available from Mouser). Any tactile switch with 2 pins spaced 0.2" can be substituted, such as the Panasonic EVQ–11 series.

Relay

See "Essential: Relay" on page 48 for information about the recommended type of 9VDC, DPDT relay. The Omron G5V-2-H1-DC9, Axicom V23105-A5006-A201, and Fujitsu RY-9W-K have all been tested for suitability.

Components for Chapter Three

Components for Chapter Three, other than resistors and capacitors, are listed in Figure 6-6.

Many of the components for projects in Chapter Three have already been mentioned for Chapters One and Two; see above.

Components for Chapter Three	Experiments			Total
	13	14	15	
LED (generic)	2	1	1	4
Power source 9VDC	1	1	1	3
Breadboard			1	1
Transistor 2N2222		3	1	4
Diode 1N4001			1	1
Relay 9VDC DPDT			1	1

Figure 6-6 *Components other than resistors and capacitors used in the third chapter of this book.*

AC Adapter

This must have an output of 9VDC. It may have additional outputs providing different voltages. See "Essential: Power Supply" on page 99 for a discussion of the options. Minimum output should be 500mA (0.5A) DC.

If you want a multi-voltage adapter, finding one can be tricky, because if you search for "ac adapter" you will find hundreds or even thousands of single-voltage units. Your answer is to search a source such as eBay for:

```
ac adapter 6v 9v
```

This should provide you with several affordable multi-voltage options. Make sure the photograph of the unit shows a little switch to select the various voltages.

Diode

The 1N4001 switching diode is cheap and generic. Buy 8 or 10, and buy a similar quantity of 1N4148 signal diodes at the same time.

Headers

These miniature plugs and sockets are an optional item. Examples are Mill-Max part numbers 800-10-064-10001000 and 801-93-050-10-001000, or 3M part numbers 929974-01-36RK and 929834-01-36-RK.

Components for Chapter Four

Components for Chapter Four, other than resistors and capacitors, are listed in Figure 6-7.

Slide Switch

The recommended slide switch is SPDT with three pins spaced 0.1" apart, and is shown in Figure 4-5. I suggest the EG1218 made by E-switch. If you buy an alternative, it must terminate in solder pins for insertion in a breadboard. An example is the NKK CS12ANW03, but if you search eBay for

```
slide switch breadboard
```

you will find some that are much cheaper. The type of contact plating, voltage rating, and current rating are unimportant for the projects in this book.

Components for Chapter Four	Experiments									Total
	16	17	18	19	20	21	22	23	24	
LED (generic)	1	4	3	2		1	2		1	14
LED (low-current)					2	1	1	3	15	22
Power source 9VDC	1	1	1	1	1	1	1	1	1	9
Breadboard	1	1	1	1	1	1	1	1	1	9
Trimmer 20K or 25K	1			1						2
Trimmer 100K		1								1
Trimmer 500K	1									1
Tactile switch	2	1	1	3	2	8	2		1	20
SPDT Slide switch			2		1		2	2	2	9
Diode 1N4001			1			1				2
Diode 1N4148		1				3				4
Timer 555 TTL type	1	4	4	3		1	2		1	16
Speaker (small)		1	1							2
7-seg LED display				3						3
4026B counter				3						3
74HC00 2-input NAND					1			1		2
74HC08 2-input AND					1	1			1	3
LM7805 regulator					1	1	1	1	1	5
Relay 9V DPDT			1			1				2
Transistor 2N2222			2			1				3
74HC32 2-input OR							1		1	2
74HC02 2-input NOR								1		1
74HC27 3-input NOR									1	1
74HC393 counter									1	1

Figure 6-7 *Components other than resistors and capacitors used in the fourth chapter of this book.*

Integrated Circuit Chips

See "Fundamentals: Choosing Chips" on page 142 for a discussion of chips. While all the chips you will need are listed in Figure 6-7 (with the exception of one more 555 timer required for Experiment 29), it's a good idea to buy an extra chip of each type, as they are easily damaged by incorrect voltage, reversed polarity, overloaded outputs, or static electricity.

Any manufacturer is acceptable. The "package" of a chip refers to its physical size, and this attribute should be checked carefully when ordering. All logic chips must be in a DIP package (meaning a dual-inline package with two rows of pins that have 0.1" spacing). This may also be referred to as PDIP (meaning a plastic dual-inline package). They are also described as "through hole." The DIP and PDIP descriptors may be appended with the number of pins, as in DIP-14 or PDIP-16. This number can be ignored.

Surface-mount chips will have packaging descriptors beginning with S, as in SOT or SSOP. Do not buy any chips with "S" type packages.

The chip family used exclusively in this book is HC (high-speed CMOS), as in 74HC00, 74HC08, and similar generic identifiers. These numbers will have additional letters or numbers added by individual manufacturers as prefixes or suffixes, as in SN74HC00DBR (a Texas Instruments chip) or MC74HC00ADG (from On Semiconductor). These versions are functionally identical. Look carefully, and you will see the 74HC00 generic number embedded in each proprietary number.

Old TTL logic chips, such as the 74LS00 series, have compatibility issues. They are not used or recommended for any of the projects in this book.

555 Timer

Unlike the logic chips, you do want the TTL version of the timer (also known as the bipolar version), not the CMOS version. Here are some guidelines:

The TTL version (which you want) often states "TTL" or "bipolar" in its datasheet, specifies a minimum power supply of 4.5V or 5V, specifies an inactive current consumption of at least 3mA, and will source or sink 200mA. Part numbers often begin with LM555, NA555, NE555, SA555, or SE555. If you search by price, the TTL versions of the 555 timer are the cheapest.

The CMOS versions (which you don't want) always state "CMOS" on the first page of their datasheets, allow a minimum power supply of 2V in most cases, claim an inactive current consumption in microamps (not milliamps), and will not source or sink more than 100mA. Part numbers include TLC555, ICM7555, and ALD7555. If you search by price, the cheapest CMOS version of the 555 timer still costs almost twice as much as the cheapest TTL version.

Seven-Segment Display

The display used in Experiment 19 must be an LED device, height 0.56", low-current red preferred, able to function at 2V forward voltage and 5mA forward current. The Avago HDSP-513A is preferred, or Lite-On LTS-546AWC, or Kingbright SC56–11EWA, or similar.

Components for Chapter Five

Components for Chapter Five, other than resistors and capacitors, are listed in Figure 6-8.

Neodymium Magnets

I suggest K&J Magnetics (*http://www.kjmagnetics.com/neomaginfo.asp*) as a source of supply, as the site maintains a very informative primer on magnets.

In Europe, supermagnete.de is a popular source.

16-Gauge Wire

This is only required for the antenna in Experiment 31. If the cost is prohibitive, try 50 or 100 feet of 22-gauge wire. If you live relatively close to an AM radio station, it should be adequate.

High-Impedance Earphone

Only required for Experiment 31, this can be ordered from the the Scitoys catalog (*http://www.scitoyscatalog.com*).

You may also find them on Amazon. On eBay, search for:

```
crystal radio earphone
```

If you search eBay for crystal radio headphones instead of earphones, you'll find some antique items from the early days of radio.

Components for Chapter Five	Experiments 25	26	27	28	29	30	31	32	33	34	Total
LED (generic)								1		7	8
LED (low current)		1		2							3
Power source 9V	1					1					2
Breadboard				1		1			1	1	4
Paperclip	1										1
Diode 1N4001		1									1
Neodyminum magnet		1									1
Loudspeaker, cheap			1								1
Tactile switch				1	1						2
Slider switch					4						4
Trimmer 10K					1						1
Trimmer 1M					1						1
555 timer TTL type					1						1
Plastic storage bin					1						1
Loudspeaker 4" min.					1						1
LM386 amplifier chip					1						1
Transistor 2N2222						1					1
16 gauge wire 50 ft							1				1
Poly/nylon rope 10 ft							1				1
High-imp earphone							1				1
Germanium diode							1				1
Arduino Uno								1	1	1	3
USB cable A-to-B type								1	1	1	3
NTC Thermistor 10K									1		1

Figure 6-8 *Components other than resistors and capacitors used in the fifth chapter of this book.*

Germanium Diode

Available from the same sources as the high-impedance earphone, above. Some may also be available from dgk, mou, or nwk.

Arduino Uno Board

For a discussion of sources, see "Beware of Imitations?" on page 280

Thermistor

The recommended thermistor for Experiment 33 is the Vishay 01-T-1002-FP. If you make a substitution, use a 10K NTC-type thermistor rated at 1% or 5% accuracy with wire leads.

Buying Tools and Equipment

See "Components" on page 317 for a list of components and see "Supplies" on page 316 for a list of supplies.

Photographs and general information about tools and equipment are provided at the beginning of each chapter of this book. See "Necessary Items for Chapter One" on page 1, "Necessary Items for Chapter Two" on page 41, and "Necessary Items for Chapter Three" on page 99. No additional tools are required for Chapters Four and Five.

Because products come and go, I have not included stock numbers or names of manufacturers of tools and equipment. The specifications and photographs at the beginning of each chapter should provide you with sufficient guidance, and if you restrict your search to large sites such as amazon.com or ebay.com, you may find everything you need fairly quickly, all in one place.

While it's true that expensive tools may be manufactured with greater precision and durability, the cheapest products should be satisfactory for the purposes of this book.

See "Suppliers" on page 326 for URLs that are referenced here with three-letter abbreviations

Tools and Equipment for Chapter One

For photographs and discussion of these items, see Chapter 1.

Only one of each item is required, unless otherwise specified.

Multimeter

For a discussion of multimeter features, see "The Multimeter" on page 1. Good sources include all, amz, eby, and jam.

Test Leads

Double-ended test leads should terminate in alligator clips approximately 1" long. The wire connecting them should measure 12" to 15" (not longer). You need at least three red and three black. Additional colors are useful.

You don't want the kind of test leads that have a plug at each end. Those are sometimes known as "jumper wires." Just search a site such as eBay for

```
test leads double ended alligator
```

and you should find what you want. Buy 10. Sources include all, eby, jam, or spk.

Safety Glasses

Try amz, eby, har, hom, or wal. Ideally, look for glasses with ANSI Z87 rating (you can use this as a search term). Avoid tinted glasses.

Tools and Equipment for Chapter Two

For photographs and discussion of these items, see Chapter 2.

Long-Nosed Pliers

They should measure approximately 5" end-to-end, with a flat inside jaw, not round. From amz, eby, mcm, mic.

Wire Cutters

Also known as "side cutters," and should measure approximately 5" end-to-end. From amz, eby, har, hom, nor, or mcm.

Flush Cutters

These are optional. From amz, eby, har, hom, nor, or mcm.

Wire Strippers

You need the type with specifically sized holes for numbered wire gauges, but the most common range (10 to 20 gauge) is not suitable.

I think you should buy wire strippers that have a specific provision for 22-gauge wire, because there's no point in making your task more difficult than it needs to be. Search online for:

```
wire strippers 20 30
```

This should find tools with holes for 20–, 22–, 24–, 26–, 28–, and 30–gauge wire. Alternatively, look on amz, eby, elg, jam, and spk.

Tools and Equipment for Chapter Three

For photographs and discussion of these items, see Chapter 3.

Low-Power Soldering Iron

This should be rated for 15W with a plated, slender, conical tip. Try all, amz, eby, jam, and mcm.

General-Duty Soldering Iron

Should be rated at 30W or 40W. Try amz, eby, har, hom, mcm, nor, or srs.

Helping Hand

Can be found at ada, amz, eby, jam, or spk.

For a small, close-up *magnifying lens* try amz, eby, or wal. It may be listed as a *magnifier* or a *loupe*.

Minigrabbers

The Pomona model 6244-48-0 is available from amz, dgk, mou, and nwk. For cheaper alternatives try eby, which is also your first choice for meter probes terminating in alligator clips.

Heat Gun

Usually sold as a general-purpose tool, and therefore is available from hardware stores. Try amz, har, hom, or nor. For a miniature heat gun, try eby.

Desoldering Equipment

Various options are available from amz, elg, jam, spk, and eby.

Soldering Iron Stand

This will be found in the same places that sell soldering irons.

Miniature Saw

My personal favorite is the #15 X-Acto blade. You also need the handle that it fits in. It is available online from Tower Hobbies, Hobbylinc, ArtCity, and many other arts/crafts sources. Also look for the larger X-Acto saw blade, #234 or #239, which you can use for cutting perforated board.

Deburring Tool

If your local hardware store does not stock this item, it is inexpensively available from amz, eby, mcm, nor, srs, and some specialty sources. The standard blade in this tool is intended for right-handed use. Left-handed blades are made, but can be hard to find. Some blades are harder than others; an E300 means that it is intended for soft metals and most plastics.

Calipers

I like Mitutoyo calipers, although many cheaper brands exist and will be sufficient for everyday use. The Mitutoyo website will show you all their available models, after which you can Google "Mitutoyo" to find retail sources. Many people prefer calipers with a digital display, switchable between metric and inches. I prefer calipers that do not require a battery.

Copper Alligator Clips

Available cheaply and in small quantities from the big general electronics suppliers such as dgk, mou, or nwk.

Suppliers

The three-letter abbreviations preceding each supplier are used throughout the text to suggest appropriate sources.

ada: Adafruit (*http://www.adafruit.com/*)

all: All Electronics (*http://www.allelectronics.com/*)

amz: Amazon (*http://www.amazon.com/*)

dgk: Digi-Key (*http://www.digikey.com/*)

eby: eBay (*http://www.ebay.com/*)

elg: Electronic Goldmine (*http://www.goldmine-elec-products.com/*)

evl: Evil Mad Scientist (*http://www.evilmadscientist.com/*)

har: Harbor Freight (*http://www.harborfreight.com/*)

hom: Home Depot (*http://www.homedepot.com/*)

ins: Instructables (*http://www.instructables.com/*)

jam: Jameco (*http://www.jameco.com/*)

mcm: McMaster-Carr (*http://www.mcmaster.com/*)

mic: Michaels crafts stores (*http://www.michaels.com/*)

mou: Mouser Electronics (*http://www.mouser.com/*)

nwk: Newark Electronics (*http://www.newark.com/*)

nor: Northern Tool (*http://www.northerntool.com/*)

plx: Parallax (*http://www.parallax.com/*)

spk: Sparkfun (*https://www.sparkfun.com/*)

srs: Sears (*http://www.sears.com/tools/*)

tay: Tayda Electronics (*http://www.taydaelectronics.com/*)

Many of these sites also host extensive tutorials and other helpful information. You can learn a lot by browsing through their pages.

Index

Numbers & Symbols

A

B

C

D

E

F

G

H

I

J

K

L

M

N

O

P

R

S

T

U

About the Author

Charles Platt became interested in computers when he acquired an Ohio Scientific C4P in 1979. After writing and selling software by mail order, he taught classes in BASIC programming, MS-DOS, and eventually in Adobe Illustrator and Photoshop. He wrote five computer books during the 1980s.

He has also written science-fiction novels such as *The Silicon Man* (published originally by Bantam and later by Wired Books) and *Protektor* (from Avon Books). He stopped writing science fiction when he started contributing to *Wired* in 1993, and became one of its three senior writers a couple of years later.

Charles began contributing to Make: Magazine in its third issue and is currently a contributing editor. Currently he is designing and building prototypes of medical equipment in his workshop in the northern Arizona wilderness.

Colophon

The cover photograph is by Marc de Vinck.

The heading and cover font are BentonSans, the text font is Myriad Pro, and the constant-width font is TheSansMonoCondensed.